Model Organisms in Biological Research

Prasun Kumar • Vijay Kothari
Editors

Model Organisms in Biological Research

Springer

Editors
Prasun Kumar
Department of Biotechnology
Sharda University
Greater Noida, Uttar Pradesh, India

Vijay Kothari
Institute of Science
Nirma University
Ahmedabad, Gujarat, India

ISBN 978-981-92-0037-5 ISBN 978-981-92-0038-2 (eBook)
https://doi.org/10.1007/978-981-92-0038-2

This Springer imprint is published by the registered company Springer Nature Singapore Pte Ltd.
The registered company address is: 152 Beach Road, #21-01/04 Gateway East, Singapore 189721, Singapore

We dedicate this work to:
Our family members, who made it possible for us to devote time to the compilation of this volume
Our teachers, who imparted the skills and knowledge enabling us to undertake such projects
Our students, who are an essential part of our journey as teachers and researchers

Preface

Various branches of biological, clinical, and environmental research depend on the availability of a suitable model organism, which can mimic the actual application platform, while allowing a reasonable pace of work. Research on some of the pathogens (e.g., some obligate intracellular parasites, including viruses) has suffered owing to nonavailability of a good model organism. On the other hand, some organisms fitted so well the criteria of being an acceptable model that they became indispensable for research in specific domains. For example, *Escherichia coli* has, for decades, remained a handy model for bacteriological research and one of the most popular platforms for expressing cloned genes. *Saccharomyces cerevisiae* has been used widely as a model yeast, and also as a platform for bioprocess intensification for industrial alcohol production. *Candida albicans* serves as a model for pathogenic yeasts. *Mycobacterium smegmatis* has offered researchers the opportunity to develop useful insights into the genus *Mycobacterium* without always taking the risk of working with *M. tuberculosis.* The preclinical research in the field of medicine has relied heavily on pigs, mice, rats, and primates as model organisms.

However, in recent years, there has been serious ethical concerns raised over use of higher animals in research labs, and strategies are being envisaged to reduce their sacrifice. In this view, three lower organisms have gained popularity, at least for preliminary in vivo studies. They are *Caenorhabditis elegans*, *Drosophila melanogaster*, and Zebrafish. Though they are not new to biology labs, their use has gained wider popularity in last two decades, and availability of their transgenic strains has contributed notably towards their widespread use. On the other hand, introduction of humanized models of higher animals has also been started. While use of conventional model organisms still continues, the alternative models (nematode, fly, zebrafish) are gaining popularity, new models are being proposed, and the regulatory agencies globally are trying to reduce the animal sacrifice in biological and pharmaceutical research.

"Organ-on–chip" type of novel approaches are also contributing their bit to reducing the use of model animals in research. The field is changing and evolving continuously and dynamically, and there is a need for researchers in the field of biology, pharmacy, cosmetics, nutraceuticals, toxicology, and environmental sciences to keep themselves updated. In the abovementioned context, this contributory volume features manuscripts from leading researchers working on different model organisms. This collection showcases the advantages and limitations of different model organisms for preclinical research, high-throughput screening for drug discovery, environmental toxicology, basic biological research, etc.

While we made extensive efforts to include chapters on as many model organisms as possible, owing to various reasons beyond the control of the editors, quality chapters on some useful model organisms like *E. coli, D. melanogaster, P. berghei*, etc. could not be made available for inclusion in this book. Nevertheless, the present collection is to our satisfaction, and we hope it will prove to be a useful resource for teaching and research. We sincerely acknowledge all the contributing authors and the publisher for making this work reach the scientific community.

Greater Noida, Uttar Pradesh, India — Prasun Kumar
Ahmedabad, Gujarat, India — Vijay Kothari

Contents

Part I Microbes as Model Organisms in Biological Research

1 **Engineering *Bacillus subtilis* Vegetative Cells for Recombinant Protein Production and Surface Display: Biotechnological and Biomedical Applications** 3
Saghi Nooraei, Mohammad Amin Ayati, Leila Badiefar, Ali Akbari, and Gholamreza Ahmadian

2 ***Bacillus subtilis* Spores as a Robust Platform for Surface Display: Applications in Vaccines Development, Whole-Cell Biocatalysts, and Environmental Biotechnology** 41
Howra Bahrulolum, Seyede Nafise Tabatabaei, Parisa Beyranvand, Sina Ekhlasi, and Gholamreza Ahmadian

3 ***Vibrio cholerae*: An Emerging Model Organism for Bacteria and Pathogens with Wide Ecological Habitats** 69
Varun R. Bavda, Ritu Rani Archana Kujur, Subrata Mishra, and Sushanta Deb

4 **Archaeal Model Organisms: Understanding Genetics, Metabolic Diversity, and Biotechnology Applications** 107
Manuela Tripepi, Rachel Denniston, Ka'alea Rennie, Daud Tariq, and Jason Yang

Part II Conventional Animal Models for Research in Health and Medicine

5 **Laboratory Rat as an Animal Model for Preclinical Studies** 135
Subramani Parasuraman, Jia Jun Chuah, and Sherly Deborah George

6 **The Mouse as a Cornerstone of Biomedical Research: A Biological Model for Human Health and Disease** 171
Priya Mori, Hina Maniya, Mehul Chauhan, Swati Goswami, and Vijay Kumar

7 **Humanized Animal Models in Biological Research** 199
Stanzin Nurboo, Sakshi Yadav, Sonam Ranwa, and Aamir Nazir

Part III Alternative Model Organisms for Research in Health and Medicine

8 Working with Worms: Protocols and Some Practical Considerations for a *C. elegans* Laboratory . . . 227
Nidhi Thakkar, Gemini Gajera, Hirva Shah, Adarshi Bangiawala, Dhyey Nagariya, Pranshu Dixit, and Vijay Kothari

9 *Caenorhabditis elegans* as a Model Organism for Proteinopathies . . . 259
Priyanka Vimal, Deepanshu Verma, Prasun Kumar Bhunia, and Prasad Kasturi

10 *Caenorhabditis elegans* as a Model Organism in Neurological Research . . . 305
Sakshi Yadav, Sonam Ranwa, Stanzin Nurboo, and Aamir Nazir

11 Zebrafish (*Danio rerio*): A Transparent Window into Biomedical Research . . . 327
Shaivee Chokshi, Keya Jantrania, and Ashutosh Kumar

12 Zebrafish in Toxicology . . . 357
Ayelen M. Sosa, M. Jimena Prieto, and Carolina S. Martinez

13 Tardigrades as an Emerging Model Animal for Astrobiology Research: A Case Study . . . 371
Soumyadeep Chakraborty, Subhasree Ray, and Soumya Pandit

Part IV Novel Approaches to Reduce Use of Animals in Research

14 Alternatives to Animal Testing and the Future of Preclinical Research . . . 387
Subramani Parasuraman, Vijeepallam Kamini, and Aaseer Thamby Sam

15 Ethical Challenges and Barriers in Animal Experimentation . . . 407
Thi Phuong Truc Nguyen and Thi Thu Hoai Nguyen

16 Non-animal Models in Research and Teaching: Organoids and Organs-on-Chips . . . 431
Sarah Teakel, Anna Walduck, Joanne H. Connolly, and Thiru Vanniasinkam

About the Editors

Prasun Kumar holds a Ph.D. in Biotechnology from CSIR-Institute of Genomics and Integrative Biology, Delhi, India. He is presently working as an Assistant Professor at the Department of Biotechnology, Sharda University, Greater Noida, India. Earlier, he was working as an Assistant Professor at the Department of Chemical Engineering, Yeungnam University, Republic of Korea. He has over seven years of experience in applied microbiological research including about two years of experience in industrial R&D. His main areas of research are bioinformatics, medical microbiology, microbial biodiversity, bioenergy, microbial biofilms, quorum sensing, quorum quenching, and genomics. His present research is oriented toward valorizing lignocellulosic biowastes into value-added products such as biopolymer, 2G ethanol, bioenergy, and antibiofilm compounds. To his credit, there are over 35 articles in SCI journals, 7 books, and 18 chapters with international publishers. He has been serving the scientific society by reviewing articles for several SCI journals and delivering guest lectures. Publons awarded him the peer review award in the year 2018. He also serves as the editorial board member of a few international journals.

Vijay Kothari is a distinguished microbiologist and faculty member at the Institute of Science, Nirma University, Ahmedabad, India, with over two decades of experience in teaching and research. His research interests include Antimicrobial Resistance (AMR), Validation of Traditional Medicine, and the intriguing field of microbial response to sonic stimulation. He has authored over 100 publications and edited 5 books. His contributions to the scientific community have been recognized with numerous accolades, such as the *Sentinel of Science* Award (2016) by Publons in recognition of his voluminous contribution as a peer-reviewer, the *Artificial Intelligence Molecular Screen* (AIMS) Award from Atomwise Inc. (2019), and the *ASM Future Leaders Mentoring Fellowship* (2023–27) by the American Society for Microbiology. A central pillar of his work is the extensive use of the nematode *Caenorhabditis elegans* as a versatile in vivo model to study host-pathogen interactions, and evaluate the efficacy of potential nutraceuticals and adaptogens. By leveraging this nematode model, Dr. Kothari has significantly advanced the understanding of how natural products and sonic stimulation can modulate microbial virulence and host immunity.

Part I

Microbes as Model Organisms in Biological Research

Engineering *Bacillus subtilis* Vegetative Cells for Recombinant Protein Production and Surface Display: Biotechnological and Biomedical Applications

1

Saghi Nooraei, Mohammad Amin Ayati, Leila Badiefar, Ali Akbari, and Gholamreza Ahmadian

1.1 Introduction

1.1.1 The Vegetative Form of *B. subtilis*: A Model for Recombinant Protein Production and Surface Display

For many years, *B. subtilis* has been used as a model microbe for Recombinant Protein Production and Surface Display. This gram-positive bacterium has been used as a model organism for cellular research investigations for a number of significant reasons. First, *B. subtilis* has gained widespread acceptance as a model organism for research in the lab due to its strong genetic manipulation potential.

Authors Saghi Nooraei and Mohammad Amin Ayati have equally contributed to this chapter.

S. Nooraei · G. Ahmadian (✉)
Department of System Biotechnology, National Institute of Genetic Engineering and Biotechnology (NIGEB), Tehran, Iran
e-mail: ahmadian@nigeb.ac.ir

M. A. Ayati
Department of Bioprocess Engineering, National Institute of Genetic Engineering and Biotechnology (NIGEB), Tehran, Iran

L. Badiefar
Laboratory Affairs, Water Research Institute-Ministry of Energy, Tehran, Iran

A. Akbari
Department of System Biotechnology, National Institute of Genetic Engineering and Biotechnology (NIGEB), Tehran, Iran

School of Life Science Engineering, College of Interdisciplinary Science and Technology, University of Tehran, Tehran, Iran

P. Kumar, V. Kothari (eds.), *Model Organisms in Biological Research*,
https://doi.org/10.1007/978-981-92-0038-2_1

Second, it is regarded as the gram-positive counterpart of the extensively researched Gram-negative bacterium *Escherichia coli*, and both microbes have been used as models for bacterial cell division research in recent decades. Third, unlike *E. coli*, it can form endospores, and spore formation is the most straightforward process of cell differentiation. In addition, *B. subtilis* has become a remarkable model for studying a variety of topics, including motility, chromosome segregation, competence, host system of bacteriophages, transcriptional regulation, biofilm formation, and the production of different enzymes like amylase, proteases, and others [11].

Several aspects of *B. subtilis* system improvement have been the subject of research in recent years. These aspects include strain optimization and improvement, regulation and enhancement of gene expression, protein surface display, enhancement of protein secretion, and optimization of the fermentation process. Techniques like genetic modification, selection of advantageous strains, and non-directional chemical or physical mutagenesis have all been applied in the field of strain improvement. Further enhancement of gene expression is pursued by utilizing plasmid-independent or genomic integration-based expression systems, regulation and engineering of promoters, as well as control of FNE gene expression through proteases and chaperone molecules. Improvement of the secretion process is mainly based on engineering of secretion pathways and identification of effective signal peptides. In surface display applications, target proteins can be placed on the surface of spores or vegetative cells. Finally, optimization of fermentation process performance involves modifying the composition of the culture medium, adjusting operating conditions, and designing effective nutritional strategies [121].

Because it lacks an outer membrane, *B. subtilis* is better than *E. coli* at producing extracellular proteins. Recombinant protein secretion in *B. subtilis* can be increased by optimizing signal peptides and secretion pathways. Furthermore, to enhance the durability and functionality of *B. subtilis* spores and vegetative cells, recombinant proteins can be attached to them [134].

Surface Display of Proteins

In fields such as biotechnology and microbiology, one of the common methods for delivering recombinant proteins is to link them to natural surface proteins of vegetative cells. This approach is more efficient and practical than other methods. In *B. subtilis*, some of the sortase enzyme substrates include YfkN and YhcR, which are encoded by their respective genes (yfkN and yhcR) and recognized on the cell wall by YwpE and/or YhcS. In one study, researchers used YhcS and YwpE to attempt to link YhcR123 or YfkN123 sequences (as a 123-amino acid spacer) to α-amylase, a reporter enzyme. The findings demonstrated that YhcS, a sortase found in the cell wall of *B. subtilis*, can identify and stabilize the YhcR anchor ([4]; [74]).

1.1.2 Why Focus on the Vegetative Form? Key Advantages Over Sporulation

The vegetative state is the active, metabolically active form of *B. subtilis*. In this state, the bacteria grow, divide rapidly by binary fission, and perform normal cellular functions such as DNA replication, RNA transcription, and protein synthesis. It is also during this stage that the bacteria produce important metabolites, enzymes, and antibiotics. However, vegetative cells are sensitive to heat, desiccation, and other environmental stresses that limit their survival under harsh conditions. When *B. subtilis* is exposed to nutrient limitation or stress, it can differentiate into spores, a dormant and highly resistant form. These spores are metabolically inactive but can withstand extreme temperatures, ultraviolet radiation, desiccation, and chemical damage. This state allows the bacterium to survive for long periods until favorable conditions return, at which point the spore can germinate again into a vegetative cell. The advantages of the vegetative state of *B. subtilis* over its spore state include rapid growth and division, active metabolism, biotechnological usefulness and responsive to the environment [121]. In the vegetative state, *B. subtilis* can grow rapidly by binary fission. Under optimal batch-culture conditions, *B. subtilis* can reach a dry-cell biomass of around 6 g/L in approximately 48 h [95]. The vegetative cells are metabolically active, meaning they can synthesize proteins, produce energy through glycolysis and the TCA cycle, and carry out normal cellular functions, such as DNA replication and transcription. Active *Bacillus* cells are used extensively in biotechnology, including the production of enzymes such as amylases and proteases, antibiotics like bacitracin, and other metabolites, because vegetative cells are metabolically active and continuously synthesizing these products. The vegetative form is highly adaptable and responsive to changes in nutrient availability, temperature, pH, and other environmental factors. Cells can adjust their metabolic processes to survive under changing conditions.

Table 1.1 shows some of the advantages and disadvantages of the vegetative form of *B. subtilis* compared to its spore state.

1.1.3 GRAS Status and Safety: A Model for Industrial and Biomedical Applications

Because *B. subtilis* does not create any harmful substances, the US Food and Drug Administration has approved it as a GRAS bacterium. Consequently, *B. subtilis* is a GRAS strain and can be readily employed in biochemical production processes that need to be highly biocompatible, such as those in the food, pharmaceutical, and cosmetic sectors [98]. The European Food Safety Authority (EFSA) uses a

Table 1.1 Some of the advantages and disadvantages of the vegetative form of *B. subtilis* compared to its spore state

Advantages	
High metabolic activity and synthesis of biomolecules	Vegetative cells actively grow, divide, and produce metabolites, while spores remain dormant [98]
Easier genetic engineering and manipulation capabilities	Vegetative cells are suitable for transformation and expression, while spores are resistant and less suitable [98]
Widely used in biotechnology and industry	Vegetative cells are used in many industrial productions, including the production of enzymes, probiotics, antibiotics, and metabolites, because only they are capable of synthesizing and secreting such compounds [98]
Disadvantages	
Sensitive to environmental stress	Vegetative cells are metabolically active but highly sensitive to environmental stresses and die quickly under unfavorable conditions [75]
Limited survival during starvation	Vegetative cells cannot survive long without nutrients and lack the ability to enter a dormant, resistant state [38]
Short lifespan outside optimal conditions	Without optimal conditions (e.g., sufficient nutrients and temperature), vegetative cells cannot persist for long. Their lifespan is limited to environments where conditions are constantly favorable

status called QPS (Qualified Presumption of Safety) that is similar to GRAS, except it is given to the product, whereas QPS is given to the microorganisms that produce it. The QPS idea is a generic, harmonized premarket safety assessment approach for microorganisms that are utilized as sources of food additives, enzymes, and feed, as well as microorganisms that are purposefully added to the food chain as live organisms. The four pillars of QPS are (i) taxonomic unit identification, at least based on rRNA sequences and other housekeeping genes; (ii) body of knowledge; (iii) potential pathogenicity; and (iv) end use. For every bacterial taxonomic unit in the QPS-recommended list, it is generally required that strains [44]. Since oligosaccharides have been shown to have a variety of biological activities and functions, they can be used in industries like food, medicine, and cosmetics that demand high stability and quality. One of the most promising prebiotics is oligosaccharides, and the need for functional food additives is predicted to steadily rise on the global market. On the other hand, because of their excellent stability, oligosaccharides produced by the GRAS bacterium *B. subtilis* have a wide range of applications [80]. Among biosugars, *B. subtilis* is capable of producing acetyl-D-glucosamine (GlcNAc), an amino monosaccharide with anti-inflammatory and anti-sarcoma properties that is frequently employed in food additives and medications. There have been numerous documented uses for GlcNAc derivatives. Therefore, there has been a lot of interest in ecologically friendly GlcNAc biosynthesis, such as employing a GRAS strain of *B. subtilis*. The peptidoglycan that makes up the cell wall of *B. subtilis* is synthesized using GlcNAc as a precursor [80].

1.2 Genetic and Molecular Tools for Engineering Vegetative *B. subtilis* Cells

Since molecular biology and genetic engineering have advanced so quickly, *B. subtilis* strains have been modified using a variety of genetic modification techniques, such as clustered regularly interspaced short palindromic repeats (CRISPR) and homologous recombinant methods, to increase their capacity to produce chemicals and proteins. To aid in the genetic engineering of *B. subtilis,* a variety of genetic tools have been created and described, including plasmids, signal peptides, regulatory elements, and promoters. Gene insertions, mutations, and deletions are the main types of genome editing procedures, while site-specific recombination systems are the main type of genome editing tools [121].

Although they are easy to use and efficient, counter-selectable marker systems limit antibiotic selection markers and are an excellent gene editing tool since they do not leave markers behind. For instance, single-gene inactivation, target gene insertion or deletion, and point mutations have all been accomplished using the blaI and upp genes [117].

1.2.1 Promoters, Vectors, and Expression Systems: Tailoring Vegetative Cells for Diverse Applications

The expression of recombinant proteins in *B. subtilis* is influenced by several factors, among which the promoter plays a key role. The level of protein production can be controlled by regulating the transcription process at the promoter level. In addition to initiating transcription, promoters also allow transcription factors to bind and facilitate its initiation, thus enabling optimal use of the cellular transcription machinery to express open reading frames. Expression systems in *B. subtilis* are usually developed in two ways: either via independent plasmids or via integration of the expression cassette into the chromosome. However, comparison with the *E. coli* expression system shows that the *B. subtilis*-based system has limitations such as plasmid instability, low plasmid diversity, and low plasmid transfer efficiency. Promoters employed for protein expression in *B. subtilis* can be broadly categorized into several types, including inducible promoters, stress-responsive promoters, self-inducible promoters, and constitutive promoters [82]. Inducible expression systems in *B. subtilis* can be activated by a wide range of compounds, including sugars such as sucrose, maltose, and xylose, the synthetic inducer IPTG (isopropyl-β-D-thiogalactopyranoside), cumate (p-isopropyl benzoate), and certain antibiotics, particularly those targeting the bacterial cell wall [92]. However, there are drawbacks to sugar-inducible promoters, including the host's use of sugars as a carbon source and control over the promoters by the carbon metabolism [92]. IPTG is one of the most frequently applied inducers of the T7 promoter for driving protein expression in *B. subtilis*. A well-known example is the construction of the IPTG-inducible promoter Pgrac, which was generated by fusing the strong groES–groEL operon (encoding heat shock proteins) with the lac operator. Using this system, the recombinant penicillin-binding protein Pbp4* was expressed at a level corresponding to approximately 16% of the total cellular protein. In another study, the engineered promoter Pgrac212, also responsive to IPTG, was employed to overproduce HRV3C protease (human rhinovirus 3C protease) in *B. subtilis*, achieving a similar yield of around 16% of total protein content. Likewise, expression of the heat-stable β-galactosidase (BgaB) using plasmids pHT1655 or pHT01 reached about 14% of the total protein when induced with IPTG [104]. Nevertheless, IPTG has significant drawbacks, as it is costly, cytotoxic, and can negatively affect host cell growth. To overcome these limitations, several IPTG-inducible promoters, such as Pgrac01, have been adapted for the development of inducer-free plasmid systems. BgaB production from plasmid pHT1655 occurred without IPTG induction and was 37 times higher compared to plasmid pHT01-bgaB, which required IPTG addition. For comparison, in plasmid pHT01-bgaB, BgaB production reached approximately 14% of total cellular protein in the presence of 1 mM IPTG. These data indicate that plasmid pHT1655 can achieve high-level BgaB production without IPTG induction, whereas plasmid pHT01-bgaB requires IPTG to reach the same production level [104].

1.2.2 CRISPR-Cas and Genome Editing: Precision Engineering in Vegetative Cells

Two approaches are commonly used to build expression systems for *B. subtilis* recombinant protein production: plasmid-based expression for larger yields but lower genetic stability, or integration of the target gene into the chromosome for stable expression (Fig. 1.1). Researchers have long been interested in the key methods used to produce recombinant proteins in *B. subtilis*, such as site-directed mutagenesis, genetic modification of *B. subtilis* strains, expression regulation and enhancement, increased levels of protein secretion, protein surface display, and fermentation optimization. Bioinformatics (such as understanding the entire genome), future strains of *B. subtilis* can be modified and engineered to simultaneously alter multiple loci or regulate multiple pathways, which significantly increases the host's capacity to produce recombinant proteins and display proteins on their surface. CRISPR-associated (Cas) genome editing systems are a cutting-edge method for altering the genome and identifying significant genetic and phenotypic effects [30, 45]. CRISPR–Cas-based genome editing is particularly advantageous because it combines ease of design with remarkable target specificity and high editing efficiency [9, 76]. CRISPR systems, as an adaptive immunity in prokaryotes, are capable of dealing with mobile elements of the systems, which include CRISPR sequences and a set of modifications from Cas genes. In addition, CRISPR–Cas genome editing systems have been used in genetic engineering to optimize and enhance protein expression in *B. subtilis*. During protein production in this bacterium, several factors can affect protein production performance, including extracellular proteases, foam formation in the fermentation process, and spore production [41]. To improve protein fermentation performance in *B. subtilis* ATCC 6051a, a

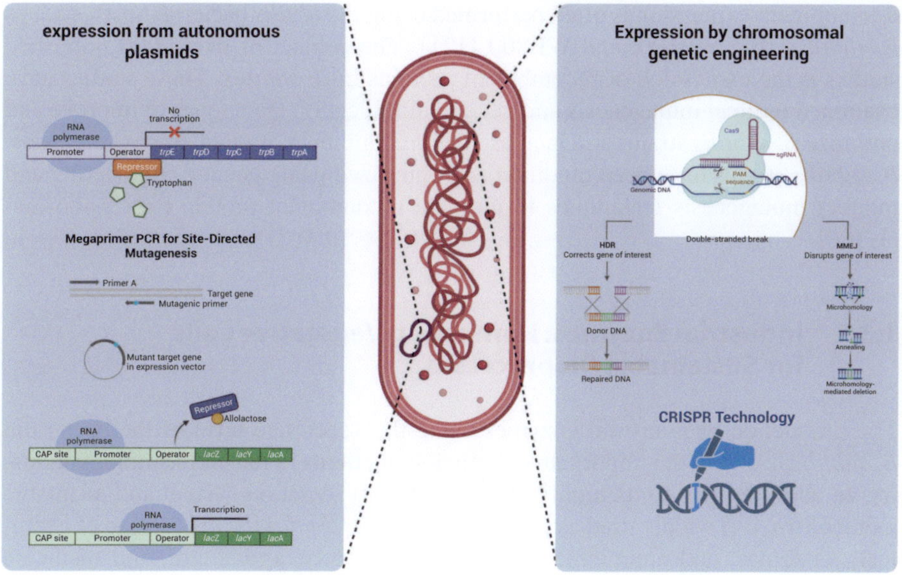

Fig. 1.1 There are two methods for creating expression systems that allow *B. subtilis* to produce recombinant proteins

CRISPR–Cas genome editing system from Streptococcus pyogenes was used to disrupt the amyE, aprE, nprE, spoIIAC, and srfC genes. Mutant strains of *B. subtilis* produce more (2.5-fold) β-cyclodextrin glycosyltransferase than the wild-type strain, and also form and produce spores in a reduced production process. These findings suggest that the *B. subtilis* genome can be modified using the CRISPR–Cas genome editing method to improve production efficiency [125].

However, when it comes to editing many genes, CRISPR/Cas9 is generally inefficient where Cas9n is more effective than CRISPR/Cas9 for large-scale genomic deletion and multi-gene editing, natural Cas9 has been replaced with Cas9n to create a CRISPR/Cas9n-mediated genome editing system for repeated editing of the *B. subtilis* genome, which is more efficient than CRISPR/Cas9 for multi-gene editing and large genomic deletion [63]. Meanwhile, the novel CRISPR/Cpf1 system is the most efficient editing tool for editing and regulating multiple genes, showing higher targeting specificity than CRISPR/Cas9 [133].

1.3 Recombinant Protein Production in Vegetative *B. subtilis* Cells

The Gram-positive and non-pathogenic bacterium *B. subtilis* is considered an excellent host for the synthesis of recombinant proteins due to its exceptional characteristics [121]. Functional peptides and oligopeptides have also been produced by *B. subtilis* because they possess an effective protein secretion mechanism.

Recombinant expression is often performed using *B. subtilis* bacterial hosts, such as *B. subtilis* 168, WB600, and WB800 [121]. The subject of over 2000 published studies is the expression of recombinant proteins in *B. subtilis*. These studies have examined indirect mutagenesis and genetic modification techniques to improve and optimize *B. subtilis* strains and increase their recombinant protein expression levels. *B. subtilis* strains have been modified and optimized using genetic engineering and indirect mutagenesis techniques to increase recombinant protein expression levels [90].

1.3.1 Industrial Enzymes: Harnessing Vegetative Cells for Sustainable Bioprocesses

The production of industrial enzymes using the vegetative form of the bacterium *B. subtilis* is one of the important and efficient methods in the biotechnology industry. In addition, due to its highly efficient protein secretion system and adaptable metabolism, it is widely used as a microbial cell factory for the production of chemicals, enzymes, and antimicrobial agents for industry, agriculture, and medicine [98]. The vegetative form of the *B. subtilis* produces the desired enzymes during its growth stages and secretes them into the surrounding environment. This bacterium is used for the production of various enzymes due to its special characteristics, including high growth rate and the ability to produce different enzymes under different environmental conditions, such as temperature, pH, and different salt concentrations, and the lack of production of toxic substances. This bacterium produces a wide range of enzymes, including proteases, amylases, catalases, lipases, and cellulases [98, 126].

The *B. subtilis* plays a major role in the production of various industrial enzymes and is therefore well known. According to statistics, this bacterium is responsible for the production of nearly 50% of enzymes available on the market. These enzymes include amylases, lipases, xylanases, lichenase, and beta-galactosidase [89, 113], cellulases [14], alkaline serine proteases, and many others [109].

1.3.1.1 Alpha-Amylase

α-1,4 glycosidic linkages are broken down by alpha-amylase (EC 3.2.1.1), which releases glucose from starch. The textile and paper industries make extensive use of this enzyme, and *B. subtilis* is a key host for the synthesis of heterologous alpha-amylase [118]. The production process of this enzyme involves growing *B. subtilis* in liquid or solid fermentation media, where it secretes the enzyme into the medium. The enzymes are then extracted from the medium. For proteins produced in *B. subtilis*, PrsA is a potent folding catalyst. Elevated PrsA expression in *B. subtilis* can increase amylase output. However, in heterologous amylase, simultaneous expression of the homologous prsA gene had a better effect. Quesada Ganuza et al. developed a new recombinant variant of PrsA that not only increased amylase yield in *B. subtilis* but also reduced the secretory stress of the strain [83]. Maximum α-amylase yields in *B. subtilis* have been reported under different fermentation conditions. In *solid-state fermentation* using wheat bran as a substrate, a study reported enzyme yields ranging between 145 and 1239 U/g [5]. In *liquid fermentation*, a study on *B. subtilis* Y25 reported the highest α-amylase production at a concentration of 95.25 ± 2.03 U/ml when starch was used as the carbon source [3]. Furthermore, co-expression of heterologous PrsA enhanced α-amylase production, resulting in up to a 2.5-fold increase in yield [83].

1.3.1.2 Xylanases

A class of enzymes known as xylanases (EC 3.2.1.8) is in charge of breaking down the β-1,4-glycosidic linkages found in xylan. Through this hydrolytic process, they release smaller sugar units such as xylose and a variety of oligosaccharides and disaccharides containing reducing sugars [122]. Xylanases are widely used in industries such as animal feed production, paper processing, textiles, and biofuel production. These enzymes are usually produced on an industrial scale using *B. subtilis* [6]. Maximum xylanase yields in *B. subtilis* have been reported under different fermentation conditions. In *solid-state fermentation*, using wheat bran as a substrate, a study achieved a maximum xylanase activity of 141.28 U/mL [27]. Additionally, under optimal fermentation conditions (5% inoculum, 20 g/L xylan, pH 6.0, for 24 h), a maximum xylanase yield of 15.51 ± 2.08 U/mL was attained [129].

1.3.1.3 Proteases

The breakdown of proteins into smaller peptides and amino acids is carried out by proteases. There are numerous uses for proteases in the food, detergent, and pharmaceutical sectors. In the detergent industry, they are responsible for the removal of stains of a protein nature, including blood and food stains. In the pharmaceutical industry, they play a vital role in the synthesis of specialized drugs and many hormones. In the food industry, they play a very important and practical role, including helping to produce cheese and fermented products. *B. subtilis* is a common microbial source for protease production due to the characteristics mentioned earlier [100]. Maximum protease yields in *B. subtilis* have been reported under various fermentation conditions. In submerged fermentation, using wheat bran as a substrate, a study reported a maximum protease production of 70.21 U/mL [93]. In solid-state fermentation, employing soy cake as a substrate, another study achieved a maximum protease activity of 960 U/g [94]. Additionally, under optimized fermentation conditions (39 °C, pH 7.5, fermentation time of 60 h, inoculum size of 5%, and rotational speed at 180 rpm), a protease activity of 174.87 U/mL was achieved [100].

1.3.1.4 Lipases

The lipase enzyme is responsible for the breakdown of fats into glycerol and fatty acids. Lipases are widely used in various industries, including the food, pharmaceutical, and detergent industries. In the food industry, lipases are used to hydrolyze fats during the production of vegetable oils. In the detergent industry, they are most widely used to remove oily stains. In addition, lipases are essential in the production of biodiesel. Today, the use of the bacterium *B. subtilis* as a factory for the production of this enzyme has received much attention. This bacterium produces lipase under certain conditions, especially in lipid-rich environments [73].

Despite all the above-mentioned features, using the vegetative form of *B. subtilis* in industrial enzyme production poses challenges and problems, including the fact that precise control of environmental conditions such as temperature, pH, and nutrient concentration is essential for optimal enzyme production. This is very difficult and makes working with this bacterium challenging. In addition, the process of extracting enzymes from the culture medium can be time-consuming and costly. Another very important thing is that some enzymes produced by *B. subtilis* may have less activity under industrial conditions and need to be optimized [66]. Maximum lipase yields in *B. subtilis* have been reported under different fermentation conditions. In submerged fermentation, using cheese whey as a substrate, a study reported a maximum lipase activity of 1314 U/mL under optimized conditions (48 h incubation at 30 °C) [28]. In solid-state fermentation, employing groundnut oil cake as a substrate, another study achieved a maximum extracellular lipase activity of 4.5 U/g dry substrate after 48 h of fermentation at pH 8 [2].

1.3.2 Therapeutic Proteins: Overcoming Challenges in Secretion and Yield

Therapeutic proteins, including growth factors, antibodies, and therapeutic enzymes, play an important role in medicine. For the production of such proteins, *B. subtilis* stands out as a particularly promising candidate. *B. subtilis* has many advantages, some of which have been mentioned so far in the text. Other advantages that make this bacterium ideal for the production of therapeutic proteins include its lack of endotoxin production, strong secretion system, ability to grow in inexpensive culture media, high genetic stability, and excellent biosafety [120].

B. subtilis is currently used in the production of certain therapeutic proteins, including:

- Interferon production
- Growth factor production
- Therapeutic enzyme production, such as lysozyme
- Recombinant antibody production

1.3.2.1 Interferon Alpha (IFN-α)

IFN-α is an important cytokine in the immune system used to treat viral diseases such as Hepatitis B and C, and some cancers such as Hairy Cell Leukemia (HCL) [34]. To produce interferon alpha in the vegetative form of *B. subtilis,* the human IFN-α gene is first placed in an expression vector and cloned into *B. subtilis*. To enhance expression, a strong promoter such as PspoVG is used. The Sec secretion system is used to export interferon from the cell to the environment. Finally, the secreted protein is collected from the culture medium and purified using methods such as chromatography. Research has shown that *B. subtilis* can produce IFN-α with high yield and without endotoxin [15]. Maximum IFN-α yields in *B. subtilis* have been reported under different expression strategies. In secretory expression, using a hybrid gene with the α-amylase signal peptide and the human IFN-α1 gene, recombinant *B. subtilis* strains achieved a maximum secretion level of 3×10^6 IU/mL [79]. In intracellular expression, a separate study reported yields of 6×10^4 IU/mL for human IFN-α1 [16]. Additionally, in a study focusing on grouper IFN-α, recombinant *B. subtilis* produced 1.3 mg/L of IFN-α in the culture supernatant, which was then purified using affinity chromatography [61].

1.3.2.2 Epidermal Growth Factor (EGF)

EGF is a therapeutic protein that has been widely used in the treatment of skin wounds, burns, and diabetic ulcers [132]. To produce this protein, the human EGF gene is cloned into a suitable plasmid and expressed in *B. subtilis*. To produce this human protein in this bacterium, a strong promoter such as PamyQ is required for high expression. The Tat secretion system is usually used to transport the protein into the extracellular environment [56]. The EGF protein is then harvested from the culture medium and purified using techniques such as ion exchange chromatography. Experimental results show that *B. subtilis* is capable of producing biologically active EGF, which has significant efficacy in wound healing [46]. Using an IPTG-inducible promoter, *B. subtilis* achieved a yield of approximately 31 mg/L of fibroblast EGF (which closely relates to EGF production data) [52]. A follow-up strategy involving secretion signal peptide optimization and the use of two copies of the hEGF expression cassette (fused with the *xynDSP* signal peptide) significantly improved yield, achieving up to 360 ± 9.41 mg/L of human EGF in the culture medium [97].

1.3.2.3 Lysozyme

Lysozyme is used as a biological preservative in the food and pharmaceutical industries [81] [31, 72]. To produce this enzyme in *B. subtilis*, lysozyme genes, which are derived from natural sources such as egg whites or lysozyme-producing bacterial strains, are typically cloned into *B. subtilis* to enable recombinant expression of the enzyme [71]. To improve gene expression, strong promoters such as Pspac are used. Typically, the Sec secretion system is used to efficiently deliver lysozyme to the culture medium. The secreted enzyme is isolated and purified using techniques such as size-exclusion chromatography. Research has shown that lysozyme produced by *B. subtilis* has potent antimicrobial activity and has the potential to be used in the treatment of antibiotic-resistant infections [40].

Reported fermentative titers for lysozyme vary substantially by host. In *Komagataella phaffii* (Pichia pastoris), fed-batch processes have reached 3.5 g/L of hen egg-white lysozyme (HEWL) with 1.5 × 10^5 U/mL activity, and up to 5.79 g/L for human lysozyme under optimized methanol-feeding strategies [21]. In *B. subtilis*, engineered protease-deficient secretion strains have achieved recombinant human lysozyme titers of approximately 0.5 g/L, whereas native *Bacillus* lysozyme production is often reported in terms of enzymatic activity, for example, around 40–55 U/mL under optimized culture conditions [70].

These examples highlight the potential of *B. subtilis* as an efficient expression host for the production of a wide range of therapeutic proteins. Through the application of genetic engineering techniques, the yield and quality of these proteins can be significantly enhanced, enabling their broader use in the pharmaceutical industry.

1.3.3 Vaccine Antigens: Leveraging Vegetative Cells for Immunization Strategies

Recombinant vaccine production plays an important role in infectious diseases. One of the best strategies for producing antigens is to use safe and efficient bacterial hosts such as *B. subtilis* for antigen expression.

1.3.3.1 Production of *B. subtilis* Anthracis Protective Antigen

To produce an effective vaccine to prevent anthrax infection, the gene encoding the protective antigen (PA) of Bacillus anthracis, which is a critical element in the formation of anthrax toxin and a major immunogenic target, has been successfully cloned and expressed in *B. subtilis*.

B. subtilis serves as an attractive system for the production of recombinant antigens due to its non-pathogenic nature, ease of cultivation, and ability to secrete extracellular proteins. The use of this bacterium as a host allows for the efficient synthesis and recovery of the PA protein, which holds significant promise for development as a vaccine component aimed at conferring immunity against anthrax infection [36]. Findings from various studies have shown that PA produced in *B. subtilis* is capable of eliciting a strong immune response, highlighting its potential as a suitable alternative to conventional anthrax vaccines [10].

1.3.3.2 Production of F1 and V Antigens of Yersinia Pestis (Plague)

A study outlines the engineering of a *Bacillus subtilis*-based expression system for a novel plague vaccine targeting the F1 and V antigens of *Yersinia pestis*. The F1 and V antigen genes are cloned into *B. subtilis*, leveraging the Tat secretion pathway for efficient extracellular release. Post-expression, antigens are purified by standard chromatography techniques and evaluated for immunogenicity through preclinical animal studies [87]. Although the exact *B. subtilis*-based system is novel and not yet evaluated in vivo, robust preclinical data from alternative expression platforms provide a valuable benchmark.

It has been designed to produce a vaccine against plague using the *B. subtilis* expression system. The genes encoding the F1 and V antigens of Yersinia pestis (the causative agent of plague) are first cloned for expression in *B. subtilis*. In this system, the Tat secretion pathway is used for efficient transport of the expressed proteins into the extracellular environment. After expression, a purification process similar to that described above is performed. After purification, the immunogenic properties of these antigens are evaluated through preclinical studies in animal models. Vaccines derived from these antigens show strong immunogenicity and induce a strong immune response capable of protecting against. This approach offers a promising alternative to traditional methods of plague vaccine production and takes advantage of the safety and efficacy of *B. subtilis* as a recombinant protein expression host [87].

1.3.3.3 Production of CagA and VacA Antigens for *Helicobacter pylori* Vaccine

Infections caused by *Helicobacter pylori* remain a major global health problem, and the development of a vaccine is therefore of high priority. The two major virulence factors of H. pylori, CagA and VacA, have been identified as promising vaccine antigens. Although Figueiredo et al. [32] demonstrated the immunogenic potential of CagA and VacA in experimental models, the use of *B. subtilis* as an expression and delivery platform for these antigens has not yet been fully developed or evaluated in large-scale animal trials. To date, no conclusive studies have reported successful prophylactic efficacy or protection rates in animal models using recombinant *B. subtilis* expressing CagA or VacA. Nevertheless, the generally regarded as safe (GRAS) status of *B. subtilis*, its strong secretion capacity, and its proven use for other vaccine antigens strongly suggest that it could be an excellent host for future development of a safe, orally deliverable *H. pylori* vaccine [32]. This strategy not only presents a promising approach to vaccine development for *Helicobacter pylori*-associated diseases but also benefits from the advantages of using *B. subtilis* as a host organism, which is known for its ease of genetic modification, high-efficiency protein production, and suitability for large-scale industrial applications. Ultimately, the creation of a vaccine based on these antigens could play a pivotal role in preventing the widespread impact of *H. pylori* infections globally.

Infections caused by Helicobacter *pylori* are a major cause of peptic ulcers and gastric cancer, and the development of a vaccine to prevent them is therefore of significant global health importance. The two major antigens of *Helicobacter pylori*, namely CagA (cytotoxin A-associated gene) and VacA (vacuolating cytotoxin A), play a key role in the pathogenicity and ability of this bacterium to induce gastric damage. To facilitate the production of these antigens for vaccine development, the respective genes encoding CagA and VacA were cloned into the expression system of *B. subtilis*, a well-known and safe bacterial host known for its ability to produce recombinant proteins. For this purpose, strong bacterial promoters, such as PspoVG, are used to drive high-level expression of CagA and VacA. Upon expression, the antigens are secreted into the surrounding culture medium. After secretion of the proteins, they are purified from the culture supernatant by various chromatographic methods. After purification, the immunological properties of the antigens are tested in animal models, and their ability to induce an immune response is assessed.

1.4 Protein Secretion and Stability in Vegetative *B. subtilis* Cells

Recombinant protein synthesis in microbial hosts is a crucial component of contemporary biotechnology and bio-based economy. High amounts of native proteins can be secreted into the culture medium by a variety of species. As a result, a lot of work has gone into creating an effective secretion mechanism for recombinant protein production. Compared to cytoplasmic expression, secretory expression offers numerous benefits, including easier product identification and purification, a simpler biological process, less cell-associated proteolytic degradation, and better protein folding and quality [18]. *B. subtilis* has several positive characteristics that make it suitable for use as a host for the production of recombinant proteins and make it widely used in the production of industrial enzymes. For example, *B. subtilis* does not produce endotoxins, has a cheap substrate, its genome is known, we have a lot of information about it, and it is GRAS (Generally Regarded As Safe). *B. subtilis* also has a strong secretion system that is directed to the culture medium. Since Gram-positive bacteria lack an outer membrane, it is thought that their thick cell wall plays a role similar to the periplasm of Gram-negative bacteria. However, the lack of an outer membrane allows for the secretion of large amounts of protein directly into the extracellular medium [114].

1.4.1 Secretion Pathways in Vegetative Cells: A Blueprint for Efficient Protein Export

All phases of life experience the phenomena of protein export from the cytoplasm to extracellular locations. There are four recognized protein secretion mechanisms in *B. subtilis*: ABC transporters, the twin-arginine translocation (Tat) system, the pseudopilin pathway, and the secretory protein translocation (Sec) pathway [51]. The Sec-dependent system can export a large range of recombinant proteins due to its high secretion efficiency and relaxed substrate specificity. It has a 20 g/L secretory capacity. The specificity of the signal peptide and its compatibility with the Sec translocase are the main factors influencing secretory capacity and efficiency. The Sec-dependent system can export a large range of recombinant proteins due to its high secretion efficiency and relaxed substrate specificity. It has a 20 g/L secretory capacity. The specificity of the signal peptide and its compatibility with the Sec translocase are the main factors influencing secretory capacity and efficiency [20]. For preproteins to be transported to the transport machinery in the cell membrane, soluble targeting factors must first recognize them. A translocation motor that binds and hydrolyzes nucleotide triphosphates, then propels the polypeptide chain across a protein channel in the membrane. The mature protein is finally freed from the membrane when the signal peptide is eliminated. Unless it is translocated in the folded state, the mature protein quickly returns to its initial shape after being released from the translocase. The majority of prokaryotic and eukaryotic species use these fundamental ideas of protein transport across membranes [103]. Proteins that are already folded in the cytoplasm and cannot be moved by the Sec pathway are moved by the twin-arginine translocation, or Tat, pathway. However, only a small number of proteins (such as PhoD and YwbN) are transported by the Tat pathway, ABC transporters, and the pseudopilin export pathway, which are specialized pathways [107]. Signal peptides, which are found in the majority of secretory proteins, may be categorized into five classes based on their structure and export pathway: lipoprotein, ABC, Com, Sec, and Tat signal peptides. The ABC signal peptide lacks an H domain, whereas the previous four kinds of signal peptides are typically made up of three different domains (H, N, and C). At least one lysine or arginine residue in the N domain, which is typically positively charged, can interact with the negatively charged anionic polymeric membrane components and the translocation machinery. The hydrophobic H domain has a hairpin-like shape because of a glycine or proline residue in the center of a helical segment. A signal peptidase cleavage site for signal peptide removal is located in the C domain [127]. A crucial component of the *B. subtilis* secretion system is signal peptides. Several times, specific signal peptides have been successful in secreting industrial proteins. For instance, the signal peptide nprB successfully secreted a lipoxgenase and a methyl parathion hydrolase, while amyQ released a pullulanase, sacB or nprE secreted an α-amylase, LipA secreted a β-agarase, YncM produced amino peptides, and so on. Prior research has demonstrated that secretion capacity is dependent on particular target proteins in addition to the signal peptide. Therefore, another practical method for secretory protein expression in *B. subtilis* is to search for the appropriate signal peptides for the target protein [53].

Non-classically Secreted Proteins Despite the numerous channels for secreting heterologous proteins with signal peptides, these techniques are not very effective. Additionally, hundreds of transfer components may be involved in each step of the Sec or Tat pathway process, which reduces efficiency and significantly limits the use of *B. subtilis* for large-scale protein synthesis. Although many secreted proteins have known secretory signals, some cytoplasmic proteins can be identified in the extracellular space without any known secretory patterns or signals. Since their secretion mechanism is still unknown, these proteins are known as non-classical secreted proteins [111]. The supernatant of some bacteria contains an increasing number of typical cytoplasmic proteins, which are non-classically secreted proteins that lack any signal peptides or secretion patterns, even though many proteins with known secretion signals are secreted into the extracellular region. For instance, the extracellular environment of certain bacteria, fungi, and even protozoa contains glyceraldehyde-3-phosphate dehydrogenase (GAPDH), which is also found on the cell surface of dangerous Streptococcus species. About half of the extracellular proteins of *B. subtilis* that were identified by proteome analysis were not anticipated in earlier genome-based predictions [130]. *B. subtilis*, one of the most well-researched Gram-positive bacteria, has the ability to secrete proteins via one or more non-classical secretion pathways. Numerous proteins lacking traditional signal peptides have been found in the extracellular environment, indicating that bacteria may secrete more proteins without signal peptides than previously believed [103]. Cell lysis is usually the cause of the discovery of non-classically released proteins in the extracellular environment. It has been established that *B. subtilis* secretes several cytoplasmic proteins devoid of a signal peptide in addition to a heterologous protein known as Est55. Cell lysis is not the cause of this common phenomenon. Furthermore, a non-classical secretion system is involved in protein production, even though the mechanisms underlying non-classical secretion are unknown. This is confirmed by a number of studies conducted by various groups in a number of bacterial species [110]. D-psicose 3-epimerase (RDPE) was demonstrated to be a non-classical secreted protein in *B. subtilis* after it was successfully produced and released into the extracellular medium without the need for a signal peptide. RDPE was then evaluated as a signal for the transfer of recombinant protein to the culture medium. Eighteen reporter proteins screened from various sources were attached to RDPE and linked by a flexible 21-base pair linker to maintain the distance between the two individual proteins. Five of eleven homologous proteins were secreted into the culture medium after binding to RDPE. In addition, two of the five heterologous proteins from different bacteria may be directed by the non-classical secreted protein RDPE. Furthermore, all fusions still showed RDPE activity, and the majority of fusion proteins retained their respective enzymatic or biological function. According to these findings, the non-classically secreted protein RDPE can export recombinant proteins to the environment through the non-classical secretion pathway. This is a novel approach to protein creation that makes use of this new potential secretion channel [17]. An essential enzyme for the biocatalytic synthesis or modification of PLs is phospholipase D (PLD). In *B. subtilis*, an eco-friendly reaction system, a PLD expression and secretion system has been developed. A superfolder green fluorescent protein serves as an N-terminal guide protein in this system's nonclassical secretory pathway. This expression system has the ability to secrete the target proteins in addition to quickly screening high-level expression strains. The culture medium's enzyme activity was 0.35 U/mL under ideal fermentation conditions, which was 2.05 times greater than that of strains of the Sec secretion pathway [128].

1.4.2 Enhancing Protein Stability: Strategies for Robust Production in Vegetative Cells

Cell autolysis drastically lowers *B. subtilis's* performance in liquid cultures, despite the fact that it is widely utilized in industrial biotechnology and agriculture. The lysis of *B. subtilis* is mediated by a number of proteins, including peptidoglycan (PG) hydrolases, prophages, and cannibalism factors. In addition to increasing the production and stability of recombinant proteins, researchers have developed a revolutionary way to enhance biomass and prevent cell lysis through the use of molecular genetic techniques. Five genes or genetic elements representing three distinct functional classes were examined: the prophage genes xpf and yqxG-yqxH-cwlA (yGlA), the skfA and sdpC encoding cannibalism factors, and lytC encoding PG hydrolases. Cell lysis was decreased and biomass was raised when skfA, sdpC, xpf, and lytC were deleted individually [128]. Improving the displayed proteins' immobilization on vegetative *B. subtilis* cells is one method of boosting protein stability. Heterologous protein-displaying microbes can be helpful biotechnological tools for lignocellulose breakdown, protein engineering, biocatalysis, bioremediation, and bioassays. Because *B. subtilis* is a model Gram-positive bacterium that is already employed in industry to generate enzymes and is genetically tractable, it is a viable host for protein display. The reporter protein Cel8A-LysM is present in *B. subtilis* strains (WB800, BRB07, BRB08, and BRB14), where the endoglucanase Cel8A from *Clostridium thermocellum* is attached to LysM cell wall binding modules. Genetic deletion of exocytoplasmic bacterial proteases improves the expression level of LysM-A8Cel, an improvement demonstrated using whole-cell cellulase measurements and isolation experiments. However, after entering the stationary phase, the expression level of the reporter decreases dramatically for all protease-deficient strains, likely as a result of cellular autolysis [47].

1.4.3 Minimizing Proteolytic Degradation: Insights from Vegetative Cell Physiology

In order to prevent proteolytic degradation after secreted proteins exit the cytoplasmic membrane, extracellular chaperones must quickly fold into their correct form. A number of extracellular chaperones, such as PrsA, metal cations, thiol-disulfide oxidoreductases, and propeptides, exist in *B. subtilis* to assist in the folding of exported proteins.

PrsA PrsA is a major lipoprotein with 270 amino acids in *B. subtilis* that is bound to the outer surface of the cytoplasmic membrane [57]. Overexpression of PrsA increased the expression of α-amylases AmyQ, AmyL, and AmyS by 2.5-, 4.7-, and 7.1-fold, respectively [48]. Overexpression of PrsA also increased the expression of the lipase homolog LipA by 1.5-fold [65]. According to these findings, elevated PrsA levels may decrease the probability of secretory protein misfolding and speed up their rate of refolding, which should lessen extracellular proteolytic degradation and boost the synthesis of exported proteins.

Metal Cations The cell wall of *B. subtilis* is made up of proteins, lipoteichoic acid, peptidoglycan, and covalently bound anionic polymers (mostly teichoic acid and teichuronic acid). Although the cell wall occasionally contains partly deacylated forms, the cytoplasmic membrane is where lipoteichoic acids are more commonly found [50]. The dlt operon (dltA–dltE) regulates the esterification of teichoic acid and lipoteichoic acid with D-alanine, which can lessen the negative charge of the cell wall and the membrane-cell wall interface. The degree of D-alanylation is decreased, and the degree of negative charge in the cell wall is increased when any of the dltA–dltD genes are inactivated. The cell wall can bind more cationic folding factors by raising the degree of negative charge. These factors can function as efficient folding agents and enhance the post-translational folding of exported proteins [42]. The peptidase ClpP or the heat stress response-related ATPases ClpC and ClpX must be present in the cell for *B. subtilis* to be able to withstand a variety of stressors. ClpC, ClpP, and ClpX appear to be directly involved in the total proteolysis of misfolded proteins, according to assays used to identify abnormalities in protein breakdown. Mutant cells for clpC, clpP, and clpX have markedly lower turnover rates for aberrant puromycyl peptides [58].

1.5 Surface Display Technologies Using Vegetative *B. subtilis* Cells

Surface expression refers to the attachment of proteins to the outer surface of the cell wall of viruses, fungi, and eukaryotic and prokaryotic cells. The first surface expression system was presented by George Smith et al. in 1980 on the surface of a bacteriophage. They used the attachment of genes encoding heterologous proteins (proteins that are expressed in a host other than the original host that produced them). With genes encoding bacteriophage capsid proteins, they succeeded in expressing heterologous proteins on the surface of bacteriophage; the first surface expression system in bacteria was presented by Frodel et al. in 1986 [69]. This system is based on the genetic fusion of the protein of interest to an outer membrane protein in *Escherichia coli* called OmpA. Surface display technology, using vegetative cells of *B. subtilis*, is a novel approach in biotechnology. In this method, proteins or antigens of interest are placed directly on the surface of bacterial cells. This allows them to be used in biological applications such as vaccines, biosensors, and biocatalysts without the need for complex extraction or purification [8]. Furthermore, unlike gram-negative bacteria, *B. subtilis* lacks an outer membrane, which facilitates easier access and manipulation of surface proteins. Furthermore, its genetic tractability, with a wide range of molecular tools and well-defined regulatory elements, allows for the precise and flexible engineering of display systems [35]. *B. subtilis* possesses remarkable cellular robustness and environmental flexibility, making it ideal for use in controlled laboratory environments and a suitable host for surface display in the development of recombinant vaccines, biocatalysts, and biosensor platforms [8]. Typically, anchor proteins are used to display surface molecules in this bacterium. These anchor proteins can bind antigens or target proteins to the cell wall. This system could be used in the design of oral vaccines, targeted drug delivery, biosequestration of substances, or even the degradation of pollutants in the environment [62].

In recent years, the surface expression of heterologous peptides and proteins on bacterial walls has gained considerable importance.

The advantages of expressing proteins on the wall over their intracellular and extracellular expression include the following: (1) By expressing proteins inside the cell, in cases where the produced protein requires a substance as a substrate that is available outside the cell for activity, protein systems for transporting the substrate into the cell must be developed simultaneously with the expression of the desired protein inside the cell. Using the surface expression method, the desired protein is expressed on the outer surface of the cell and there is no longer a need to transport the substrate into the cell. (2) Surface expression of enzymes and proteins facilitates their purification steps; applications of this method include the production of recombinant vaccines, the preparation of antibody libraries, and the preparation of biosensors. One of the methods of changing the surface of bacterial cells is the surface expression of recombinant proteins on the surface of these cells. Many surface expression systems have been studied in both Gram-positive and Gram-negative bacteria.

1.5.1 Principles of Surface Display in Vegetative Cells

In surface display in *B. subtilis*, a protein, such as an antigen, enzyme, or peptide, is usually linked to a carrier protein or an anchor. The aim is to make the target protein accessible, and this technique allows for the exogenous presentation of recombinant proteins directly onto the surface of living cells, thus eliminating the need for protein extraction [62]. Several approaches have been developed for this purpose, which are discussed below.

- Utilizing Natural Cell Wall Anchor Proteins
  This method utilizes proteins naturally present in the bacterium *B. subtilis* that can attach to its cell wall. These proteins facilitate the attachment of the desired proteins to the bacterial surface. Some examples of these proteins are LytC, WapA, and YhcS, which bind to a specific part of the cell wall called peptidoglycan. These proteins possess domains that aid in their attachment to the wall, ensuring the stable anchoring of the new proteins on the bacterial surface [74].
- Anchoring via the Surface Layer (S-layer) Proteins
  The S-layer is a highly ordered, crystalline proteinaceous layer that naturally covers the surface of many bacterial species, including certain strains of Bacillus. Composed of self-assembling proteins, the S-layer provides a robust structural platform that can be exploited for surface display. By engineering fusion constructs in which the target protein is genetically linked to functional domains of S-layer proteins, it becomes possible to localize and stably present the recombinant protein on the exterior of the bacterial cell. This strategy leverages the inherent self-assembly and surface localization properties of S-layer components for efficient and uniform display [49].
- Membrane Anchor-Based Display
  In this method, parts called membrane anchors are used to attach the protein of interest to the inner membrane of the bacterium *B. subtilis*. These anchors usually contain parts that are hydrophobic or act as signals, and they easily fit into the cell's lipid-like membrane. When the target protein combines with these anchors, it is positioned in the membrane in such a way that its active part faces the outside of the cell. This allows the protein to easily interact with foreign substances, antibodies, or other cells [108].
- Enzymatic Display via Covalent or Non-Covalent Anchoring Mechanisms
  In this method, the target proteins are attached to the bacterial surface and immobilized by chemical or enzymatic interactions. This attachment can be covalent (i.e., permanent and strong) or non-covalent (temporary and removable). In the covalent type, enzymes such as sortase cause strong bonds between a specific part of the target protein and the bacterial cell wall. These bonds are usually very stable.

In contrast, non-covalent methods use moieties that are able to recognize and bind to the cell wall without forming a permanent bond. These moieties can include domains such as CBM (carbohydrate binding module), LysM domains, or antibody-based tags. These methods have high precision, good stability, and great flexibility, and are suitable for various applications such as vaccine development, disease detection systems, and catalysis of reactions on the cell surface [60].

In general, the choice of the best method for displaying a protein on the cell surface depends on the type of target protein, the intended application (e.g., vaccine development, use as a biocatalyst or biosensor), and the characteristics of the host cell. Also, factors such as environmental conditions and cell culture method can affect the success of this process and therefore need to be carefully considered in the design of the system.

1.6 Scaling Up Vegetative *B. subtilis* for Industrial Applications

Nowadays, *B. subtilis* has attracted considerable interest for its diverse industrial applications. It is clear that the great characterizations of *B. subtilis,* including robust growth and adaptability to the various environmental conditions, have made it a popular species for large-scale fermentation and successful industrial production [1]. The first step for such an industrial process is the production of good-quality biomass with desired cell density via microbial cultivation, which is often performed as a batch or fed-batch. The initial experiments for the optimization of fermentation are mainly carried out on a laboratory scale. However, it is necessary to ensure that the same results can be obtained on a larger scale in a pilot plant. After further optimizations of the process at the pilot level, the system can be scaled up to start industrial production, considering economic aspects.

1.6.1 Fermentation Optimization: Lessons from Vegetative Cell Cultivation

For a long time, *B. subtilis* was considered to be an obligate aerobe that can only use molecular oxygen as a terminal electron acceptor [19]. But more investigations showed that *B. subtilis* is able to grow anaerobically too, either by using nitrate ammonification or by fermentation [68]. Due to excellent physiological characteristics and highly adaptable metabolism of this bacterial species, it easily and quickly grows on inexpensive substrates. Therefore, *B. subtilis* fermentation cycle is usually short [98]. *B. subtilis* cultivation using both batch and fed-batch methods produces high vegetative cell concentrations, although the results of some studies recommend the use of a fed-batch system. The process of fermentation optimization is cost-effectively started in Erlenmeyer flasks and then continued step by step to scale up in bioreactors. To characterize optimized conditions of fermentation, microbial culture, type of fermentation, growth medium, and growth conditions should be optimized [96].

1.6.2 Media and Process Engineering: Maximizing Yield and Efficiency in Vegetative Cells

B. subtilis is a microbial cell factory that is commonly used in industrial fermentation as a model strain [86, 105]. The medium components for *B. subtilis* cultivation are usually a combination of a carbon source and a nitrogen source, mineral salts, microelements, and vitamins. In both flask and bioreactor, the composition and concentration of these components should be adjusted to maximize yield in the vegetative cells [39]. Moreover, general properties of different *B. subtilis* strains (such as thermostability, pH prolife, and pH stability) and thereby their optimized fermentation conditions may vary [23]. According to the involvement of multiple parameters in bioprocess technologies and complications that inhibit the application of engineering principles, effective statistical methods for designing optimization experiments are also required. In this way, Researchers succeeded in optimizing a fermentation medium for the production of Nattokinase by *B. subtilis* at pH 7.5 using Response surface methodology and central composite rotary design (CCRD). For this purpose, they evaluated the effects of Glucose, Peptone, $CaCl_2$, and $MgSO_4$ in the medium as variable factors. Their results showed that the optimized medium containing 1% of glucose, 5.5% of peptone, 0.2% of MgSO4, and 0.5% of $CaCl_{2.}$ After 10 h of fermentation, the amount of nattokinase produced (3194.25 U/ml) has doubled in comparison to the original level (1599.09 U/ml). Therefore, the medium composition can greatly increase the yield of the product [26]. Researchers might also use cheap substrates like orange peels, soy beans, and diluted seawater to enhance *B. subtilis* SPB1's production of biosurfactants. They discovered that only orange peels (15.5 g/L), soy beans (10 g/L), and 30% diluted seawater could be included in the ideal medium for the creation of biosurfactants. Such an organic medium doubled the level of production [37]. In a study, the best inoculation conditions for maximum biosurfactant production by *B. subtilis* SPB1 in submerged fermentation were described. It was shown that a two-stage inoculation strategy increased lipopeptide production to approximately 3.4 g/L. It is worth noting that the first inoculation age was 23 h, followed by a second inoculation age of 4 h and 0.01, respectively [67]. In the other study, researchers isolated a new *B. subtilis* source of c-PGA and optimized its culture conditions and medium composition using composite design and response surface methodology. They also investigated the influence of bioreactor stirring rates on the production of c-PGA and established the oxygen volumetric mass transfer coefficients (k_La). At the first step, they could increase the c-PGA production under optimized conditions of 37 °C, pH 6.9, and 1.22 mM Zn^{2+} threefold, compared to the conventional medium. Then, they indicated that ɣ-PGA production was 70% higher in bioreactor than shaker cultures, and c-PGA production had high dependency on the oxygen transfer [22]. It was shown that the number of effective *B. subtilis* cells increased as a result of optimized two-stage fermentation. In order to improve the solid medium qualities, researchers used an ideal two-stage fermentation process with a temperature of 47 °C, a fermentation time of 48 h, and a Mn^{2+} content of 4.9 mg $MnSO_4$/g dry medium. In those circumstances, *B. subtilis*'s effective cell population after 72 h was 1.79 × 1010/g dry medium, which was 29.7% and 8.48% greater than that of 24- and 48-h traditional fermentation, respectively [131].

In addition to the above, there is a real challenge to obtain a strain of *B. subtilis* with high potential for the specific industrial demands [23]. To achieve such a goal, systematic metabolic engineering is a viable option. In fact, a better biosynthetic pathway for *B .subtilis* that meets the needs of the research community and industry can be constructed by combining systems biology, synthetic biology, and evolution-based engineering, which are distinct aspects of metabolic engineering (Fig. 1.2). With this explanation that system biology can provide a full understanding of *B. subtilis* via genome-scale information, synthetic biology can be used to construct new synthetic pathways or to optimize existing biosynthetic pathways and evolution-based engineering screens mutants which meet production needs [117]. Researchers used metabolic engineering techniques to create *B. subtilis* strains that produce Menaquinone-7 (MK-7) at high titers. They discovered that the MK-7 titer can be positively enhanced by overexpressing the glycerol dissociation pathway (GlpK and GlpD), the shikimate pathway (AroGD146N), the pyrimidine metabolic pathway (PyrGE156K), the methylerythritol phosphate pathway (HepS), and VHb. Additionally, scientists were able to create a recombinant strain BSMK_11 by knocking down the mgsA and araM genes and simultaneously overexpressing the genes glpK, glpD, aroGfbr, pyrGfbr, hepS, and vgb. In a 5 L fermenter, the recombinant strain produced MK-7 titer was 281.4 ± 5.0 mg/L [119]. In another study, A group of researchers identified a metabolic engineering strategy based on transcriptome analysis that effectively increases riboflavin production. Since microbial riboflavin synthesis requires large amounts of oxygen, the researchers aimed to reduce the oxygen limitation of *B. subtilis* during the industrial production process. In this way, they could demonstrate the effect of hypoxia on purine and nitrogen metabolism according to transcriptional profiling. Then, they succeeded in enriching the

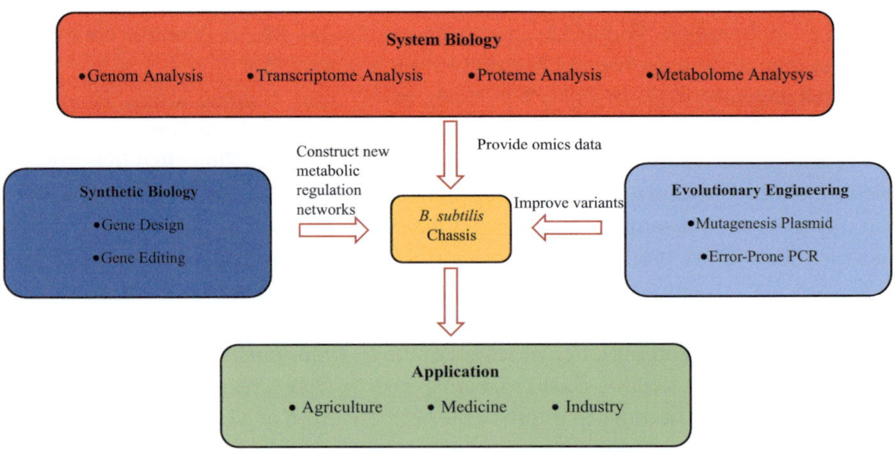

Fig. 1.2 Systems biology, synthetic biology, and evolution-based engineering are used to build *B. subtilis* chassis cells

precursor supply pool by purR knockout and tnrA and glnR knockdown to balance intracellular nitrogen metabolism. At the end, further improvement of the riboflavin yield was performed through dynamic control of the exogenous expression of vgb gene. Fed-batch fermentation of recombinant strain RF1-aPaGaTgV in a 5- liter bioreactor increased the riboflavin yield by 45.51% that was higher than that produced in non-recombinant *B. subtilis* (RF1) [123]. Researchers reduced the disadvantage of *B. subtilis* during fermentative production of acetoin by blocking the spore synthesis pathway and deleting the essential genes for controlling cell autolysis. Such modifications increased biomass and the efficiency of substrate utilization by the strain. After that, they optimized intracellular redox balance through the Rex protein to reduce the harmful effects of NADH produced by the glycolytic metabolic pathway on the process of acetoin production. Subsequently, they blocked competing pathways to increase acetoin yield and enhanced the acetoin pathway using the population cell density-induced promoter. By such combinatorial attempts, they succeeded in producing acetoin in a final titer of 64.3 g/L, which was 89.5% of the theoretical yield [112]. During the other study, a group of researchers enhanced biotin biosynthesis by metabolic engineering of *B. subtilis*. For this purpose, they screened suitable *B. subtilis* strains with the potential to synthesize desthiobiotin (DTB) and biotin. Then, they relieved the feedback repression of biotin synthesis, deleted biotin uptake protein YhfU, and engineered the access tunnel to the active site of biotin synthase (BioB) for reactants and regulated its expression. In this way, they succeeded in increasing DTB and biotin titer to 8.3 mg/L and 11.2 mg/L, which had been improved 37.4-fold and 1130-fold compared to the wild type strain. Their findings suggested a novel strategy for improving DTB production and enhancing its conversion efficiency to biotin [116].

1.6.3 Monitoring and Quality Control: Ensuring Consistency in Large-Scale Production

In addition to the optimization of the fermentation process to maximize growth, yield, and efficiency in vegetative cells of *B. subtilis*, it is necessary to monitor and control all fermentative parameters, including the optimum carbon and nitrogen sources, inoculum volume, and environmental conditions such as moisture content, pH, temperature, agitation, oxygenation, and incubation period, continuously. Because the large-scale industrial bioreactors are different from small lab-scale fermenters. Gradients in large industrial fed-batch fermenters, which are mainly formed by the feeding strategy and insufficient mixing, can change the local medium composition, oxygen, or pH. Accordingly, any change in one of those parameters can dramatically affect the yield of cells and protein product stabilization. Thus, a quantification method is needed for quality control of fermentative processes [24, 85, 91]. For example, the BioPAT® Spectro NIR sensor was directly connected to a bioreactor and *Bacillus* fermentation was monitored in 50-liter bioreactors under harsh industrial conditions. Their method was reported to provide rapid, sensitive, non-destructive, and robust measurements without interfering with the metabolism of the microorganisms [7]. Ensuring good quality control during large-scale bioprocesses is also possible by *B. subtilis* marker gene analysis [91].

1.7 Environmental Applications of Vegetative *B. subtilis* Cells

B. subtilis has also been regarded for its remarkable role in agriculture and aiding environmental cleanup efforts. In fact, the metabolic versatility of this bacterium allows it to improve sustainable agricultural practices and restore polluted ecosystems [1].

1.7.1 *B. subtilis* Benefits in Agriculture

B. subtilis is a soil-inhabiting bacterium with biocontrol and biofertilizer functions. This bacterium, as a plant growth-promoting rhizobacterium (PGPR), is the most commonly used in agricultural production [99]. *B. subtilis has* defenses against phytopathogens in two direct and indirect biocontrol mechanisms, making it very valuable in agriculture. In direct mechanism, *B. Subtilis* synthesizes many secondary metabolites with antibiotic properties, phytohormones, cell-wall-degrading enzymes, and antioxidants that help plant to suppress disease caused by pathogen attack. The promotion of plant growth and the stimulation of plant-induced systemic resistance (ISR) is related to the indirect mechanism. *B. subtilis* can also promote its growth and inhibit the growth of pathogens by solubilizing soil phosphate, enhancing nitrogen fixation, and producing siderophores [43]. Nitrogen fixation and phosphate solubilization activity of *B. subtilis* allow it to be used as a biofertilizer to help plants take in certain nutrients from their environments [1]. Moreover, *B. subtilis* significantly enhances stress tolerance in its plant hosts by activating stress–response genes, phytohormones, and stress-related metabolites [43].

1.7.2 Use of *B. subtilis* in Bioremediation: Water and Wastewater Management

Population growth, as well as an increase in urbanization and industrialization, has led to the release of an extensive volume of hazardous waste into our surrounding environment [124]. Given the importance of water resources management throughout the world, contamination of surface waters and groundwater is one of the major challenges of today's societies. Municipal and industrial wastewaters containing many toxic inorganic and organic matters are also a potential hazard to the natural water system [54]. Microbial bioremediation is a promising technique for the recovery of such contaminated water resources *B. subtilis*, a microorganism with unique characteristics and metabolic pathways, is an efficient and eco-friendly candidate in the bioremediation of many organic pollutants, including industrial chemicals, pesticides, petroleum compounds, hydrocarbons, and polycyclic aromatic hydrocarbons (PAHs). This capability stems from its capacity to synthesize a variety of enzymes, bio-surfactants, and bio-degradative substances [88]. To lower the levels of metals such as lead, cadmium, mercury, chromium, arsenic, and nickel in the environment, *B. subtilis* can employ a variety of bioremediation mechanisms, including biosorption, bioaccumulation, bioprecipitation, and extracellular polymeric substance (EPS)-mediated biosorption [115]. In 2019, Velupillaimani and Muthaiyan identified protocatechuate 3,4-dioxygenase (3,4-PCD) enzyme in the β-ketoadipate pathway of one *B. subtilis* isolate that efficiently degraded protocatechuate (PCA) up to 20 mM concentration. According to the enzyme assay, they could indicate that 3,4-PCD was produced and cleaved the benzene ring at the ortho position and formed keto compounds of the betaketoadipate pathway. Their results also indicated that PCA metabolism in that *B. subtilis* isolate is encoded by chromosomal genes. Since protocatechuate is a common intermediate in the biodegradation of several aromatic compounds, they concluded that isolated *B. subtilis* has potential for bioremediation of aromatic pollutants in marine environments [106]. In another study, researchers found that co-culturing *B. subtilis* with the microbial communities present in wastewater samples containing high concentrations of nitrate or ammonium enabled rapid nitrate removal. They reported that under these conditions, *nasD*, *nasE*, *narG*, *narH,* and *narI* genes involved in assimilatory and respiratory of nitrogen metabolism of *B. subtilis* were overexpressed. It was arguably the reason for the observed boost in denitrifi. This was suggested to be the reason for the observed enhancement in denitrification [84]. The results of some experiments have also shown that *B. subtilis* is capable of removing 100% of 500 ppm of lead, 92.3% of 100 ppm of cadmium, and 89% of 100 ppm of copper from water. Thus, it can efficiently biosorb heavy metals from water and can be employed for industrial wastewater treatment [88].

1.8 Challenges and Future Perspectives: Advancing Vegetative *B. subtilis* as a Model Organism

Today, *B. subtilis* has become one of the most important bacteria in scientific research. It is known as a basic model in microbiology, and many studies have been conducted on its genetic, physiological, and biochemical properties. Using advanced technologies such as the CRISPR/Cas9 system and modern computer methods, this bacterium has found many applications in fields such as biotechnology, agriculture, medicine, and even environmental cleanup. However, for its wider use, we must pay attention to important issues such as safety, ethics, and how to use laboratory results in the real world. As we learn more about this bacterium, it is clear that it can play an important role in solving global problems and creating sustainable solutions in the future. In recent years, studies on *B. subtilis* have made great progress and have yielded new and interesting results in various fields. These new achievements have increased our understanding of the complex properties of this bacterium and shed light on its diverse applications. In particular, important steps have been taken in the development of genetic tools specific to this bacterium [1]. Despite the great progress made in the field of omics studies, which demonstrate the high ability of this bacterium to produce recombinant proteins, there are still challenges in better understanding the secretion pathways. To solve these problems and improve performance, it is necessary to design and implement new and more effective strategies [25, 77, 101].

In addition, the use of proteomics and metabolomics technologies has greatly contributed to our better understanding of the proteins and metabolites present in *B. subtilis*. These methods have also helped us understand how this bacterium regulates its complex metabolic pathways and protects itself against harsh conditions [12]. With the advent of genetic engineering and synthetic biology, there are now exciting opportunities to modify *B. subtilis* strains and improve their characteristics to meet a range of biotechnological objectives [64]. In addition to biotechnology, *Bacillus subtilis* has also garnered significant attention in the field of agriculture, where its application as a biofertilizer and biocontrol agent promises to increase crop yields and combat plant diseases in a sustainable and environmentally friendly manner. These genetically engineered strains have shown promising potential as effective cell factories, facilitating the production of enzymes, pharmaceuticals, and biofuels, thereby making a significant contribution to the bioeconomy and improving environmentally friendly industrial processes [78]. New studies in the medical profession suggest that *B. subtilis* may be used as a probiotic and therapeutic agent to treat infections and digestive issues and to support gut health. Furthermore, in reaction to the worldwide coronavirus disease of 2019 (COVID-19) [102], the potential of *B. subtilis* as a therapeutic agent for the treatment of COVID-19 has been the subject of inquiry. This new field of study has potential since *B. subtilis* has a number of intriguing features that make it a desirable subject for additional research [29, 55].

Potential Applications That Have Not Yet Been Explored

- *B. subtilis*'s possible contribution to the fight against infectious illnesses and other new and developing worldwide issues.
- Applying its biotherapeutic qualities to environmental pollution in ways other than BTEX chemicals.
- Examining its potential for producing bioplastics or other environmentally friendly substitutes for the industrial.

1.8.1 Limitations of Vegetative Cells as a Protein Production Host

Despite being a promising model organism, the biology of *B. subtilis* cannot be used to characterize all bacterial species. Even while they offer insightful information, the genetic processes and mechanisms unique to each bacterium might not be immediately applicable to other bacteria. Despite developments in proteomics, genomics, and other omics methods, a fundamental technology in *B. subtilis*'s cellular machinery and regulatory networks remains to be fully understood. Possible ethical and safety issues with synthetic biology and genetic engineering. Public participation and thorough scientific communication are necessary to resolve concerns about genetically modified organisms (GMOs) and gain social acceptability. It is still difficult to translate scientific results into practical uses. Although *B. subtilis* exhibits considerable promise across a range of industries, its effectiveness and scalability in practical contexts, such as agricultural fields [1].

1.8.2 Overcoming Host-Product Incompatibility: Lessons from Vegetative Cell Engineering

The study of *B. subtilis* today lies at the nexus of engineering, biology, and computer science. Multidisciplinary approaches to comprehending the intricate relationships within this bacterium are made possible by the development of systems biology and computational modeling. These methods can be applied to the development of prediction models of bacterial behavior in various environmental settings, the selection of bacterial strains with desired features, and the optimization of biofuel production routes. Additionally, developments in synthetic biology allow for the construction of unique genetic circuits in *B. subtilis*, which would allow it to carry out a range of functions, from complex chemical manufacturing to environmental sensing. These developments could revolutionize industries and shift them toward ecologically friendly and sustainable methods [1]. Microorganisms must have stable and steady heritable features in order to further improve their performance and turn microbial strains into production hosts for valuable chemicals. Changes are made to their genome's DNA sequence for this reason. Epigenetic methods in production strain creation have not yet been documented. Metabolic engineering is the preferred technique for modifying the genetic composition of bacteria in a focused and intentional way, especially for species like *B. subtilis* that possess a well-developed genetic toolkit. The first stage involves increasing the metabolic capacity of a particular final pathway or pathways towards a useful metabolite in comparison to the wild-type ancestor. This is frequently accomplished by genetic engineering techniques, including recombinant DNA. This straightforward reductionist method, which adheres to a local idea [13], is predicated on the idea that important metabolic network nodes are adaptable enough to handle the occasionally significant shifts in flux distribution that come with an excess of a desirable chemical. For instance, metabolic flux through the highly controlled purine pathway must rise at high growth rates and even treble at low growth rates in order for *B. subtilis* to reach the GTP requirements for optimal riboflavin production. Even still, in vivo flow study revealed that the *B. subtilis* metabolic network was surprisingly rigid, at least when it came to its main metabolic pathways [33]. Without intentional manipulations, the network offers the necessary flexibility to meet the GTP requirement of artificially high riboflavin production. After genetic engineering produces a transformant with the desired genomic structure, the clonal founder cell must proliferate for at least 50 generations in order to produce a culture from which the inoculum for large-scale fermentation can be obtained and from which the desired phenotype can be tested. If the manufactured genetic modification has a negative effect on cell physiology, it is possible that cells with adaptive mutations will be enriched during this multiplication. These adaptive mutations boost the fitness of cells with the mutant allele and provide the network the flexibility it needs. However, a global or systems view of bacterial metabolism is required when focusing on strains that transform their carbon substrate into a desired product with carbon yields near the stoichiometric achievable [59]. This includes core metabolism, which provides a few key metabolic building blocks from which all cellular components are formed, and pathways, which supply precursors from which chemicals are biosynthesised. The ratio of the phosphorylated to unphosphorylated form of the nucleotide redox carrier NAD(P)H/H+, energy load, competitive metabolic pathways, redox balance, and product storage inside the cell or product excretion/secretion are all taken into account. For a multitude of reasons, such as acquiring nutrients, competing with other species, and adapting to environmental changes, *B. subtilis* spontaneously secretes a diverse spectrum of proteins. N-terminal signal peptides found in the majority of these secreted proteins guide their export from the cytoplasm. As a result, effective signal peptides are also necessary for industrial fermentations to secrete protein products. In addition to signal peptides, a complex cellular machinery is required to export proteins from the cytoplasm, transfer them through the cytoplasmic membrane, and subsequently release them into the culture medium in a fully folded, physiologically active, and protease-resistant form. This implies that it is possible to improve the secretory production of high-value proteins by engineering the secretion mechanism as well as the characteristics of the proteins that are released [44].

1.9 Conclusion

The vegetative form of *B. subtilis* serves as a practical and well-studied model system for recombinant protein production and surface display applications. Its genetic tractability, efficient protein secretion capability, and non-pathogenic nature make it a reliable choice for both research and industrial purposes.

This bacterium's ability to secrete proteins directly into the culture medium has proven valuable for producing industrial enzymes such as proteases and amylases. In medical applications, it has shown potential for manufacturing therapeutic proteins and vaccine components. The surface display system using vegetative cells offers additional possibilities for biocatalysis and vaccine development. While *B. subtilis* presents clear advantages, certain limitations need consideration. Issues like protein degradation by endogenous proteases and genetic instability during large-scale cultivation require ongoing attention. Current genetic engineering approaches, including CRISPR-based modifications, continue to improve strain performance. Looking ahead, further optimization of protein production and secretion systems in *B. subtilis* could enhance its utility. The organism's established role in research and industry suggests it will remain relevant for developing biological solutions to practical problems. Continued study of its basic biology will likely yield additional applications while addressing current limitations.

This chapter has outlined the established uses and realistic potential of vegetative *B. subtilis*, providing a balanced perspective on its current and future applications in biological research and biotechnology.

References

1. Akinsemolu AA, Onyeaka H, Odion S, Adebanjo I. Exploring Bacillus subtilis: ecology, biotechnological applications, and future prospects. J Basic Microbiol. 2024;64(6):2300614. https://doi.org/10.1002/jobm.202300614.
2. Aktar L, Saha ML. Optimization of lipase production by bacillus subtilis 20B. Bangladesh J Bot. 2025;54(1):129–36. https://doi.org/10.3329/bjb.v54i1.80585.
3. Aladejana OM, Oyedeji O, Omoboye O, Bakare M. Production, purification and characterization of thermostable alpha amylase from Bacillus subtilis Y25 isolated from decaying yam (Dioscorea rotundata) tuber. Notulae Sci Biol. 2020;12(1):154–71. https://doi.org/10.15835/nsb12110521.
4. Al-Aneed B, Nguyen HD, Masoudi AA, Nooraei S, Hajizade A, Tarrahimofrad H, Ahmadian G, Razmyar J, Khatami M. Sortase-mediated surface display of a chimeric immunogen on Bacillus subtilis: a chicken necrotic enteritis vaccine candidate. Microb Cell Fact. 2025;24(1):147. https://doi.org/10.1186/s12934-025-02762-5.
5. Almanaa TN, Vijayaraghavan P, Alharbi NS, Kadaikunnan S, Khaled JM, Alyahya SA. Solid state fermentation of amylase production from Bacillus subtilis D19 using agro-residues. J King Saud Univ Sci. 2020;32(2):1555–61. https://doi.org/10.1016/j.jksus.2019.12.011.
6. Alponti JS, Maldonado RF, Ward RJ. Thermostabilization of Bacillus subtilis GH11 xylanase by surface charge engineering. Int J Biol Macromol. 2016;87:522–8. https://doi.org/10.1016/j.ijbiomac.2016.03.003.
7. Alves-Rausch J, Bienert R, Grimm C, Bergmaier D. Real time in-line monitoring of large scale Bacillus fermentations with near-infrared spectroscopy. J Biotechnol. 2014;189:120–8. https://doi.org/10.1016/j.jbiotec.2014.09.004.
8. Bahrulolum H, Ahmadian G. Bacillus subtilis surface display technology: applications in bioprocessing and sustainable manufacturing. Biotechnol Biofuels Bioprod. 2025;18(1):34.
9. Bahrulolum H, Tarrahimofrad H, Rouzbahani FN, Nooraei S, Sameh MM, Hajizade A, Ahmadian G. Potential of CRISPR/Cas system as emerging tools in the detection of viral hepatitis infection. Virol J. 2023;20(1):91. https://doi.org/10.1186/s12985-023-02048-5.
10. Baillie L, Moir A, Manchee R. The expression of the protective antigen of Bacillusanthracis in Bacillus subtilis. J Appl Microbiol. 1998;84(5):741–6. https://doi.org/10.1046/j.1365-2672.1998.00405.x.
11. Barák I. Special issue "Bacillus subtilis as a model organism to study basic cell processes". Microorganisms. 2021;9:2459. https://doi.org/10.3390/microorganisms9122459.
12. Bate AR, Bonneau R, Eichenberger P. Bacillus subtilis systems biology: applications of-omics techniques to the study of endospore formation. In: The bacterial spore: from molecules to systems; 2016. p. 129–44. https://doi.org/10.1128/9781555819323.ch6.
13. Becker J, Wittmann C. Systems and synthetic metabolic engineering for amino acid production–the heartbeat of industrial strain development. Curr Opin Biotechnol. 2012;23(5):718–26. https://doi.org/10.1016/j.copbio.2011.12.025.
14. Bien TLT, Tsuji S, Tanaka K, Takenaka S, Yoshida K-i. Secretion of heterologous thermostable cellulases in Bacillus subtilis. J Gen Appl Microbiol. 2014;60(5):175–82. https://doi.org/10.2323/jgam.60.175.

15. Breitling R, Gerlach D, Hartmann M, Behnke D. Secretory expression in Escherichia coli and Bacillus subtilis of human interferon α genes directed by staphylokinase signals. Mol Gen Genet MGG. 1989;217:384–91. https://doi.org/10.1007/BF02464908.
16. Breitling R, Gase K, Behnke D. Intracellular expression of hIFNα genes in Escherichia coli and Bacillus subtilis directed by staphylokinase signals. J Basic Microbiol. 1990:655–62. https://doi.org/10.1002/jobm.362030090830(9).
17. Chen J, Zhao L, Fu G, Zhou W, Sun Y, Zheng P, Sun J, Zhang D. A novel strategy for protein production using non-classical secretion pathway in Bacillus subtilis. Microb Cell Fact. 2016;15:1–16. https://doi.org/10.1186/s12934-016-0469-8.
18. Choi J, Lee S. Secretory and extracellular production of recombinant proteins using Escherichia coli. Appl Microbiol Biotechnol. 2004;64:625–35. https://doi.org/10.1007/s00253-004-1559-9.
19. Cruz Ramos H, Hoffmann T, Marino M, Nedjari H, Presecan-Siedel E, Dreesen O, Glaser P, Jahn D. Fermentative metabolism of Bacillus subtilis: physiology and regulation of gene expression. J Bacteriol. 2000;182(11):3072–80. https://doi.org/10.1128/jb.182.11.3072-3080.2000.
20. Cui W, Han L, Suo F, Liu Z, Zhou L, Zhou Z. Exploitation of Bacillus subtilis as a robust workhorse for production of heterologous proteins and beyond. World J Microbiol Biotechnol. 2018;34:1–19. https://doi.org/10.1007/s11274-018-2531-7.
21. Cui L, Huang H, Zhang H, Wang X, Qin X, Tu T, Zhang J, Su X, Yu H, Bai Y. Recombinant expression of hen egg white lysozyme with the assistance of xylanase fusion partner in Pichia pastoris. Bioengineered. 2022;13(5):13860–71. https://doi.org/10.1080/21655979.2022.2084496.
22. Da Silva SB, Cantarelli VV, Ayub MAZ. Production and optimization of poly-γ-glutamic acid by Bacillus subtilis BL53 isolated from the Amazonian environment. Bioprocess Biosyst Eng. 2014;37:469–79. https://doi.org/10.1007/s00449-013-1016-1.
23. Dash BK, Rahman MM, Sarker PK. Molecular identification of a newly isolated Bacillus subtilis BI19 and optimization of production conditions for enhanced production of extracellular amylase. Biomed Res Int. 2015;2015(1):859805. https://doi.org/10.1155/2015/859805.
24. de Souza CG, Martins FICC, Zocolo GJ, Figueiredo JEF, Canuto KM, de Brito ES. Simultaneous quantification of lipopeptide isoforms by UPLC-MS in the fermentation broth from Bacillus subtilis CNPMS22. Anal Bioanal Chem. 2018;410:6827–36. https://doi.org/10.1007/s00216-018-1281-6.
25. de Souza CC, Guimarães JM, do Pereira SS, Mariúba LAM. The multifunctionality of expression systems in Bacillus subtilis: emerging devices for the production of recombinant proteins. Exp Biol Med. 2021;246(23):2443–53. https://doi.org/10.1177/15353702211030189.
26. Deepak V, Kalishwaralal K, Ramkumarpandian S, Babu SV, Senthilkumar S, Sangiliyandi G. Optimization of media composition for Nattokinase production by Bacillus subtilis using response surface methodology. Bioresour Technol. 2008;99(17):8170–4. https://doi.org/10.1016/j.biortech.2008.03.018.
27. Devi S, Dwivedi D, Bhatt AK. Utilization of Agroresidues for the production of Xylanase by bacillus safensis XPS7 and optimization of production parameters. Fermentation. 2022;8(5):221. https://doi.org/10.3390/fermentation8050221.
28. El-Naga MYA, Khan MA, Abu-Hussien SH, Mahdy SM, Al-Farga A, Hegazy AA. Optimizing lipase production by Bacillus subtilis on cheese whey and evaluating its antimicrobial, antibiofilm, anti virulence and biosafety properties. Sci Rep. 2025;15(1):11087. https://doi.org/10.1038/s41598-025-92181-8.
29. Elshaghabee FM, Rokana N, Gulhane RD, Sharma C, Panwar H. Bacillus as potential probiotics: status, concerns, and future perspectives. Front Microbiol. 2017;8:1490. https://doi.org/10.3389/fmicb.2017.01490.
30. Farsani AM, Mokhtari N, Nooraei S, Bahrulolum H, Akbari A, Farsani ZM, Khatami S, Sadat Ebadi M, Ahmadian G. Lipid nanoparticles: the game-changer in CRISPR-Cas9 genome editing. Heliyon. 2024;10(2) https://doi.org/10.1016/j.heliyon.2024.e24606.

31. Ferraboschi P, Ciceri S, Grisenti P. Applications of lysozyme, an innate immune defense factor, as an alternative antibiotic. Antibiotics. 2021;10(12):1534. https://doi.org/10.3390/antibiotics10121534.
32. Figueiredo C, Quint W, Nouhan N, van den Munckhof H, Herbrink P, Scherpenisse J, de Boer W, Schneeberger P, Perez-Perez G, Blaser MJ. Assessment of Helicobacter pylori vacA and cagA genotypes and host serological response. J Clin Microbiol. 2001;39(4):1339–44. https://doi.org/10.1128/jcm.39.4.1339-1344.2001.
33. Fischer E, Sauer U. Large-scale in vivo flux analysis shows rigidity and suboptimal performance of Bacillus subtilis metabolism. Nat Genet. 2005;37(6):636–40. https://doi.org/10.1038/ng1555.
34. Fish EN, Platanias LC. Interferon receptor signaling in malignancy: a network of cellular pathways defining biological outcomes. Mol Cancer Res. 2014;12(12):1691–703. https://doi.org/10.1158/1541-7786.MCR-14-0450.
35. Fu LL, Xu ZR, Li WF, Shuai JB, Lu P, Hu CX. Protein secretion pathways in Bacillus subtilis: implication for optimization of heterologous protein secretion. Biotechnol Adv. 2007;25(1):1–12. https://doi.org/10.1016/j.biotechadv.2006.08.002.
36. Gat O, Inbar I, Aloni-Grinstein R, Zahavy E, Kronman C, Mendelson I, Cohen S, Velan B, Shafferman A. Use of a promoter trap system in Bacillus anthracis and Bacillus subtilis for the development of recombinant protective antigen-based vaccines. Infect Immun. 2003;71(2):801–13.
37. Ghribi D, Mnif I, Boukedi H, Kammoun R, Ellouze-Chaabouni S. Statistical optimization of low-cost medium for economical production of Bacillus subtilis biosurfactant, a biocontrol agent for the olive moth Prays oleae. Afr J Microbiol Res. 2011;5(27):4927–36. https://doi.org/10.5897/AJMR11.1125.
38. Gloria G, Guerrero M. Sporulation, structure assembly, and germination in the soil bacterium Bacillus thuringiensis: survival and success in the environment and the insect host. Microbiol Res. 2023;14(2):466–91. https://doi.org/10.3390/microbiolres14020035.
39. Grigs O, Didrihsone E, Bolmanis E. Investigation of a broad-bean based low-cost medium formulation for Bacillus subtilis MSCL 897 spore production. Fermentation. 2023;9(4):390. https://doi.org/10.3390/fermentation9040390.
40. Guariglia-Oropeza V, Helmann JD. Bacillus subtilis σV confers lysozyme resistance by activation of two cell wall modification pathways, peptidoglycan O-acetylation and D-alanylation of teichoic acids. J Bacteriol. 2011;193(22):6223–32. https://doi.org/10.1128/jb.06023-11.
41. Gupta M, Rao KK. Phosphorylation of DegU is essential for activation of amyE expression in Bacillus subtilis. J Biosci. 2014;39:747–52. https://doi.org/10.1007/s12038-014-9481-5.
42. Harwood CR, Cranenburgh R. Bacillus protein secretion: an unfolding story. Trends Inmicrobiol. 2008;16(2):73–9. https://doi.org/10.1016/j.tim.2007.12.001.
43. Hashem A, Tabassum B, Abd_Allah EF. Bacillus subtilis: a plant-growth promoting rhizobacterium that also impacts biotic stress. Saudi J Biol Sci. 2019;26(6):1291–7. https://doi.org/10.1016/j.sjbs.2019.05.004.
44. Hohmann HP, van Dijl JM, Krishnappa L, Prágai Z. Host organisms: Bacillus subtilis. Ind Biotechnol Microorg. 2017;1:221–97. https://doi.org/10.1002/9783527807796.ch7.
45. Hong K-Q, Liu D-Y, Chen T, Wang Z-W. Recent advances in CRISPR/Cas9 mediated genome editing in Bacillus subtilis. World J Microbiol Biotechnol. 2018;34:1–9. https://doi.org/10.1007/s11274-018-2537-1.
46. Huang R, Lam E, Chen Y, Hackett J, Lam T, Liu D, Ma M, Siu K, Sivakesava S, Xu Z. Human epidermal growth factor excreted by recombinant Escherichia coli K-12 has the correct N-terminus and is fully bioactive. Process Biochem. 1999;35(1–2):1–5. https://doi.org/10.1016/S0032-9592(99)00024-2.
47. Huang GL, Gosschalk JE, Kim YS, Ogorzalek Loo RR, Clubb RT. Stabilizing displayed proteins on vegetative Bacillus subtilis cells. Appl Microbiol Biotechnol. 2018;102:6547–65. https://doi.org/10.1007/s00253-018-9062-x.
48. Hyyryläinen HL, Marciniak BC, Dahncke K, Pietiäinen M, Courtin P, Vitikainen M, Seppala R, Otto A, Becher D, Chapot-Chartier MP. Penicillin-binding protein folding is depen-

dent on the PrsA peptidyl-prolyl cis-trans isomerase in Bacillus subtilis. Mol Microbiol. 2010;77(1):108–27. https://doi.org/10.1111/j.1365-2958.2010.07188.x.
49. Ilk N, Egelseer EM, Sleytr UB. S-layer fusion proteins—construction principles and applications. Curr Opin Biotechnol. 2011;22(6):824–31. https://doi.org/10.1016/j.copbio.2011.05.510.
50. Iwasaki H, Shimada A, Ito E. Comparative studies of lipoteichoic acids from several Bacillus strains. J Bacteriol. 1986;167(2):508–16. https://doi.org/10.1128/jb.167.2.508-516.1986.
51. Jongbloed JD, Antelmann H, Hecker M, Nijland R, Bron S, Airaksinen U, Pries F, Quax WJ, Van Dijl JM, Braun PG. Selective contribution of the twin-arginine translocation pathway to protein secretion in Bacillus subtilis. J Biol Chem. 2002;277(46):44068–78. https://doi.org/10.1074/jbc.M203191200.
52. Jun J-S, Jeong H-E, Hong K-W. Exploring and engineering novel strong promoters for high-level protein expression in Bacillus subtilis DB104 through transcriptome analysis. Microorganisms. 2023;11(12):2929. https://doi.org/10.3390/microorganisms11122929.
53. Kang X-M, Cai X, Huang Z-H, Liu Z-Q, Zheng Y-G. Construction of a highly active secretory expression system in Bacillus subtilis of a recombinant amidase by promoter and signal peptide engineering. Int J Biol Macromol. 2020;143:833–41. https://doi.org/10.1016/j.ijbiomac.2019.09.144.
54. Kehinde LI, Adekanmi AA, Ahmad LK, Akinkunmi OO, Tomi OB, Ajewole OM. Bioremediation of waste water from pharmaceutical industry by bacteria (Bacillus subtilis). J Environ Issues Clim Change. 2022;1(1):38–50. https://doi.org/10.59110/jeicc.v1i1.64.
55. Khodavirdipour A, Chamanrokh P, Alikhani MY, Alikhani MS. Potential of Bacillus subtilis against SARS-CoV-2–a sustainable drug development perspective. Front Microbiol. 2022;13:718786. https://doi.org/10.3389/fmicb.2022.718786.
56. Kolkman MA, van der Ploeg R, Bertels M, van Dijk M, van der Laan J, van Dijl JM, Ferrari E. The twin-arginine signal peptide of Bacillus subtilis YwbN can direct either Tat-or Sec-dependent secretion of different cargo proteins: secretion of active subtilisin via the B. subtilis Tat pathway. Appl Environ Microbiol. 2008;74(24):7507–13. https://doi.org/10.1128/AEM.01401-08.
57. Kontinen V, Saris P, Sarvas M. A gene (prsA) of Bacillus subtilis involved in a novel, late stage of protein export. Mol Microbiol. 1991;5(5):1273–83.
58. Krüger E, Witt E, Ohlmeier S, Hanschke R, Hecker M. The clp proteases of Bacillus subtilis are directly involved in degradation of misfolded proteins. J Bacteriol. 2000;182(11):3259–65. https://doi.org/10.1128/jb.182.11.3259-3265.2000.
59. Lee JW, Na D, Park JM, Lee J, Choi S, Lee SY. Systems metabolic engineering of microorganisms for natural and non-natural chemicals. Nat Chem Biol. 2012;8(6):536–46. https://doi.org/10.1038/nchembio.970.
60. Liew PX, Wang CL, Wong S-L. Functional characterization and localization of a Bacillus subtilis sortase and its substrate and use of this sortase system to covalently anchor a heterologous protein to the B. subtilis cell wall for surface display. J Bacteriol. 2012;194(1):161–75. https://doi.org/10.1128/JB.05711-11.
61. Lin H-J, Xiao Joe JT, Lu W-J, Huang M-Y, Sun T-H, Lin S-P, Li Y-C, Tsui Y-C, Lu M-W, Victor Lin H-T. Secretory production of functional grouper type I interferon from Epinephelus septemfasciatus in Escherichia coli and Bacillus subtilis. Int J Mol Sci. 2020a;21(4):1465. https://doi.org/10.3390/ijms21041465.
62. Lin P, Yuan H, Du J, Liu K, Liu H, Wang T. Progress in research and application development of surface display technology using Bacillus subtilis spores. Appl Microbiol Biotechnol. 2020b;104:2319–31.
63. Liu D, Huang C, Guo J, Zhang P, Chen T, Wang Z, Zhao X. Development and characterization of a CRISPR/Cas9n-based multiplex genome editing system for Bacillus subtilis. Biotechnol Biofuels. 2019a;12:1–17. https://doi.org/10.1186/s13068-019-1537-1.

64. Liu Y, Liu L, Li J, Du G, Chen J. Synthetic biology toolbox and chassis development in Bacillus subtilis. Trends Biotechnol. 2019b;37(5):548–62. https://doi.org/10.1016/j.tibtech.2018.10.005.
65. Ma RJ, Wang YH, Liu L, Bai LL, Ban R. Production enhancement of the extracellular lipase LipA in Bacillus subtilis: effects of expression system and Sec pathway components. Protein Expr Purif. 2018;142:81–7. https://doi.org/10.1016/j.pep.2017.09.011.
66. Majeed T, Lee CC, Orts WJ, Tabassum R, Shah TA, Jardan YAB, Dawoud TM, Bourhia M. Characterization of a thermostable protease from Bacillus subtilis BSP strain. BMC Biotechnol. 2024;24(1):49. https://doi.org/10.1186/s12896-024-00870-5.
67. Mnif I, Ellouze-Chaabouni S, Ghribi D. Optimization of inocula conditions for enhanced biosurfactant production by Bacillus subtilis SPB1, in submerged culture, using Box–Behnken design. Probiotics Antimicrob Proteins. 2013;5:92–8. https://doi.org/10.1007/s12602-012-9113-z.
68. Nakano MM, Zuber P. Anaerobic growth of a "strict aerobe" (Bacillus subtilis). Ann Rev Microbiol. 1998;52(1):165–90. https://doi.org/10.1146/annurev.micro.52.1.165.
69. Navarre WW, Schneewind O. Surface proteins of gram-positive bacteria and mechanisms of their targeting to the cell wall envelope. Microbiol Mol Biol Rev. 1999;63(1):174–229. https://doi.org/10.1128/mmbr.63.1.174-229.1999.
70. Naveed M, Wang Y, Yin X, Chan MWH, Aslam S, Wang F, Xu B, Ullah A. Purification, characterization and bactericidal action of lysozyme, isolated from Bacillus subtillis BSN314: a disintegrating effect of lysozyme on gram-positive and gram-negative bacteria. Molecules. 2023;28(3):1058. https://doi.org/10.3390/molecules28031058.
71. Naveed M, Wen S, Chan MWH, Wang F, Aslam S, Yin X, Xu B, Ullah A. Expression of BSN314 lysozyme genes in Escherichia coli BL21: a study to demonstrate microbicidal and disintegarting potential of the cloned lysozyme. Braz J Microbiol. 2024;55(1):215–33. https://doi.org/10.1007/s42770-023-01219-4.
72. Nawaz N, Wen S, Wang F, Nawaz S, Raza J, Iftikhar M, Usman M. Lysozyme and its application as antibacterial agent in food industry. Molecules. 2022;27(19):6305. https://doi.org/10.3390/molecules27196305.
73. Ng WZ, Chan E-S, Gourich W, Adiiba SH, Liow MY, Ooi CW, Tey BT, Song CP. Unveiling the role of mechanical process intensifications and chemical additives in boosting lipase-catalyzed hydrolysis of vegetable oil for fatty acid production: a comprehensive review. Int J Biol Macromol. 2024; https://doi.org/10.1016/j.ijbiomac.2024.138144.
74. Nguyen HD, Phan TTP, Schumann W. Analysis and application of Bacillus subtilis sortases to anchor recombinant proteins on the cell wall. AMB Express. 2011;1:1–11. https://doi.org/10.1186/2191-0855-1-22.
75. Nicholson WL, Munakata N, Horneck G, Melosh HJ, Setlow P. Resistance of Bacillus endospores to extreme terrestrial and extraterrestrial environments. Microbiol Mol Biol Rev. 2000;64(3):548–72. https://doi.org/10.1128/mmbr.64.3.548-572.2000.
76. Nishida K, Arazoe T, Yachie N, Banno S, Kakimoto M, Tabata M, Mochizuki M, Miyabe A, Araki M, Hara KY. Targeted nucleotide editing using hybrid prokaryotic and vertebrate adaptive immune systems. Science. 2016;353(6305):aaf8729. https://doi.org/10.1126/science.aaf8729.
77. Nooraei S, Bahrulolum H, Katalani C, Ajorloo MB, Ahmadian G, Hajizade A. Virus-like particles: applications in vaccine development and drug delivery for infectious agents. In: Microbiology in the era of artificial intelligence. CRC Press; 2025. p. 63–82. https://www.taylorfrancis.com/books/mono/10.1201/9781003410164.
78. Nowocień K, Sokołowska B. Bacillus spp. as a new direction in biocontrol and deodorization of organic fertilizers. AIMS Environ Sci. 2022;9(2) https://doi.org/10.3934/environsci.2022007.
79. Palva I, Lehtovaara P, Kääriäinen L, Sibakov M, Cantell K, Schein CH, Kashiwagi K, Weissmann C. Secretion of interferon by Bacillus subtilis. Gene. 1983;22(2–3):229–35. https://doi.org/10.1016/0378-1119(83)90107-5.

80. Park SA, Bhatia SK, Park HA, Kim SY, Sudheer PD, Yang Y-H, Choi K-Y. Bacillus subtilis as a robust host for biochemical production utilizing biomass. Crit Rev Biotechnol. 2021;41(6):827–48. https://doi.org/10.1080/07388551.2021.1888069.
81. Proctor VA, Cunningham F, Fung DY. The chemistry of lysozyme and its use as a food preservative and a pharmaceutical. Crit Rev Food Sci Nutr. 1988;26(4):359–95. https://doi.org/10.1080/10408398809527473.
82. Promchai R, Promdonkoy B, Tanapongpipat S, Visessanguan W, Eurwilaichitr L, Luxananil P. A novel salt-inducible vector for efficient expression and secretion of heterologous proteins in Bacillus subtilis. J Biotechnol. 2016;222:86–93. https://doi.org/10.1016/j.jbiotec.2016.02.019.
83. Quesada-Ganuza A, Antelo-Varela M, Mouritzen JC, Bartel J, Becher D, Gjermansen M, Hallin PF, Appel KF, Kilstrup M, Rasmussen MD. Identification and optimization of PrsA in Bacillus subtilis for improved yield of amylase. Microb Cell Factories. 2019;18:1–16. https://doi.org/10.1186/s12934-019-1203-0.
84. Rahimi S, Modin O, Roshanzamir F, Neissi A, Alam SS, Seelbinder B, Pandit S, Shi L, Mijakovic I. Co-culturing Bacillus subtilis and wastewater microbial community in a bio-electrochemical system enhances denitrification and butyrate formation. Chem Eng J. 2020;397:125437. https://doi.org/10.1016/j.cej.2020.125437.
85. Ray AK, Bairagi A, Ghosh KS, Sen SK. Optimization of fermentation conditions for cellulase production by Bacillus subtilis CY5 and Bacillus circulans TP3 isolated from fish gut. Acta Ichthyol Piscat. 2007;37:47–53.
86. Ren K, Wang Q, Chen J, Zhang H, Guo Z, Xu M, Rao Z, Zhang X. Design-build-test of recombinant Bacillus subtilis chassis cell by lifespan engineering for robust bioprocesses. Synth Syst Biotechnol. 2024;9(3):470–80. https://doi.org/10.1016/j.synbio.2024.04.004.
87. Robinson C, Heather Z, Slater J, Potts N, Steward KF, Maskell DJ, Fontaine MC, Lee J-J, Smith K, Waller AS. Vaccination with a live multi-gene deletion strain protects horses against virulent challenge with Streptococcus equi. Vaccine. 2015;33(9):1160–7. https://doi.org/10.1016/j.vaccine.2015.01.019.
88. Rocco DH, Freire BM, Oliveira TJ, Alves PLM, de Oliveira Júnior JM, Batista BL, Grotto D, Jozala AF. Bacillus subtilis as an effective tool for bioremediation of lead, copper and cadmium in water. Discov Appl Sci. 2024;6(8):430. https://doi.org/10.1007/s42452-024-06101-y.
89. Schallmey M, Singh A, Ward OP. Developments in the use of Bacillus species for industrial production. Can J Microbiol. 2004;50(1):1–17. https://doi.org/10.1139/w03-076.
90. Schumann W. Production of recombinant proteins in Bacillus subtilis. Adv Appl Microbiol. 2007;62:137–89. https://doi.org/10.1016/S0065-2164(07)62006-1.
91. Schweder T. Bioprocess monitoring by marker gene analysis. Biotechnol J. 2011;6(8):926–33. https://doi.org/10.1002/biot.201100248.
92. Seo S-O, Schmidt-Dannert C. Development of a synthetic cumate-inducible gene expression system for Bacillus. Appl Microbiol Biotechnol. 2019;103(1):303–13. https://doi.org/10.1007/s00253-018-9485-4.
93. Shad AA, Ahmad T, Iqbal MF, Asad MJ, Nazir S, Mahmood RT, Wajeeha AW. Production, partial purification and characterization of protease through response surface methodology by Bacillus subtilis K-5. Braz Arch Biol Technol. 2024;67:e24210355. https://doi.org/10.1590/1678-4324-2024210355.
94. Soares VF, Castilho LR, Bon EP, Freire DM. High-yield Bacillus subtilis protease production by solid-state fermentation. Appl Biochem Biotechnol. 2005;121(1):311–9. https://doi.org/10.1385/ABAB:121:1-3:0311.
95. Stamenković SS, Karabegović I, Beškoski V, Nikolić N, Lazić M. Bacillus subtilis NCIM2063 batch cultivation: the influence of the substrate concentration and oxygen transfer rate on the biomass yield. Adv Technol. 2020;9(1):44–9. https://doi.org/10.5937/savteh2001044S.
96. Stamenković-Stojanović S, Karabegović I, Beškoski V, Nikolić N, Lazić M. Bacillus based microbial formulations: optimization of the production process. Hemijska Industrija. 2019;73(3):169–82.

97. Su H-H, Chen J-C, Chen P-T. Production of recombinant human epidermal growth factor in Bacillus subtilis. J Taiwan Inst Chem Eng. 2020a;106:86–91. https://doi.org/10.1016/j.jtice.2019.10.024.
98. Su Y, Liu C, Fang H, Zhang D. Bacillus subtilis: a universal cell factory for industry, agriculture, biomaterials and medicine. Microb Cell Fact. 2020b;19:1–12. https://doi.org/10.1186/s12934-020-01436-8.
99. Sun B, Bai Z, Bao L, Xue L, Zhang S, Wei Y, Zhang Z, Zhuang G, Zhuang X. Bacillus subtilis biofertilizer mitigating agricultural ammonia emission and shifting soil nitrogen cycling microbiomes. Environ Int. 2020;144:105989. https://doi.org/10.1016/j.envint.2020.105989.
100. Sun B, Zou K, Zhao Y, Tang Y, Zhang F, Chen W, Tang X, Chang C, Zheng Y. The fermentation optimization for alkaline protease production by Bacillus subtilis BS-QR-052. Front Microbiol. 2023;14:1301065. https://doi.org/10.3389/fmicb.2023.1301065.
101. Syed MA, Ali S, Hussain T. Microbiology in the era of artificial intelligence: nanotechnology, quantum, and next generation sequencing. CRC Press; 2025. https://www.taylorfrancis.com/books/mono/10.1201/9781003410164.
102. Tabatabaei SN, Keykhaee Z, Nooraei S, Ayati MA, Behzadmand M, Azimi S, Eskati F, Ahmadian G. SARS-CoV-2 and coronaviruses: understanding transmission, impact, and strategies for prevention and treatment. Drugs Drug Candidates. 2025;4(1):5. https://doi.org/10.3390/ddc4010005.
103. Tjalsma H, Antelmann H, Jongbloed JD, Braun PG, Darmon E, Dorenbos R, Dubois J-YF, Westers H, Zanen G, Quax WJ. Proteomics of protein secretion by Bacillus subtilis: separating the "secrets" of the secretome. Microbiol Mol Biol Rev. 2004;68(2):207–33. https://doi.org/10.1128/mmbr.68.2.207-233.2004.
104. Tran DTM, Phan TTP, Huynh TK, Dang NTK, Huynh PTK, Nguyen TM, Truong TTT, Tran TL, Schumann W, Nguyen HD. Development of inducer-free expression plasmids based on IPTG-inducible promoters for Bacillus subtilis. Microb Cell Fact. 2017;16:1–10. https://doi.org/10.1186/s12934-017-0747-0.
105. van Tilburg AY, Cao H, van der Meulen SB, Solopova A, Kuipers OP. Metabolic engineering and synthetic biology employing Lactococcus lactis and Bacillus subtilis cell factories. Curr Opin Biotechnol. 2019;59:1–7. https://doi.org/10.1016/j.copbio.2019.01.007.
106. Velupillaimani D, Muthaiyan A. Potential of Bacillus subtilis from marine environment to degrade aromatic hydrocarbons. Environ Sustain. 2019;2:381–9. https://doi.org/10.1007/s42398-019-00080-2.
107. Voigt B, Antelmann H, Albrecht D, Ehrenreich A, Maurer K-H, Evers S, Gottschalk G, Van Dijl JM, Schweder T, Hecker M. Cell physiology and protein secretion of Bacillus licheniformis compared to Bacillus subtilis. J Mol Microbiol Biotechnol. 2009;16(1–2):53–68. https://doi.org/10.1159/000142894.
108. Wang J, Wu K, Zhang D, Tang H, Xie H, Hong L, Pan Y, Lan M, Hu S, Ning X. Expressions and clinical significances of angiopoietin-1,-2 and Tie2 in human gastric cancer. Biochem Biophys Res Commun. 2005;337(1):386–93. https://doi.org/10.1016/j.bbrc.2005.09.051.
109. Wang J-P, Yeh C-M, Tsai Y-C. Improved subtilisin YaB production in Bacillus subtilis using engineered synthetic expression control sequences. J Agric Food Chem. 2006;54(25):9405–10. https://doi.org/10.1021/jf061982f.
110. Wang G, Chen H, Zhang H, Song Y, Chen W. The secretion of an intrinsically disordered protein with different secretion signals in Bacillus subtilis. Curr Microbiol. 2013;66:566–72. https://doi.org/10.1007/s00284-013-0315-8.
111. Wang G, Xia Y, Song X, Ai L. Common non-classically secreted bacterial proteins with experimental evidence. Curr Microbiol. 2016;72:102–11. https://doi.org/10.1007/s00284-015-0915-6.
112. Wang Q, Zhang X, Ren K, Han R, Lu R, Bao T, Pan X, Yang T, Xu M, Rao Z. Acetoin production from lignocellulosic biomass hydrolysates with a modular metabolic engineering system in Bacillus subtilis. Biotechnol Biofuels Bioprod. 2022;15(1):87. https://doi.org/10.1186/s13068-022-02185-z.

113. Watzlawick H, Altenbuchner J. Multiple integration of the gene ganA into the Bacillus subtilis chromosome for enhanced β-galactosidase production using the CRISPR/Cas9 system. AMB Express. 2019;9(1):158. https://doi.org/10.1186/s13568-019-0884-4.
114. Westers L, Westers H, Quax WJ. Bacillus subtilis as cell factory for pharmaceutical proteins: a biotechnological approach to optimize the host organism. Biochim Biophys Acta (BBA)-Mol Cell Res. 2004;1694(1–3):299–310. https://doi.org/10.1016/j.bbamcr.2004.02.011.
115. Wróbel M, Śliwakowski W, Kowalczyk P, Kramkowski K, Dobrzyński J. Bioremediation of heavy metals by the genus Bacillus. Int J Environ Res Public Health. 2023;20(6):4964. https://doi.org/10.3390/ijerph20064964.
116. Wu Y, Du G-Q, Ma D-H, Li J-L, Fang H, Dong H-N, Jin Z-X, Zhang D-W. Pathway and protein channel engineering of Bacillus subtilis for improved production of desthiobiotin and biotin. Synth Syst Biotechnol. 2025;10(1):307–13. https://doi.org/10.1016/j.synbio.2024.11.005.
117. Xiang M, Kang Q, Zhang D. Advances on systems metabolic engineering of Bacillus subtilis as a chassis cell. Synth Syst Biotechnol. 2020;5(4):245–51.
118. Yan S, Wu G. Bottleneck in secretion of α-amylase in Bacillus subtilis. Microb Cell Fact. 2017;16:1–8. https://doi.org/10.1186/s12934-017-0738-1.
119. Yang S, Wang Y, Cai Z, Zhang G, Song H. Metabolic engineering of Bacillus subtilis for high-titer production of menaquinone-7. AICHE J. 2020a;66(1):e16754. https://doi.org/10.1002/aic.16754.
120. Yang M, Zhu G, Korza G, Sun X, Setlow P, Li J. Engineering Bacillus subtilis as a versatile and stable platform for production of nanobodies. Appl Environ Microbiol. 2020b;86(8):e02938–19. https://doi.org/10.1128/AEM.02938-19.
121. Yang H, Qu J, Zou W, Shen W, Chen X. An overview and future prospects of recombinant protein production in Bacillus subtilis. Appl Microbiol Biotechnol. 2021;105(18):6607–26. https://doi.org/10.1007/s00253-021-11533-2.
122. Yardimci GO, Cekmecelioglu D. Assessment and optimization of xylanase production using co-cultures of Bacillus subtilis and Kluyveromyces marxianus. 3 Biotech. 2018;8(7):290. https://doi.org/10.1007/s13205-018-1315-y.
123. You J, Yang C, Pan X, Hu M, Du Y, Osire T, Yang T, Rao Z. Metabolic engineering of Bacillus subtilis for enhancing riboflavin production by alleviating dissolved oxygen limitation. Bioresour Technol. 2021;333:125228. https://doi.org/10.1016/j.biortech.2021.125228.
124. Yusuf RO, Noor ZZ, Hassan MAA, Agarry SE, Solomon BO. A comparison of the efficacy of two strains of Bacillus subtilis and pseudomonas fragii in the treatment of tannery wastewater. Desalin Water Treat. 2013;51(16–18):3189–95. https://doi.org/10.1080/19443994.2012.751155.
125. Zhang K, Duan X, Wu J. Multigene disruption in undomesticated Bacillus subtilis ATCC 6051a using the CRISPR/Cas9 system. Sci Rep. 2016;6(1):27943. https://doi.org/10.1038/srep27943.
126. Zhang X, Al-Dossary A, Hussain M, Setlow P, Li J. Applications of Bacillus subtilis spores in biotechnology and advanced materials. Appl Environ Microbiol. 2020a;86(17):e01096–20.
127. Zhang K, Su L, Wu J. Recent advances in recombinant protein production by Bacillus subtilis. Annu Rev Food Sci Technol. 2020b;11(1):295–318. https://doi.org/10.1146/annurev-food-032519-051750.
128. Zhang H, Li X, Liu Q, Sun J, Secundo F, Mao X. Construction of a super-folder fluorescent protein-guided secretory expression system for the production of phospholipase D in Bacillus subtilis. J Agric Food Chem. 2021;69(24):6842–9. https://doi.org/10.1021/acs.jafc.1c02089.
129. Zhang G, Li Z, Chen G, Zhang L, Cai W, Deng S, Zhang H, Wu L, Li H, Liu H. Purification and characterization of the low molecular weight xylanase from Bacillus cereus L-1. Braz J Microbiol. 2023;54(4):2951–9. https://doi.org/10.1007/s42770-023-01129-5.
130. Zhao L, Chen J, Sun J, Zhang D. Multimer recognition and secretion by the non-classical secretion pathway in Bacillus subtilis. Sci Rep. 2017;7(1):44023. https://doi.org/10.1038/srep44023.

131. Zhao Z-M, Xi J-T, Xu J-F, Ma L-T, Zhao J. Enhancement of Bacillus subtilis growth and sporulation by two-stage solid-state fermentation strategy. Processes. 2019;7(10):644. https://doi.org/10.3390/pr7100644.
132. Zheng S-Y, Wan X-X, Kambey PA, Luo Y, Hu X-M, Liu Y-F, Shan J-Q, Chen Y-W, Xiong K. Therapeutic role of growth factors in treating diabetic wound. World J Diabetes. 2023;14(4):364. https://doi.org/10.4239/wjd.v14.i4.364.
133. Zhu H, Liang C. CRISPR-DT: designing gRNAs for the CRISPR-Cpf1 system with improved target efficiency and specificity. Bioinformatics. 2019;35(16):2783–9. https://doi.org/10.1093/bioinformatics/bty1061.
134. Zweers JC, Barák I, Becher D, Driessen AJ, Hecker M, Kontinen VP, Saller MJ, Vavrová LU, van Dijl JM. Towards the development of Bacillus subtilis as a cell factory for membrane proteins and protein complexes. Microb Cell Fact. 2008;7(1):10. https://doi.org/10.1186/1475-2859-7-10.

Bacillus subtilis Spores as a Robust Platform for Surface Display: Applications in Vaccines Development, Whole-Cell Biocatalysts, and Environmental Biotechnology

2

Howra Bahrulolum, Seyede Nafise Tabatabaei, Parisa Beyranvand, Sina Ekhlasi, and Gholamreza Ahmadian

2.1 Introduction

2.1.1 Overview of *B. subtilis* as a Model Organism

B. subtilis is the rod-shaped, gram-positive, aerobic bacterium that have the ability to form endospores. This bacterium is classified as a generally regarded as safe (GRAS) organism by the US Food and Drug Administration (FDA) [1]. Originally recognized by Christian Gottfried Ehrenberg, *B. subtilis* has become a key model organism in microbiology over the course of its significant role in scientific studies. *B. subtilis* presents several important benefits as a model organism, merging robust research potential with practical uses [2]. There are several key reasons why this bacterium became a focus for studying fundamental cellular processes. *B. subtilis* is highly suitable for genetic manipulation. Its well-characterized genome facilitates genetic engineering and manipulation. Researchers can quickly introduce mutations or plasmids to study gene function. These key features make *B. subtilis* a popular choice for laboratory research. *B. subtilis* is regarded as the gram-positive counterpart of *Escherichia coli*, a model gram-negative bacterium. While both are used in cell division studies, *B. subtilis* is unique in its ability to form endospores, providing a simple model of cellular differentiation. Moreover, this bacterium is an excellent candidate for investigating motility, chromosome segregation, competence, serving as a host for bacteriophages, transcription regulation, and biofilm formation [3, 4].

H. Bahrulolum · S. N. Tabatabaei · P. Beyranvand · S. Ekhlasi · G. Ahmadian (✉)
Department of Industrial and Environmental Biotechnology, National Institute of Genetic Engineering and Biotechnology (NIGEB), Tehran, Iran
e-mail: ahmadian@nigeb.ac.ir

P. Kumar, V. Kothari (eds.), *Model Organisms in Biological Research*,
https://doi.org/10.1007/978-981-92-0038-2_2

The total genome length of *B. subtilis* is approximately 4.2 Mbp and encodes around 4100 genes. The laboratory strain *B. subtilis* 168 is commonly used as the reference strain. The pan-genome of *B. subtilis*, which encompasses genes from various isolates, includes roughly 6250 genes. Within this, the core genome comprises about 2500 conserved genes. Additionally, approximately 300 genes are dedicated to the process of endospore development [1]. Thanks to its remarkably efficient protein secretion system and versatile metabolism, *B. subtilis* has become a widely used bacterial cell factory for the microbial production of many chemicals, enzymes, and antimicrobial agents across different industrial, agricultural, and medical sectors. Key research areas involving *B. subtilis* include studies on cell development, such as research on endospore formation, biofilm development, cell differentiation processes, and morphogenesis. In addition, *B. subtilis* plays an important role in industrial applications, including investigations into its protein export mechanisms, production of secondary metabolites, and a significant producer of enzymes, promotion of plant growth, and broader biotechnological uses [5, 6].

B. subtilis offers several advantages that make it highly suitable for vaccine development and biotechnological applications. Its rapid growth and ease of cultivation enable its use across various experimental approaches. Moreover, certain cellular processes in *B. subtilis* resemble those of eukaryotic organisms, allowing insights gained from its study to be applied to more complex biological systems. Industrially, *B. subtilis* is favored due to its single cell membrane, which simplifies protein secretion. It also benefits from a short fermentation time of around 48 hours, efficient metabolic activity, and adaptability to diverse substrates [2].

2.1.2 Overview of Spore Formation, Structure, and Resistance Mechanisms

The spore formation in *B. subtilis* explains one of nature's most intricate survival strategies, characterized by a complex interplay of cellular differentiation, structural development, and mechanisms of resistance. This tightly controlled process changes a vegetative cell into a dormant spore that is highly resistant and can endure extreme environmental conditions.

2.1.2.1 Spore Formation Process of *B. subtilis*

The *B. subtilis* is capable of starting the sporulation process when nutrients are limited [7]. The sporulation process begins with the asymmetric division of rod-shaped *B. subtilis*, resulting in the formation of a polar septum. This division generates two compartments with identical genetic content but distinct morphological features: the larger one, called the "mother cell," and the smaller one, referred to as the "forespore." The mother cell and the forespore temporarily reside side by side, connected by the outer cell wall, until the initially flat polar septum begins to curve as the mother cell engulfs the forespore. As a result, the forespore transforms into a roughly spherical structure, bounded by two membranes, within the cytosol of the mother cell. Subsequently, the forespore matures into a dormant, partially

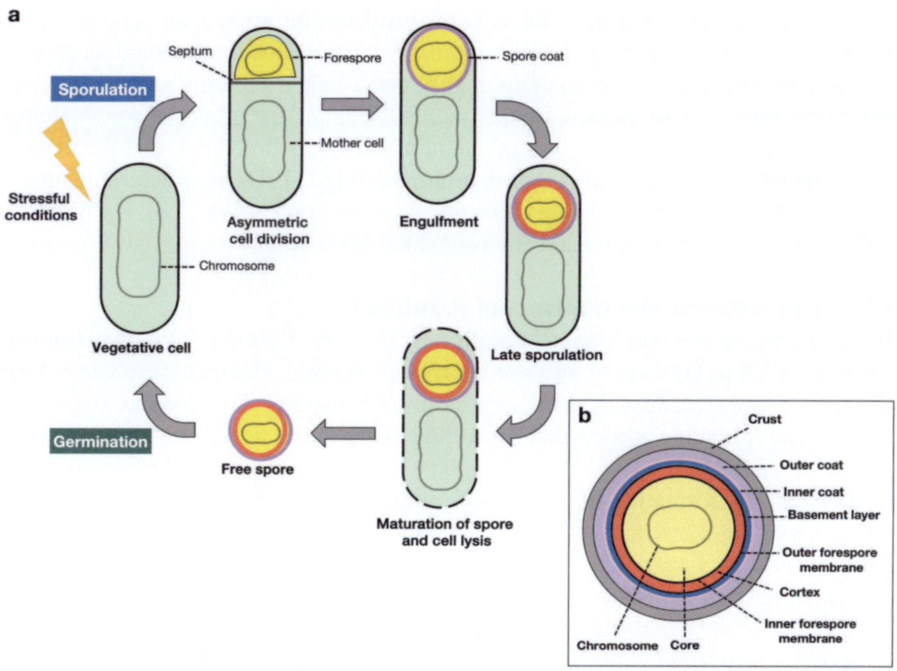

Fig. 2.1 (**a**) The *B. subtilis* sporulation cycle process. Under stressful environmental conditions sporulation begins. The asymmetric division of the sporulating cell forms a structure which comprises the mother cell and a smaller forespore, which later develops into the spore. The next stage involves the engulfment of the forespore by the mother cell. Once the engulfment is complete, the forespore appears as a double-membraned cell within the mother cell. Subsequently, specific protective structures such as spore coat, cortex, and outer layers begin to form. Finally, the mother cell undergoes lysis, resulting in the formation of a mature spore. In the presence of an appropriate stimulus, such as the availability of nutrients in the environment, dormant spores can regain viability through the process of germination. (**b**) Structure of bacterial spore

dehydrated cell, which is released when the mother cell undergoes programmed lysis [8, 9]. This carefully controlled process converts a vegetative cell into a dormant spore through several distinct phases which depicted in Fig. 2.1a [10, 11].

2.1.2.2 Spore Structure of *B. subtilis*

The *B. subtilis* spore possesses a highly advanced multi-layered design that offers remarkable protection against environmental challenges. This complex arrangement consists of outer layers, protective coats, and core elements. In *B. subtilis*, the spore's outermost structure is known as the crust, a layer that bears some resemblance to the exosporium found in other *Bacillus* species such as *B. anthracis* and *B. cereus*. While the exosporium is a glycoprotein-rich layer separated from the spore coat by an interspace, the crust of *B. subtilis* is believed to provide chemical resistance, though direct experimental confirmation is lacking. Interestingly, some proteins found in the *B. subtilis* crust have orthologs in the exosporium of other species, suggesting potential functional similarities. Beneath the crust lies the spore coat, a protein-rich layer that acts as the primary defense against environmental threats such as chemicals and enzymes. The coat contains highly cross-linked proteins and carotenoids that contribute to UV resistance. Deeper within the spore, the structural core includes the inner membrane, a cortex rich in calcium and dipicolinic acid, and the central core, which houses DNA and is tightly protected [11–13]. These structural features, including the crust, coat, and core elements, are illustrated in Fig. 2.1b, highlighting their roles in spore protection and environmental interactions.

2.1.2.3 Resistance Mechanisms of *B. subtilis*

To endure extreme environmental conditions, *B. subtilis* utilizes various resistance strategies such as endospore formation, DNA protection, chemical resistance, heat resistance, stress response systems, and biofilm formation [2]. Endospore formation includes the following aspects that contribute to resistance against harsh conditions:

Dormancy: The formation of endospores allows the bacterium to enter a dormant state, making it resistant to heat, desiccation, and chemical damage.

Thick Protective Layers: The spore coat and cortex provide physical protection against environmental stresses.

Spores demonstrate significantly enhanced resilience against numerous agents, including desiccation, freezing, thawing, high temperatures in both wet and dry conditions, UV and γ-radiation, elevated pressures, and a wide range of harmful chemicals, through a coordinated system of protective barriers and molecular adaptations [14].

2.1.3 Key Features That Make Spores Ideal for Surface Display and Biotechnological Applications

The concept and methodology of *B. subtilis* spore surface display (BSSD) is straightforward to implement; yet, the efficiency of the display can fluctuate based on two key factors. One is the quantity of protein exposed on the spore surface, and the other is its functionality and stability. The effectiveness of BSSD is influenced by several factors, including the location and properties of the anchor protein, the nature of the target protein, the method of attachment to the spore coat, the selection of the expression vector and host strain, employing a flexible peptide linker between spore coat and target protein, and other experimental conditions [15].

Presenting recombinant proteins on *B. subtilis* spores provides numerous benefits. The challenges related to protein folding that occur when unfolded polypeptide chains pass through membranes are avoided. Unlike other display systems, the target proteins located on the spore's coat do not need to cross membranes. Additionally, chaperones available in the mother cell can aid in the proper folding of proteins. Since spores are capable of enduring a variety of physical and chemical stresses, the displayed protein sequences remain accessible even under harsh screening conditions. Conversely, vegetative cells cannot withstand such treatments. By leveraging the resilient characteristics of spores, the displayed proteins are pre-immobilized. Unlike traditional immobilization techniques, where proteins are expressed and purified before being attached to a matrix, immobilization occurs naturally during the sporulation process. Immobilized proteins typically confer both economic and technological advantages, as they can be conveniently isolated after reactions and maintain enhanced stability in harsh environments [16].

Finally, the robustness and safety of *B. subtilis* spores, along with their ease of manipulation in the lab and the absence of certain limitations found in other systems, render these spores an effective platform for presenting heterologous proteins [17].

2.2 Molecular Mechanisms of Spore Surface Display

The display of proteins on spore surfaces is an advanced biotechnological approach that involves attaching proteins or peptides to the surfaces of bacterial spores, predominantly using *B. subtilis* as the host organism. This technique takes advantage of the distinct characteristics of bacterial spores to develop stable and adaptable platforms for a range of biological applications.

2.2.1 Role of Spore Coat Proteins in Recombinant and Non-Recombinant Approach of Spore Surface Display

The spore coat proteins serve as a durable foundation for surface display, making it suitable for extended applications. The outer spore coat demonstrates exceptional durability, offering excellent protection against various environmental adversities, such as high temperatures, dehydration, and exposure to chemicals. This property renders spore display systems particularly advantageous for uses that require prolonged stability, including vaccine delivery, enzyme immobilization, or biosensing in harsh conditions. In addition, the spore display system preserves the activity of target proteins while keeping them exposed on the spore's outer surface [18, 19]. Spore coat proteins serve as versatile platforms for molecular attachment through both recombinant and non-recombinant approaches, offering distinct mechanisms and advantages for different applications.

2.2.1.1 Recombinant Approach of *BSSD* Technique

The spore coat proteins play a crucial role in the recombinant approach of protein display in *B. subtilis* spores. Inner and outer coat layers of the spore contain specific proteins that can be utilized for surface display. CotB, CotC, CotG, and CotZ are commonly used outer coat proteins as anchoring motifs for displaying foreign proteins because they are highly exposed on the spore surface [20]. Inner coat proteins (e.g., CotE, CotH) help in the proper assembly of the outer coat and can also serve as fusion partners [21].

The mechanisms of the recombinant approach involve genetic modification of bacterial genomes to create fusion proteins that combine target proteins with spore coat proteins, presenting specific target molecules on the surface of the spores and thus eliminating the need for protein purification steps. This technique is currently the predominant approach used in both research and industry [22]. The process of recombinant display approach in *B. subtilis* involves several steps, starting with the integration of the fused foreign and spore coat protein genes into the *amyE* locus of the *B. subtilis* chromosome by double-crossover recombination. Expression is carefully timed to align with sporulation, facilitating the assembly of fusion proteins into the spore coat structure. The creation of functional protein complexes and the final surface exposure of the recombinant protein enable recognition of the display protein by antibodies or other molecules (Fig. 2.2a).

There are numerous possible benefits to adopting a surface display system that utilizes bacterial spores: (i) The expression system is guaranteed to be highly stable due to the well-established resilience and stability of bacterial spores. (ii) A solid safety profile is provided by employing spores from species like *B. subtilis*, *B. clausii*, *B. coagulans*, *B. cereus*, and *B. natto*, which are used as food additives in both human and animal food products, as well as in various prescription and nonprescription remedies for gastrointestinal issues; (iii) The production of large quantities of spores is straightforward and cost-effective, thanks to the already existing and widely employed methods for industrial-scale manufacturing and marketing of numerous spore-based items [20].

2.2.1.2 Non-Recombinant Approach of *BSSD* Technique

The non-recombinant display system entails the direct attachment of a heterologous protein/peptide (e.g., enzymes, antigens, antibodies) to the surface of bacterial spores, eliminating the need to engineer recombinant strains and thereby reducing the environmental risks associated with the release of genetically modified organisms. Spore coat proteins enable non-recombinant attachment through their surface chemistry, charge, and structural properties, making *B. subtilis* spores a flexible platform for biotechnological applications without genetic engineering.

In this method, spores are incubated with the target antigen under defined conditions that favor binding, typically using a buffer optimized for adsorption. Following incubation, separation methods such as centrifugation are used to recover spores that have successfully bound the antigen, separating them from unbound fractions [23–26] (Fig. 2.2b). This strategy enhances the stability and protection of the adsorbed protein by utilizing its interaction with the spore surface, which helps prevent degradation and preserves activity even in challenging environmental conditions [27, 28].

Altogether, each method has its own advantages and disadvantages which are summarized in Table 2.1.

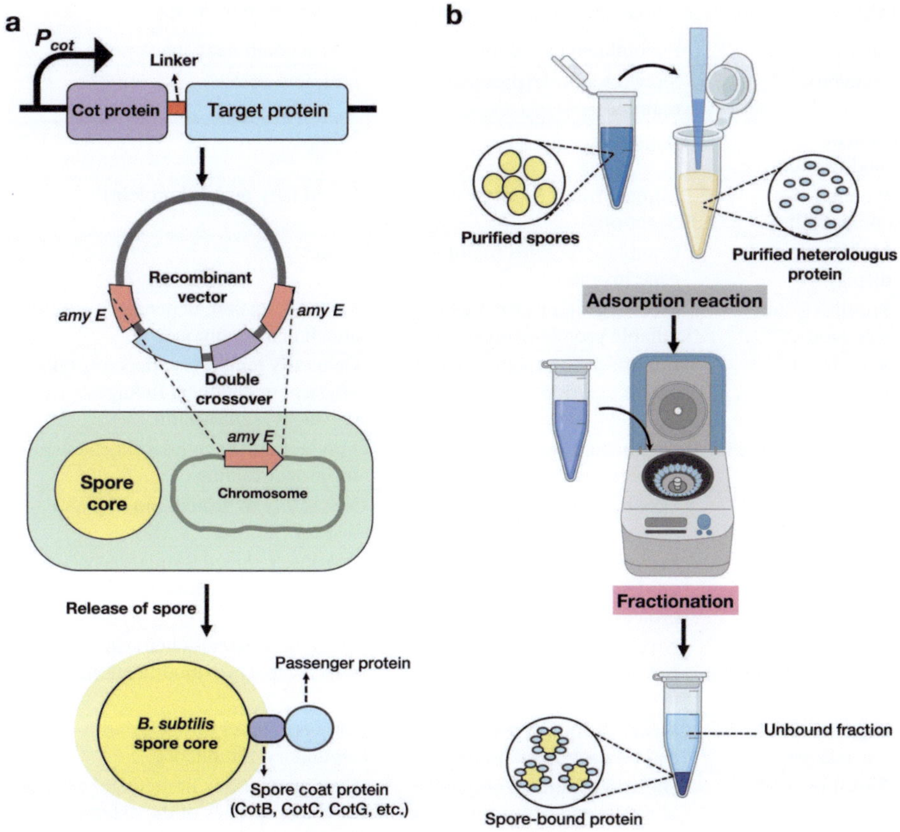

Fig. 2.2 (**a**) Recombinant approach of surface display of protein on *B. subtilis* spore. The target protein is genetically fused to spore coat proteins, resulting in the construction of a recombinant plasmid that carries the gene encoding the fusion protein. This ensures proper expression and integration into spore coat during the sporulation process. Following transformation, the recombinant plasmid is integrated into the genome of *B. subtilis*. During the sporulation process, Cot proteins are embedded in the spore's coat, effectively displaying the fused antigen on the spore surface. (**b**) non-recombinant approach of surface display of protein on *B. subtilis* spore. Purified proteins can spontaneously attach to the *B. subtilis* spore surface via chemical interactions. This adsorption takes place at 25 °C in an acidic buffer with a pH of 4.0. After the binding process, centrifugation is employed to separate spores with attached antigens from those that remain unbound

Table 2.1 Comparison of recombinant and non-recombinant approach of *BSSD*

Characteristic	Recombinant attachment	Non-recombinant attachment
Time/cost	Slow (cloning/expression required)	Fast, low-cost
Genetic modification	Required	None
Protein compatibility	Limited (must be expressible in *B. subtilis*)	Broad (any purified protein)
Multi-enzyme display	Complex (requires multiple gene fusions)	Easier
Specificity and precision	Precise gene insertion and reliable protein expression	Lower precision; depends on external attachment techniques
Stability and integration	Long-lasting genetic stability and integration	Generally temporary; depends on physical or chemical linkages (covalent > adsorption
Protein folding and functionality	Augmented by the molecular chaperones of the host	May not utilize cellular mechanisms for proper folding
Consistency and reproducibility	Consistent and repeatable protein presentation on spores	Variability in protein binding and display
High throughput potential	Scalable for large-scale production with uniform protein expression	Possibly reduced throughput due to inconsistent attachment
Avoidance of cross-linking agents	No requirement for cross-linking agents, minimizing toxicity risks	Often necessitates chemical agents, which can cause modifications
Regulatory advantages	Easier to comply with safety and efficacy regulations	May encounter obstacles with regulatory compliance
Versatility and customization	High adaptability for genetic alterations and customizations	Customization is limited and depends on the properties of the external protein

2.3 Applications of BSSD Technology in Vaccine Development

Pathogenic microorganisms such as bacteria, viruses, and parasites are responsible for causing infectious diseases. Sometimes their spread leads to pandemics, causing substantial increases in morbidity and mortality across a wide geographic area [29]. Vaccination has significantly reduced the burden of infectious diseases and mortality among different populations [30]. Vaccines have significantly improved public health since national immunization programs became widespread in the 1960s. In countries with high coverage, many childhood diseases have nearly vanished [31].

Traditional vaccines like live attenuated, inactivated, and subunit types offer lasting protection against many serious diseases. Despite their success against some infectious diseases, traditional vaccines continue to face major challenges in addressing pathogens that can evade the adaptive immune response. Additionally, for emerging viruses, the primary barrier is not vaccine efficacy but the urgency of rapid development and widespread distribution [32].

Furthermore, while traditional vaccines effectively reduce disease severity and complications, they often fail to generate strong mucosal immunity, limiting their ability to block viral transmission. This is especially concerning for highly contagious respiratory viruses, where preventing spread is crucial for outbreak control. Moreover, the immunity provided by these vaccines wanes over time; therefore, booster doses may be required to maintain protection against the pathogen. Rapid mutations in respiratory viruses such as influenza and SARS-CoV-2 lead to immune evasion. Without robust mucosal immunity, these variants can still infect mucosal surfaces despite the presence of systemic antibodies [33]. Such limitations underscore the urgent need for alternative strategies, such as mucosal vaccines, which target mucosal surfaces to elicit both local and systemic immune responses.

Spore surface display-based vaccine is emerging as a promising platform for mucosal immunization, leveraging the natural stability and resilience of bacterial spores to display antigens on their surface. Among spore-forming bacteria, *B. subtilis* spores have attracted considerable attention in this technology due to the advantages offered by the bacterium. These spores offer several key advantages, including high thermostability, and low production costs. Their stability at room temperature eliminates the need for cold-chain storage, making them ideal for use in low-resource settings [33]. Numerous studies have also demonstrated that mucosal immunization of *B. subtilis* spores display antigens (administrated either intranasally and/or orally) can elicit robust antibody responses and confer protection against infectious diseases [34]. Furthermore, *B. subtilis* spores possess inherent adjuvant properties that enhance the immunogenicity of weak antigens. Upon reaching mucosal tissues, spores can either germinate and present antigens to antigen-presenting cells (APCs) or trigger immune responses without germination [35]. With GRAS status and frequent use in food and probiotic products, *B. subtilis* spores benefit from well-established production systems that support rapid and scalable vaccine manufacturing. Unlike viral vectors, they pose no risk of pathogenicity, pre-existing immunity, or tumorigenesis. The availability of advanced genetic tools further strengthens their potential as a safe, affordable, and thermostable system for developing next-generation mucosal vaccines [36].

2.3.1 Strategies of BSSD Systems for Antigen Display

The concept of employing *B. subtilis* spores for antigen display on their surface was initially introduced by Isticato and colleagues in 2001. They used the CotB as an anchor protein to display the C-terminal 459-amino-acid fragment of tetanus toxin (TTFC) [20]. The successful outcomes of this study led to the widespread use of this technique as a suitable vaccine platform for other infectious diseases, including bacterial, viral, and etc.,

As described in previous section, there are two main strategies for antigen display on *B. subtilis* spore surfaces; a recombinant, and non-recombinant approach.

In the recombinant *BSSD* approach, a fusion gene is constructed between a DNA sequence encoding spore coat proteins (e.g., CotB, CotC, or CotG) and the antigen of interest, with the expression driven by the native transcriptional and translational signals of spore coat gene. During the sporulation process, expression of the fusion construct takes place in mother cell, leading to the assembly of the chimeric protein around the forming spore. During the late phase of sporulation, the mother cell undergoes lysis, leading to the release of the antigen-displaying spore [35]. Spore-based vaccines provide several advantages such as stability and the capacity to endure passage through the gastrointestinal tract, where they can germinate. Their resilience in the harsh acidic environment of the stomach makes *B. subtilis* spores an efficient platform for oral vaccine development [17]. However, the release of genetically modified microorganism into the environment is the major drawback of this approach. Such issues are especially important when the system is intended for applications involving humans or animals [37]. In the non-recombinant *BSSD* approach (spore adsorption), antigens are directly bound to the spore antigens are directly bound to the spore surface, thereby avoiding the requirement for genetic modification of the host strain. The adsorption process between spores and antigens is thought to rely on a combination of electrostatic and hydrophobic forces. The reaction typically done under controlled conditions (e.g., pH 4.0, 25 °C) to optimize adsorption [38]. Following incubation, spores that display the antigen are separated from unbound spores using centrifugation or an alternative separation technique [39]. Compared with recombinant spore display, the non-recombinant technique provides higher antigen display efficiency. Furthermore, it offers the unique capability to present multimeric antigens in their native form, thereby promoting accurate receptor interaction and strong immune activation [40].

To date, a wide range of *B. subtilis* spore-based vaccines have been developed, utilizing different spore coat proteins to display target antigens (either through recombinant approach or non-recombinant approach) for combating diverse pathogens. These studies are summarized in Table 2.2.

2.3.2 Immune Responses Induced by *B. subtilis* Spore-Based Vaccines

B. subtilis spore-based vaccines exhibit the ability to induce both innate and adaptive immune response through different mucosal vaccination route (e.g., oral, nasal, and sublingual) [42]. Both recombinant and non-recombinant display approaches have been observed to function as immunostimulatory agents, capable of acting as adjuvants. These vaccines can promote the production of pro-inflammatory cytokines and facilitate the dendritic cell (DCs) activation and maturation, which are essential for antigen presentation and T-cell activation [43]. De Souza et al. (2014) investigated the adjuvant properties of *B. subtilis* spores employing HIV gag p24 protein as antigen. In BALB/c and C57BL/6 mice, subcutaneous co-administration (but not adsorption onto the spore surface) markedly increased antigen-specific immune responses. Spores induced activation of APCs, evidenced by MHC and CD40 upregulation and enhanced pro-inflammatory cytokine production by DCs.

Table 2.2 Summary of antigen display strategies using *B. subtilis* spores for vaccine development [41]

Virus/bacterium/parasite	Target antigen (s)	Anchor protein
SARS-CoV-2	RBD (Wuhan-Hu-1 & Omicron B.1.1.529), HR1-HR2 domain	CotZ/CotY
Influenza	M2e (Matrix 2 protein)	CotB/CotC
Porcine reproductive and respiratory syndrome virus (PRRSV)	GP5 protein	CotB
Transmissible gastroenteritis virus (TGEV)	Spike protein	CotB
Rabies virus (RABV)	G protein	Cot G
Duck hepatitis A virus type 1 (DHAV-1)	VP1 (capsid protein)	CotG-E
White spot syndrome virus	VP28 protein	CotB
Feline coronavirus (FCoV)	HR2-derived peptide (HR2P)	CotB/CotC
Enterovirus 71 (EV71)	EV71 antigen	CotB
Grass carp reovirus (GCRV)	Vp7	CotB/CotC
Nervous necrosis virus (NNV)	Major capsid protein (MCP)	CotC
Bacillus anthracis	Secreted protective antigen (PA)	CotB
Mycobacterium tuberculosis	MPT64 (secretory antigen)	CotB
Clostridium difficile	FliD (flagellin protein)	CotB
Clostridium difficile	Toxin A (A26-39) or Toxin B (B15-24)	CotB/CotG
Clostridium tetani	TTFC (C-terminal tetanus toxin fragment)	CotB
Clostridium perfringens	Cpa247–370 fused to GST	CotB
Helicobacter acinonychis	UreA (urease subunit alpha)	CotG/CotZ
Helicobacter pylori	CagA (vacuolating cytotoxin A)	CgeA
Salmonella serovar Pullorum	OmpC (outer membrane protein)	CotG
Vibrio vulnificus	OmpK (outer membrane protein)	CotY
Mannheimia haemolytica	Chimeric proteins MhCP1, MhCP2	Non-recombinant
Haemonchus contortus	Glyceraldehyde-3-phosphate dehydrogenase (HcGAPDH)	CotB
Clonorchis sinensis	Tegumental protein (CsTP22.3)	CotG
Clonorchis sinensis	Serpin (CsSerpin)	CotC
Clonorchis sinensis	Enolase (Csenolase)	CotC
Clonorchis sinensis	Paramyosin (CsPmy)	CotC
Schistosoma japonicum	GST protein	CotC
Plasmodium falciparum	Circumsporozoite protein (rPfCSP)	Non-recombinant

Importantly, these adjuvant effects were absent in MyD88^{-}/$^{-}$ and TLR2^{-}/$^{-}$ knockout mice, highlighting the involvement of innate immune pathways.

Importantly, immunization with *B. subtilis* spore-based vaccines has been shown to induce strong mucosal and systemic immune responses. High levels of antigen-specific immunoglobulin A (IgA) at mucosal sites and immunoglobulin G (IgG) in serum have been detected, indicating robust humoral immunity. In a recent study, *B. subtilis* spores were used as a mucosal vaccine platform against *Mannheimia haemolytica*. Two spore-bound antigens (Spore-MhCP1 and Spore-MhCP2) administered intranasally to mice elicited strong mucosal IgA responses in bronchoalveolar lavage, saliva, and feces, along with significant bactericidal activity, whereas intramuscular immunization mainly induced serum IgG [34].

Moreover, these vaccines can enhance cellular immune responses by activating $CD4^+$ and $CD8^+$ T cells specific to the displayed antigen, contributing to long-term protection (Fig. 2.3). In a murine TB model, *B. subtilis* spores coated with a fusion protein of Mtb antigens (Spore-FP1) were used as mucosal boosters following BCG priming. Spore-FP1 reduced pulmonary bacterial burden, enhanced antigen-specific IgG/IgA responses, and promoted robust $CD4^+/CD8^+$ memory T-cell proliferation with Th1/Th17/Treg cytokine production. Importantly, mucosal delivery induced lung-resident Trm cells, demonstrating superior protection compared with BCG alone [44].

Several studies have further suggested that *B. subtilis* spores are recognized by Toll-like receptors, particularly TLR2 and TLR4, leading to the activation of

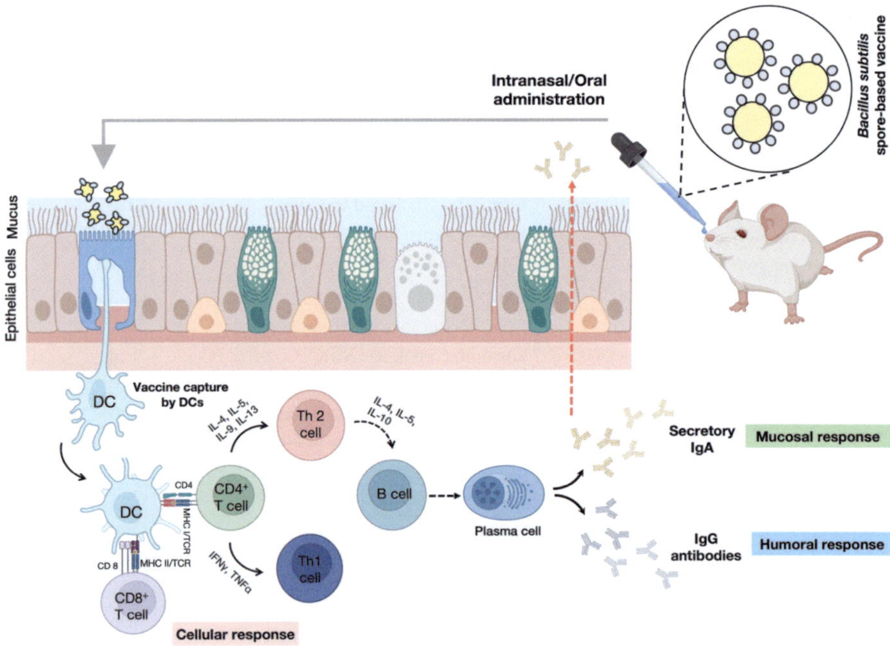

Fig. 2.3 Overview of immune responses induced by spore-based vaccines. Oral/internasal administration of spore-based vaccine activates both innate and adaptive immune responses at mucosal sites. It stimulates APCs such as DCs. DCs present the antigen using MHC I and MHC II molecules to $CD4^+$ and $CD8^+$ T cells. Activation of $CD8^+$ and $CD4^+$ Th1leads to cellular immune response. Activation of $CD4^+$ Th2 leading to the induction of humoral and mucosal immunity through the production of secretory IgA (sIgA) at mucosal surfaces (mucosal response) and IgG systemically (humoral response) [47]

downstream signaling pathways involved in immune modulation [45]. This interaction promotes the differentiation of T helper cells and supports a balanced Th1/Th2 response. The presence of IgG2a (a Th1-associated isotype) and IgG1 (a Th2-associated isotype) in the sera of immunized mice provides evidence of this dual activation pattern [46].

Together, these findings suggest that *B. subtilis* spore-based vaccine platforms are capable of eliciting comprehensive immune responses, making them promising candidates for mucosal immunization strategies against infectious diseases.

2.4 Applications of BSSD Technology as Whole-Cell Biocatalysts

Whole-cell biocatalysts are microbial systems that contain all the necessary enzymes and cofactors for the transformation of specific substrates into targeted products [48]. These systems often utilize living microorganisms, such as *B. subtilis*, that have been genetically modified to express specific enzymes or engineered metabolic pathways for enhanced catalytic performance [49]. Whole-cell biocatalysts provide numerous advantages, such as superior selectivity, improved catalytic effectiveness, operation under mild conditions, and minimal environmental impact [50].

Whole-cell biocatalysts streamline downstream processing [51], adhering to the principles of green chemistry [52]. Moreover, their reliance on cost-effective raw materials, such as renewable feedstocks, reduces production expenses, thereby enhancing their economic feasibility for large-scale industrial applications [53].

Besides *B. subtilis*, other microorganisms such as yeasts (*Saccharomyces cerevisiae* [54, 55], *Kluyveromyces marxianus*, and *Candida tropicalis*), bacteria (*E. coli* [56], *Pseudomonas putida* [57, 58], *Lactobacillus plantarum* [59, 60], *Zymomonas mobilis* [61], and *Methylosinus trichosporium* [62, 63]), and fungi (*Aspergillus niger* [64]) can act as whole-cell biocatalysts, producing enzymes and converting raw materials into valuable products like bioethanol and chemicals.

In the field of biocatalysis, spores act as robust platforms for the immobilization or presentation of enzymes [65], thereby enhancing their catalytic efficiency in chemical reactions across a wide range of industries, such as pharmaceutical manufacturing and food production [66, 67]. Their exceptional stability enables them to endure elevated temperatures and harsh chemical environments, making them particularly suitable for industrial applications [65]. They offer key advantages as genetically engineered platforms for enzyme incorporation [68], enhancing reaction efficiency and cofactor regeneration for cost-effective production [69]. Spores exhibit remarkable robustness, enabling them to persist in harsh environments, including intense heat and organic solvents, making them valuable for multiple uses and prolonged preservation [70].

Enzyme immobilization on spore surfaces involves their attachment to the outer layer of bacterial spores, such as those derived from *B. subtilis*. These spore-enzyme complexes function as spore-based biocatalysts, utilizing the entire spore structure to enhance catalytic efficiency, making them suitable for various industrial applications [27].

2.4.1 Industrial Applications of Spore Surface Display

Food Industry: Enzymes such as α-amylase, lipase, and β-galactosidase have been successfully immobilized on *B. subtilis* spores for various applications in food processing. This immobilization strategy greatly improves the durability and recyclability of the enzymes, making them especially effective for processes such as starch hydrolysis and lipid modification [27, 70].

Bioremediation: The expression of the AtzA enzyme on *B. thuringiensis* spores facilitates the detoxification of atrazine, a widely used herbicide, by converting it into less harmful substances [71]. The spore-based system promotes long-term persistence in soil and aquatic environments, improving the efficiency of pollutant degradation while eliminating the necessity for costly purification processes [72].

Pharmaceuticals: Immobilized enzymes are employed in pharmaceutical synthesis to facilitate selective reactions and enhance product yields. Although specific examples involving spores are relatively infrequent, recent studies have shown their potential. A notable example is the adsorption of bromelain, a therapeutic protease with anti-inflammatory properties, onto *Bacillus* spores. The immobilized enzyme exhibited enhanced thermal and storage stability, as well as improved catalytic efficiency (higher Vmax and lower Km) compared to the free enzyme. These findings suggest that edible spore carriers can provide a stable and biocompatible platform for the oral delivery of therapeutic enzymes, broadening their potential use in drug formulations [73, 74].

2.4.2 Advantages of Using BSSD Technology as Biocatalyst

As mentioned in previous sections, *B. subtilis* spores exhibit exceptional stability, enabling them to endure extreme temperatures and pH conditions that would typically lead to the denaturation of free enzymes [65]. When enzymes are immobilized on these spores, they benefit from the protective spore coat, which serves as a barrier against unfolding or loss of activity under adverse environmental conditions [27]. The T7 RNA polymerase-enabled high-density protein display (TIED) approach enables high loading density protein display on *B. subtilis* spores, with 10^6–10^7 enzymes displayed per spore .This method retains robust catalytic activities comparable to free enzymes while enhancing kinetic stability under methanol and elevated temperature conditions [65].

The *B. subtilis* spore coat is a complex, multilayered structure consisting of over 70 distinct proteins. This protective barrier plays a crucial role in shielding the spore from various environmental stressors, including elevated temperatures, chemical agents, and enzymatic degradation by lysozyme [13].

The spore coat is essential for providing resistance against UV radiation, hydrogen peroxide, and other sporicidal agents. Genetic mutations that alter the composition or integrity of the coat layers can markedly diminish this protective capability [75].

Heterologous proteins can be presented on the surface of engineered *B. subtilis* spores by fusing target enzymes with spore coat proteins, such as CotA, CotB, and CotC. This strategy facilitates high-density enzyme display while preserving the inherent properties of the spores [76]. The recombinant spores displaying exochitinase have demonstrated remarkable stability, retaining >70% activity after three reuse cycles and 62.7% after the fourth [77]. Enzymes such as lipase and tyrosinase have also been successfully immobilized on *B. subtilis* spores. Notably, spore-displayed lipase exhibited reusability over three reaction cycles without a significant decline in enzymatic performance. Similarly, spore-immobilized tyrosinase retained approximately 62% of its activity after six consecutive reaction cycles [70]. The application of genetically engineered spores streamlines the production of biocatalysts by eliminating the requirement for protein purification and subsequent immobilization on synthetic matrices [78].

2.5 Applications of BSSD Technology in Bioremediation and Environmental Biotechnology

One of the most notable traits of *B. subtilis* spores is their extreme resistance to environmental stressors, allowing them to survive under harsh and polluted conditions. Their ability to be produced at an industrial scale, coupled with their metabolically dormant state, enhances their stability, safety, and shelf life. These features collectively make *Bacillus* spores a versatile and reliable biological tool across diverse branches of environmental biotechnology [4].

Enzymes sensitive to environmental conditions can be displayed on the surface of stable spores, gaining stability and becoming less vulnerable, thus operating effectively in harsh environments. Therefore, BSSD gives spores an essential and key advantage, making them highly applicable in the field of environmental biotechnology [25, 76].

2.5.1 Use of Spore-Displayed Enzymes for Degrading Pollutants

Rapid urbanization, industrialization, and population growth have intensified environmental pollution through unsustainable practices and excessive waste generation [79]. Globally, over 80% of wastewater is released untreated or partially treated, introducing toxic substances such as heavy metals (Pb^{2+}, Hg^{2+}, Cu^{2+}, Cd^{2+}) that can cause severe health effects, including the development of abnormalities in fetuses [79]. Other harmful pollutants, such as halogenated compounds and phenolic dyes, are also commonly released [80]. These challenges highlight the urgent need for effective environmental remediation and waste management. Highly resistant and easily produced *B. subtilis* spores offer a cost-effective and efficient platform for tackling rising environmental contamination. Their use in biotechnological applications supports the removal of industrial pollutants, contributes to waste reduction, and promotes ecosystem restoration, ultimately aiding in the rehabilitation of the environment and the health of our planet [2]. Various functional proteins and enzymes can be displayed on the surface of *B. subtilis* spores, enhancing their stability and functional efficiency, making them suitable for use in bioremediation, which usually involves tolerating harsh environments [81] .

In a study using the BSSD system, a His12 tag was fused to the C-terminal of CotE (chosen for its abundance in the outer layer of the spore) to enable surface display. The recombinant spore expressing pCotE-His12 was able to bind and absorb Ni^{2+} and Cd^{2+} ions, successfully removing them from the environment [82]. Ni^{2+} ions were absorbed by bacillus spore using the same method described in Hinc et al. (2010). In this study, his-tag was genetically fused to the Cot B of the spore. Recombinant spores showed enhanced activity in binding to the Ni^{2+} ions in comparison with wild-type spores [83].

In another study, Laccase enzymes were successfully displayed on *B. subtilis* spores and applied for the decolorization of indigo carmine, a common textile dye, by utilizing Cot A as an anchoring protein. The recombinant spores exhibited high enzymatic activity under extreme pH and temperature conditions and retained stability in organic solvents. Kinetic analysis confirmed their catalytic efficiency, with significant dye removal achieved in both small- and large-scale setups. These findings highlight the potential of spore-displayed laccases as a robust biocatalyst for environmental bioremediation of dye-contaminated effluents [84].

In 2015, engineered *B. subtilis* spores were used for the first time to express and immobilize organophosphorus hydrolase (OPH) on their surface coat, using EDC-NHS as a coupling reagent. The displayed enzyme remained functional and exhibited enhanced tolerance to heat and pH variations. OPH was employed to detect and degrade organophosphorus pesticides commonly used in agriculture, demonstrating the effectiveness of engineered spores as a robust biocatalyst for bioremediation and the elimination of toxic environmental pollutants [25].

Halogenated compounds, particularly organo-haloacids, are toxic pollutants that are strictly regulated in drinking water due to their carcinogenic effects and potential to impair fetal development. Using BSSD technology, researchers engineered spores to present L-2-haloacid dehalogenase on their surface, enhancing the enzyme's stability and activity. The enzyme-coated spores were immobilized on a membrane to create a reusable bioreactor capable of degrading 2-chloropropionic acid (CPA), as well as other α-haloacids such as 2-bromopropionic acid (BPA), monochloroacetic acid (MCA), and dichloroacetic acid (DCA), under harsh conditions. This study represents another successful application of BSSD in bioremediation for the removal of environmental pollutants [85].

Tyrosinases, copper-containing monooxygenases, are particularly promising for the treatment of phenol-contaminated environments. In a study, tyrosinase was successfully displayed on *B. subtilis* spore surfaces using CotE as an anchor protein. Enzyme activity was confirmed in the presence of L-tyrosine and $CuSO_4$, and the recombinant spores retained functionality after multiple uses. These findings suggest that BSSD is an effective and reusable platform for the bioremediation of phenol, highlighting its potential for sustainable environmental cleanup [86].

To illustrate the successful application of BSSD in heavy metal removal, Diaco (2017) demonstrated its effectiveness in extracting lead from wastewater. Lead, a toxic heavy metal, accumulates in the body and can impair brain, liver, kidney, and bone function. In this study, researchers genetically fused the lead-binding protein PbrR to spore coat protein CotC, creating a recombinant protein (CotC-PbrRs) displayed on the spore surface. This recombinant spore effectively binds lead, demonstrating significant promise for the effectiveness and applicability of BSSD in bioremediation [87].

The haloalkane dehalogenase DhaA degrades sulfur mustard (HD) efficiently but is limited by its sensitivity to pH and temperature. To enhance stability, DhaA from *Rhodococcus rhodochrous* was fused to spore coat protein *CotG* and displayed on *B. subtilis* spores using BSSD technology. The recombinant strain expressed CotG-DhaA on the spore surface, as confirmed by dot blotting and immunofluorescence. The displayed enzyme retained activity, achieving 1.74 ± 0.06 U/mL against an HD analog, demonstrating the potential of CotG-mediated BSSD for enzyme stabilization and bioremediation in harsh environments [88].

Microbial metabolites like 2-hydroxy-6-oxohexa-2,4-dienoic acids (HODAs), produced from the meta-cleavage of aromatic compounds, can hinder mineralization. Spore-displayed enzymes, specifically meta-cleavage product hydrolases (MfphA and BphD), fused to CotG, can effectively transform HODAs. These fusions exhibit optimal activity at pH 7 and 70 °C, maintain about 80% activity at 80 °C and pH 10, and can be recycled up to ten times, demonstrating their potential in bioremediation applications [89].

In another study focused on environmental bioremediation, BSSD was combined with directed evolution to enhance the catalytic performance of lignin peroxidase, an enzyme that facilitates the degradation of lignin in agricultural waste and the breakdown of azo dyes with similarly complex aromatic structures. A total of 2700 mutants were screened, resulting in the identification of a superior variant exhibiting a 196.36% increase in catalytic activity. Structural modeling and functional analysis clarified the molecular basis for the enhanced activity. These findings underscore the potential of BSSD as an effective platform for engineering robust, immobilized enzymes capable of mitigating environmental pollution through the degradation of industrial dyes and agricultural biomass [90].

2.5.2 Spores as Biosensors for Environmental Monitoring

Biosensors are devices that turn a biological response into an electrical signal. They need to be very specific, work well under different conditions, like changes in pH and temperature, and be reusable [91]. Creating biosensors involves a mix of knowledge from chemistry, biology, and engineering. The materials in biosensors fall into three main categories based on how they work:

Biocatalytic Group: This includes enzymes that help speed up reactions.
Bioaffinity Group: This includes antibodies and nucleic acids that can specifically bind to other molecules.
Microbe-Based: This includes microorganisms used for their biological properties [92].

The first microbe-based or cell-based sensor was actualized by Diviès in 1975 [93], marking a foundational step in the advancement of biosensors. Building on this concept, the BSSD system uses robust spores to securely hold sensing elements, allowing for real-time monitoring and detection of environmental pollutants. Furthermore, when engineered with catalytic or binding proteins, these biosensors can actively contribute to pollutant degradation. The ability of engineered spores to detect and remediate contaminants positions BSSD as a powerful tool for both environmental monitoring and bioremediation [76]. Another key advantage of BSSD in biosensor-based environmental surveillance is its high tolerance to stressful and harsh conditions that typically prevail in polluted environments containing toxic elements [22].

This system enables the surface display of receptor proteins, enzymes, or antibodies capable of detecting specific biological or chemical signals. As a result, engineered spores can be tailored for the detection of a wide variety of environmental pollutants or disease-related biomarkers. These biosensors have shown potential for real-time, continuous monitoring in air or water, offering sensitive, efficient, and reusable solutions for both industrial and biomedical applications. When compared to traditional cell-based biosensors, spore-based systems offer superior resilience under harsh environmental conditions such as high temperature, extreme pH, oxidizing agents, and osmotic stress [76, 94].

Emphasizing their remarkable resilience to extreme environmental stresses, *B. subtilis* spores have been incorporated into 3D-printed materials [95]. Unlike vegetative cells, spores can remain viable for extended periods, making them ideal for long-term applications. When embedded into 3D-printed structures, these spore-based living materials retain their functionality over time, offering a durable and practical solution. Such materials hold significant potential for use in environmental sensing, chemical and material production, and bioelectronics. The spores are evenly distributed throughout the 3D material and can germinate to carry out their engineered functions, such as responding to chemicals like IPTG, xylose (Xyl), vanillic acid (Van), and cuminic acid (Cum) [95, 96].

Recent advances in fabrication techniques, including electrospinning and bioprinting, further enhance the incorporation of spore-based systems into porous substrates without compromising enzymatic activity. This has enabled the development of embedded biosensing and bioremediation platforms capable of continuous pollutant detection and degradation. Moreover, engineering spores to co-display functional enzymes with binding domains such as cellulose-binding domains allows for targeted immobilization on support materials, facilitating their use in structured bioreactors like fiber brushes or spiral sheets. These innovations position BSSD as a robust and versatile platform for environmental monitoring and remediation [80].

Fantino et al. (2009) developed a whole-cell bacterial sensor, termed sposensor, using immobilized engineered *B. subtilis* spores. The system uses two spores: reporting spores carrying a reporter gene for target detection, and control spores to confirm germination and viability. The system works in a single step by incubating spores in a medium containing the target compound. β-Galactosidase was used as the reporter, and its activity was measured using a color change. The sposensor effectively detected both a metal ion (Zn^{2+}) and the antibiotic bacitracin. Thanks to the spores' stability and strength, sposensor is a simple, reliable tool for detecting toxic compounds in the environment [97].

Another notable example of BSSD in environmental monitoring is a green fluorescent protein-based whole-cell biosensor (WCB-GFP) developed using *B. subtilis* to detect arsenic contamination. In this system, the reporter gene gfpmut3a was placed under the control of the arsenic-responsive promoter Pars and introduced into *B. subtilis* via plasmid transformation. The resulting strain, BsWCB-GFP, was specifically activated by inorganic arsenic compounds As (III) and As (V), but not by organic arsenic such as DMA(V). The biosensor demonstrated high tolerance to arsenic toxicity and could operate effectively even under concentrations considered harmful in water and soil. Significantly, dormant spores of the engineered strain were able to detect arsenic concentrations ranging from 0.1 to 1000 μM within 4 hours after germination, indicating its potential for use in rapid, field-deployable monitoring of arsenic pollution in environmental samples [98].

A recent study demonstrated the scalable production of spore-based biosensors by integrating genetically engineered *B. subtilis* spores into a PEG-PVA hydrogel matrix. The spores were modified to express superfolder GFP (sfGFP) in response to IPTG, serving as a model analyte. After sporulation, the spores were embedded into the hydrogel, which was then cast onto a PET support film using a roll-to-roll coating process. This method allowed the continuous fabrication of sensor strips (5 × 20 mm), capable of producing visible fluorescence upon analyte detection. The process achieved a production rate of one million biosensor units in less than 4 days, with each sensor costing only 0.0009 USD, significantly cheaper than commercial biosensors. This example highlights the potential of spore-based systems for low-cost, large-scale biosensor fabrication, further supporting their use in environmental monitoring and field applications [99].

2.6 BSSD as a Model for Other Spore-Forming Bacteria

The BSSD system has been extensively studied and utilized due to its stability, safety, and ease of genetic manipulation. Its well-characterized genetics, stability, and ease of surface engineering make it a valuable model for other spore-forming bacteria.

Although various *Clostridium* and *Bacillus* strains form spores, *B. subtilis* is preferred for display due to its GRAS status and safety for humans [100]. However, other spore-forming bacteria, such as *Bacillus thuringiensis*, have also been explored for spore surface display applications. *B. thuringiensis* is a gram-positive, spore-forming microorganism commonly found in soil, sericulture habitats, and both marine and freshwater environments [101]. The proteins employed for surface display in *B. thuringiensis* spores differ from those utilized in *B. subtilis.* In a study by Du et al. (2005), an SSD system was developed using *B. thuringiensis* spores, in which the N-terminal region of the insecticidal protoxin (Cry1Ac) was replaced with a heterologous protein. By employing a sporulation-specific promoter, the system successfully displayed GFP and a single-chain antibody (scFv) against phOx (anti-phOx) on the spore surface. The engineered spores were able to specifically bind their target [102]. In another study, Park et al. (2009) have demonstrated the display of recombinant proteins, such as GFP and β-galactosidase, on the spores of *B. thuringiensis* via fusion with InhA, an exosporium protein. Surface expression was confirmed through flow cytometry, confocal microscopy, enzyme assays, and immunogold electron microscopy [103]. These examples demonstrate the flexibility of spore surface display systems and suggest that the methods developed for *B. subtilis* could serve as a foundation for similar applications in other spore-forming bacteria. Therefore, selecting appropriate bacteria that possess the key features required for spore surface display will broaden the scope of this technology in various industrial and medical fields.

2.7 Challenges and Future Direction of BSSD Technology

As described in previous sections, to date, numerous BSSD systems have been successfully developed and utilized in the field of biotechnological manufacturing. Despite significant advances in this technology, challenges still remain that need to be addressed for broader application. This section outlines several key challenges along with potential approaches to mitigate them.

The majority of these studies are performed on a small scale in the laboratory. Hence, the importance of the scaling-up of this platform and the strategies to achieve it should be considered in upcoming studies. Scaling up the production of this platform while ensuring quality and consistency poses further challenges. This demands well-designed fermentation and purification processes, combined with rigorous quality control and validation [104].

One of the existing challenges in recombinant display systems is the low expression level of passenger proteins. Using *a* strong promoter can lead to increased transcription of the genes of interest and enhance the protein display efficiency. In this demand, Nguyen et al. (2014) explored the use of IPTG-inducible promoters to modulate the anchor protein CotB for α-amylase BSSD. Their study indicated that the level of passenger protein displayed on the spore surface was influenced by the concentration of IPTG, although the optimal inducer amount differed according to the carrier and passenger proteins. This demonstrates that passenger proteins can be expressed in a tunable manner, allowing customization of their amount for particular applications [105]. However, IPTG is inappropriate for large-scale production of foreign proteins because it is costly and toxic [106]. Future research should focus on engineering strong promoters to enhance expression efficiency, particularly those derived from probiotic or GRAS bacteria, which hold promise for applications in the food and medical industries. Another strategy for regulating gene expression is using engineered riboswitches, which are synthetic RNA devices that can modulate gene expression at transcriptional and post-transcriptional levels. They are constructed using artificial aptamers or modified natural riboswitches through directed evolution. When integrated into recombinant plasmids, they can be used alongside spore surface display technology to control target protein expression on spores [107]. Furthermore, the efficiency of surface display of the protein of interest can be improved using advanced tools like CRISPR/Cas9 systems for precise genome-level gene integration. Owing to its high accuracy and efficiency, CRISPR/Cas9 has gained significant attention in metabolic engineering [108]. Studies on gene knock-in and knock-out demonstrate its potential for enhancing target protein overexpression in spore surface display systems [109].

Selecting suitable carrier proteins to ensure the protective properties of the spore coat and the intrinsic stability of the displayed protein for the spore surface display system is another obstacle in this system. Although the spore's outer-coat protein provides considerable protection against environmental stresses, some coat proteins may still undergo structural instability or lose their function, especially if they are not effectively anchored. Thus, consideration of this aspect is essential for achieving protein activity and durability. In a recent study by Potot et al., they investigated the potential use of inner-coat protein oxalate decarboxylase (OxdD) to display an active passenger enzyme. Their study shows that OxdD can effectively display active phytase, a widely used feed enzyme, on the spore surface. Interestingly, despite the higher abundance of the outer-coat protein CotG, the OxdD-Phy fusion showed greater surface representation. Moreover, the capability of OxdD as a carrier protein was further confirmed by displaying the functional tetrameric β-glucuronidase from *E. coli* on the spore surface [23]. Given the potential of this protein in surface display, future studies should focus on investigating other inner spore coat proteins for their applicability as effective carriers in spore surface display systems. Additionally, a multilayered structure of *B. subtilis* spores can hinder the efficient display of heterologous proteins [13]. A major limitation is the high density of these proteins on the spore, which affects the efficiency of protein display and the functional versatility of the BSSD system. To conquer this issue, multiple anchor proteins have been studied to improve display efficiency. In a study by Liu et al. (2019), they utilized CotC and CotG coat proteins in combination to display trehalose synthase (TreS). The successful functional expression of TreS on the *B. subtilis* spore indicated the benefits of multi-anchor strategies for improving both performance and reusability [110].

For effective fusion protein construction, it is essential to study the protein's molecular dynamics and select an appropriate linker. Flexible linkers allow structural flexibility, while rigid linkers help prevent domain interference. Thus, choosing an optimal linker is critical for successful surface display [100]. Rigid linkers, such as EAAAK, help maintain protein structure but may restrict the mobility and flexibility of the displayed proteins, whereas highly flexible linkers, like GGGGS, might reduce protein stability. To overcome this limitation, a combinatorial approach combining rigid and flexible linkers has been designed, with intermediate flexibility (e.g., FFFRR) demonstrating excellent performance in fusion proteins such as PhoC-GFP [111]. Implementing similar strategies in spore surface display systems can improve protein stability, accessibility, and functionality, thereby expanding their biotechnological potential applications. Using linker libraries with adjustable flexibility in spore surface display systems can also improve protein stability, accessibility, and function, offering promising applications in various industrial sectors.

Moreover, future advancements in spore display systems may focus on protein adsorption onto a nonrecombinant spore surface display system, which requires deeper insight into the underlying adhesion mechanisms. While hydrophobic interactions appear significant, especially in cases like β-galactosidase binding, electrostatic forces seem less influential. The possible roles of other forces, such as van der Waals and capillary effects, and their sensitivity to environmental factors and protein properties, remain open questions for further research [17].

By addressing current limitations and leveraging emerging advancements, spore surface display technology can be extended to a broad range of biotechnological applications, offering sustainable and efficient solutions across diverse sectors.

2.8 Conclusions

According to existing studies, BSSD technology can be extended to a wide range of industrial and environmental applications. BSSD technology presents a versatile platform with advantages such as stability, cost-effectiveness, and multifunctionality. Beyond mucosal vaccines and oral drug delivery, its applications extend to biocatalysis and environmental biotechnology, enabling efficient enzyme display and sustainable solutions for bioremediation and biosensing. By overcoming the current challenges, this technology holds great promise for commercialization and broader industrial applications. As the applications of *B. subtilis* spore surface display continue to grow across biotechnology, medicine, and environmental science, it will be exciting to witness the innovative solutions this promising technology will bring in the years ahead.

Competing Interests The authors have no conflicts of interest to declare that are relevant to the content of this chapter.

References

1. Errington J, Aart LTV. Microbe profile: Bacillus subtilis: model organism for cellular development, and industrial workhorse. Microbiology (Reading). 2020;166(5):425–7. https://doi.org/10.1099/mic.0.000922.
2. Akinsemolu AA, et al. Exploring Bacillus subtilis: ecology, biotechnological applications, and future prospects. J Basic Microbiol. 2024;64(6):e2300614. https://doi.org/10.1002/jobm.202300614.

3. Barák I. Special Issue "Bacillus subtilis as a model organism to study basic cell processes". Microorganisms. 2021;9(12):2459. https://doi.org/10.3390/microorganisms9122459.
4. Stülke J, et al. Bacillus subtilis, a Swiss Army knife in science and biotechnology. J Bacteriol. 2023;205(5):e0010223. https://doi.org/10.1128/jb.00102-23.
5. Su Y, et al. Bacillus subtilis: a universal cell factory for industry, agriculture, biomaterials and medicine. Microb Cell Factories. 2020;19(1):173. https://doi.org/10.1186/s12934-020-01436-8.
6. Saadati F, et al. Advances and principles of hyaluronic acid production, extraction, purification, and its applications: a review. Int J Biol Macromol. 2025;312:143839. https://doi.org/10.1016/j.ijbiomac.2025.143839.
7. Higgins D, Dworkin J. Recent progress in Bacillus subtilis sporulation. FEMS Microbiol Rev. 2012;36(1):131–48. https://doi.org/10.1111/j.1574-6976.2011.00310.x.
8. Tan IS, Ramamurthi KS. Spore formation in Bacillus subtilis. Environ Microbiol Rep. 2014;6(3):212–25. https://doi.org/10.1111/1758-2229.12130.
9. Ghaedmohammadi S, Bahrulolum H, Ahmadian G. A systemic review on bacillus subtilis spore structure and its applications in the development of mucosal vaccines and adjuvants. J Arak Univ Med Sci. 2022;25(3):338–53. https://doi.org/10.32598/jams.25.3.6859.1.
10. Piggot PJ, Hilbert DW. Sporulation of Bacillus subtilis. Curr Opin Microbiol. 2004;7(6):579–86. https://doi.org/10.1016/j.mib.2004.10.001.
11. Koopman N, et al. Mechanisms and applications of bacterial sporulation and germination in the intestine. Int J Mol Sci. 2022;23(6) https://doi.org/10.3390/ijms23063405.
12. Henriques AO, Moran CP Jr. Structure, assembly, and function of the spore surface layers. Ann Rev Microbiol. 2007;61:555–88. https://doi.org/10.1146/annurev.micro.61.080706.093224.
13. McKenney PT, Driks A, Eichenberger P. The Bacillus subtilis endospore: assembly and functions of the multilayered coat. Nat Rev Microbiol. 2013;11(1):33–44. https://doi.org/10.1038/nrmicro2921.
14. Setlow P. Spore resistance properties. Microbiol Spectrum. 2014;2(5) https://doi.org/10.1128/microbiolspec.TBS-0003-2012.
15. Wang H, Wang Y, Yang R. Recent progress in Bacillus subtilis spore-surface display: concept, progress, and future. Appl Microbiol Biotechnol. 2017;101(3):933–49. https://doi.org/10.1007/s00253-016-8080-9.
16. Mahmoodi A, Farinas ET. Applications of Bacillus subtilis protein display for medicine, catalysis, environmental remediation, and protein engineering. Microorganisms. 2024;12(1):97. https://doi.org/10.3390/microorganisms12010097.
17. Isticato R, Ricca E. Spore surface display. Microbiol Spectrum. 2014;2(5) https://doi.org/10.1128/microbiolspec.TBS-0011-2012.
18. Rostami A, et al. Display of B. pumilus chitinase on the surface of B. subtilis spore as a potential biopesticide. Pestic Biochem Physiol. 2017;140:17–23. https://doi.org/10.1016/j.pestbp.2017.05.008.
19. Leggett MJ, et al. Bacterial spore structures and their protective role in biocide resistance. J Appl Microbiol. 2012;113(3):485–98. https://doi.org/10.1111/j.1365-2672.2012.05336.x.
20. Isticato R, et al. Surface display of recombinant proteins on Bacillus subtilis spores. J Bacteriol. 2001;183(21):6294–301. https://doi.org/10.1128/jb.183.21.6294-6301.2001.
21. Yang H, et al. An overview and future prospects of recombinant protein production in Bacillus subtilis. Appl Microbiol Biotechnol. 2021;105(18):6607–26. https://doi.org/10.1007/s00253-021-11533-2.
22. Lin P, et al. Progress in research and application development of surface display technology using Bacillus subtilis spores. Appl Microbiol Biotechnol. 2020;104(6):2319–31. https://doi.org/10.1007/s00253-020-10348-x.
23. Potot S, et al. Display of recombinant proteins on Bacillus subtilis spores, using a coat-associated enzyme as the carrier. Appl Environ Microbiol. 2010;76(17):5926–33. https://doi.org/10.1128/aem.01103-10.

24. Hinc K, et al. Expression and display of UreA of helicobacter acinonychis on the surface of Bacillus subtilis spores. Microb Cell Factories. 2010a;9(1):2. https://doi.org/10.1186/1475-2859-9-2.
25. Falahati-Pour SK, et al. Covalent immobilization of recombinant organophosphorus hydrolase on spores of Bacillus subtilis. J Appl Microbiol. 2015;118(4):976–88. https://doi.org/10.1111/jam.12744.
26. Gashtasbi F, Ahmadian G, Noghabi KA. New insights into the effectiveness of alpha-amylase enzyme presentation on the Bacillus subtilis spore surface by adsorption and covalent immobilization. Enzym Microb Technol. 2014;64-65:17–23. https://doi.org/10.1016/j.enzmictec.2014.05.006.
27. Ugwuodo CJ, Nwagu TN. Stabilizing enzymes by immobilization on bacterial spores: a review of literature. Int J Biol Macromol. 2021;166:238–50. https://doi.org/10.1016/j.ijbiomac.2020.10.171.
28. Li S, et al. Display of lignin peroxidase on the surface of Bacillus subtilis. Appl Biochem Biotechnol. 2024;196(10):6849–63. https://doi.org/10.1007/s12010-024-04869-8.
29. Poorolajal J. The global pandemics are getting more frequent and severe. J Res Health Sci. 2021;21(1):e00502. https://doi.org/10.34172/jrhs.2021.40.
30. Shattock AJ, et al. Contribution of vaccination to improved survival and health: modelling 50 years of the expanded Programme on immunization. Lancet. 2024;403(10441):2307–16. https://doi.org/10.1016/S0140-6736(24)00850-X.
31. Pollard AJ, Bijker EM. A guide to vaccinology: from basic principles to new developments. Nat Rev Immunol. 2021;21(2):83–100. https://doi.org/10.1038/s41577-020-00479-7.
32. Pardi N, et al. mRNA vaccines — a new era in vaccinology. Nat Rev Drug Discov. 2018;17(4):261–79. https://doi.org/10.1038/nrd.2017.243.
33. Bahrulolum H, Beyranvand P, Ahmadian G. Harnessing Bacillus subtilis Spore Surface Display (BSSD) technology for mucosal vaccines and drug delivery: innovations in respiratory virus immunization. Drugs Drug Candidates. 2024;3(4):774–95. https://doi.org/10.3390/ddc3040044.
34. Uddin MS, et al. Mucosal immunization with spore-based vaccines against Mannheimia haemolytica enhances antigen-specific immunity. Vaccines (Basel). 2024;12(4) https://doi.org/10.3390/vaccines12040375.
35. Amuguni H, Tzipori S. Bacillus subtilis: a temperature resistant and needle free delivery system of immunogens. Hum Vaccin Immunother. 2012;8(7):979–86. https://doi.org/10.4161/hv.20694.
36. Desvaux M, et al. Protein cell surface display in gram-positive bacteria: from single protein to macromolecular protein structure. FEMS Microbiol Lett. 2006;256(1):1–15. https://doi.org/10.1111/j.1574-6968.2006.00122.x.
37. Detmer A, Glenting J. Live bacterial vaccines--a review and identification of potential hazards. Microb Cell Factories. 2006;5:23. https://doi.org/10.1186/1475-2859-5-23.
38. Ricca E, et al. Mucosal vaccine delivery by non-recombinant spores of Bacillus subtilis. Microb Cell Factories. 2014;13(1):115. https://doi.org/10.1186/s12934-014-0115-2.
39. Isticato R, Ricca E, Baccigalupi L. Spore adsorption as a nonrecombinant display system for enzymes and antigens. J Vis Exp. 2019;145 https://doi.org/10.3791/59102.
40. Huang J-M, et al. Mucosal delivery of antigens using adsorption to bacterial spores. Vaccine. 2010;28(4):1021–30. https://doi.org/10.1016/j.vaccine.2009.10.127.
41. Bahrulolum H, Ahmadian G. Bacillus subtilis surface display technology: applications in bioprocessing and sustainable manufacturing. Biotechnol Biofuels Bioprod. 2025;18(1):34. https://doi.org/10.1186/s13068-025-02635-4.
42. Saggese A, et al. The bacterial spore as a mucosal vaccine delivery system. Int J Mol Sci. 2023;24(13) https://doi.org/10.3390/ijms241310880.
43. de Almeida MEM, et al. Bacillus subtilis spores as delivery system for nasal Plasmodium falciparum circumsporozoite surface protein immunization in a murine model. Sci Rep. 2022;12(1):1531. https://doi.org/10.1038/s41598-022-05344-2.

44. Copland A, et al. Mucosal delivery of fusion proteins with Bacillus subtilis spores enhances protection against tuberculosis by Bacillus Calmette-Guérin. Front Immunol. 2018;9:346. https://doi.org/10.3389/fimmu.2018.00346.
45. Liao B, et al. Disruption of spore coat integrity in Bacillus subtilis enhances macrophage immune activation. Curr Issue Mol Biol. 2025;47(5) https://doi.org/10.3390/cimb47050378.
46. Barnes AG, et al. Bacillus subtilis spores: a novel microparticle adjuvant which can instruct a balanced Th1 and Th2 immune response to specific antigen. Eur J Immunol. 2007;37(6):1538–47. https://doi.org/10.1002/eji.200636875.
47. Cho C-S, et al. Mucosal vaccine delivery using Mucoadhesive polymer particulate systems. Tissue Eng Regener Med. 2021;18(5):693–712. https://doi.org/10.1007/s13770-021-00373-w.
48. Lin B, Tao Y. Whole-cell biocatalysts by design. Microb Cell Factories. 2017;16:1–12. https://doi.org/10.1186/s12934-017-0724-7.
49. Bao T, et al. Efficient whole-cell biocatalyst for acetoin production with NAD+ regeneration system through homologous co-expression of 2, 3-butanediol dehydrogenase and NADH oxidase in engineered Bacillus subtilis. PLoS One. 2014;9(7):e102951. https://doi.org/10.1371/journal.pone.0102951.
50. Anteneh YS, Franco CMM. Whole cell actinobacteria as biocatalysts. Front Microbiol. 2019;10:77. https://doi.org/10.3389/fmicb.2019.00077.
51. Pinto A, Contente ML, Tamborini L. Advances on whole-cell biocatalysis in flow. Curr Opin Green Sustain Chem. 2020;25:100343. https://doi.org/10.1016/j.cogsc.2020.04.004.
52. Wachtmeister J, Rother D. Recent advances in whole cell biocatalysis techniques bridging from investigative to industrial scale. Curr Opin Biotechnol. 2016;42:169–77. https://doi.org/10.1016/j.copbio.2016.05.005.
53. Molinari F, Pollegioni L, Rosini E. Whole-cell bioconversion of renewable biomasses-related aromatics to cis, cis-muconic acid. ACS Sustain Chem Eng. 2023;11(6):2476–85. https://doi.org/10.1021/acssuschemeng.2c06534.
54. Zhang C, et al. Saccharomyces cerevisiae cell surface display technology: strategies for improvement and applications. Front Bioeng Biotechnol. 2022;10:1056804. https://doi.org/10.3389/fbioe.2022.1056804.
55. Ye M, et al. Cell-surface engineering of yeasts for whole-cell biocatalysts. Bioprocess Biosyst Eng. 2021;44:1003–19. https://doi.org/10.1007/s00449-020-02484-5.
56. Hu B, et al. Whole-cell P450 biocatalysis using engineered Escherichia coli with fine-tuned heme biosynthesis. Adv Sci. 2023;10(6):2205580. https://doi.org/10.1002/advs.202205580.
57. Wang Y, et al. Reconstruction of lactate utilization system in Pseudomonas putida KT2440: a novel biocatalyst for l-2-hydroxy-carboxylate production. Sci Rep. 2014;4(1):6939. https://doi.org/10.1038/srep06939.
58. Nikel PI, de Lorenzo V. Pseudomonas putida as a functional chassis for industrial biocatalysis: from native biochemistry to trans-metabolism. Metab Eng. 2018;50:142–55. https://doi.org/10.1016/j.ymben.2018.05.005.
59. Gobinath D, Prapulla SG. Transgalactosylating β-galactosidase from probiotic Lactobacillus plantarum MCC2156: production and permeabilization for use as whole cell biocatalyst. J Food Sci Technol. 2015;52:6003–9. https://doi.org/10.1007/s13197-014-1656-4.
60. Nguyen H-M, et al. Display of a β-mannanase and a chitosanase on the cell surface of Lactobacillus plantarum towards the development of whole-cell biocatalysts. Microb Cell Factories. 2016;15:1–14. https://doi.org/10.1186/s12934-016-0570-z.
61. Yang S, et al. Zymomonas mobilis as a model system for production of biofuels and biochemicals. Microb Biotechnol. 2016;9(6):699–717. https://doi.org/10.1111/1751-7915.12408.
62. Baldo H, et al. Methane biohydroxylation into methanol by Methylosinus trichosporium OB3b: possible limitations and formate use during reaction. Front Bioeng Biotechnol. 2024;12:1422580. https://doi.org/10.3389/fbioe.2024.1422580.
63. Hwang IY, et al. Batch conversion of methane to methanol using Methylosinus trichosporium OB3b as biocatalyst. J Microbiol Biotechnol. 2015;25(3):375–80. https://doi.org/10.4014/jmb.1412.12007.

64. Xiao M, Qi C, Obbard JP. Biodiesel production using Aspergillus Niger as a whole-cell biocatalyst in a packed-bed reactor. GCB Bioenergy. 2011;3(4):293–8. https://doi.org/10.1111/j.1757-1707.2010.01087.x.
65. Hui Y, Cui Z, Sim S. Stress-tolerant, recyclable, and renewable biocatalyst platform enabled by engineered bacterial spores. ACS Synth Biol. 2022;11(8):2857–68. https://doi.org/10.1021/acssynbio.2c00256.
66. Sun Z, et al. Artificially sporulated Escherichia coli cells as a robust cell factory for interfacial biocatalysis. Nat Commun. 2022;13(1):3142. https://doi.org/10.1038/s41467-022-30915-2.
67. Fernandez-Lafuente R. Enzyme immobilization. Molecules. 2023;28(3):1373. https://doi.org/10.3390/molecules28031373.
68. Chappell TC, et al. Advancing protein display on bacterial spores through an extensive survey of coat components. bioRxiv [Preprint]. 2024;24 https://doi.org/10.1101/2024.11.22.624950.
69. Intasian P, et al. Enzymes, in vivo biocatalysis, and metabolic engineering for enabling a circular economy and sustainability. Chem Rev. 2021;121(17):10367–451. https://doi.org/10.1021/acs.chemrev.1c00121.
70. Zhang X, et al. Applications of Bacillus subtilis spores in biotechnology and advanced materials. Appl Environ Microbiol. 2020;86(17):e01096–20. https://doi.org/10.1128/AEM.01096-20.
71. Hsieh H-Y, et al. A Bacillus spore-based display system for bioremediation of atrazine. Appl Environ Microbiol. 2020;86(18):e01230–20. https://doi.org/10.1128/AEM.01230-20.
72. Yamaguchi H, Miyazaki M. Bioremediation of hazardous pollutants using enzyme-immobilized reactors. Molecules. 2024;29(9):2021. https://doi.org/10.3390/molecules29092021.
73. Fernandez-Lafuente R. Enzyme immobilization 2016. Molecules. 2017;22(4):601. https://doi.org/10.3390/molecules22040601.
74. Nwagu TN, Ugwuodo CJ. Stabilizing bromelain for therapeutic applications by adsorption immobilization on spores of probiotic Bacillus. Int J Biol Macromol. 2019;127:406–14. https://doi.org/10.1016/j.ijbiomac.2019.01.061.
75. Riesenman PJ, Nicholson WL. Role of the spore coat layers in Bacillus subtilis spore resistance to hydrogen peroxide, artificial UV-C, UV-B, and solar UV radiation. Appl Environ Microbiol. 2000;66(2):620–6. https://doi.org/10.1128/AEM.66.2.620-626.2000.
76. Guoyan Z, et al. Bacillus subtilis spore surface display technology: a review of its development and applications. J Microbiol Biotechnol. 2019;29(2):179–90. https://doi.org/10.4014/jmb.1807.06066.
77. Ullah M, et al. Display of bacterial Exochitanase on Bacillus subtilis spores improved enzyme stability and recyclability. Molecules. 2024;29(18):4302. https://doi.org/10.3390/molecules29184302.
78. Hui Y. Applications of genetically engineered Bacillus subtilis in biocatalysis and functional materials [Ph.D. Dissertation]. California Institute of Technology; 2023. https://doi.org/10.7907/cdja-ck19.
79. Briffa J, Sinagra E, Blundell R. Heavy metal pollution in the environment and their toxicological effects on humans. Heliyon. 2020;6(9):e04691. https://doi.org/10.1016/j.heliyon.2020.e04691.
80. Davenport B, Hallam SJ. Emerging enzyme surface display systems for waste resource recovery. Environ Microbiol. 2023;25(2):241–9. https://doi.org/10.1111/1462-2920.16284.
81. Huang GL, et al. Stabilizing displayed proteins on vegetative Bacillus subtilis cells. Appl Microbiol Biotechnol. 2018;102:6547–65. https://doi.org/10.1007/s00253-018-9062-x.
82. Kim W, et al. Removal of Ni 2+ and Cd 2+ by surface display of polyhistidine on Bacillus subtilis spore using CotE anchor protein. Biotechnol Bioprocess Eng. 2019;24:375–81. https://doi.org/10.1007/s12257-018-0467-2.
83. Hinc K, et al. Efficient binding of nickel ions to recombinant Bacillus subtilis spores. Res Microbiol. 2010b;161(9):757–64. https://doi.org/10.1016/j.resmic.2010.07.008.
84. Cho E-A, et al. Decolorization of indigo carmine by laccase displayed on Bacillus subtilis spores. Enzym Microb Technol. 2011;49(1):100–4. https://doi.org/10.1016/j.enzmictec.2011.03.005.

85. Roshid MHO, et al. Development of a renewable endospore-displayed enzyme bioreactor for the bioremediation of α-haloacids. J Water Process Eng. 2024;65:105789. https://doi.org/10.1016/j.jwpe.2024.105789.
86. Hosseini-Abari A, et al. Surface display of bacterial tyrosinase on spores of Bacillus subtilis using CotE as an anchor protein. J Basic Microbiol. 2016;56(12):1331–7. https://doi.org/10.1002/jobm.201600203.
87. Diaco SM. Bioremediation of lead using spore surface displayed proteins. Honors theses and capstones. University of New Hampshire; 2017. p. 360. Available online: https://scholars.unh.edu/honors/360.
88. Wang F, et al. Bacillus subtilis spore surface display of haloalkane dehalogenase DhaA. Curr Microbiol. 2019;76(10):1161–7. https://doi.org/10.1007/s00284-019-01723-7.
89. Qu Y, et al. Catalytic transformation of HODAs using an efficient meta-cleavage product hydrolase-spore surface display system. J Mol Catal B Enzym. 2014;102:204–10. https://doi.org/10.3390/microorganisms12010097.
90. Shi N, et al. High-throughput screening and identification of lignin peroxidase based on spore surface display of Bacillus subtilis. J Sci Food Agric. 2025;105(4):2179–89. https://doi.org/10.1002/jsfa.13988.
91. Cammann K. Bio-sensors based on ion-selective electrodes. Fresen Z Anal Chem. 1977;287(1):1–9. https://doi.org/10.1007/BF00539519.
92. Mehrotra P. Biosensors and their applications–a review. J Oral Biol Craniofac Res. 2016;6(2):153–9. https://doi.org/10.1016/j.jobcr.2015.12.002.
93. Divies C. Remarques sur l'oxydation de l'éthanol par une électrode microbienne d'Acetobacter xylinum. [Remarks on ethanol oxidation by an "Acetobacter xylinum" microbial electrode (author's transl)]. Ann Microbiol (Paris). 1975;126(2):175–86.
94. Park M. Surface display technology for biosensor applications: a review. Sensors. 2020;20(10):2775. https://doi.org/10.3390/s20102775.
95. González LM, Mukhitov N, Voigt CA. Resilient living materials built by printing bacterial spores. Nat Chem Biol. 2020;16(2):126–33. https://doi.org/10.1038/s41589-019-0412-5.
96. Sahin O, et al. Physical basis for the adaptive flexibility of Bacillus spore coats. J R Soc Interface. 2012;9(76):3156–60. https://doi.org/10.1098/rsif.2012.0470.
97. Fantino J-R, Barras F, Denizot F. Sposensor: a whole-bacterial biosensor that uses immobilized Bacillus subtilis spores and a one-step incubation/detection process. J Mol Microbiol Biotechnol. 2009;17(2):90–5. https://doi.org/10.1159/000206634.
98. Valenzuela-García LI, et al. Design of a whole-cell biosensor based on Bacillus subtilis spores and the green fluorescent protein to monitor arsenic. Microbiol Spectrum. 2023;11(4):e00432–23. https://doi.org/10.1128/spectrum.00432-23.
99. Schulz-Schönhagen K. Bacillus subtilis biosensors: engineering a living material sensor platform [Ph.D. Dissertation]. ETH Zurich; 2019. https://doi.org/10.3929/ethz-b-000389191.
100. Chen H, Ullah J, Jia J. Progress in Bacillus subtilis spore surface display technology towards environment, vaccine development, and biocatalysis. J Mol Microbiol Biotechnol. 2017;27(3):159–67. https://doi.org/10.1159/000475177.
101. Belousova ME, et al. Dissecting the environmental consequences of Bacillus thuringiensis application for natural ecosystems. Toxins. 2021;13(5):355. https://doi.org/10.3390/toxins13050355.
102. Du C, et al. Surface display of recombinant proteins on Bacillus thuringiensis spores. Appl Environ Microbiol. 2005;71(6):3337–41. https://doi.org/10.1128/AEM.71.6.3337-3341.2005.
103. Park TJ, et al. Spore display using Bacillus thuringiensis exosporium protein InhA. J Microbiol Biotechnol. 2009;19(5):495–501. https://doi.org/10.4014/jmb.0802.163.
104. Ulmer JB, Valley U, Rappuoli R. Vaccine manufacturing: challenges and solutions. Nat Biotechnol. 2006;24(11):1377–83. https://doi.org/10.1038/nbt1261.
105. Nguyen QA, Schumann W. Use of IPTG-inducible promoters for anchoring recombinant proteins on the Bacillus subtilis spore surface. Protein Expr Purif. 2014;95:67–76. https://doi.org/10.1016/j.pep.2013.11.018.

106. Khow O, Suntrarachun S. Strategies for production of active eukaryotic proteins in bacterial expression system. Asian Pac J Trop Biomed. 2012;2(2):159–62. https://doi.org/10.1016/s2221-1691(11)60213-x.
107. Wittmann A, Suess B. Engineered riboswitches: expanding researchers' toolbox with synthetic RNA regulators. FEBS Lett. 2012;586(15):2076–83. https://doi.org/10.1016/j.febslet.2012.02.038.
108. Bahrulolum H, et al. Potential of CRISPR/Cas system as emerging tools in the detection of viral hepatitis infection. Virol J. 2023;20(1):91. https://doi.org/10.1186/s12985-023-02048-5.
109. Wiull K, et al. CRISPR/Cas9-mediated genomic insertion of functional genes into Lactiplantibacillus plantarum WCFS1. Microbiol Spectrum. 2025;13(2):e0202524. https://doi.org/10.1128/spectrum.02025-24.
110. Liu H, et al. Production of trehalose with trehalose synthase expressed and displayed on the surface of Bacillus subtilis spores. Microb Cell Factories. 2019;18(1):100. https://doi.org/10.1186/s12934-019-1152-7.
111. Huang Z, et al. A study on the effects of linker flexibility on acid phosphatase PhoC-GFP fusion protein using a novel linker library. Enzym Microb Technol. 2016;83:1–6. https://doi.org/10.1016/j.enzmictec.2015.11.002.

Vibrio cholerae: An Emerging Model Organism for Bacteria and Pathogens with Wide Ecological Habitats

3

Varun R. Bavda, Ritu Rani Archana Kujur, Subrata Mishra, and Sushanta Deb

3.1 Introduction

The members of the *Vibrio cholerae* species group share a wide range of ecological habitats and are the causative agent for deadly cholera infection in humans. The ongoing seventh cholera pandemic (active since 1961) is a major concern for health monitoring agencies; outbreaks are reported from various regions across the globe. The World Health Organization (WHO) is tracking and reporting current outbreaks and upsurges worldwide (https://www.who.int/emergencies/situations/cholera-upsurge). Researchers have prioritized *V. cholerae* as a model organism to explore microbial genetics, evolution, ecology and several other important aspects to get deeper insights into the bacterial physiology and metabolism [215]. *V. cholerae* possess multipartite genome (ChrI and ChrII), it is challenging to correlate observed phenotype to the distinct chromosome, surprisingly plethora of vibrio genome studies have revolutionized our understanding of genes encoded to each chromosome, thus provide insightful model to better understand other dual chromosomal bacteria [56, 63, 71, 213, 214]. Earlier reports have demonstrated *V. cholerae* O139 induced intestinal infection in fruit fly as novel human oral infection model [25, 171]. It has also been reported that upregulation and downregulation of certain *V. cholerae*

V. R. Bavda
Division of Genetics, Wadsworth Center, New York State Department of Health, Albany, NY, USA

R. R. A. Kujur
Institute of Life Sciences, Bhubaneswar, Odisha, India

S. Mishra
Department of Chemical Engineering, IIT Bombay, Mumbai, India

S. Deb (✉)
Veterinary Microbiology and Pathology, Washington State University, Pullman, USA

P. Kumar, V. Kothari (eds.), *Model Organisms in Biological Research*,
https://doi.org/10.1007/978-981-92-0038-2_3

genes established a model for quorum sensing and biofilms formation of intestinal pathogens [185]. Bacterial genetics and infection are not the sole models; a cholera model has been proposed to predict climate-sensitive infectious diseases, which provided various means to develop advanced methods for outbreak prediction [128]. Additionally, in the perspective of microbial ecology, *Vibrio cholerae* can be considered a crucial model organism for heterotrophic bacteria thriving in natural ecosystems [49]. An in-depth information is available on *V. cholerae*, an extensively studied pathogen, that can be leveraged to get ecological insights of other bacterial members [49]. *V. cholerae* strains inhabit diverse ecological settings and host types, and the availability of complete genome sequences from various ecological niches makes this bacterium a perfect model for exploring the potential genetic factors associated with ecologically driven bacterial phenotypes [93]. This chapter aims to discuss the crucial discoveries in the field of *V. cholerae* infection, with in-depth information on Vibrio ecology that could potentially provide a window to develop a model system for studying bacterial genetics and virulence. Multiple factors, such as shorter generation time, easy requirements for survival, streamlined genetic machinery, and well-studied biology, make *V. cholerae* an excellent candidate for a model microbe. This bacterium continues to be explored as a model in other fields, including multi-chromosome organization, host–pathogen interactions, and the evolutionary arms race driving pathogen evolution. Various microbes have demonstrated their potential to be considered model organisms in microbiology. In this context, *V. cholerae* still has a long way to go before it can be placed among the great model microbes.

3.2 *Vibrio cholerae*: Genome Structure and Evolution

Widespread adoption of genomics and metagenomics approach have revolutionized our understanding of adaptive capability [71, 72], evolution [64], metabolic function [40], and pathogenic determinants of various microbes from diverse ecological settings [56, 57, 71, 73, 188, 189]. The evolution of *V. cholerae's* genome is marked by the acquisition of integrative mobile genetic elements (IMGE's) that differentiate the pathogenic strains; however, the core genome remains relatively similar. This indicates a strong contribution of the core genome in the bacterium's adaptability [51, 134]. *V. cholerae* comprises two circular chromosomes (the larger with 2,961,146 bp and the smaller with 1,072,314 bp), a feature shared among Vibrionaceae species. Both chromosomes together encode for 3885 ORFs [93]. The larger chromosome harbours genes responsible for supporting essential cellular housekeeping functions, including processes involved in the central dogma, such as DNA replication, transcription, and translation. It also codes for antigens involved in pathogenicity, such as cholera toxin genes. The smaller chromosome has a significant number of genes with origins outside the γ-Proteobacteria (indicated by the ParA homologs, where the ones in the smaller chromosome map to plasmids and phages), suggesting it may have evolved from a megaplasmid acquired by an ancestor of *Vibrio* species. It is known for its integron island that helps in the acquisition and shuffling of foreign genes by encoding for the enzyme integrase, its toxin–antitoxin system, and repetitive sequences [93, 214]. While it is not understood yet why the smaller chromosome has not integrated into the larger one, it is hypothesized that this dual nature of the chromosomes in *V. cholerae* has helped it to adapt to various environmental conditions, such as human intestines, as well as aquatic environments. This could be a result of the smaller chromosome's contribution in acquiring foreign genes, leading to adaptation.

3.2.1 Mechanism of Horizontal Gene Transfer (HGT) in *Vibrio cholerae*: Significance

Microbial genome sequencing gained popularity to infer genetic, metabolic, and evolutionary trajectories of bacteria isolated from various ecological habitats across the globe [55, 62, 67–70]. One major mechanism of gene transfer other than the vertical mode of gene transfer is the horizontal gene transfer (HGT) that helps *V. cholerae* colonize and survive due to the transfer of plasmids, transposons, and phages. HGT occurs by conjugation, transformation, and transduction. In particular, HGT enhances pathogenicity [54]. Due to the integron island present on the smaller chromosome, genes acquired from an HGT event are used for synthesizing new genetic combinations leading to antibiotic resistance and higher virulence [93]. If the genes acquired from an HGT event are beneficial, they are retained; if not, they may decay or be lost. For instance, genome comparisons indicate that environmental non-O1/non-O139 strains have frequently lost certain virulence islands (e.g., the cholera toxin prophage, TCP island) and instead gained other islands (like those encoding type III/VI secretion systems), reflecting adaptation to alternate niches.

3.2.2 Three Strategies of HGT Occurrence Within *V. cholerae* Species

1. Natural transformation: This occurs in aquatic environments where *V. cholerae* is found on exoskeleton surfaces. *V. cholerae* becomes naturally competent, induced by chitin found on the exoskeleton surfaces [138]. It then takes up DNA from lysed cells in the marine environment. For example, the ancestral O1 El Tor strain probably acquired a DNA segment coding for a new O-antigen via an HGT event [79].
2. Conjugation: This occurs through the transfer of integrative conjugative elements (ICE's), which are DNA segments that can transfer through a pilus and are resistant to multiple antibiotics [207]. The most cited ICE is the SXT element, which is 100 kb DNA segment (provided resistant to trimethoprim, sulfamethoxazole, streptomycin, and chloramphenicol) and was identified in different strains of *V. cholerae* responsible for the 7th pandemic worldwide [219].
3. Transduction: The acquisition of the CTXΦ phage by *V. cholerae* is the most significant cause of cholerae. The lysogenic phage integrates into the VC

genomes of one of the chromosomes or both of them and produces the cholerae toxin (ctxA and ctxB) [93, 176]. Only the strains with a TCP pilus can allow the entry of the CTXΦ phage.

However, HGT acts as a double-edged sword, where the occurrence of HGT is regulated by VC. Many VC strains from the seventh pandemic carry CRISPR spacers and restriction modification systems to limit the consequences from HGT events [3].

3.2.3 Pathogenicity Islands *Vibrio cholerae*: Significance

PAIs are large chunks of DNA clusters encoding for virulence factors and are acquired through an HGT event. The most notable ones include the TCP island (encoding for the type 4 pilus), the VPI-1island (encoding for toxins), and the VSP islands.

These are A/T rich regions and vary most among pathogenic strains [174].

The two major PAIs that help the bacterium cause cholera are:

1. Vibrio Pathogenicity Island 1 (VPI-1): It is a 41 kb island located on chromosome 1, which encodes for the toxin co-regulated pilus (TCP) that not only helps in colonizing within the intestine but also acts as a receptor for the CTXΦ phage to enter, thereby allowing for the cholerae toxins to act [150].
2. Vibrio Pathogenicity Island 2 (VPI-2): It is a 57 kb island also located on chromosome 1, it contains genes for sialic acid scavenging and metabolism—including a sialidase (nanH) that cleaves sialic acids, transport genes (dctPQM), and catabolic enzymes (NanA, NanK in the nan-nag cluster) to use sialic acid as source of carbon. Both VPI-1 and 2 can excise itself from the chromosome to form a circular DNA under specific conditions [150].

Nearly every evolutionary milestone in *V. cholerae* had toxin production capability, becoming pandemic, developing drug resistance, which can be traced to HGT of a mobile element. The two-chromosome architecture itself may be a relic of an ancient HGT (plasmid acquisition) that became fixed in the lineage [216].

In summary, while HGT provides the raw genetic material for adaptation, from the spread of the SXT segment during the pandemic to processes like biofilm formation and phage resistance, the core dual genome architecture provides mechanisms to shuffle and form novel genes and establish a regulation system.

Fluoroquinolone Resistance in *V. cholerae:*

Non-synonymous nucleotide substitution in gyrA and parC genes contributed to the development of fluoroquinolone resistance phenotype in *V. cholerae* with reduced drug binding affinity. Additionally, plasmid-integrated qnr gene have been reported to be associated with this resistance [23]. A steady increase in ciprofloxacin resistance in clinical *V. cholerae* O1/O139 strains was also reported from India [20]. Global surveillance suggests variable resistance trends, emphasizing the need for continued monitoring [104]. To date, studies on fluoroquinolone resistance have been largely restricted to a limited number of bacterial species and are not broadly representative of diverse ecological habitats. Therefore, employing *V. cholerae* as a model organism to study fluoroquinolone resistance requires further validation and comprehensive research across varied environmental contexts.

3.3 *Vibrio cholerae* as Model for Studying Bacterial Ecology

V. cholerae (VC) is a gram-negative, facultative anaerobic bacteria causes the severe diarrheal disease of cholera, which is a significant global health threat. From the tropics to the temperate waters, VC thrives and can be found in vast geographical regions worldwide. The population residing particularly in these regions with inadequate sanitation and low quantities of clean water is at more risk of cholera infections. The species of VC is further divided into 200 serogroups based on the surface antigen presented on their lipopolysaccharide layer, out of which only two serotypes, O1 and O139 can cause cholera disease. These strains harbour the genes that code for cholera toxin (CT) and the toxin co-regulated pilus (TCP), primarily involved in disease development [39, 110]. Since most of the isolated VC strains do not infect and cause cholera in humans, it seems that humans as a host are only a minor part of the VC life cycle and not necessarily important for its persistence in the environment [39, 158].

3.3.1 Environmental Reservoirs and Survival Strategies

Its existence in diverse aquatic environments is absolutely necessary for its long-term survival and spread. The capability of VC to thrive in diverse and challenging aquatic conditions is largely because of its elaborate balance of both intra- and intercellular adaptive mechanisms. Adaptive responses of VC include stress-responsive gene regulation, biofilm formation, and interaction with other microbes in the reservoirs. This allows VC to survive a variety of stresses, such as nutrient limitation, salinity, and temperature fluctuations of the aquatic reservoirs, and to evade predation by bacteriophages and protists [132].

3.3.1.1 Nutrition Acquisition

VC is a prototrophic microorganism and relies on acquiring nutrients from the surrounding environment. Different conditions within the aquatic environment and intra- and interspecies competition can limit the available nutrients. VC employs a range of strictly controlled mechanisms for nutrient uptake as a survival and reproduction strategy, which are also used when it is in such a condition inside a host.

Chitin Utilization by VC

In an aquatic environment, VC utilizes various carbon sources. One of these is chitin, an insoluble biopolymer found in abundant quantities in the exoskeleton of many marine animals, including zooplanktons [98, 143, 202]. Chitin serves as a crucial nitrogen and carbon source for VC in aquatic environments. Chitin is made up of β-1,4-linked N-acetylglucosamine (GlcNAc) monomers. Chitin serves not only as a source of nutrients but also functions as a signalling molecule, provides a surface for biofilm formation, and helps in bacterial dissemination. VC utilizes the mechanism of chemotaxis to migrate towards the chitin molecules and uses its MSHA Pilus and protein GbpA to attach to the surface [117, 143]. Upon attachment, VC activates its chitin catabolic response, which enables the degradation and metabolism of chitin residues and induces natural competence that allows horizontal gene transfer [144]. Within the biofilm, the bacteria are in close proximity to extracellular DNA, which facilitates the gene uptake and adaptation to different environmental stresses. Biofilm formation on zooplankton with chitin exoskeleton helps with bacterial dissemination across aquatic bodies through passive locomotion or mechanical transfer, thus helping in VC survival transmission and evolution [24, 33, 155].

VC produces two extracellular chitinase enzymes, ChiA-1 and ChiA-2, that break down the insoluble chitin. The shorter GlcNAc oligomers produced due to chitinase activity are then transported inside the cell via the ABC transporter [125, 143]. The sensor kinase ChiS of the two-component system (TCS) initiates the chitin signalling cascade. In the absence of chitin, the chitin-binding protein (CBP) binds to ChiS and inactivates it. The degraded chitin subunits enter the periplasm via chitoporins, where they bind to CBP and dissociate it from ChiS. Thus, activated ChiS initiates the signalling cascade through either autophosphorylation or an alternative mechanism which is yet unknown [125].

ChiS sensor kinase regulates 41 different genes induced by chitin, including those involved in chitin degradation, GlcNAc metabolism, and natural competence. It regulates two different regulatory pathways: one controls catabolic pathways and the other regulates cellular competence. The catabolic pathway is regulated by the response regulator protein, ChiR, which controls the transcription of genes in the Chitobiose (*chb*) operon [143, 231]. On the other hand, ChiS activates the transmembrane regulator, TfoS, responsible for the induction of natural competence. TfoS induces the transcription of for, a small regulatory RNA, which aids the translation of an mRNA of a key regulator tfoX, responsible for chitin metabolism and natural competence. Further, TfoX and HapR activate the downstream regulator QstR, which can induce a subset of competence genes [53, 129, 230, 231].

Presence of alternative carbon source in the surrounding environment downregulates the chitin utilization via the mechanism called carbon catabolite repression (CCR), which allows VC to prioritize the utilization of sugars over chitin [24]. Utilization of glucose and other PTS sugars interferes with the cAMP accumulation and, in turn, represses the chitin metabolism and natural competence. Since cAMP-bound CRP initiate colonization, chitin degradation, and competence, utilization of PTS sugar suppresses the CMP synthesis and inhibits these processes. Interestingly, cAMP-CRP complex also inhibits biofilm formation, which is characteristically associated with chitin utilization. This complex regulation highlights the intricate regulatory network of VC environmental adaptation. Further studies need to elucidate how biofilm development bypasses cAMP-CRP inhibition during growth on chitin [24].

The chitin utilization system plays a crucial role in VC's ability to adapt as it alleviates the nutrient limitations, promotes biofilm formation and facilitates the exchange of genetic material. Additionally, its components contribute to host survival and play a role in pathogenesis. GbpA, chitin-binding protein, also facilitates the colonization of VC in the host intestine mucus layer, as chitin and mucin share structural similarities [21, 224]. ChiA2 aids survival in the host and virulence as it allows mucin degradation and releases GlcNAc and oligomers that provide the nutrients to VC during the host gut colonization [149]. The connection between the environment and nutrient acquisition by the host is key in the evolution of VC, allowing it to evolve into a successful pathogen where it can use the same system to survive in two different environments.

Glycogen Storage, Utilization, and Virulence

In addition to chitin as a carbon source, VC overcomes the carbon limitation condition by synthesizing and storing intracellular glycogen when carbon resources are abundant. Glycogen is a polysaccharide molecule composed of glucose monomers, which are interlinked by α-1,4 glycosidic bonds with branching at α-1,6 positions [28]. Glycogen metabolism is tightly controlled by the availability of nitrogen and carbon in VC. Under conditions with high nitrogen concentration, the glycogen synthesis and degradation occur at a base level; however, the nitrogen limitation condition upregulates the synthesis, leading to intracellular glycogen accumulation [28].

Glycogen synthesis in VC is initiated by two ADP-glucose phosphorylases, GlgC1 and GlgC2 enzymes, which produce ADP-glucose from glucose-1-phosphate and ATP. The glycogen is then made by the polymerization of ADP-glucose monomers. The glycogen synthase GlgA is responsible for forming α-1,4 linkages, while the branching enzyme GlgB facilitates the creation of α-1,6 linkages [28]. When carbon and nitrogen levels fluctuate in the surrounding environment, VC initiates glycogen degradation by utilizing three key enzymes: GlgX, a debranching enzyme, GlpG, a maltodextrin phosphorylase; and MalQ, a 4-α-glucanotrasferase [28, 107].

The regulation of glycogen metabolism in the VC is controlled by carbon storage regulator CsrA, which is thought to repress *glg* genes in VC, like *E. coli* [13, 107]. CrsA is post-transcriptionally regulated by the two-component system (TCS) VarSA, which activates three small regulatory RNAs (sRNAs) in response to unknown environmental stimuli. Deletion of sensor histidine kinase varS significantly reduces the glycogen storage compared to WT, indicating its importance in glycogen storage in VC [107, 210]. The activation of the VarSA pathway and CrsA repression also cause inhibition of biofilm formation by activation of HapR, which suggests a potential link between carbon storage, biofilm regulation, and quorum sensing [210].

The capability to metabolize glycogen enhances the survival and dissemination of VC in various environments. Glycogen storage granules have been found in cells isolated from rice water stool, which indicates the preparation of VC to store the glycogen pool to transit from a nutrient-rich to a nutrient-poor host environment like pond water, stool samples, etc. Interestingly, the mutants incapable of glycogen degradation survived like wild-type bacteria compared to mutants deficient in storing glycogen. This itself suggests the protective role of glycogen against environmental stress independent of its utilization. However, the glycogen degradation is absolutely required in carbon-poor environments, as mutants deficient in glycogen breakdown showed a significant reduction in viability compared to wild-type VC [28, 107].

In the infant mouse model of transmission, glycogen-rich VC has displayed increased virulence. Mutants that cannot metabolize glycogen were also impaired for transmission following incubation in pond water, mimicking the passage from host to environment before infecting a new host. This attenuation of bacteria indicates that both glycogen storage and utilization are crucial for VC's ability to thrive in the environment and successfully infect the new hosts [28].

Thus, in the VC lifecycle, glycogen metabolism plays a fundamental role in its survival, transmission, and host infection. The ability to store glycogen during nutrient-rich environments and utilize it under poor nutrient conditions enhances VC's resilience under various environmental stresses, prolongs survival post-shedding from the host, and promotes its chances of successful host-to-host transmission.

Inorganic Phosphate Acquisition and Regulation

Inorganic phosphate (Pi) is a crucial and essential nutrient, and has limited presence in aquatic settings, VC must adapt to the limited amount of Pi. Phosphate homeostasis is finely regulated by the two-component system, PhoBR, which also plays a crucial role in regulating biofilm development, cell motility, and virulence [166, 167]. Under the low Pi concentration, the sensor histidine kinase, PhoR (transmembrane protein), phosphorylates its cytoplasmic cognate response regulator PhoB, which activates the Pho regulon to increase Pi acquisition for the surrounding environments. It induces the production of PhoA, alkaline phosphatase, that provides the exogenous phosphate to VC cells and allows to survive in Pi-limiting environments. On the other hand, when Pi is in abundance, the phosphate-specific transport system (Pts) represses the PhoBR activity and maintains the cellular balance [120, 121].

The PhoBR system plays a crucial role during infection and survival of VC as *phoB* mutants exhibit reduced survival in low-Pi conditions in the host intestine and pond waters after the passage through the host [120, 167]. PhoBR system also negatively influences the virulence master regulator *tcpP*; both strains lacking or over expressing phoBR system attenuates the colonization of VC in the host. During the late stage of infection, PhoB upregulates the genes linked to c-di-GMP metabolism, which promotes the motility and dissemination of VC in the environment. This indicates that Pi limitation in the host environment induces the adaptations that facilitate the bacterial spread [166, 167].

In surplus Pi conditions, VC synthesizes long chains of Pi—polyphosphate (poly-P), which enhances the resistance to various environmental stresses, including osmotic, acidic, and oxidative stress. While poly-P is not directly involved in motility, biofilm formation, or virulence, its loss sensitizes the VC to stressors in Pi-limited conditions [100]. The precise molecular mechanisms underlying the protection of VC remain unclear, although the ability to dynamically control the Pi acquisition and storage is crucial for VC survival, transmission, and persistence in the environment.

Iron Homeostasis in VC

VC employs multiple iron acquisition and regulatory mechanisms to survive in limiting iron conditions. Furthermore, the iron-dependent regulator, controls the iron homeostasis by downregulating the iron transporters, Feo and Fbp, under sufficient intracellular iron concentrations. The iron-bound complex, Fur-Fe2+, also enhances the expression of genes responsible for iron storage, metabolism, and antioxidant defence. On the other hand, unbound Fur under low iron conditions represses the iron-dependent nonessential proteins and activates alternative pathways like manganese- dependent superoxide dismutase. Beyond iron regulation, Fur also regulates the virulence genes in the pathogenicity island of VC, as *fur* deletion mutants showed reduced colonization in mice [6, 147, 229].

VC synthesizes the siderophore vibriobactin, the ultimate product of the VibBDEFH operon [50, 115, 126], to scavenge the iron efficiently from the environment, which binds to ferric iron and transports it via outer membrane transporter ViuA. Iron transport across the membrane is further powered by the TonB-ExbBD complexes and ABC transporters ViuPDGC and VctPDGC. ViuB facilitates the processing of vibriobactin, allowing the sequestered iron molecule to be released into bioavailable form [34, 35, 126, 146, 159, 184, 228]. VC can also utilize the siderophores made by other bacterial species, like enterobactin, which is identified by the IrgA and VctA receptors, and saves on the metabolic cost of synthesis [146, 228]. Vibriobactin is not essential for virulence, however iron acquisition using siderophores supports survival and transmission in sewage-contaminated waters [184].

During host infection, VC overcomes iron requirements by implementing alternative strategies, like production of hemolysin to lyse the host cells to obtain heme or hemoglobin released here [196]. Three independent receptors, HutA, HutR, and HasR, classified as TonB-dependent receptors, can detect and import the iron derived from heme [94, 145]. However, mutation in all three receptors does not completely impair the colonization in the host thus suggesting the alternative mechanisms of iron acquisition in vivo. Iron-limiting conditions also upregulate the CpxAR two-component system, which upregulates the iron homeostasis and cell envelope stress responses [2]. Furthermore, iron scarcity promotes the expression of FumC, a non-iron-containing fumarate hydratase, to ensure continued metabolic activity [147]. These diverse regulatory and acquisition mechanisms underscore the crucial role of iron in VC survival, virulence, and environmental persistence.

3.4 Response to Physiological Stressors

VC thrives in aquatic environments here physiological conditions can fluctuate significantly like salinity and temperature, which can profoundly affect the growth and survival dynamics. To overcome such stressors, VC has developed various adaptive mechanisms that allow it to persist even in suboptimal environmental conditions. These stress response mechanisms are key for its stability in natural reservoirs between outbreaks, supporting its viability, persistence, and host infections under favourable conditions.

3.4.1 Salinity-Induced Stress Response

Changes in salinity and osmolarity are very frequent in the coastal and estuarine habitats of VC, especially during the transition from water to host. Although VC can grow optimally at 200 mM NaCl, it can tolerate a wide range of salt concentrations utilizing osmoadaptive strategies. To counter osmotic stress, VC synthesizes a compound named ectoine via the ectABC operon, and simultaneously imports glycine betaine from the external environment using OpuD transporter [111, 165, 182]. Biofilm formation is also regulated by osmolarity, which is crucial for environmental persistence. At low osmolarity, the transcription regulator OscR represses the genes involved in biofilm formation; on the other hand, high ionic strength promotes CosR expression, which promotes biofilm formation and reduces cell motility [182, 183]. Additionally, at optimal ionic strength, the transcription regulator CosR suppresses the biosynthesis of compatible solutes and genes coding for transporter proteins [183]. Together, they highlight the complex nature of regulatory networks connecting environmental changes and bacterial survival strategies.

3.4.2 Temperature-Induced Stress Response

To adapt to two different environments, aquatic and human hosts, VC has developed temperature-responsive mechanisms. Water temperatures above 15 °C promote the VC attachment of chitin by upregulating adhesins like MSHA and GbpA, eventually enhancing environmental persistence and host transmission [195]. Temperatures between 15 and 25 °C trigger the biofilm formation by activating diguanylate cyclases (DGCs), which elevates c-di-GMP levels, indicating a shift to environmental survival strategies when leaving the human host with 37 °C [122, 128, 130, 208]. During host infection at 37 °C, it promotes the translation of virulence regulator ToxT via temperature sensitive RNA, which links the temperature sensing and host colonization [223]. During the low temperature exposure, at 4 °C, it induces the expression of cold shock proteins CspA and CspV, which may contribute to the transformation to a viable but non-culturable state of VC, allowing long-term survival during non-favourable conditions [7, 58]. Together, these strategies enable VC to adapt dynamically to changing temperatures, promoting persistence under different ecological niches.

3.4.3 Viable but Non-culturable State in *Vibrio cholerae*

VC has been found to deplete its metabolic activity, undergo a change in cell shape from a comma to a round, coccoid shape, and enter into a state of nonculturable under laboratory conditions in response to environmental stress, also known as Viable, but non-culturable (VBNC) [46], Conditionally viable, environmental VC (CVEC), and persister cells state, based on the surrounding environmental conditions [11, 37, 81, 106]. This state is most induced under nutrient-deprived conditions, leading to a reduction in carbohydrate, lipid, protein, and RNA content of the bacterial cell [95]. Thus, depleted metabolite requirements of such cells significantly contribute to environmental survival under unfavourable conditions. Conditions like QS and transit through a human host allow these cells to restore their metabolic activities, highlighting the significance of reactivation in the transmission and outbreaks of VC [47, 108, 160, 217].

3.4.4 Response to Predatory Micrograzers

In the aquatic environments, VC faces strong predation pressure from protists like heterotrophic nonflagellates (HNFs) and ciliates, which are key drivers of bacterial mortality [164]. Being bacterioplankton, VC always remains exposed to the persistent danger of phagotrophic zooplanktons that graze on bacteria at the rate of 600–2000 bacterial cells per hour [133]. Microcosm experiments in the Gulf of Mexico have shown that bacterial grazing by protists effectively diminished VC from environmental water samples. However, during the blooms of phytoplankton, the bacterial doubling rates can reach more than four per day, which allows VC to increase in numbers and evade the predation for a short time, potentially accumulating at infectious levels in the water environments [225].

3.5 Utilization of VPS and T6SS as Defense System

To survive against the grazing pressure, VC utilizes biofilm formation as a central defence strategy. While planktonic cells are readily consumed by predation, cells entrapped within Vibrio-polysaccharide (VPS)- enriched biofilm matrix are more resistant to predation [198]. Exposure to predators triggers the biofilm formation in VC and smooth-to-rugose colony morphology transition concomitant with increased production of VPS [141]. The VPS-enriched matrix not only provides physical defence against the grazers but also acts as a reservoir for QS-controlled anti-protozoal factors that can accumulate only at high cell densities [78]. Such properties together positively impact the survival of VC against predation in the environment. QS has been shown to be important for protection against protozoan predation as QS-deficient strains of VC are more prone to grazing compared to wild type. VPS mutants are also susceptible to grazing but outperform QS mutants, suggesting that QS provided greater protection against grazing by amoeba compared to VPS [78,

141]. QS-regulated anti-protozoal factors and protease PrtV serve as an anti-predator strategy against various predators like *T. pyriformis*, *C. bergensis*, and *C. elegans* [141, 212].

The VC type VI secretion system (T6SS) serves as a powerful defence mechanism, protecting VC from potent predators like *E. coli*, *D. discoideum*, and mammalian immune cells—macrophages [135, 169, 170, 238]. T6SS mediates the secretion of syringe-like structure-forming proteins VgrG1-3 that puncture the plasma membrane and deliver the virulence factor VasX into the target cells [16, 32, 170]. Although all VCs have the T6SS system, the expression is strain-specific; some O1-O139 strains showed constitutive expression, whereas the pandemic El Tor strain does not express T6SS under laboratory conditions [29, 135].

3.5.1 Defence Against Predatory Protozoa

In aquatic settings, VC is subject to ongoing predation by bacterivorous protozoa such as ciliates and flagellates and has evolved a number of defence mechanisms with which to contend with predators. One of its most significant mechanisms is the Type VI Secretion System (T6SS), which injects toxic effectors like VasX and TseL into predator cells upon contact, killing predators such as *D. discoideum* [74, 148, 169]. In addition to active killing, VC enhances Vibrio polysaccharide (VPS) production and switches to a rugose colony type when subjected to grazing stress, where it produces biofilms that provide physical protection against protozoan grazers [198]. Biofilms produced on chitin substrata exhibit enhanced anti-protozoal activity when compared with abiotic substrata, primarily because ammonium a degradation product of chitin, is produced, which is flagellate toxic in nature like *R. nasuta* [199]. This ammonium production is reduced in quorum-sensing (QS)-deficient strains, hence reducing their toxicity. QS also regulates the production of novel anti-protozoal compounds like the HapR-dependent protease PrtV, which is toxic to predators like *C. roenbergensis* and *T. pyriformis* [141, 212]. Interestingly, even though both VPS and HapR are involved in grazing resistance, HapR's influence is greater; yet, mutants lacking *hapR*, although more vulnerable because of loss of QS-regulated factors, are more grazing resistant in some habitats as a consequence of enhanced biofilm formation [78, 132, 198, 233]. This highlights VC's use of overlapping and independent mechanisms like direct attack via T6SS, QS-regulated toxins, and structural defences like VPS for protozoan predator evasion and environmental survival.

3.5.2 Defence Against Vibriophages

Phages, or bacteriophages, play major roles in VC and cholera epidemic dynamics and evolution. Phages of prevailing serogroups during an outbreak can end transmission by infecting and destroying VC within aquatic environments and infected hosts [80, 81, 103, 157]. This contributes to causing natural attenuation of outbreaks as observed in the demonstration of a negative correlation of phage occurrence and viable VC or incidences of reported cholera cases over a period of three years. High titres of lytic phages, often enriched within the human host and expelled into the environment in stool, correlate with the decline of seasonal epidemics and provide support to the theory that phages contribute to cholera's self-limiting nature of outbreak [81, 103]. Experiment illustrates that phages isolated in rice water stool reduce VC infectivity within pond water and can prevent transmission, and act efficiently to modulate the infectious dose and induce collapse of the epidemic [157, 235].

In addition to controlling population dynamics, phage pressure determines the evolution of VC. Phages preferentially infect the O antigen, and *V. cholerae* may change this structure to avoid recognition, leading to the emergence of new clones [178]. Environmental vibriophages vary in host specificity, some infecting only O1 or O139 strains, and others being lytic or lysogenic phages depending on the host exerting selective pressure favouring phage-resistant or lysogenic strains. Such "phage territorialism" will also influence the frequency of occurrence of particular serogroups and is potentially available for exploitation for control of outbreaks [80].

Phages are also a key player in VC evolution through horizontal transfer of genes and genomic rearrangements. Horizontal gene transfer acquiring cholera toxins by the use of the toxin-coregulated pilus (TCP) as a receptor from the CTXϕ phage is amongst the most critical examples. Such phage-transferred genes are typical of the manner in which bacteriophages not only regulate VC populations during epidemics but also sustain the causative pathotypes [218]. Phage interactions are thus at the centre of both the development and decline of cholera epidemics.

3.6 Biofilm Formation in the Aquatic Environment

V. cholerae (VC), in the aquatic environment, generally exists in the form of the microbial communities known as biofilms. These organized communities are collections of cells adhered to various surfaces and are covered by extracellular matrix produced by themselves or acquired externally. Biofilm production enhances bacterial survival within the environment by providing protection against a variety of environmental stress factors, including nutrient deprivation and protozoan and bacteriophage predation [4, 18, 99, 141, 204]. The biofilms formed by VC are found on the majority of surfaces within aquatic settings, including both biotic surfaces like zooplanktons and phytoplanktons, and abiotic surfaces like marine chitin rain [172, 202].

Other than protection, biofilm also serves as mechanism for delivering high bacterial burdens to enhance the transmission into the human hosts and cause the disease and cholera outbreaks [14, 201]. The hyper infectious phenotype of VC is also supported by the presence of biofilm-like aggregates in the stool of cholera patients, which suggests the role of biofilm in disease transmission from host to host. Given the critical importance of biofilm in environmental survival of VC and disease transmission, this section discusses the VC biofilm development, structural composition, regulation, and the impact of environmental stimulus on biofilm formation [91, 131, 201].

3.6.1 Molecular Insights into VC Biofilm Formation

The formation of VC biofilm involves multiple steps, including surface scanning, initial attachment, production of extracellular matrix components (EMC), followed by microcolony formation, and finally three-dimensional, highly organized 3D structures (Fig. 3.1).

3.6.2 Surface Attachment

The very first step of biofilm formation is surface attachment. The transition of VC from the free motile stage to the non-motile biofilm stage involves the landing of swimming bacteria onto the surface of attachment and then shifting to a sessile state. Cyclic diguanylate (c-di-GMP) plays a crucial role as a signalling molecule in this transition by repressing motility while upregulating surface attachment and biofilm matrix production. c-di-GMP signalling molecules are synthesized by the

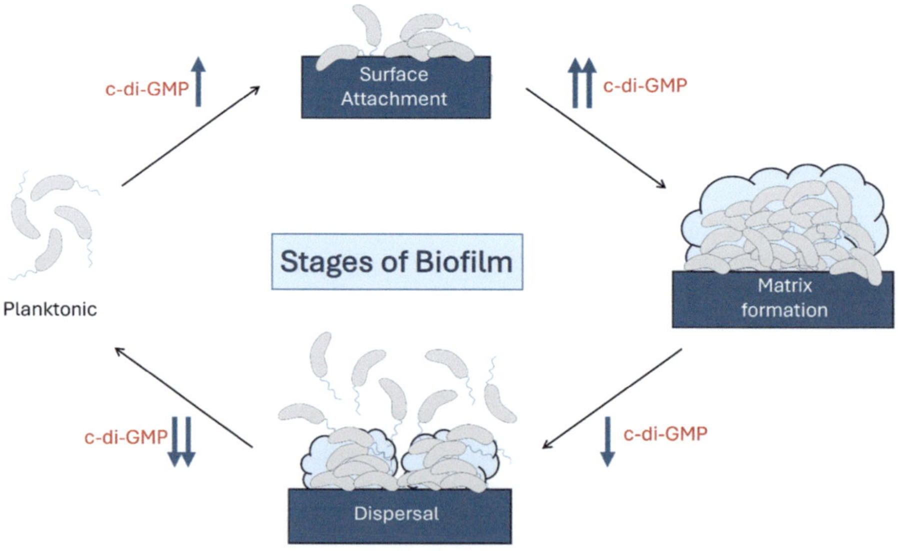

Fig. 3.1 Stages of biofilm: Intracellular levels of c-di-GMP (bis-(3-5)-cyclic dimeric guanosine monophosphate) control various developmental stages of biofilm in *Vibrio cholerae* (VC). It promotes the transition of VC from planktonic (motile) state to biofilm (sessile) state by downregulating the expression of genes involved in cell motility and upregulating gene expression responsible for surface attachment and extracellular matrix production. Increasing concentrations of c-di-GMP enhance the surface attachment and production of matrix components, whereas decreasing concentrations of c-di-GMP encourage dispersal of biofilm

enzyme called diguanylate cyclases (DGCs) and degraded by the enzyme called phosphodiesterases PDEs), which allows the tight regulation of biofilm formation. c-di-GMP can react with various types of cell surface receptors or RNAs and modulate various cell processes [204].

The initiation of surface attachment is mediated by type IV mannose-sensitive hemagglutinin (MSHA) pili, which can interact with surfaces through regular. The torque created by the polar flagellum enhances the MSHA Pili interactions, leading to two discrete near-surface motility behaviours: the stronger surface interaction called orbiting and weaker interactions called roaming. These interactions play a key role in guiding the transition to a sessile state and the formation of microcolonies [123, 204].

Based on the concentration of c-di-GMP, the Receptor protein MshE controls the dynamics of the polymerization of the MshA subunits and thus the formation of the MSHA pili. Increased levels of the second messenger c-di-GMP allow pilus elongation, while reduced levels cause retraction. ATPase PilT is also essential for both the process of motility and adhesion to surfaces since the tack of PilT decreases the initial adherence but enhances the retention after adherence. Another ATPase, PilU, also assists in the process of retraction and depends on the activity of PilT [105, 175, 204, 220].

Other than MSHA, video call produces two other types of IV Pili, namely toxin co-regulated pilus (TCP) and chitin-regulated pilus (ChiRP). TCP acts as a virulence factor and is important for VC for intestinal colonization, microcolony formation, and cell aggregation via pilus-pilus contact in the intestine. TCPs are upregulated within biofilm and thus contribute to increased infectivity of biofilms, especially important for colonization in the intestine. While MSHA aids in biofilm formation and chitin adherence, its constitutive expression negatively impacts intestinal colonization by interfering with immune evasion. Two positive regulatory elements, VC1371 and VcRfaH, were recently found and have been shown to be important for MSHA-mediated biofilm formation and in intestinal colonization. Coordinate regulation of TCP and MSHA is necessary for successful colonization of both host species and environmental reservoirs [173, 181, 203, 205].

VC cell harbours a single Na + dependent flagellum on one of the cell poles, which it utilizes for the purpose of swimming motility. During the early stages of the development of the biofilm, the rotation of the flagella is inhibited to facilitate adhesion on surfaces and prevent detachment. The intracellular c-di-GMP concentrations inversely regulate the mutual inhibition between the processes of motility and the development of the biofilm [48, 204, 205]. Disrupting flagellin (flaA) or flagellar biogenesis in VC leads to elevated c-di-GMP levels, enhancing biofilm formation through the flagellum-dependent biofilm regulatory (FDBR) response. This response, partially dependent on the flagellar stator and Na^+ flux, upregulates c-di-GMP and promotes biofilm development. Three diguanylate cyclases—CdgA, CdgL, and CdgO are also essential for the higher levels of cellular c-di-GMP and signalling pathway that give rise to FDBR response and thus ultimately triggering biofilm formation [226]. Single-cell analysis reveals an increase in intracellular c-di-GMP upon surface attachment, which may play a role in the reduction of flagellar activity. At the transcriptional level, c-di-GMP inhibits the transcription of flagellar genes by interacting with FlrA, the master regulator of flagellar formation, and FlrC, destabilizing their complex. Such multi-layered repression enables the switch from motile to sessile growth, thus inducing biofilm formation and surface colonization [38, 194, 226].

The proteins from the repeats-in-toxin family play a crucial role in different interactions mediated by cell-surface attachment, cell–cell interactions, and cell–matrix interactions. VC contains two such adhesins called FrhA and CraA that are thought to localize on the outer membrane by the type I secretion system. Their maintenance at the cell surface is controlled by a well-characterized and conserved, posttranslational proteolytic mechanism dependent on c-di-GMP concentration. FrhA and CraA contribute to biofilm formation and initial attachment in a strain- and substrate-specific manner. In *V. cholerae* O1 classical strains, the absence of these adhesins reduces biofilm formation and surface attachment. In contrast, *V. cholerae* El Tor strains exhibit enhanced biofilm formation on chitin-rich surfaces when utilizing this system [90, 118]. Since O1 classical strains lack functional MSHA pili, RTX adhesins may play a compensatory role in adhesion. In addition, VC produces the N-acetylglucosamine (GlcNAc)-specific adhesion protein GbpA, which is controlled by a c-di-GMP riboswitch. GbpA enables attachment to chitin and mucin substrates, demonstrating the organism's versatility in using different adhesion proteins for attachment in different environments [21, 113, 117, 195, 197].

3.6.3 Extracellular Matrix Production

During the biofilm formation, the initial cell surface attachment of VC is followed by another key step of production of extracellular matrix (ECM). ECM of VC is composed of various complex molecules, including polysaccharides, proteins, nucleic acids, and other biomolecules required for cells to attach to each other and surfaces [205].

Vibrio polysaccharide (VPS) is an essential component of the VC biofilm matrix, which is important for the integrity and stability of biofilms. VC produces two different variants of VPS: a major variant that is 80% and a minor variant that is 20%, each with a different sugar composition. The major variant is made up of [−4)-α-L-GulpNAcAGly3OAc-(1–4)-β-D-Glcp-(1–4)-α-D GlcpNAc-(1–4)-α-D-Galp-(1-] whereas in the minor variant α-D-Glc is replaced with α-D-GlcNAc [234]. Upon cell surface attachment, VPS is secreted extracellularly and remains throughout the process of biofilm maturation [17, 204]. A recent study suggests that the regulation of VPS biosynthesis is performed by a tyrosine phosphoregulatory system located within the VPS operon. Autophosphorylation of tyrosine kinase VpsO at the C-terminal tail inhibits its own activity which prevents the VPS production. Dephosphorylation of VpsO by the tyrosine phosphatase VpsU allows the formation of high-oligomeric structures of VpsO, which promotes the VPS production. Furthermore, the phospho-state of VpsO also controls proteolysis significantly, leading to the fast turnover of inactive VpsO. Regulation of both post-transcriptional and transcriptional VPS biosynthesis precisely underscores the high-energy demand involved in matrix synthesis, so that the establishment of the biofilm is tightly controlled according to the environment [177, 205].

The ECM of VC biofilm also contains many important proteins that are important for biofilm architecture and stability. Among these, several are RbmA, RbmC, and Bap1, which play different but complementary roles in biofilm formation. These proteins are secreted through the type II secretion system (T2SS), and the deletion of T2SS leads to inhibited biofilm formation. RbmA ranks among the earliest proteins secreted into the ECM after adhesion onto surfaces. It plays a significant function in microcolony formation by regulating cell orientation, as well as in the attachment of daughter cells onto mother cells [17, 87]. Biofilms that do not contain RbmA have reduced density and enhanced susceptibility to external perturbations, including invasion and predation. At the structural level, RbmA contains two fibronectin type III (FnIII) modules (FNIII-1 and FNIII-2) that dimerize and are essential in ensuring the stability of the biofilm. In addition, RbmA is capable of binding Vibrio polysaccharide (VPS) directly and undergoes conformational changes among different structural states, affecting the biofilm structure. Proteolytic cleavage of RbmA further affects the biofilm formation [75, 87, 154, 227]. Three proteases, HapA, PrtV, and IvaP, that proteolytically generate a C-terminal fragment RbmA* harbouring the FnIII-2 domain, retain the VPS binding capability. While full-length RbmA mainly interacts with VPS-producing cells, RbmA* can also bind to both VPS-producing and non-producing cells, suggesting its role in the recruitment of all VC cells to biofilm [17, 85, 190].

RbmC and Bap1 are other important ECM proteins that share structural similarities and play complementary roles in biofilm formation. During the early stages of biofilm formation, Bap1 is present at the interface between the surface and founder cell, whereas RbmC is absent at the early stage but surrounds biofilm aggregates in more mature biofilms. Both RbmC and Bap1 proteins are crucial for biofilm attachment; in their absence, the poorly attached biofilms are prone to detachment. Both proteins share similar β-prism domains that bind glycans, contributing towards cellular adherence and providing structural integrity to biofilms. Together, these ECM proteins perform independent and specific roles in *V. cholerae* biofilm development that provide stability, robustness, and versatility in various environmental settings [1, 17, 83].

Extracellular DNA (eDNA) and outer membrane vesicles (OMVs) are other very important components found in the ECM of biofilms, although its regulation and contribution to the biofilm architecture is still under study. eDNA acts as a structural component in biofilms by interacting with Vibrio polysaccharide (VPS) and being a potential source of nutrients and genetic material via horizontal gene transfer [27, 89, 109, 179]. The eDNA-encoding extracellular nucleases—Dns (an endonuclease) and Xds (an exonuclease)—degrade eDNA to nucleoside monophosphates that are usable by *V. cholerae* as a source of phosphate. Thus, deletion of dns and xds enhances biofilm formation and reduces detachment, indicating eDNA degradation here is important in biofilm dispersal. There are three NupC-family transporters used to uptake nucleosides, and their deletion increases biofilm formation, perhaps due to a negative feedback loop resulting from nucleoside accumulation in the matrix. The eDNA release pathways are still not exactly clear, but cell lysis, autolysis, active secretion, and transport using OMVs are possible pathways [60, 89, 142, 179]. Conversely, OMVs comprised of lipopolysaccharides, phospholipids, and outer membrane proteins (OMPs) also play roles in biofilm formation. The vesicles carry important proteins of the biofilm matrix, such as RbmA, RbmC, and Bap1, described above. Moreover, the outer membrane protein OmpT was shown to interact with Bap1, while the function of the OMVs in the context of VC biofilm formation is apparently not clear [5, 76]. Now, more investigations are required to determine the specific mechanisms through which OMVs influence biofilm stability and structural organization.

3.6.4 Biofilm Formation and Dispersal: Gene Regulation

Optimum gene expression in biofilm provides maximum fitness advantage to the community members at both the single-cell level and population level as it enables the adaptation to stable or fluctuating environments. The main regulatory circuitry for biofilm formation involves transcriptional regulators that regulate the expression of biofilm matrix synthesis genes and accordingly modify production of key matrix components such as Vibrio polysaccharide (VPS) or biofilm-associated proteins, which support proper development of the biofilm. Secondary regulatory pathways integrate into the primary network, meaning that accumulation of biofilm gene expression in response to environmental signals by *V. cholerae* occurs through activation or repression and therefore occurs when any of the conditions are influenced by the substrate, such as stress, nutrient availability, or quorum-sensing signals. Besides, biofilm dispersal also has regulatory mechanisms that enable bacteria to turn again into remaining planktonic when conditions become adverse. Such a switch further facilitates the association of new surfaces and adds to the transmission and survival of *V. cholerae* in diverse environments [205].

3.6.5 Primary Regulatory Pathways

VC biofilm formation is tightly controlled by various transcription regulators. It can integrate environmental signals and modulate the pathways to ensure biofilm formation and dispersal dynamics (F. H. [232]; F. H. [233]). VpsR and VpsT serve as the master regulators of biofilm formation. VpsR is an ATPase, the principal positive regulator of biofilm genes, which can trigger the type II secretion system (T2SS) for secreting major biofilm proteins (F. H. [232]). Disruption of VpsR abolishes the expression of biofilm genes. The VpsR activity is controlled through phosphorylation. At physiological pH, it can bind to c-di-GMP, which allows dimerization, which is required for its DNA binding activity to activate the transcription of genes important for biofilm formation, such as vpsU, vpsL, bap1, and the rbm operon. VpsT is also involved in biofilm formation as it binds to c-di-GMP and activates vpsL, vpsA, and rbmA ([119]; F. [96, 234, 236]). VpsT also controls ci-di-GMP homeostasis by activating the vpvABC operon that codes for diguanylate cyclase (DGC) [88] (Fig. 3.2).

H-NS acts as a key inhibitor of biofilm genes in VC, as in the absence of HNS, VC shows enhanced biofilm formation. H-NS negatively regulates the biofilm gene

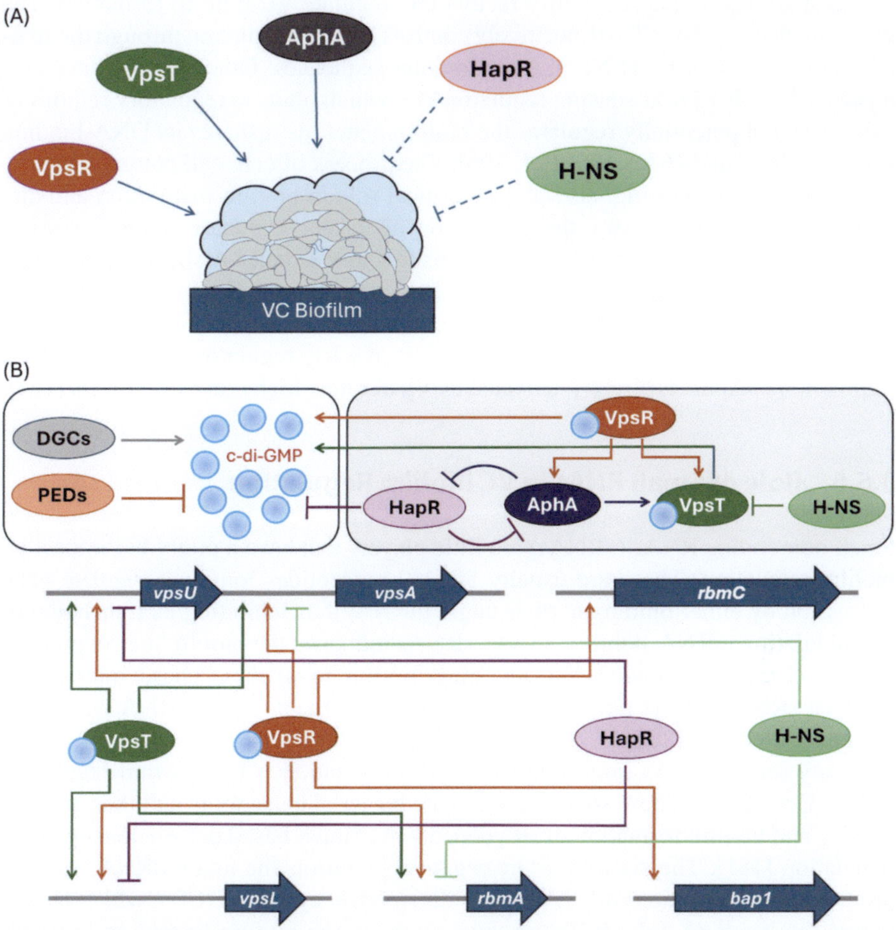

Fig. 3.2 Key regulatory factors of biofilm in *Vibrio cholerae*. (**a**) Biofilm development in Vibrio cholerae is controlled by a complex array of positive and negative regulators. The positive regulators are VpsR, VpsT, and AphA, while the negative regulators are HapR and H-NS. (**b**) The signal molecule c-di-GMP plays a significant role by binding to VpsR and VpsT and controlling biofilm development. c-di-GMP is produced by diguanylate cyclase enzymes (DGCs) and broken down by phosphodiesterase enzymes (PDEs). Master biofilm regulator VpsR activates AphA and VpsT, along with genes for major biofilm matrix components. AphA activates VpsT and represses HapR, whereas HapR represses AphA, VpsT, and c-di-GMP synthesis. Furthermore, H-NS represses VpsT and VPS (Vibrio polysaccharide) genes, which are essential for biofilm architecture. All these regulators compete for binding sites within the promoter regions of VPS-associated genes. VpsR and VpsT bind to their target promoters in a c-di-GMP-dependent manner to induce biofilm formation. They also directly activate rbmA, a key protein of biofilm matrix. VpsR also induces the expression of rbmC and bap1, additional matrix proteins, by binding to their promoters

expression by binding to AT-rich sequence and changing the nucleoid topology, it represses the transcription [8, 10, 220, 236]. VpsT promotes the biofilm gene expression by displacing the H-NS from vpsA and vpsL promoters in a c-di-GMP manner, alleviating repression. VpsR also enhances the expression of vpsT and reduces the occupancy of H-NS at its promoter, indicating coordinated action of biofilm gene activation [9, 193].

There are additional regulatory factors that regulate VC biofilm formation. The key virulence regulator ToxR positively controls biofilm formation through the inhibition of repression by H-NS at key promoter sequences. Other studies have also reported the role of iron-sensing regulator, Fur, which binds to regulatory regions of vps genes and potentially regulates the biofilm genes, as it shares its DNA-binding site with VpsT and H-NS [86, 114, 205]. The process of cell–cell communication, known as quorum sensing, also plays a critical role in biofilm maturation and dispersal. At high cell density, the expression of master regulator HapR increases, which acts as a repressor at high cell densities and reduces the activity of vpsL, vpsT, and c-di-GMP-degrading enzymes [61, 209, 221, 233]. On the other hand, AphA is defined as a low cell density master QS regulator. Opposite to HapR, AphA positively regulates biofilm formation and acts as a key regulator of virulent genes. Expression of AphA is directly activated by VpsR under high c-di-GMP levels [127].

3.6.6 Role of Small RNAs in VC Biofilm Regulation

Small non-coding RNAs (sRNAs) are little players that have a pretty big impact on biofilm regulation. These are usually 50–250 nucleotides long and regulate gene expression by either binding to their target mRNA, thus inhibiting its translation or by modulating mRNA stability. In VC, sRNAs influence the biofilm matrix production by directly regulating the key transcription regulators. These interactions between sRNA and the target mRNA are often facilitated by the RNA chaperone proteins called Hfq [205].

A key set of sRNAs important for biofilm regulation is the quorum regulatory RNAs (Qrr1-Qrr4). These sRNAs coordinate bacterial behaviours and thus quorum sensing and biofilm formation in Hfq dependent manner based on cell density in the population [151]. The sRNA for Qrr negatively controls the luxO mRNA, while it promotes the translation of VCA0939, a diguanylate cyclase (DGC), which in turn boosts the levels of c-di-GMP and thus ramps up the biofilm formation independently of HapR. In VC, the QS autoinducer (AI) molecule DPO (5-dimethylpyrazin-2-ol) binds to its receptor VqmA and positively regulates another QS-regulated sRNA, VqmR. The function of VqmR is Hfq dependent and represses the vpsT translation and thus reduce biofilm formation. VC lacking vqmA or vqmR shows increased vpsT expression, however overexpression of VqmR or supplementing DPO in the culture inhibits biofilm formation [163, 200, 205, 237].

Two other VC Hfq dependent sRNA, VadR, and VrrA, have been reported to control the biofilm formation. VrxAB system activates VadR, which then represses the biofilm genes, (vpsU, vpsL, rbmA), although VrxAB system is known to induce the biofilm production, which suggests the complexity of regulatory system. While VadR controls the VC cell curvature, its overexpression reduces the biofilm formation. The expression of VrrA is controlled by sigma factor RpoE and it represses the rbmC, ompA and ompT, in turn reduces the biofilm formation. Overall, these sRNAs integrate the environmental signals and collectively act to fine-tune the development of biofilm in VC [36, 101, 124, 206, 210].

3.6.7 Dispersal of Biofilm

Biofilm dispersion can be achieved through two different processes, which can be either active regulation or passive process such as shear force of fluid flow. Active dispersion refers to response to microenvironmental cues or exogenous stimulation. Multiple concurrent regulatory pathways operate to achieve the change from a biofilm towards a planktonic state under a controlled mechanism [205].

The process of active biofilm dispersion in VC is regulated and influenced by the intracellular c-di-GMP concentration, which controls many key cellular mechanisms such as pilus retraction, cellular motility, and extracellular matrix degradation [102, 112]. MSHA pilus retraction is induced by low c-di-GMP levels; this process is crucial for the detachment of biofilm from surfaces to initiate the early stage of biofilm dispersal. The reduction in the intracellular c-di-GMP concentrations is achieved through the activity of the enzyme CdpA, which is nitric oxide (NO)-responsive phosphodiesterase (PDE), under the microaerobic conditions. NO has been shown to induce biofilm dispersion in multiple bacterial species, while its role as a biofilm dispersal agent in VC is still being investigated [15, 82, 97, 205, 211].

Additionally, the LapDG system, a conserved signalling module involved in surface adhesion and detachment, also plays an important role in VC biofilm dispersal. LapD, the receptor for c-di-GMP, regulates the activity of LapG, which is a periplasmic protease. LapG is responsible for the degradation of adhesins such as CraA and FrhA and thus promotes biofilm dispersal. However, LapG bound to c-di-GMP, sequesters the LapG, in turn prevents the cleavage of adhesins and maintains the biofilm stability. This system is also regulated at the transcription level by c-di-GMP-dependent regulators FlrA, VpsT, and VpsR, which regulate the expression of FrhA and CraA, which acts as an additional layer of biofilm dispersal regulation [19, 31, 45, 118, 156].

In addition to the above mechanisms described, there is another well-studied c-di-GMP signalling module, NspS/MbaA, which also influences the biofilm dispersal by modulating the levels of cellular c-di-GMP. It responds to host-derived signals such as polyamines like spermidine. NspS bound to spermidine in the periplasmic space does not bind to MbaA, which then exhibits its PDE activity, reducing the c-di-GMP levels and thus triggering the biofilm dispersal. This regulatory mechanism highlights complex control of biofilm dispersal in VC, where there are multiple signalling pathways that combines environmental cues and control the transition process of VC cells from biofilm-associated to dispersed planktonic lifestyle [26, 30, 44, 191].

The decrease in the production of ECM components alone is not enough for biofilm dispersal; it also requires the degradation of the ECM to detach the cells and enter the planktonic state. Several enzymes have been identified in VC involved in the degradation of ECM, including proteases such as IvaP, PrtV, and HapA, which process a major matrix protein RbmA capable of autoproteolysis. rbmB locus codes for a putative polysaccharide lyase that can potentially hydrolyze VPS. Overexpression of RbmB reduces the accumulation of VPS and thus the biomass of biofilm, while its deletion impairs the dispersion of biofilm. Another crucial component of biofilm matrix, extracellular DNA (eDNA), is processed by the nucleases Xds and Dns, which impact the biofilm dispersion, although it appears condition-dependent [83, 92, 136, 168, 180, 190, 205].

Quorum sensing (QS) also plays a crucial role in biofilm dispersion of VC by controlling the production of ECM components and cellular detachment. The biofilm ECM components synthesis is repressed by the QS regulators HapR and sRNA VqmR at high cell density, which promotes the dispersion, especially under the nutrient starvation condition. Together, HapR and sigma factor RpoS enhance the motility and detachment upon biofilm maturation [30, 162, 163, 186, 187, 233]. Additionally, under carbon starvation conditions, the expression of VbdR is induced, which retracts MSHA pilus and upregulates serine protease IvaP, which processes RbmA, and thus facilitates the cellular detachment from biofilm. Interestingly, the leftover ECM components of the mature biofilm suggest that complete degradation is not mandatory for the release of cells [92, 187]. Multiple two-component systems are reported to be involved in spontaneous biofilm dispersion in static cultures such as CarRS, PhoBR, NtrBC, VieSA, and DbfSR. Among all, DbfSR plays a crucial role in the downregulation of matrix components production, and deletion of this system leads to defective biofilm dispersion. The DbfSR system is regulated by the CpxR, which is an envelope stress regulator, and is involved in antibiotic resistance against polymyxin B. These enzymes collectively regulate the biofilm dispersion, enhancing the transition from the sessile to motile lifestyle of VC [22, 30, 31, 42, 137, 139, 140, 153].

3.6.8 Model for Defence Against Protozoan Predation and Biofilm Development

Previous studies have demonstrated the role of *V. cholerae* emphasizing biofilm formation a defence against protozoan predation, could be a potential mechanism represent by persistent aquatic bacterial pathogens for defence [141]. *V. cholerae* produces the hapR factor that regulates cell density involved in avoiding protozoan predation. This antiprotozoal factor has been reported in various other members of vibrio group as well as in non-vibrio bacteria [41]. This indicates *V. cholerae* antiprotozoal defence mechanism shared with other bacterial member making *V. cholerae* as first choice to select as target. *V. cholerae,* being relatively easy to modify at the genetic level, provides flexibility to study the mechanism of biofilm development. Various steps of biofilm development, including Attachment, maturation, and dispersal, are well studied and standardized [84]. Additionally, *V. cholera-associated* biofilm studies have been well characterized from various aquatic ecosystems as well as from human infection. Biofilms play a role in transmission and virulence, linking environmental biology to pathogenesis [239]. Multiple signalling pathways; quorum sensing, c-di-GMP, and stress responses are involved, providing rich biological insight [222, 239]. The widespread study of *V. cholerae* isolated from diverse habitats and hosts represents a promising reference for biofilm study.

3.7 *V. cholerae* in the Context of Global Health

3.7.1 The Seventh Pandemic and Ongoing Challenges

Since 1817, *V. cholerae* has caused seven pandemics. In 1961, EI tor biotype caused cholera's seventh pandemic, which claimed millions of lives even though the disease was treatable and preventable. It originated in Indonesia and spread globally, particularly to South Asia, Africa, Latin America, and the Caribbean (i.e., Haiti) from 1963 to 2010. And thereafter sero and bio variants of O1 emergence have been reported from various regions of the world [12]. The O1 biotype (further classified as "classical" and "EI tor" biotypes) of *V. cholerae* is known to contain pathogenicity islands (VPI1 & VPI2) and pandemic islands (VSP1 & VSP2) along with a genetic element known as *ctx,* which includes core and RS2 region 8 [161].

For modern scientific study, the strains are available only from the 6th (1899–1923) and 7th pandemics (1961 - present). The causing agent for the 6th pandemic was classical biotype strains, whereas, for the ongoing 7th pandemic, it was EI tor biotype strains. An exhaustive investigation of the 7th pandemic epidemiology and phylogeny has been done. A study conducted by Mutreja et al. [152], suggested that strains (comprising >190 genomes for the analysis) of the 7th pandemic have spread around the world in at least three independent waves and are monophyletic. With the ongoing pandemic, the changes in the pattern of individual genes have also been reviewed, suggesting that it is still going on possibly concerning human immune response [116].

The recently occurred global cholera resurgence has been classified as a grade 3 emergency by WHO in January 2023—an event requiring the utmost response. Accounting for 4007 deaths in 2023, reported cases of more than 500,000, with the highest counts in the Democratic Republic of the Congo, Haiti, and Afghanistan, an increase of 71% in deaths was seen in comparison to the previous year. Still, this issue seems to be ignored on political platforms and health agendas, thus leaving this pandemic to persist for so long. Regardless of global commitments to Sustainable Development Goal 6, we fail to execute effective water, sanitation, and hygiene (WASH) practices to the core of negligence. The program aimed to provide worldwide access to sustainable water and sanitation. WHO report (2020) stipulates that about two billion people lack access to safe drinking water.

In addition, the failure to provide sufficient vaccination for cholera and treatment for those in need further elevates the problem. Presently, millions of doses are supplied by Gavi, but any holdups in the production of vaccine and distribution mean that supply remains fragile, with the only manufacturer—EuBiologics. In 2013, a global stockpile of the oral cholera vaccine was established with two million doses globally available for an outbreak emergence. The production has been expanded to thirty-six million doses by 2023. Still, in early 2024, a depletion was seen in the stockpile despite the October 2022 decision to maximize the reserves by introducing a single-dose regimen instead of a two-dose regimen as an outbreak response. Euvichol-S a new simplified vaccine, is prequalified to offer efficacy similar to vaccines already existing and will help lessen the shortage of vaccines due to its cost-effective and easier production by 2025.

The cholera resurgence is an extreme threat to health. By 2030, the Global Task Force on Cholera Control aims to lessen 90% of deaths by cholera from 2015. That seems to be increasingly unlikely without coordinated and immediate action. For a very long period, cholera's high mortality rate has been ignored, most likely because of the countries and people it most affects. The cholera disease is mainly driven by extreme poverty, displacement, and conflict. Until or unless we focus on the root causes of cholera, it will continue to thrive.

3.7.2 Vaccination, Prevention, and Control Strategies

It has been observed that cholera is more prevalent in low-resource communities lacking basic infrastructural facilities. For instance, regions prone to war, poverty, earthquakes, flood-like natural disasters, insurgency, and other factors lead to people's displacement from their homes. Thus, in this scenario, the basic problem people are facing is access to clean and portable water and sanitation facilities, leading to open defecation problems, and they are also forced to share the limited available resources. These problems can be commonly seen in Nigeria, BAY states (north-eastern states of Borno, Adamawa, and Yobe), etc. [77].

The global climate crisis is another serious problem resulting in the rise in sea levels, extreme weather conditions, and temperatures in small island states that are less developed. The conditions favouring the outbreak and spread of waterborne diseases such as cholera. To completely eradicate cholera, the most appropriate and effective way revolves around case management, provision of clean and adequate water, sanitation, prevention, and surveillance activities consisting of health education and control employing antibiotic drugs and Oral Cholera Vaccine (OCV) [77]. With proper treatment and rapid fluid replacement therapy, the mortality rate can be reduced to 1% from 50%, especially in the case of left untreated severe cases [43].

Recently, OCV has become one of the cost-effective and essential tools for preventing and controlling cholera in non-emergency and emergencies. OCV provides protection for up to 6 months and extended protection for up to 3 years in case of single and double doses, respectively.

Depending on the feasibility of conducting campaigns and local epidemiology, WHO includes OCVs as a comprehensive cholera control strategy in pre-emptive situations. And it can be used along with other proven public health interventions and not as a stand-alone intervention, regarding it as a "supplementary" tool for long-term prevention and cholera outbreak control.

In order to prevent and control cholera deaths, it is necessary to adopt prompt intervention strategies, which include clean and potable water, facilities for sanitation and hygiene, Oral Rehydration Therapy (ORS), Oral Cholera Vaccine (OCV), Antibiotic treatment, and community-based education systems. Still, this practice requires continual investment and a commitment to improving the services and infrastructure.

The further essential intervention that could mitigate the cholera outbreak situation, when persons already have accessibility to treatment whenever there is any sign of symptoms, is "real-time surveillance," which includes proper reporting, on-time, and quality data by electronic systems deployed, strengthening laboratories, and case management [192]. Mainly accomplished by establishing staffed and well-equipped treatment centres within the current community centres and health facilities. Adequate knowledge and proper training among health workers are critical in the assessment of patients suffering from dehydration, recognizing clinical signs and symptoms, following appropriate treatment protocols, and lastly protecting themselves while helping others, thereby preventing the further spreading of cholerae infection [59].

The effective elimination of cholera has been observed in some developing as well as developed countries, simply by investment in proper hygiene, which includes adequate water treatment and proper management of human waste disposal system [52]. Therefore, a combination of OCV campaigns and other proven interventions must be promoted by the WHO and other international partners to control or prevent outbreaks of cholera in non-emergency and emergencies as soon as possible.

3.8 Limitations of *Vibrio cholerae* as a Model Organism

Vibrio cholerae exhibits limited ecological versatility in the context of soil-associated microbes. *V. cholerae* is sporadic in terrestrial or soil-based environments, and may not effectively model microbial interactions in plant-associated or non-aquatic ecosystems. As a human-specific pathogen, the virulence factors and disease progression mechanisms of *V. cholerae* are often not generalizable to other hosts or infection models. Furthermore, commonly used animal models for cholera, such as infant mice and rabbits, do not fully recapitulate the complexity of human infection dynamics, which may limit the translational relevance of in vivo studies. Due to its pathogenic nature, research on *V. cholerae* biofilms or host immune responses may be inherently influenced by its virulence-associated traits, potentially skewing interpretations when applied to broader microbial systems. Consequently, the use of *V. cholerae* as a model for studying commensal or non-pathogenic biofilm communities may be inappropriate or yield limited ecological insight.

3.9 Conclusion

V. cholerae, due to its distinct genomic features, has gained attention to be considered as a model for biological study. Its capability to acquire exogenous genetic material by various means of horizontal gene transfer enhances pathogenic potential and ecological versatility. The widespread distribution of this bacterium across diverse ecological habitats and human hosts highlights its importance in the study of microbial ecology and evolution. Moreover, highly regulated ability to form biofilms through multiple signalling pathways has fascinated researchers to include *V. cholerae* for benchmark study. Well-explored mechanism of *V. cholerae-mediated* biofilm formation provides a robust framework for investigating biofilm-associated pathogenicity and resilience. Its involvement in climate-sensitive disease patterns further highlights its importance for exploring how environmental factors influence microbial virulence. Collectively, these attributes support the use of *V. cholerae* as a novel and versatile model organism for investigating bacteria, bacterial pathogenicity, and ecological dynamics.

References

1. Absalon C, Dellen K, Watnick PI. A communal bacterial adhesin anchors biofilm and bystander cells to surfaces. PLoS Pathog. 2011;7(8):e1002210.
2. Acosta N, Pukatzki S, Raivio TL. The Vibrio cholerae Cpx envelope stress response senses and mediates adaptation to low iron. J Bacteriol. 2015;197:262–76. https://doi.org/10.1128/jb.01957-14.
3. Adade NE, Aniweh Y, Mosi L, Valvano MA, Duodu S, Ahator SD. Comparative analysis of Vibrio cholerae isolates from Ghana reveals variations in genome architecture and adaptation of outbreak and environmental strains. Front Microbiol. 2022;13:998182. https://doi.org/10.3389/fmicb.2022.998182.
4. Alam M, Sultana M, Nair GB, Siddique AK, Hasan NA, Sack RB, Sack DA, et al. Viable but nonculturable *Vibrio cholerae* O1 in biofilms in the aquatic environment and their role in cholera transmission. Proc Natl Acad Sci. 2007;104(45):17801–6. https://doi.org/10.1073/pnas.0705599104.
5. Altindis E, Fu Y, Mekalanos JJ. Proteomic analysis of Vibrio cholerae outer membrane vesicles. PNAS. 2014;111(15):1548–56.
6. Andrews SC, Robinson AK, Rodríguez-Quiñones F. Bacterial iron homeostasis. FEMS Microbiol Rev. 2003;27:215–37. https://doi.org/10.1016/S0168-6445(03)00055-X.
7. Asakura H, Ishiwa A, Arakawa E, Makino S, Okada Y, Yamamoto S, Igimi S. Gene expression profile of vibrio Cholerae in the cold stress-induced viable but non-Culturable state. Environ Microbiol. 2007;9:869–79. https://doi.org/10.1111/j.1462-2920.2006.01206.x.
8. Ayala JC, Wang H, Benitez JA, Silva AJ. RNA-Seq analysis and whole genome DNA-binding profile of the Vibrio cholerae histone-like nucleoid structuring protein (H-NS). Genom Data. 2015a;5:147–50.
9. Ayala JC, Wang H, Silva AJ, Benitez JA. Repression by H-NS of genes required for the biosynthesis of the Vibrio cholerae biofilm matrix is modulated by the second messenger cyclic diguanylic acid. Mol Microbiol. 2015b;97(4):630–45.
10. Ayala JC, Silva AJ, Benitez JA. H-NS: an overarching regulator of the Vibrio cholerae life cycle. Res Microbiol. 2017;168(1):16–25.
11. Ayrapetyan M, Williams TC, Oliver JD. Bridging the gap between viable but non-Culturable and antibiotic persistent bacteria. Trends Microbiol. 2014;23:7–13.

12. Ayyappan MV, Kishore P, Panda SK, Kumar A, Uchoi D, Nadella RK, Priyadarshi H, Obaiah MC, George D, Hamza M, Ramannathan SK. Emergence of multidrug resistant, ctx negative seventh pandemic Vibrio cholerae O1 El Tor sequence type (ST) 69 in coastal water of Kerala, India. Sci Reports. 2024;14:2031.
13. Baker CS, Morozov I, Suzuki K, Romeo T, Babitzke P. CsrA regulates glycogen biosynthesis by preventing translation of glgC in Escherichia coli. Mol Microbiol. 2002;44:1599–610.
14. Bari SM, Nayeemul MK, Roky M, Mohiuddin MK, Mekalanos JJ, Faruque SM. Quorum-sensing autoinducers resuscitate dormant *Vibrio Cholerae* in environmental water samples. Proc Natl Acad Sci. 2013;110(24):9926–31. https://doi.org/10.1073/pnas.1307697110.
15. Barraud N, Kelso M, Rice S, Kjelleberg S. Nitric oxide: a key mediator of biofilm dispersal with applications in infectious diseases. Curr Pharm Des. 2014;21(1):31–42.
16. Basler M, Pilhofer M, Henderson GP, Jensen GJ, Mekalanos JJ. Type VI secretion requires a dynamic contractile phage tail-like structure. Nature. 2012;483:182–6.
17. Berk V, Fong JCN, Dempsey GT, Develioglu ON, Zhuang X, Liphardt J, Yildiz FH, Chu S. Molecular architecture and assembly principles of Vibrio cholerae biofilms. Science. 2012;337:236–9.
18. Beyhan S, Yildiz FH. Smooth to rugose phase variation in Vibrio cholerae can be mediated by a single nucleotide change that targets C-Di-GMP signalling pathway. Mol Microbiol. 2007;63:995–1007.
19. Beyhan S, Tischler AD, Camilli A, Yildiz FH. Transcriptome and phenotypic responses of Vibrio cholerae to increased cyclic Di-GMP level. J Bacteriol. 2006;188:3600–13.
20. Bhattacharya D, Sugunan AP, Ghosh AR, Yadav RN. Antibiotic resistance pattern of Vibrio cholerae strains isolated from cholera cases in Andaman and Nicobar Islands during 2000–2006. Indian J Med Res. 2011;133(3):225–30. https://www.ijmr.org.in/text.asp?2011/133/3/225/78394.
21. Bhowmick R, Ghosal A, Das B, Koley H, Saha DR, Ganguly S, Nandy RK, Bhadra RK, Chatterjee NS. Intestinal adherence of Vibrio cholerae involves a coordinated interaction between colonization factor GbpA and mucin. Infect Immun. 2008;76:4968–77.
22. Bilecen K, Yildiz FH. Identification of a calcium-controlled negative regulatory system affecting Vibrio cholerae biofilm formation. Environ Microbiol. 2010;11:2015–29.
23. Bina XR, Provenzano D, Nguyen N, Bina JE. Vibrio cholerae RND family efflux systems are required for antimicrobial resistance, optimal virulence factor production, and colonization of the infant mouse small intestine. Infect Immun. 2008;76(6):2559–67. https://doi.org/10.1128/IAI.00126-08.
24. Blokesch M. Chitin colonization, chitin degradation and chitin-induced natural competence of Vibrio cholerae are subject to catabolite repression. Environ Microbiol. 2012;14:1898–912.
25. Blow NS, Salomon RN, Garrity K, Reveillaud I, Kopin A, Jackson FR, et al. Vibrio cholerae infection of Drosophila melanogaster mimics the human disease cholera. PLoS Pathog. 2005;1:e8. https://doi.org/10.1371/journal.ppat.0010008.
26. Bomchil N, Watnick P, Kolter R. Identification and characterization of a Vibrio cholerae gene, mbaA, involved in maintenance of biofilm architecture. J Bacteriol. 2003;185(4):1384–90.
27. Borgeaud S, Metzger LC, Scrignari T, Blokesch M. The type VI secretion system of Vibrio cholerae fosters horizontal gene transfer. Science. 2015;347:63–7.
28. Bourassa L, Camilli A. Glycogen contributes to the environmental persistence and transmission of Vibrio cholerae. Mol Microbiol. 2009;72:124–38.
29. Breen P, Winters AD, Theis KR, Withey JH. The Vibrio cholerae type six secretion system is dispensable for colonization but affects pathogenesis and the structure of zebrafish intestinal microbiome. Infect Immun. 2021;89(9):e0015121.
30. Bridges AA, Bassler BL. The intragenus and interspecies quorum-sensing autoinducers exert distinct control over Vibrio cholerae biofilm formation and dispersal. PLoS Biol. 2019;17(11):e3000429.
31. Bridges AA, Fei C, Bassler BL. Identification of signaling pathways, matrix-digestion enzymes, and motility components controlling Vibrio cholerae biofilm dispersal. PNAS. 2020;117(51):32639–47.

32. Brooks TM, Unterweger D, Bachmann V, Kostiuk B, Pukatzki S. Lytic activity of the Vibrio cholerae type VI secretion toxin VgrG-3 is inhibited by the antitoxin TsaB. J Biol Chem. 2013;288:7618–25.
33. Broza M, Gancz H, Halpern M, Kashi Y. Adult non-biting midges: possible windborne carriers of Vibrio cholerae. Environ Microbiol. 2005;7:576–85.
34. Butterton JR, Calderwood SB. Identification, cloning, and sequencing of a gene required for ferric vibriobactin utilization by Vibrio cholerae. J Bacteriol. 1994;176:5631–8.
35. Butterton JR, Payne JA, Calderwoodl SM. Viability of the nonculturable Vibrio cholerae O1 and O139. System Appl Microbiol. 1992;24:331–41.
36. Butz HA, Mey AR, Ciosek AL, Payne SM. Vibrio cholerae CsrA directly regulates varA to increase expression of the three nonredundant Csr small RNAs. 2019;10:10–1128.
37. Chaiyanan S, Huq A, Maugel T, Colwell RR. Viability of the nonculturable Vibrio cholerae O1 and O139. System Appl Microbiol. 2001;24:331.
38. Chakraborty S, Biswas M, Dey S, Agarwal S, Chakrabortty T. The heptameric structure of the flagellar regulatory protein FlrC is indispensable for ATPase activity and disassembled by cyclic-diGMP. J Biol Chem. 2020;295(50):16960.
39. Charles RC, Ryan ET. Cholera in the 21st century. Curr Opin Infect Dis. 2011;24:472–7.
40. Chaudhari M, Verma S, Deb S. Insights into molecular interplay in tuberculosis–COVID-19 co-infection via integrated multi-omics strategies. J. 2025;8(4):41.
41. Chavez-Dozal A, Gorman C, Erken M, Steinberg PD, McDougald D, Nishiguchi MK. Predation response of Vibrio fischeri biofilms to bacterivorus protists. Appl Environ Microbiol. 2013;79 https://doi.org/10.1128/AEM.02710-12.
42. Cheng AT, Zamorano-Sánchez D, Teschler JK, Wu D, Yildiz FH. NtrC adds a new node to the complex regulatory network of biofilm formation and Vps expression in Vibrio cholerae. J Bacteriol. 2018;200(15):10–1128.
43. Clemens JD, Nair GB, Ahmed T, Qadri F, Holmgren J. The effectiveness of one or two doses of either Shanchol (Shantha biotechnics; India) or ORC-vax (Vabiotech; Vietnam) OCVs showed a combined vaccine effectiveness of 75% (95% CI, 61-84). Lancet. 2017;390(10101):1539–49.
44. Cockerell, Rutkovsky SR, Zayner AC, Cooper JP, Porter RE, Pendergraft LR, Parker SS, McGinnis ZM, Karatan MW. Vibrio cholerae NspS, a homologue of ABC-type periplasmic solute binding proteins, facilitates transduction of polyamine signals independent of their transport. Microbiology. 2014;160:832–43.
45. Collins AJ, Smith TJ, Sondermann H, O'Toole GA. From input to output: the lap/c-Di-GMP biofilm regulatory circuit. Ann Rev Microbiol. 2020;74:607–31 38.
46. Colwell RR, Brayton PR, Grimes DJ, Roszak DB, Huq SA, Palmer LM. Viable but non-culturable Vibrio cholerae and related pathogens in the environment: implications for release of genetically engineered microorganisms. Nat Biotechnol. 1985;3:817–20. https://doi.org/10.1038/nbt0985-817.
47. Colwell RR, Brayton P, Herrington D, Tall B, Huq A, Levine MM. Viable but non-culturable Vibrio cholerae O1 revert to a cultivable state in the human intestine. World J Microbiol Biotechnol. 1996;12:28–31.
48. Conner JG, Zamorano-Sánchez D, Park JH, Sondermann H, Yildiz FH. The ins and outs of cyclic Di-GMP signaling in Vibrio cholerae. Curr Opin Microbiol. 2017;36(20):39.
49. Cottingham KL, Chiavelli DA, Taylor RK. Environmental microbe and human pathogen: the ecology and microbiology of Vibrio cholerae. Front Ecol Environ. 2003;1(2):80–6. https://doi.org/10.1590/s1020-49891997000100002.
50. Crosa JH, Walsh CT. Genetics and assembly line enzymology of Siderophore biosynthesis in bacteria. Microbiol Mol Biol Rev. 2002;66:223–49. https://doi.org/10.1128/MMBR.66.2.223-249.2002.
51. Cuervo MP, Castillo A, Santiago-Connolly LM. Overview of biological hazards and foodborne diseases. In: Encyclopedia of food safety; 2024. p. 1–18.
52. Cutler D, Miller G. The role of public health improvements in health advances: the twentieth-century United States. Demography. 2005;42(1):1–22.

53. Dalia AB, Lazinski DW, Camilli A. Identification of a membrane-bound transcriptional regulator that links chitin and natural competence in Vibrio cholerae. 2014:01028-13. https://doi.org/10.1128/JB.01028-13.
54. Das B, Pazhani GP, Sarkar A, Mukhopadhyay AK, Nair GB, Ramamurthy T. Molecular evolution and functional divergence of Vibrio cholerae. Curr Opin Infect Dis. 2016;29:520–7. https://doi.org/10.1097/QCO.0000000000000272.
55. Das L, Deb S, Das SK. Description of *Acinetobacter kanungonis* sp. nov., based on phylogenomic analysis. Int J Syst Evol Microbiol. 2021;71 https://doi.org/10.1099/ijsem.0.004833.
56. Das L, et al. Pufferfish (Tetraodon cutcutia) sampled from a freshwater river serves as an intermediate reservoir of a sucrose nonfermenting variant of Vibrio cholerae PS-4. Microbiol Spectrum. 2022;10(1):e01221–21. https://doi.org/10.1128/spectrum.01221-21.
57. Dash J, et al. Biochemical, functional and genomic characterization of a new probiotic Ligilactobacillus salivarius F14 from the gut of tribes of Odisha. World J Microbiol Biotechnol. 2023;39(7):171.
58. Datta PP, Bhadra RK. Cold shock response and major cold shock proteins of vibrio Cholerae. Appl Environ Microbiol. 2003;69:6361–9. https://doi.org/10.1128/AEM.69.11.6361-6369.2003.
59. Davis W, Narra R, Mintz ED. Cholera. Curr Epidemiol Rep. 2018;5:303–15.
60. de Aldecoa I, Alejandra L, Zafra O, González-Pastor JE. Mechanisms and regulation of extracellular DNA release and its biological roles in microbial communities. Front Microbiol. 2017;8:1390. https://doi.org/10.3389/fmicb.2017.01390.
61. De Silva RS, Kovacikova G, Lin W, Taylor RK, Skorupski K, Kull FJ. Crystal structure of the Vibrio cholerae quorum-sensing regulatory protein HapR. J Bacteriol. 2007;189(15):5683–91. https://doi.org/10.1128/JB.01807-06.
62. Deb S. Pan-genome evolution and its association with divergence of metabolic functions in Bifidobacterium genus. World J Microbiol Biotechnol. 2022;38:231. https://doi.org/10.1007/s11274-022-03430-1.
63. Deb S, Basak S. Comparative study of Codon usage pattern and compositional distribution between whole genome and virulence gene set of Vibrio cholerae N16961. Comput Mol Biol. 2015;5 https://doi.org/10.5376/cmb.2015.05.0006.
64. Deb S, Basak S. Variation in synonymous codon usage in Paenibacillus sp. 32O-W genome. Bioinformation. 2016;12(11):396.
65. Deb S, Das SK. Draft genome sequence of Idiomarina sp. strain W-5T, isolated from the Andaman Sea. Microbiol Resour Announc. 2020a;9:e01510–9. https://doi.org/10.1128/MRA.01510-19.
66. Deb S, Das SK. Draft genome sequence of *Microbacterium oryzae* strain MB-10, isolated from a Rice field in India. Microbiol Resour Announc. 2020b;9:e01532–19. https://doi.org/10.1128/MRA.01532-19.
67. Deb S, Das L, Das SK. Phylogenomic analysis reveals that Arthrobacter mysorens Nand and Rao 1972 (approved lists 1980) and Glutamicibacter mysorens Busse 2016 are later heterotypic synonyms of Arthrobacter nicotianae Giovannozzi-Sermanni 1959 (approved lists 1980) and Glutamicibacter nicotianae Busse 2016. Curr Microbiol. 2020a;77(11):3793–8.
68. Deb S, Badhai J, Das SK. Draft genome sequences of Vibrio alginolyticus strain S6-61 and Vibrio diabolicus strain S7-71, isolated from corals in the Andaman Sea. Microbiol Resour Announc. 2020b;9:e01465–19. https://doi.org/10.1128/MRA.01465-19.
69. Deb S, Badhai J, Das SK. Draft genome sequences of two Vibrio fortis strains isolated from Coral (*Fungia* sp.) from the Andaman Sea. Microbiol Resour Announc. 2020c;9:e01225–19. https://doi.org/10.1128/MRA.01225-19.
70. Deb S, Basu J, Choudhary M. An overview of next generation sequencing strategies and genomics tools used for tuberculosis research. J Appl Microbiol. 2024;2:135. https://doi.org/10.1093/jambio/lxae174.
71. Deb S, et al. Genomic insights and functional adaptation of Mycobacterium intracellulare strains isolated from patients with inflammatory bowel disease. Microbiol Spectrum. 2025;13(12):e03414–24.

72. Deb S, Kumari L, Singh UB. Comparative population genomic analysis of Brevibacterium casei isolated from a tuberculosis patient. Folia Microbiol. 2026:1–12.
73. Debnath T, Deb S, Das SK. Influence of geochemistry in the tropical hot springs on microbial community structure and function. Curr Microbiol. 2023;80(1):4.
74. Dong TG, Ho BT, Yoder-Himes DR, Mekalanos JJ. Identification of T6SS-dependent effector and immunity proteins by Tn-Seq in Vibrio cholerae. Proc Natl Acad Sci USA. 2013;110 https://doi.org/10.1073/pnas.1310238110.
75. Drescher K, Dunkel J, Nadell CD, Teeffelen S, Grnja I. Architectural transitions in Vibrio cholerae biofilms at single-cell resolution. PNAS. 2016;113(14):2066–72.
76. Duperthuy M, Sjöström AE, Sabharwal D, Damghani F, Uhlin BE, Wai SN. Role of the Vibrio cholerae matrix protein Bap1 in cross-resistance to antimicrobial peptides. PLoS Pathog. 2013;9(10):e1003620.
77. Ebob T. Cholera prevention and control strategies; A global overview. J Adv Med Med Res. 2020;32(12):27–53.
78. Erken M, Weitere M, Kjelleberg S, McDougald D. In situ grazing resistance of Vibrio cholerae in the marine environment. FEMS Microbiol Ecol. 2011;76(3):504–12. https://doi.org/10.1111/j.1574-6941.2011.01067.x.
79. Faruque SM, Saha MN, Asadulghani, Sack DA, Sack RB, Takeda Y, Nair GB. The O139 serogroup of *Vibrio cholerae* comprises diverse clones of epidemic and nonepidemic strains derived from multiple *V. cholerae* O1 or non-O1 progenitors. J Infect Dis. 2000;182:1161–8. https://doi.org/10.1086/315764.
80. Faruque SM, Naser I, Bin MJ, Islam ASG, Faruque, Ghosh AN, Nair GB, Sack DA, Mekalanos JJ. Seasonal epidemics of cholera inversely correlate with the prevalence of environmental cholera phages. Proc Natl Acad Sci USA. 2005a;102 https://doi.org/10.1073/pnas.0408992102.
81. Faruque SM, Islam MJ, Ahmad QS, Faruque ASG, Sack DA, Nair GB, Mekalanos JJ. Self-limiting nature of seasonal cholera epidemics: role of host-mediated amplification of phage. Proc Natl Acad Sci USA. 2005b;102 https://doi.org/10.1073/pnas.0502069102.
82. Floyd KA, Lee CK, Xian W, Nametalla M, Valentine A, Crair B, Zhu S, et al. C-Di-GMP modulates type IV MSHA pilus retraction and surface attachment in Vibrio cholerae. Nat Commun. 2020;11(1):1549. https://doi.org/10.1038/s41467-020-15331-8.
83. Fong JCN, Yildiz FH. The rbmBCDEF gene cluster modulates development of rugose colony morphology and biofilm formation in Vibrio cholerae. J Bacteriol. 2007;189:2319–30.
84. Fong JCN, Yildiz FH. Biofilm matrix proteins. Microbiol Spectr. 2015;3(2):MB-0001-2014. https://doi.org/10.1128/microbiolspec.MB-0001-2014.
85. Fong JC, Rogers A, Michael AK, Parsley NC, Cornell WC, Lin YC, Singh PK, Hartmann R, Drescher K, Vinogradov E, Dietrich LE. Structural dynamics of RbmA governs plasticity of Vibrio cholerae biofilms. elife. 2017;6:e26163.
86. Gao H, Ma L, Qin Q, Qiu Y, Zhang J. Fur represses Vibrio cholerae biofilm formation via direct regulation of vieSAB, cdgD, vpsU, and vpsA-K transcription. Front Microbiol. 2020;11(587159):71.
87. Giglio KM, Fong JC, Yildiz FH, Sondermann H. Structural basis for biofilm formation via the Vibrio cholerae matrix protein RbmA. J Bacteriol. 2013;195:3277–86.
88. Guest T, Haycocks JRJ, Warren GZL, Grainger DC. Genome-wide mapping of Vibrio cholerae VpsT binding identifies a mechanism for c-di-GMP homeostasis. Nucleic Acids Res. 2022;50(1):149–59.
89. Gumpenberger T, Vorkapic D, Zingl FG, Pressler K, Lackner S. Nucleoside uptake in Vibrio cholerae and its role in the transition fitness from host to environment. Mol Microbiol. 2016;99(3):470–83.
90. Guo S, Vance TDR, Stevens CA, Voets IK, Davies PL. RTX Adhesins are key bacterial surface megaproteins in the formation of biofilms. Trends Microbiol. 2019;27(5):453–67.
91. Hang L, Manohar J, Muhammad A, Bridges EA, Vanderspurt C, Kirn TJ, Taylor RK, et al. Use of *in vivo* -induced antigen technology (IVIAT) to identify genes uniquely expressed

during human infection with *Vibrio cholerae*. Proc Natl Acad Sci. 2003;100(14):8508–13. https://doi.org/10.1073/pnas.1431769100.
92. Hatzios SK, Abel S, Martell J, Hubbard T, Sasabe J. Chemoproteomic profiling of host and pathogen enzymes active in cholera. Nat Chem Biol. 2016;12(4):268–74.
93. Heidelberg JF. DNA sequence of both chromosomes of the cholera pathogen Vibrio cholerae. Nature. 2002;406:477–86. https://doi.org/10.1038/35020000.
94. Henderson DP, Payne SM. Characterization of the Vibrio cholerae outer membrane heme transport protein HutA: sequence of the gene, regulation of expression, and homology to the family of TonB-dependent proteins. J Bacteriol. 1994;176:3269–77.
95. Hood MA, Guckert JB, White DC, Deck F. Effect of nutrient deprivation on lipid, carbohydrate, DNA, RNA, and protein levels in Vibrio cholerae. Appl Environ Microbiol. 1986;52:788–93.
96. Hsieh M-L, Hinton DM, Waters CM. VpsR and cyclic Di-GMP together drive transcription initiation to activate biofilm formation in Vibrio cholerae. Nucleic Acids Res. 2018;46(17):8876–87. https://doi.org/10.1093/nar/gky606.
97. Hughes HQ, Floyd KA, Hossain S, Anantharaman S, Kysela DT, Ibáñez Aldecoa AL, Zafra O, González-Pastor JE. Nitric oxide stimulates type IV MSHA pilus retraction in Vibrio cholerae via activation of the phosphodiesterase CdpA. Proc Natl Acad Sci. 2022;119:e2108349119.
98. Huq A, Small EB, West PA, Huq MI, Rahman R, Colwell RR. Ecological relationships between Vibrio cholerae and planktonic crustacean copepods. Appl Environ Microbiol. 1983;45:275–83.
99. Islam MS, Jahid MIK, Rahman MM, Rahman MZ, Islam MS, Kabir MS, Sack DA, Schoolnik GK. Biofilm acts as a microenvironment for plankton-associated Vibrio cholerae in the aquatic environment of Bangladesh. Microbiol Immunol. 2007;51:369–79.
100. Jahid IK, Silva AJ, Benitez JA. Polyphosphate stores enhance the ability of Vibrio cholerae to overcome environmental stresses in a low-phosphate environment. Appl Environ Microbiol. 2006;72:7043–9.
101. Jang J, Jung K-T, Park J, Yoo C-K, Rhie G-E. The Vibrio cholerae VarS/VarA two-component system controls the expression of virulence proteins through ToxT regulation. Microbiology. 2011;157(5):1466–73.
102. Jenal U, Reinders A, Lori C. Cyclic Di-GMP: second messenger extraordinaire. Nat Rev Microbiol. 2017;15(5):271–84.
103. Jensen MA, Faruque SM, Mekalanos JJ, Levin BR. Modeling the role of bacteriophage in the control of cholera outbreaks. Proc Natl Acad Sci USA. 2006;103:4652.
104. Jin D, Li Y, Dai Y, Zeng Y, Liang Y, Chen S, Wang W, Huang L, Li W. Systematic review and meta-analysis on antimicrobial resistance of vibrio cholerae in environmental water and clinical samples over two decades. Antimicrob Resist Infect Control. 2022;11:81. https://doi.org/10.1186/s13756-022-01100-3.
105. Jones CJ, Utada A, Davis KR, Thongsomboon W, Zamorano Sanchez D. C-Di-GMP regulates motile to sessile transition by modulating MshA pili biogenesis and near-surface motility behavior in Vibrio cholerae. PLoS Pathog. 2015;11(10):95.
106. Jubair M, Morris JG, Ali A. Survival of vibrio Cholerae in nutrient-poor environments is associated with a novel "persister". Phenotype. PLoS One. 2012;7:E45187. https://doi.org/10.1371/journal.pone.0045187.
107. Kamp HD, Patimalla-Dipali B, Lazinski DW, Wallace-Gadsden F, Camilli A. Gene fitness landscapes of Vibrio cholerae at important stages of its life cycle. PLoS Pathog. 2013;9:e1003800. https://doi.org/10.1371/journal.ppat.1003800.
108. Kamruzzaman M, Udden SMN, Cameron DE, Calderwood SB, Nair GB, Mekalanos JJ, Faruque SM. Quorum-regulated biofilms enhance the development of conditionally viable, environmental Vibrio cholerae. Proc Natl Acad Sci USA. 2010;107 https://doi.org/10.1073/pnas.0913404107.
109. Kanampalliwar A, Singh DV. Extracellular DNA builds and interacts with vibrio polysaccharide in the biofilm matrix formed by Vibrio cholerae. Environ Microbiol Rep. 2020;12(5):594–606.

110. Kaper JB, Morris JG, Levine MM. Cholera. Clin Microbiol Rev. 1995;8:48–86. https://doi.org/10.1128/CMR.8.1.48.
111. Kapfhammer D, Karatan E, Pflughoeft KJ, Watnick PI. Role for glycine betaine transport in Vibrio cholerae osmoadaptation and biofilm formation within microbial communities. Appl Environ Microbiol. 2005;71:3840–7. https://doi.org/10.1128/AEM.71.7.3840-3847.2005.
112. Karatan E, Watnick P. Signals, regulatory networks, and materials that build and break bacterial biofilms. Microbiol Mol Biol Rev. 2009;73(2):310–47.
113. Kariisa AT, Weeks K, Tamayo R. The RNA domain Vc1 regulates downstream gene expression in response to cyclic diguanylate in Vibrio cholerae. PLoS One. 2016;11(2):98.
114. Kazi MI, Conrado AR, Mey AR, Payne SM, Davies BW. ToxR antagonizes H-NS regulation of horizontally acquired genes to drive host colonization. PLoS Pathog. 2016;12(4):e1005570.
115. Keating TA, Marshall CG, Walsh CT. Reconstitution and characterization of the Vibrio cholerae vibriobactin synthetase from VibB, VibE, VibF, and VibH. Biochemistry. 2000;39:15522–30. https://doi.org/10.1021/bi0016523.
116. Kim EJ, Lee CH, Nair GB, Kim DW. Whole-genome sequence comparisons reveal the evolution of vibrio cholerae O1. Trends Microbiol. 2015;23(8):479–89.
117. Kirn TJ, Jude BA, Taylor RK. A colonization factor links Vibrio cholerae environmental survival and human infection. Nature. 2005;438:863–6. https://doi.org/10.1038/nature04241.
118. Kitts G, Giglio KM, Zamorano-Sánchez D, Park JH, Townsley L. A conserved regulatory circuit controls large adhesins in Vibrio cholerae. 2019;10:10–1128.
119. Krasteva PV, Fong JCN, Shikuma NJ, Beyhan S, Navarro MVAS, Yildiz FH, Sondermann H. Vibrio cholerae VpsT regulates matrix production and motility by directly sensing cyclic Di-GMP. Science. 2010;327:866–8.
120. Kruger WMA, Humphreys S, Ketley JM. A role for the PhoBR regulatory system homologue in the Vibrio cholerae phosphate-limitation response and intestinal colonization. Microbiology. 1999, 145:2463–75.
121. Krüger WMA, LMS L, Soares MR, Neves-Manta FS, Batista SCM, AGDC N-F, Perales J, Bisch PM. The phosphate-starvation response in Vibrio cholerae O1 and phoB mutant under proteomic analysis: disclosing functions involved in adaptation, survival and virulence. Proteomics. 2006;6:1495–511.
122. Lama JR, Seas CR, León-Barúa R, Gotuzzo E, Sack RB. Environmental temperature, cholera, and acute diarrhoea in adults in Lima, Peru. J Health Popul Nutr. 2004;22:399–403.
123. Lauga E, DiLuzio WR, Whitesides GM, Stone HA. Swimming in circles: motion of bacteria near solid boundaries. Biophys J. 2006;90:400–12.
124. Lenz DH, Miller MB, Zhu J, Kulkarni RV, Bassler BL. CsrA and three redundant small RNAs regulate quorum sensing in Vibrio cholerae. Mol Microbiol. 2005;58(4):1186–202.
125. Li X, Roseman S. The Chitinolytic Cascade in Vibrios is regulated by chitin oligosaccharides and a two-component chitin catabolic sensor/kinase. Proc Natl Acad Sci USA. 2004;101:627.
126. Li N, Zhang C, Li B, Liu X, Huang Y, Xu S, Gu L. Unique iron coordination in iron-chelating molecule vibriobactin helps Vibrio cholerae evade mammalian siderocalin-mediated immune response. J Biol Chem. 2012;287:8912–9.
127. Lin W, Kovacikova G, Skorupski K. The quorum sensing regulator HapR downregulates the expression of the virulence gene transcription factor AphA in vibrio Cholerae by antagonizing Lrp- and VpsR-mediated activation. Mol Microbiol. 2007;64(4):953–67.
128. Lipp EK, Huq A, Colwell RR. Effects of global climate on infectious disease: the cholera model. Clin Microbiol Rev. 2002;15(4):757–70. https://doi.org/10.1128/CMR.15.4.757-770.2002.
129. Lo Scrudato M, Blokesch M. A transcriptional regulator linking quorum sensing and chitin induction to render Vibrio cholerae naturally transformable. Nucleic Acids Res. 2013;41:3644–58.
130. Lobitz B, Beck L, Huq A, Wood B, Fuchs G, Faruque ASG, Colwell R. Climate and infectious disease: use of remote sensing for detection of Vibrio cholerae by indirect measurement. Proc Natl Acad Sci. 2000;97:1438–43.
131. Lombardo M-J, Michalski J, Martinez-Wilson H, Morin C, Hilton T, Osorio CG, Nataro JP, Tacket CO, Camilli A, Kaper JB. An *in vivo* expression technology screen for *Vibrio cholerae*

genes expressed in human volunteers. Proc Natl Acad Sci. 2007;104(46):18229–34. https://doi.org/10.1073/pnas.0705636104.
132. Lutz C, Erken M, Noorian P, Sun S, McDougald D. Environmental reservoirs and mechanisms of persistence of vibrio Cholerae. Front Microbiol. 2013;4:375. https://doi.org/10.3389/fmicb.2013.00375.
133. Macek M, Carlos G, Memije P, Ramírez P. Ciliate-vibrio Cholerae interactions within a microbial loop: an experimental study. Aquat Microb Ecol. 1997;13(3):257–66. https://doi.org/10.3354/ame013257.
134. Maciel-Guerra A, Babaarslan K, Baker M, Rahman A, Hossain M, Sadique A, Alam J, Uzzaman S, Ferdous Rahman Sarker M, Sultana N, Islam Khan A, Ara Begum Y, Hassan Afrad M, Senin N, Hossain Habib Z, Shirin T, Qadri F, Dottorini T. Core and accessory genomic traits of Vibrio cholerae O1 drive lineage transmission and disease severity. Nat Commun. 2024;15(1):8231.
135. MacIntyre DL, Miyata ST, Kitaoka M, Pukatzki S. The Vibrio cholerae type VI secretion system displays antimicrobial properties. Proc Natl Acad Sci U S A. 2010;107 https://doi.org/10.1073/pnas.1012931107.
136. Maestre-Reyna M, Huang W-C, Wu W-J, Singh PK, Hartmann R. Vibrio cholerae biofilm scaffolding protein RbmA shows an intrinsic, phosphate-dependent autoproteolysis activity. IUBMB Life. 2021;73(2):418–31.
137. Martinez-Wilson HF, Tamayo R, Tischler AD, Lazinski DW, Camilli A. The vibrio Cholerae hybrid sensor kinase vieS contributes to motility and biofilm regulation by altering the cyclic diguanylate level. J Bacteriol. 2008;190(19):6439–47.
138. Marvig RL, Blokesch M. Natural transformation of Vibrio cholerae as a tool—optimizing the procedure. BMC Microbiol. 2010;10:155. https://doi.org/10.1186/1471-2180-10-155.
139. Matson JS, Yoo HJ, Hakansson K, Dirita VJ. Polymyxin B resistance in El Tor Vibrio cholerae requires lipid acylation catalyzed by MsbB. J Bacteriol. 2010;192:2044–52.
140. Matson JS, Livny J, DiRita VJ. A putative Vibrio cholerae two-component system controls a conserved periplasmic protein in response to the antimicrobial peptide polymyxin B. PLoS One. 2017;12(10):e0186199.
141. Matz C, McDougald D, Moreno AM, Yung PY, Yildiz FH, Kjelleberg S. Biofilm formation and phenotypic variation enhance predation-driven persistence of Vibrio cholerae. Proc Natl Acad Sci U S A. 2005;102(46):16819–24. https://doi.org/10.1073/pnas.0505350102. Epub 2005 Nov 2. PMID: 16267135; PMCID: PMC1283802.
142. McDonough E, Lazinski DW, Camilli A. Identification of in vivo regulators of the Vibrio cholerae Xds gene using a high-throughput genetic selection. Mol Microbiol. 2014;92(2):302–15.
143. Meibom KL, Li XB, Nielsen AT, Wu C-Y, Roseman S, Schoolnik GK. The Vibrio cholerae chitin utilization program. Proc Natl Acad Sci USA. 2004;101:2524–9.
144. Meibom KL, Blokesch M, Dolganov NA, Wu C-Y, Schoolnik GK. Chitin induces natural competence in Vibrio cholerae. Science. 2005;310:1824–7. https://doi.org/10.1126/science.1120096.
145. Mey AR, Payne SM. Haem utilization in Vibrio cholerae involves multiple TonB-dependent Haem receptors. Mol Microbiol. 2001;42:835–49.
146. Mey AR, Wyckoff EE, Oglesby AG, Rab E, Taylor RK, Payne SM. Identification of the Vibrio cholerae enterobactin receptors VctA and IrgA: IrgA is not required for virulence. Infect Immun. 2002;70:3419–26.
147. Mey AR, Wyckoff EE, Kanukurthy V, Fisher CR, Payne SM. Iron and fur regulation in Vibrio cholerae and the role of fur in virulence. Infect Immun. 2005;73:8167–78.
148. Miyata ST, Kitaoka M, Brooks TM, McAuley SB, Pukatzki S. Vibrio cholerae requires the type VI secretion system virulence factor VasX to kill Dictyostelium Discoideum. Infect Immun. 2011;79:2941–9.
149. Mondal M, Nag D, Koley H, Saha DR, Chatterjee NS. The Vibrio cholerae extracellular chitinase ChiA2 is important for survival and pathogenesis in the host intestine. PLoS One. 2014;9:e103119.

150. Montero DA, Vidal RM, Velasco J, George S, Lucero Y, Gómez LA, Carreño LJ, García-Betancourt R, O'Ryan M. Vibrio cholerae, classification, pathogenesis, immune response, and trends in vaccine development. Front Med. 2023;10:1155751. https://doi.org/10.3389/fmed.2023.1155751.
151. Mukherjee S, Bassler BL. Bacterial quorum sensing in complex and dynamically changing environments. Nat Rev Microbiol. 2019;17(6):371–82.
152. Mutreja A, Kim DW, Thomson NR, Connor TR, Lee JH, Kariuki S, Croucher NJ, Choi SY, Harris SR, Lebens M, Niyogi SK. Evidence for several waves of global transmission in the seventh cholera pandemic. Nature. 2011;477(7365):462–5.
153. Nadell CD, Drescher K, Wingreen NS, Bassler BL. Extracellular matrix structure governs invasion resistance in bacterial biofilms. ISME J. 2015;9(8):1700–9.
154. Nadell CD, Drescher K, Foster KR. Spatial structure, cooperation and competition in biofilms. Nat Rev Microbiol. 2016;14(9):589–600.
155. Nahar S, Sultana M, Naser MN, Nair GB, Watanabe H, Ohnishi M, Yamamoto S, et al. Role of shrimp chitin in the ecology of toxigenic Vibrio cholerae and cholera transmission. Front Microbiol. 2011;2:260.
156. Navarro MVAS, Newell PD, Krasteva PV, Chatterjee D, Madden DR. Structural basis for C-Di-GMP-mediated inside-out signaling controlling periplasmic proteolysis. PLoS Biol. 2011;9(2):e1000588.
157. Nelson EJ, Chowdhury A, Flynn J, Schild S, Bourassa L, Shao Y, LaRocque RC, Calderwood SB, Qadri F, Camilli A. Transmission of Vibrio cholerae is antagonized by lytic phage and entry into the aquatic environment. PLoS Pathog. 2008;4:e1000187.
158. Nelson EJ, Harris JB, Morris JG, Calderwood SB, Camilli A. Cholera transmission: the host, pathogen and bacteriophage dynamic. Nat Rev Microbiol. 2009;7 https://doi.org/10.1038/nrmicro2204.
159. Occhino DA, Wyckoff EE, Henderson DP, Wrona TJ, Payne SM. Vibrio cholerae iron transport: Haem transport genes are linked to one of two sets of tonB, exbB, exbD genes. Mol Microbiol. 1998;29:1493–507.
160. Oh YT, Park Y, Yoon MY, Bari W, Go J, Min KB, Raskin DM, Lee K-M, Yoon SS. Cholera toxin production during anaerobic trimethylamine N-oxide respiration is mediated by stringent response in vibrio Cholerae. J Biol Chem. 2014;289:13232–42.
161. Pant A, Bag S, Saha B, Verma J, Kumar P, Banerjee S, Kumar B, Kumar Y, Desigamani A, Maiti S, Maiti TK. Molecular insights into the genome dynamics and interactions between core and acquired genomes of vibrio cholerae. Proc Natl Acad Sci. 2020;117(38):23762–73.
162. Papenfort K, Bassler BL. Quorum sensing signal-response systems in gram-negative bacteria. Nat Rev Microbiol. 2016;14(9):576–88.
163. Papenfort K, Silpe JE, Schramma KR, Cong J-P, Seyedsayamdost MR, Bassler BL. A Vibrio cholerae autoinducer-receptor pair that controls biofilm formation. Nat Chem Biol. 2017;13(5):551–7.
164. Pérez M, Elena M, Macek M, Galván MTC. Do protozoa control the elimination of Vibrio choleraein brackish water? Int Rev Hydrobiol. 2004;89(2):215–27. https://doi.org/10.1002/iroh.200310644.
165. Pflughoeft KJ, Kierek K, Paula I, Watnick PI. Role of ectoine in Vibrio cholerae osmoadaptation. Appl Environ Microbiol. 2003;69(10):5919–27.
166. Pratt JT, McDonough E, Camilli A. PhoB regulates motility, biofilms, and cyclic Di-GMP in Vibrio cholerae. J Bacteriol. 2009;191:6632–42.
167. Pratt JT, Ismail AM, Camilli A. PhoB regulates both environmental and virulence gene expression in Vibrio cholerae. Mol Microbiol. 2010;77:1595–605.
168. Pressler K, Mitterer F, Vorkapic D, Reidl J, Oberer M, Schild S. Characterization of Vibrio cholerae's extracellular nuclease Xds. Front Microbiol. 2019;10:2057.
169. Pukatzki S, Ma AT, Sturtevant D, Krastins B, Sarracino D, Nelson WC, Heidelberg JF, Mekalanos JJ. Identification of a conserved bacterial protein secretion system in Vibrio cholerae using the dictyostelium host model system. Proc Natl Acad Sci USA. 2006;103:1528.

170. Pukatzki S, Ma AT, Revel AT, Sturtevant D, Mekalanos JJ. Type VI secretion system translocates a phage tail spike-like protein into target cells where it cross-links actin. Proc Natl Acad Sci USA. 2007;104:15508–13.
171. Purdy AE, Watnick PI. Spatially selective colonization of the arthropod intestine through activation of vibrio cholerae biofilm formation. Proc Natl Acad Sci USA. 2011;108:19737–42. Epub 2011/11/23. pmid:22106284; PubMed Central PMCID: PMC3241763. https://doi.org/10.1073/pnas.1111530108.
172. Rawlings TK, Ruiz GM, Colwell RR. Association of *Vibrio cholerae* O1 El Tor and O139 Bengal with the copepods *Acartia tonsa* and *Eurytemora affinis*. Appl Environ Microbiol. 2007;73(24):7926–33. https://doi.org/10.1128/AEM.01238-07.
173. Reguera G, Kolter R. Virulence and the environment: a novel role for Vibrio cholerae toxin coregulated pili in biofilm formation on chitin. J Bacteriol. 2005;187(10):3551–5.
174. Robins WP, Mekalanos JJ. Genomic science in understanding cholera outbreaks and evolution of Vibrio cholerae as a human pathogen. In: Nair GB, Takeda Y, editors. Cholera outbreaks, vol. 379. Springer Berlin Heidelberg; 2014. p. 211–29. https://doi.org/10.1007/978-3-642-38079-3_16.
175. Roelofs KG, Jones CJ, Helman SR, Shang X, Orr MW, Goodson JR, Galperin MY, Yildiz FH, Lee VT. Systematic identification of Cyclicdi-GMP binding proteins in Vibrio cholerae reveals a novel class of cyclic-Di-GMP-binding ATPases associated with type II secretion systems. PLoS Pathog. 2015;11(10):e1005232.
176. Safa A, Sultana Jime J, Shahel F. Cholera toxin phage: structural and functional diversity between vibrio cholerae biotypes. AIMS Microbiology. 2020;6:144–51. https://doi.org/10.3934/microbiol.2020.1.144.
177. Schwechheimer C, Hebert K, Tripathi S, Singh PK, Floyd KA. A tyrosine phosphoregulatory system controls exopolysaccharide biosynthesis and biofilm formation in Vibrio cholerae. PLoS Pathog. 2020;16(8):e1008745.
178. Seed KD, Faruque SM, Mekalanos JJ, Calderwood SB, Qadri F, Camilli A. Phase variable O antigen biosynthetic genes control expression of the major protective antigen and bacteriophage receptor in vibrio Cholerae O1. PLoS Pathog. 2012;8:e1002917.
179. Seper A, Fengler VHI, Roier S, Wolinski H, Kohlwein SD, Bishop AL, Camilli A, Reidl J, Schild S. Extracellular nucleases and extracellular DNA play important roles in Vibrio cholerae biofilm formation. Mol Microbiol. 2011;82:1015–37.
180. Seper A, Pressler K, Kariisa A, Haid AG, Roier S. Identification of genes induced in Vibrio cholerae in a dynamic biofilm system. Int J Med Microbiol. 2014;304(5–6):749–63.
181. Shi M, Zheng Y, Wang X, Wang Z, Yang M. Two regulatory factors of Vibrio cholerae activating the mannose-sensitive Haemagglutinin pilus expression is important for biofilm formation and colonization in mice. Microbiology. 2021;167(10):001098.
182. Shikuma NJ, Yildiz FH. Identification and characterization of OscR, a transcriptional regulator involved in osmolarity adaptation in Vibrio cholerae. J Bacteriol. 2009;191:4082–96.
183. Shikuma NJ, Davis KR, Fong JNC, Yildiz FH. The transcriptional regulator, CosR, controls compatible solute biosynthesis and transport, motility and biofilm formation in vibrio Cholerae. Environ Microbiol. 2013;15:1387–99.
184. Sigel SP, Stoebner JA, Payne SM. Iron-vibriobactin transport system is not required for virulence of Vibrio cholerae. Infect Immun. 1985;47:360–2.
185. Silva AJ, Benitez JA. Vibrio cholerae biofilms and cholera pathogenesis. PLoS Negl Trop Dis. 2016;10(2):e0004330. https://doi.org/10.1371/journal.pntd.0004330.
186. Singh PK, Bartalomej S, Hartmann R, Jeckel H, Vidakovic L. Vibrio cholerae combines individual and collective sensing to trigger biofilm dispersal. Curr Biol. 2017;27(21):3359–66.
187. Singh PK, Rode DKH, Buffard P, Nosho K, Bayer M. Vibrio cholerae biofilm dispersal regulator causes cell release from matrix through type IV pilus retraction. 2021; https://doi.org/10.1101/2021.05.02.442311.
188. Singh UB, et al. Phylogeny and evolution of SARS-CoV-2 during Delta and Omicron variant waves in India. J Biomol Struct Dyn. 2024a;42(9):4769–81.

189. Singh RP, et al. First report on in-depth genome and comparative genome analysis of a metal-resistant bacterium Acinetobacter pittii S-30, isolated from environmental sample. Front Microbiol. 2024b;15:1351161.
190. Smith DR, Maestre-Reyna M, Lee G, Gerard H, Wang AH-J, Watnick PI. In situ proteolysis of the Vibrio cholerae matrix protein RbmA promotes biofilm recruitment. PNAS. 2015;112(33):10491–6.
191. Sobe RC, Bond WG, Wotanis CK, Zayner JP, Burriss MA. Spermine inhibits Vibrio cholerae biofilm formation through the NspS-MbaA polyamine signaling system. J Biol Chem. 2017;292(41):17025–36.
192. Spiegel P, Ratnayake R, Hellman N, Ververs M, Ngwa M, Wise PH, Lantagne D. Responding to epidemics in large-scale humanitarian crises: a case study of the cholera response in Yemen, 2016–2018. BMJ Glob Health. 2019;4(4):e001709.
193. Srivastava D, Harris RC, Waters CM. Integration of cyclic Di-GMP and quorum sensing in the control of vpsT and aphA in Vibrio cholerae. J Bacteriol. 2011;193(22):6331–41.
194. Srivastava D, Hsieh M-L, Khataokar A, Neiditch MB, Waters CM. Cyclic Di-GMP inhibits Vibrio cholerae motility by repressing induction of transcription and inducing extracellular polysaccharide production. Mol Microbiol. 2013;90:1262–76.
195. Stauder M, Vezzulli L, Pezzati E, Repetto B, Pruzzo C. Temperature affects Vibrio cholerae O1 El tor persistence in the aquatic environment via an enhanced expression of GbpA and MSHA adhesins. Environ Microbiol Rep. 2010;2:140–4.
196. Stoebner JA, Payne SM. Iron-regulated hemolysin production and utilization of heme and hemoglobin by Vibrio cholerae. Infect Immun. 1988;56:2891–5.
197. Sudarsan N, Lee ER, Weinberg Z, Moy RH, Kim JN, Link KH, Breaker RR. Riboswitches in eubacteria sense the second messenger cyclic Di-GMP. Science. 2008;321:411–3.
198. Sun S, Kjelleberg S, McDougald D. Relative contributions of Vibrio polysaccharide and quorum sensing to the resistance of Vibrio cholerae to predation by heterotrophic protists. PLoS One. 2013;8(2):e56338.
199. Sun S, Tay QXM, Kjelleberg S, Rice SA, McDougald D. Quorum sensing-regulated chitin metabolism provides grazing resistance to Vibrio cholerae biofilms. ISMEJ. 2015;9:1812–20.
200. Svenningsen SL, Tu KC, Bassler BL, Tai J-SB, Mukherjee S, Nero T, Olson R, Tithof J. Gene dosage compensation calibrates four regulatory RNAs to control Vibrio cholerae quorum sensing. EMBO J. 2009;28(4):429–39 169. https://doi.org/10.1101/2021.12.16.472970.
201. Tamayo R, Patimalla B, Camilli A. Growth in a biofilm induces a hyperinfectious phenotype in Vibrio cholerae. Infect Immun. 2010;78:3560–9.
202. Tamplin ML, Gauzens AL, Huq A, Sack DA, Colwell RR. Attachment of Vibrio cholerae serogroup O1 to zooplankton and phytoplankton of Bangladesh waters. Appl Environ Microbiol. 1990;56:1977–80.
203. Taylor RK, Miller VL, Furlong DB, Mekalanos JJ, Teschler JK, Cheng AT, Yildiz FH, et al. Use of phoA gene fusions to identify a pilus colonization factor coordinately regulated with cholera toxin. PNAS. 1987;84(9):2833–7.
204. Teschler JK, Zamorano-Sánchez D, Utada AS, Warner CJA, Wong GCL, Linington RG, Yildiz FH. Living in the matrix: assembly and control of Vibrio cholerae biofilms. Nat Rev Microbiol. 2015;13(5):255–68. https://doi.org/10.1038/nrmicro3433.
205. Teschler JK, Nadell CD, Drescher K, Yildiz FH. Mechanisms underlying *Vibrio cholerae* biofilm formation and dispersion. Ann Rev Microbiol. 2022;76(1):503–32. https://doi.org/10.1146/annurev-micro-111021-053553.
206. Timmermans J, Melderen L. Post-transcriptional global regulation by CsrA in bacteria. Cell Mol Life Sci. 2010;67(17):2897–908.
207. Toma C, Nakasone N, Song T, Iwanaga M. *Vibrio cholerae* SXT Element, Laos. Emerg Infect Dis. 2005;11:346–7. https://doi.org/10.3201/eid1102.040933.
208. Townsley L, Yildiz FH Temperature affects C-Di-GMP signaling and biofilm formation in Vibrio cholerae. Epub ahead of print. 2015. https://doi.org/10.1111/1462-2920.12799.

209. Tsou AM, Cai T, Liu Z, Zhu J, Kulkarni RV. Regulatory targets of quorum sensing in Vibrio cholerae: evidence for two distinct HapR-binding motifs. Nucleic Acids Res. 2009;37(8):2747–56.
210. Tsou AM, Liu Z, Cai T, Zhu J. The VarS/VarA two-component system modulates the activity of the Vibrio cholerae quorum-sensing transcriptional regulator HapR. Microbiology. 2011;157:1620–8.
211. Utada AS, Bennett RR, Fong JCN, Gibiansky ML, Yildiz FH, Golestanian R, Wong GCL. Vibrio cholerae use pili and flagella synergistically to effect motility switching and conditional surface attachment. Nat Commun. 2014;5:4913.
212. Vaitkevicius K, Lindmark B, Ou G, Song T, Toma C, Iwanaga M, Zhu J, et al. A Vibrio cholerae protease needed for killing of Caenorhabditis elegans has a role in protection from natural predator grazing. Proc Natl Acad Sci USA. 2006;103:9280–5.
213. Val M-E, et al. Genome engineering in vibrio cholerae: a feasible approach to address biological issues. PLoS Genet. 2012;8(1):e1002472. https://doi.org/10.1371/journal.pgen.
214. Val M-E, et al. Management of multipartite genomes: the Vibrio cholerae model. Curr Opin Microbiol. 2014;22:120–6. https://doi.org/10.1016/j.mib.2014.10.003.
215. Van Kessel JC, Camilli A. Vibrio cholerae: a fundamental model system for bacterial genetics and pathogenesis research. J Bacteriol. 2024;206:e00248–24. https://doi.org/10.1128/jb.00248-24.
216. Wachsmuth K, Olsvik Ø, Evins GM, Popovic T. vibrio Cholerae and cholera: molecular to global perspective. In: Wachsmuth IK, Blake PA, Olsvik Ø, editors. . Washington DC: ASM Press; 1994. p. 357–70. https://doi.org/10.1128/9781555816404.ch26.
217. Wai SN, Moriya T, Kondo K, Misumi H, Amako K. Resuscitation of Vibrio cholerae O1 strain TSI-4 from a viable but nonculturable state by heat shock. FEMS Microiol Lett. 1996;136:187–91.
218. Waldor MK, Mekalanos JJ. Lysogenic conversion by a filamentous phage encoding cholera toxin. Science. 1996;272:1910–4.
219. Waldor MK, Tschäpe H, Mekalanos JJ. A new type of conjugative transposon encodes resistance to sulfamethoxazole, trimethoprim, and streptomycin in Vibrio cholerae O139. J Bacteriol. 1996;178:4157–65. https://doi.org/10.1128/jb.178.14.4157-4165.1996.
220. Wang H, Ayala JC, Silva AJ, Benitez JA. The histone-like nucleoid structuring protein (H-NS) is a repressor of Vibrio cholerae exopolysaccharide biosynthesis (Vps) genes. Appl Environ Microbiol. 2012;78:2482–8.
221. Waters CM, Lu W, Rabinowitz JD, Bassler BL. Quorum sensing controls biofilm formation in Vibrio cholerae through modulation of cyclic Di-GMP levels and repression of vpsT. J Bacteriol. 2008;190:2527–36.
222. Watnick PI, Kolter R. Steps in the development of a Vibrio cholerae El tor biofilm. Mol Microbiol. 1999;34(3):586–95. https://doi.org/10.1046/j.1365-2958.1999.01624.x.
223. Weber GG, Kortmann J, Narberhaus F, Klose KE. RNA thermometer controls temperature-dependent virulence factor expression in Vibrio cholerae. Proc Natl Acad Sci USA. 2014;111:14241.
224. Wong E, Vaaje-Kolstad G, Ghosh A, Hurtado-Guerrero R, Konarev P. V, Ibrahim AFM, Svergun DI, Eijsink VGH, Chatterjee NS, Aalten DMF. The Vibrio cholerae colonization factor GbpA possesses a modular structure that governs binding to different host surfaces. PLoS Pathog. 2012;8:e1002373.
225. Worden AZ, Seidel M, Smriga S, Wick A, Malfatti F, Bartlett D, Azam F. Trophic regulation of Vibrio cholerae in coastal marine waters. Environ Microbiol. 2006;8:21–9.
226. Wu DC, Zamorano-Sánchez D, Pagliai FA, Park JH, Floyd KA. Reciprocal C-Di-GMP signaling: incomplete flagellum biogenesis triggers c-Di-GMP signaling pathways that promote biofilm formation. 2020;16:e1008703.
227. Wucher BR, Elsayed M, Adelman JS, Kadouri DE, Nadell CD. Bacterial predation transforms the landscape and community assembly of biofilms. Curr Biol. 2021;31(12):2643–51.

228. Wyckoff EE, Valle A, Smith SL, Payne SM. A multifunctional ATP-binding cassette transporter system from Vibrio cholerae transports Vibriobactin and Enterobactin. J Bacteriol. 1999;181:7588–96.
229. Wyckoff EE, Mey AR, Leimbach A, Fisher CF, Payne SM. Characterization of ferric and ferrous iron transport systems in Vibrio cholerae. J Bacteriol. 2006;188:6515–23.
230. Yamamoto S, Izumiya H, Mitobe J, Morita M, Arakawa E, Ohnishi M, Watanabe H. Identification of a chitin-induced small RNA that regulates translation of the tfoX gene, encoding a positive regulator of natural competence in Vibrio cholerae. J Bacteriol. 2011;193:1953–65.
231. Yamamoto S, Mitobe J, Ishikawa T, Wai SN, Ohnishi M, Watanabe H, Izumiya H. Regulation of natural competence by the orphan two-component system sensor kinase ChiS involves a non-canonical transmembrane regulator in Vibrio cholerae. Mol Microbiol. 2014;91:326–47.
232. Yildiz FH, Dolganov NA, Schoolnik GK. VpsR, a member of the response regulators of the two-component regulatory systems, is required for expression of Vps biosynthesis genes and EPS(ETr)-associated phenotypes in Vibrio cholerae O1 El tor. J Bacteriol. 2001;183:1716–26.
233. Yildiz FH, Liu XS, Heydorn A, Schoolnik GK. Molecular analysis of Rugosity in a Vibrio cholerae O1 El tor phase variant. Mol Microbiol. 2004;53:497.
234. Yildiz F, Fong J, Sadovskaya I, Grard T, Vinogradov E. Structural characterization of the extracellular polysaccharide from Vibrio cholerae O1 El-tor. PLoS One. 2014;9:e86751. https://doi.org/10.1371/journal.pone.0086751.
235. Zahid MSH, Udden SMN, Faruque ASG, Calderwood SB, Mekalanos JJ, Faruque SM. Effect of phage on the infectivity of Vibrio cholerae and emergence of genetic variants. Infect Immun. 2008;76:5266–73.
236. Zamorano-Sánchez D, Fong JCN, Kilic S, Erill I, Yildiz FH. Identification and characterization of VpsR and VpsT binding sites in Vibrio cholerae. J Bacteriol. 2015;197:1221–35.
237. Zhao X, Koestler BJ, Waters CM, Hammer BK. Post-transcriptional activation of a diguanylate cyclase by quorum sensing Small RNAs promotes biofilm formation in Vibrio cholerae. Mol Microbiol. 2013;89(5):989–1002.
238. Zheng J, Ho B, Mekalanos JJ. Genetic analysis of anti-amoebae and anti-bacterial activities of the type VI secretion system in Vibrio cholerae. PLoS One. 2011;6:e23876.
239. Zhu J, Mekalanos JJ. Quorum sensing-dependent biofilms enhance colonization in Vibrio cholerae. J Bacteriol. 2003;185(9):2820–32. https://doi.org/10.1128/JB.01756-07.

4 Archaeal Model Organisms: Understanding Genetics, Metabolic Diversity, and Biotechnology Applications

Manuela Tripepi, Rachel Denniston, Ka'alea Rennie, Daud Tariq, and Jason Yang

4.1 Introduction

Archaea, along with Bacteria and Eukarya, constitute one of the three domains of life [36]. They have gained considerable scientific interest due to their unique evolutionary trajectories, distinct cellular architectures, and diverse metabolic pathways. Initially, archaea were described as extremophiles inhabiting harsh environments, but they are now recognized as ubiquitous, inhabiting a wide array of habitats, including moderate environments such as the human microbiome, terrestrial soils, and marine ecosystems [91]. Their ubiquitous presence underscores their ecological significance and positions them as pivotal subjects for investigating fundamental biological processes [55]. It is also intriguing to note that despite being among the most abundant and widespread life forms on Earth, archaea are non-pathogenic [34]. Four major archaeal groups have emerged as model organisms due to their genetic tractability: methanogens, halophiles, thermophilic euryarchaeota, and crenarchaeota [64]. These models have facilitated in-depth studies of archaeal genetics, physiology, and molecular biology, providing insights into their unique mechanisms and evolutionary relationships [64].

Methanogens, for instance, are the only biological producers of methane, playing a crucial role in the global carbon cycle [27]. Species such as *Methanococcus maripaludis* have become model organisms, offering a platform for genetic manipulation and functional genomics studies [94]. The genetic tools developed for these organisms have enabled detailed investigations into methanogenic pathways and regulation [27]. Halophilic archaea, thriving in high-salinity environments, have

M. Tripepi (✉) · R. Denniston · K. Rennie · D. Tariq · J. Yang
Department of Biological and Chemical Sciences, College of Life Sciences,
Thomas Jefferson University, Philadelphia, PA, USA
e-mail: Manuela.Tripepi@jefferson.edu

P. Kumar, V. Kothari (eds.), *Model Organisms in Biological Research*,
https://doi.org/10.1007/978-981-92-0038-2_4

also been established as model systems, leading to significant advancements in our understanding of archaeal DNA replication and repair mechanisms [84].

Thermophilic archaea, such as those belonging to the order Thermococcales, have thermostable enzymes that are studied for understanding protein stability and also provide insights into adaptations to high-temperature environments and biotechnology [7, 30, 43]. The availability of genetic tools in thermophiles has facilitated studies on their unique metabolic pathways and stress response mechanisms [47]. Crenarchaeota, particularly those in the order Sulfolobales, are known for their ability to thrive in acidic and high-temperature environments [97]. *Sulfolobus* species have been models for studying archaeal cell cycle regulation, DNA replication, and repair [61].

Establishing these model organisms has been pivotal in elucidating archaea's molecular and cellular biology. Advancements in genetic techniques, including the development of transformation methods, selectable markers, and reporter genes, have empowered researchers to perform precise genetic manipulations [64, 74, 94]. These tools have been essential for functional genomics studies, enabling the identification and characterization of genes involved in various metabolic and regulatory pathways [74, 94]. The study of archaeal model organisms has provided valuable evolutionary insights. Comparative analyses have revealed that many archaeal proteins involved in DNA replication, transcription, and translation share more remarkable similarities with their eukaryotic counterparts than those in bacteria [46]. These data suggest a closer evolutionary relationship between archaea and eukaryotes, offering clues about the evolution of complex cellular processes [46].

Exploring key archaeal model species has significantly advanced our understanding of microbial life, shedding light on archaea's unique physiological adaptations, genetic systems, and evolutionary relationships, enriching our comprehension of life's diversity and resilience to harsh environments.

4.2 Methanogens

Methanogens are obligate anaerobic archaea that synthesize ATP via methanogenesis [16] and are also the only group of organisms that can perform this process [112]. The process involves the reduction of single-carbon compounds, such as carbon dioxide and methanol, via various methanogenic pathways [29]. Methanogens contribute to 1.1% of all the photosynthetically produced organic carbon on Earth [113]. They can be found in a wide variety of microbiomes, from pond sediment to animal digestive tracts [16]. They thrive in more neutral pH environments, which is unlike many other archaea that have been cultured within laboratory settings [94]. Furthermore, while they are associated with oral and gastrointestinal illnesses, they are not directly pathogenic for humans [31].

This group was instrumental in the identification of archaea as a separate domain by Carl Woese [118]. Through his work and related work, archaea were determined to have unique 16S rRNA sequences, isoprenoid membrane proteins, a lack of homology to other bacteria, and a unique amino acid, pyrrolysine, that, among other factors, served as evidence of the need for a new Archaea domain [16]. Despite their value in producing methane, methanogens have been historically understudied for two reasons. The first is their anaerobic nature, as well as the fact that they need a reducing environment with a redox potential below 50 mV, making them difficult to culture [39]. Second, many methanogens form syntrophic relations with bacteria, consuming excess hydrogen or acetate produced by their bacterial partners, creating an environment where both the archaea and bacteria can survive [100]. As they were closely linked to the syntrophic bacteria, they were often discarded as waste rather than properly studied to characterize these relationships [16].

Metabolism For all methanogens, the breakdown of a carbon substrate to methane involves the Wolfe Cycle [112]. To begin, carbon dioxide is reduced to formylmethanofuran via the reduction and oxidation of ferredoxin [11]. To achieve energetically favorable circumstances, this process is done in three steps. The methyl group of methylH4MPT is transferred to CoM-SH, then CoM-SH is reduced to methane and the heterodisulfide CoM-S-S-CoB, and lastly CoM-S-S-CoB is reduced to restore the ferredoxins, completing the cycle [112]. The anaplerotic nature of this cycle is especially important, as not only does it allow for energy formation, but purine intermediaries are siphoned off in order to generate nucleic acids [9]. Without the Wolfe cycle, any hopes of energy production or DNA synthesis are cut short, which would halt essential cellular processes. To initiate energy production, methanogens rely on different metabolic strategies. Hydrogenotrophic methanogens, or those that use hydrogen gas as an electron donor, are the most common and easiest to cultivate [16]. Other metabolism categories include carboxydotrophic methanogens, which use carbon dioxide and carbon monoxide; methylotrophic methanogens, which use methanol, methylamines, and methyl sulfides; and acetoclastic methanogens, which use acetate for energy [16].

Genomic Insights Genetic tools have been instrumental in advancing the study of methanogens, particularly in exploring DNA replication and gene regulation [107]. Antibiotics, gene expression tools, and shuttle vectors are used for uncovering further insight within the methanogen species (Table 4.1) [27]. *Methanococcus voltae* is a popular model organism for methanogens, as it grows quickly in complex media and has a single-cell morphology [57]. The absence of a pseudomurein cell wall in *M. voltae* was also key, as it aided the integrity of the cells while allowing unique antibiotic selection methods to function that could not be done with bacteria [107]. *M. voltae*'s versatility was used to define the nutrient media for other *Methanococcus* species as well, such *Methanococcus maripaludis* and *Methanococcus methanosarcina* [57].

Gene Expression Tools Gene reporters have been used in *Methanococcus maripaludis* to study the regulation of *nifH*; a gene which contributes to nitrogen fixation [59]. To study how *Methanococcus maripaludis* regulates the *nifH* gene, researchers linked

Table 4.1 Representative model organisms of methanogens

Model organism	Genetic tools available	Biotechnology applications	Challenges/ considerations	References
Methanococcus voltae	Strain A3 genome has been sequenced *UidA* and *lacZ* are used as fusion reporter genes. Neomycin is used for antibiotic selection	Biofuel and energy production Naturally produces methane, which is able to be used for biofuel and energy production	Substrate range is limited Limited metabolism does not allow for more diverse studies	Argyle et al. [5]; Kessler and Leigh [59]; Leigh et al. [64]; Stepanenko and Heng [107]; Li et al. [66]
Methanococcus maripaludis	Strain S2, DSM 2067, and C7 genomes has been sequenced. *NifH* is used as a fusion reporter gene. Neomycin is used for antibiotic selection. pURB500 plasmid is transduced for plasmid replication	Biofuel and energy production Nitrogen fixation mechanism is used for energy production like methane, for preserving energy by producing ammonia as an inhibitory agent, and is used in agriculture by producing biofertilizer	Substrate range is limited A more unique and complex nitrogen regulation mechanism	Kessler and Leigh [59]; Sarmiento et al. [94]; Stepanenko and Heng [107]; Li et al. [66]

the *lacZ* reporter gene to the *nifH* promoter [94]. This reporter system revealed that *M. maripaludis* regulates nitrogen fixation using mechanisms similar to those found in other nitrogen-fixing organisms, such as bacteria [59]. *UidA* and *lacZ* are fusion reporter genes that have been used to monitor gene expression within hydrogenotrophic methanogens [94]. *M. voltae* is a hydrogenotrophic methanogen whose gene expression was studied through the utilization of *uidA* [94]. *UidA* produces a beta-glucuronidase enzyme and *lacZ* produces a beta-galactosidase enzyme, and both enzymes create a visible signal when they are active, which indicates the level of gene expression [64]. Markerless genetic exchange has also been incredibly effective in *M. methanococcus* and *M. methanosarcina.* In other systems, a marker is left behind on a gene to trace a result, but in these species of methanogens, after the change to the gene is made, the selectable marker is removed so the modified organism can be reused in future experiments without interference from the marker used [64].

Use of Antibiotics as Selection Tools Methanogens have a unique cell structure that differentiates them from bacteria and eukaryotes, as they do not have peptidoglycan in the cell wall and instead use a protein S-layer, which aids in gas exchange [79]. This unique cell structure and a unique ribosomal RNA component influence their susceptibility to antibiotics, rendering many conventional antibiotics ineffective ([107]; [27]). However, certain antibiotics, such as puromycin and neomycin, have been effectively utilized as selection tools in the genetic engineering of methanogens [79]. Puromycin is an antibiotic that inhibits protein synthesis [107]. Methanogens are generally sensitive to puromycin, and this renders this antibiotic a valuable selection agent. To confer resistance, the *pac* gene, encoding puromycin N-acetyl-transferase from *Streptomyces alboniger*, is introduced into methanogens. This enzyme inactivates puromycin, allowing only genetically modified cells harboring the *pac* gene to survive in its presence [80]. This strategy has been successfully employed in species like *Methanococcus voltae* and *Methanococcus maripaludis*. Another antibiotic used for selection in methanogens is neomycin—an aminoglycoside antibiotic [5]. In *Methanococcus maripaludis*, neomycin resistance has been achieved by introducing a neomycin resistance gene that encodes for the enzyme aminoglycoside phosphotransferase, allowing researchers to select for genetically modified strains that survive antibiotic treatment ([5]; [107]).

Shuttle Vectors Shuttle vectors have also been utilized within methanogens by transferring the plasmid between different organisms for better understanding of how archaea function (Table 4.1) [23]. *M. maripaludis* contains a small plasmid pURB500 that provides insight into how plasmids are able to replicate within methanogens through DNA loops [94]. Within further discovery of *M. maripaludis*, it was found that combining the pWLG30 plasmid with the *lacZ* enzyme, gene expression within *M. maripaludis* could be better understood due to the *lacZ* enzyme changing color when the gene was active [94].

Biofuel and Renewable Energy Synthesis Methanogens, having the ability to convert hydrogen into methane, have created sustainable advances in renewable energy by making biogas [66]. Biogas is a mixture of methane and has a larger amount of carbon dioxide and hydrogen sulfide, differentiating it from natural gases such as fossil fuels [66]. Hydrogenotrophic methanogens, such as *Methanococcus* spp., provide a scalable, biologically driven solution for renewable energy storage and are central to power-to-gas systems that convert renewable electricity into storable energy (Table 4.1) [25]. In these systems, electricity is used to electrolyze water, producing molecular hydrogen (H_2), which is then combined with carbon dioxide (CO_2) and biologically converted into methane by methanogens. The methane obtained with this process is referred to as "biomethane." This bioprocess enables long-term energy storage and offers a sustainable alternative to fossil fuels by utilizing carbon-neutral or recycled CO_2 sources, such as biogas plants or industrial emissions [25].

4.3 Halophiles

Halophilic archaea, commonly referred to as halophiles, thrive in environments with high salinity, such as natural brines, alkaline salt lakes, the Dead Sea, marine solar salterns, and rock salt deposits [106]. The term "halophiles" is derived from the Greek words meaning "salt-loving" and is associated with archaea, bacteria, and eukaryotes; however, the archaeal classification is the largest among the group [106]. The first known sample of haloarchaeal microorganisms was reported in the 1960s from Permian salt sediments and is believed to date back approximately 250 million years [106]. Halophiles have adapted to the high ionic stress, high temperatures, and scarcity of moisture and nutrients, but do not use the typical bacterial survival mechanism of spores [106]. The haloarchaea possess specialized biochemical and molecular adaptations that allow them to thrive in environments where most other organisms would not [76]. Specific methodologies and cellular features contribute to their survival, such as processing carbohydrates and proteins, accumulating compatible solutes through the organic osmolyte mechanism, and undergoing morphological changes like spherical formation to maintain osmotic balance [37, 50]. The haloarchaea cellular membranes contain unique characteristics that provide stability in hyper-saline conditions, with their protein often heat-stabilizing and enabling them to function at elevated temperatures and high salinity [76]. Additionally, haloarchaea have evolved mechanisms such as the production of protective pigments like bacteriorhodopsins, which help shield them from damaging UV radiation, further enhancing their resilience in harsh conditions [102].

As mentioned above, the organic osmolyte mechanism allows haloarchaea to generate compatible solutes within their cytoplasm, stabilizing them against the saline environment [50]. Furthermore, they can have an influx of potassium ions into the cytoplasm, facilitating the release of sodium ions from the cell and maintaining cellular balance despite external environmental stressors [68]. By maintaining cellular fluid balance, haloarchaea can remain viable or dormant for extended periods [50, 106]. Another key adaptation is the formation of spheres by haloarchaea in confined salt deposits, known as fluid inclusions. These inclusions are thought to serve as a habitat for haloarchaea. For instance, the strain *Halobacterium salinarum* was isolated from a fluid inclusion in a 97,000-year-old crystal from Death Valley [106]. In a recent study, rod-shaped *H. salinarum* adapted to a spherical shape when exposed to a lab-grown Halite crystal. Researchers have concluded that the spherical formation of haloarchaea is a resting state, unlike bacteria, which produce spores to remain dormant. This spherical formation is facilitated by the non-covalent bonding of glycoproteins in the surface layer, enabling such interactions [38].

Metabolism The majority of cultivated haloarchaea are aerobic heterotrophs that use peptides or simple carbohydrates as carbon sources and energy, with few facultative anaerobes as exceptions ([104]; [35]). Recent studies have discovered haloarchaea that are anaerobic and have novel metabolic capacities ([105]; [103]; [104]; [35]). Metabolism varies between haloarchaea species. There are carbohydrate-utilizing species and non-carbohydrate-utilizing pathways, with some species able to utilize both. Carbohydrate-utilizing pathways are able to metabolize various sugars such as glucose, fructose, arabinose, xylulose, sucrose, and lactose [37]. Haloarchaea use a modified version of the Embden-Meyerhof pathway and Entner-Doudoroff (ED) pathway variants [37]. For example, *Haloferax volcanii* use the semi-phosphorylated ED pathway for glucose degradation [109]. *H. salinarum* was found to have both the forward and reverse EM pathway, with the enzyme 6-phosphofructokinase missing [37]. Noncarbohydrate-utilizing pathways use amino acids and various compounds found in hypersaline environments [37]. An example would be *H. volcanii*, which uses glycerol and organic acids excreted by the primary halophilic producers *Dunaliella salina* and *Microcoleus chthonoplastes* [37]. There are also haloarchaea that can utilize other sources and processes for growth. *H. salinarum* three bioenergetic processes. It can grow phototrophically using bacteriorhodopsin, a light-driven proton pump [90]. *H. salinarum* can also utilize respiration to enhance membrane potential and fermentation of arginine for energy [111]. Halo(natrono)archaea utilize chitin and cellulose as growth substrates [105]. Anaerobic lithoheterotrophic haloarchaea can utilize elemental sulfur and acetate [103]. Haloarchaea are versatile in their metabolism across species, and novel pathways will continue to be discovered as haloarchaea are further studied.

Genetic Insights and Genetic Tools Haloarchaea are used in the laboratory due to their easy cultivation, fast growth kinetics, and resistance to contamination by non-halophilic microorganisms ([2]; [64]). The model organisms of haloarchaea are *H. volcanii* and *H. salinarum*. Both model organisms have genome sequences completed, as listed in Table 4.2 ([81]; [52]). *H. volcanii* has easy growth conditions, a short generation time, and facile genetics [49, 72, 73]. *H. volcanii* takes about 1.83 hours for generation time at its optimal temperature of 45 °C in aerobic conditions ([90]; [2]). It can grow on complex and defined media with a wide range of salinities from 1.8 to 3.5 M NaCl [2]. The complete genome sequence of *H. volcanii* was completed in 2010 [52]. *H. salinarum* can grow over a wide range of salinity between 2.6 and 5.1 M NaCl. It takes about 1.86 hours for generation time at its optimal temperature of 50 °C in aerobic conditions [90]. The genome of *H. salinarum* was published in 2000 and was the first haloarchaea genome sequenced [81].

Gene Expression Tools There are tools used in gene expression for *H. volcanii*, including the tryptophan-inducible promoter p.*tnaA* for protein expression and p.syn promoter for protein overexpression (Table 4.2) [49]. *H. volcanii* also has a CRISPR-Cas system that uses CRISPR interference for genetic manipulation [71]. *H. salinarum* has genome instability because of its insertion sequence rearrangements, and so *Haloferax* is preferred for genetic manipulation due to its insertion sequence elements being confined to megaplasmids [51]. However, *H. salinarum* had been used to study DNA repair due to its high resistance to ultraviolet light and ionizing radiation [64]. For gene deletion and replacement, the "pop-in, pop-out method" is used in *H. salinarum* and *H. volcanii* (Table 4.2) [51]. A ura3-based gene replacement strategy uses H. *salinarum*'s sensitivity to 5-fluoroorotic acid to select

Table 4.2 Representative model organisms of halophiles

Model organism	Genetic tools available	Biotechnology applications	Challenges/considerations	References
Haloferax volcanii	Promoters *p.tnaA* and *p.syn* CRISPR-Cas Novobiocin and mevinolin are used for antibiotic selection. Shuttle vectors pHK2, pHV2, pHV1/4 *pyrE1*- *pyrE2* gene knockout system Selectable markers available: leucine, tryptophan, uracil, and thymidine biosynthesis Strain DS2 genome has been sequenced	Biocatalysts in industry Antibiotic peptides known as halocin for the textile industry and medical field Silver and gold nanoparticle production Production of C_{50} carotenoid for pharmaceuticals and cosmetics Production of endogenous laccases for bio-based polymer synthesis and pollutant degradation	High salt concentrations can cause corrosion mediated damage to stainless-steel bioreactors Tendency to form biofilms at high volumes	Bitan-Banin et al. [14]; Hartman et al. [52]; Leigh et al. [64], Atomi et al. [6], Yin et al. [120] Amoozegar et al. [4] Haque et al. [49], Maier et al. [71]; Pérez-Arnaiz et al. [84]; Pohlschroder et al. [85]
Halobacterium salinarum	Novobiocin and mevinolin are used for antibiotic selection Shuttle vectors pGRB1, PHH1, and pNRC100 ura3-based gene replacement strategy Novel modular system called pHsal vector series Strain NRC-1 genome has been sequenced	Biocatalysts in industry Production of antibiotic peptides known as halocins for the textile industry and medical field Bacteriorhodopsin production for photochemical application	Genome instability due to insertion sequence rearrangements	Peck et al. [82]; Ng et al. [81]; Leigh et al. [64], Atomi et al. [6]; Silva-Rocha et al. [101]; Yin et al. [120]; Amoozegar et al. [4]; Haque et al. [49]; Harrison and Allers [51]

the mutant *ura3* gene as a counterselectable marker [82]. A gene knockout system was developed for *H. volcanii,* where the *pyrE1* gene is deleted while the *pyrE2* gene is used to create gene knockouts [14].

Use of Antibiotics as Selection Tools The genera *Haloferax and Halobacterium* have selection systems based on the antibiotic resistance of novobiocin, mevinolin, and simvastatin (Table 4.2) [6]. Archaea lack the peptidoglycan walls that other antibiotics target. Novobiocin inhibits DNA gyrase, an enzyme required for DNA synthesis. Mevinolin inhibits 3-hydroxy-3-methylglutaryl coenzyme A reductase (HMG-CoA), which produces precursors for isoprenoid compounds in archaeal membrane lipids. The antibiotics were used to develop shuttle vectors. For example, *H. volcanii* mutants that encode the gene for novobiocin-resistant DNA gyrase have been used as a selectable marker to develop *H. volcanii-E. coli* shuttle vectors ([6]; [84]). Mevinolin-resistant DNA strains of *H. volcanii* have been shotgun-cloned and isolated as DNA fragments to create shuttle vectors ([15]; [6]).

Shuttle Vectors *Haloferax volcanii* and *Halobacterium salinarum* both possess replicative shuttle vectors that support genetic manipulation (Table 4.2). For *H. volcanii*, commonly used vectors derive from pHK2, pHV2, and pHV1/4 plasmids ([3]; [64]). In *H. salinarum*, vectors have been developed based on plasmids originating from pGRB1, pHH1, and pNRC100 ([15]; [62]; [64]). Recently, a modular set of four vectors—known as the pHsal series—has been specifically optimized for *H. salinarum* [101].

Biocatalysts, Antibiotics Production, and Other Biotechnological Applications Haloarchaea have a variety of potential biotechnological applications in industry and also biomedicine, as seen in Table 4.2 ([26]; [33]). The unique properties and adaptation of haloarchaea enzymes allow potential use as biocatalysts in harsh industrial conditions without the need for purification or sterilization [4]. Haloarchaea also produce antibiotic peptides called halocins, which are currently used in the textile industry to prevent halophiles from growing in brine ([13]; [20]). Haloarchaea have been utilized in various industries, including bioplastics, the food industry, and biomedicine [120]. Polyhydroxyalkanoates (PHAs) are a family of biodegradable and biocompatible polyesters produced by many microorganisms, enabling the development of bioplastics [87]. The biotechnological use of halophiles offers several advantages, as it can help create a low-cost, biocompatible, and environmentally friendly product. One key benefit is that high salt concentrations prevent contamination by non-halophiles, significantly reducing costs, energy consumption, and the complexity of sterilization processes [87]. Haloarchaea are also valuable in the food industry due to their ability to secrete extracellular hydrolytic enzymes such as proteases, which function effectively in a wide range of food processing environments, including high-salt and high-pH conditions [120].

The use of haloarchaea is gaining special significance in biomedicine, especially in the search for new bioactive molecules [26]. Natural products, often referred to as bioactive molecules, are important sources of anticancer drugs and present fewer side effects than the current ones. Halophilic microorganisms show great potential in producing carotenoids such as β-carotene, bacterioruberin, and xanthophylls, and have been explored as bioactive compounds in cancer treatment in recent studies [75]. For instance, carotenoid extracts from *Halobacterium halobium* have been tested on hepatoma (HepG2) cells, demonstrating a dose-dependent inhibition of cell viability. These findings highlight the strong antioxidant potential of these extracts and suggest their promise in cancer therapy [75].

4.4 Thermococcales

Thermococcales are anaerobic, sulfur-metabolizing archaea that exhibit optimal growth at temperatures ranging from 80 °C to over 100 °C. The order of *Thermococcales*, in Euryarchaeota, includes hyperthermophilic microorganisms primarily found in deep-sea hydrothermal vents and other high-temperature environments [42, 44]. The ability of *Thermococcales* to thrive in extreme conditions provides valuable insights into molecular mechanisms of thermophily, DNA replication, and protein stability [122]. Proteins from *Thermococcales* exhibit extraordinary thermal stability. The factors contributing to this stability include increased hydrophobic interactions, ionic networks, and reduced conformational flexibility. These adaptations are crucial for maintaining protein function under extreme conditions [122]. Organisms from the genera *Thermococcus* and *Pyrococcus* have earned the title of model organisms due to their unique physiological and genetic characteristics [47, 88].

Metabolism Thermococcales archaea have several versatile metabolic pathways, with most species using a mix of peptides or carbohydrate substrates as energy sources [1]. As an example, *Pyrococcus furiosus* uses starch, glycogen, β-glucans, cellobiose, and maltose [40]. In this process, reduced ferredoxin is produced, which needs to be regenerated to continue breaking down substrates [1]. There are three pathways that can be used: S_0 reduction to H_2S, proton reduction to H_2, or alanine formation. The first of these pathways involving sulfur is by far the most common in *Thermococcles*, with species necessitating S^0 in the culture media [116]. However, some *Thermococcales* can be facultative in sulfur use and perform ferredoxin oxidation by the other two pathways. The reduction to H_2 gas is performed most prominently by *Thermococcus kodakarensis,* which is one of the most potent H_2 gas-producing *Thermococcales*, and has many uses as a genetic tool [58]. Alternatively, *P. furiosus* can produce alanine as a byproduct in exchange for ferredoxin [116]. The presence of sulfur enhances these two pathways, making it essential to the metabolism of *Thermococcales* as a whole.

Genomic Insights Within the order *Thermococcales*, several species have emerged as valuable tools for genetic manipulation and expression studies. *Thermococcus kodakarensis* thrives at 85 °C, with a growth range between 60 °C and 100 °C [7].

It was initially isolated from a solfatara field on Kodakara Island, Japan [78]. This archaeon is an obligate anaerobe and heterotroph, metabolizing various organic substrates in the presence of elemental sulfur, producing hydrogen sulfide [116]. It first gained a spot in research for the ease of laboratory growth and metabolic flexibility. However, the discovery of natural competence, allowing efficient DNA uptake, made this organism a top model organism for archaea [95]. *P. furiosus* is a well-established model organism for the genus *Pyrococcales*. *P. furiosus* is a hyperthermophile that exhibits optimal growth near 100 °C. It was isolated from geothermally heated marine sediments off the coast of Italy [40]. It is a heterotrophic, strictly anaerobic organism that metabolizes carbohydrates and peptides.

Gene Expression Tools Developing genetic systems in *Thermococcales* has been fundamental for their establishment as model organisms. Genetic manipulation techniques allow researchers to explore gene function and regulation in these archaea, since they can be used to study gene expression and vector uptake with ease. For example, *Thermococcus kodakarensis* has been extensively studied due to its natural competence for DNA uptake, facilitating targeted gene disruptions and complementation studies [110]. This has allowed for functional analyses of metabolic pathways and stress responses, particularly oxidative, osmotic, and thermal [64]. The ease of use has also expanded the fields of proteomics [56] and genetics [53], since the entire *T. kodakarensis* KOD1 genome has been sequenced (Table 4.3) [42]. There are two main routes to determine gene uptake with *T. kodakarensis*. Auxotrophic gene selection can be done using the *pyrF* gene to disrupt the *trpE* gene, inhibiting the uptake of uracil ([95]; [96]). Additionally, the *TK0664* gene can be used for counterselection of *T. kodakarensis* to measure the uptake of 6-methyl purine [93].

Other species that are used as model organisms are *Pyrococcus furiosus* JCM8422 [64] and *Pyrococcus abyssi* GE5 [24], whose genomes have facilitated the development of sophisticated gene expression tools such as shuttle vectors, inducible promoters, and reporter systems. For example, *P. furiosus* was used by scientists to establish a shuttle vector-based transformation system that enables the overexpression of target proteins and the study of gene function in this archaeon [114]. This genomic data has provided insights into its metabolic pathways and enzymatic machinery, contributing to its status as a model organism for studying hyperthermophilic archaea.

Use of Antibiotics as Selection Tools The antibiotic simvastatin can be used with *T. kodakarensis* to determine gene uptake. 3-Hydroxy-3-methylglutaryl coenzyme A (HMG-CoA) reductase can be inhibited with simvastatin [77]. Resistance to the simvastatin with overexpression of HMG-CoA via the hmg_{Tk} gene can then be used as a marker gene for DNA uptake [77]. This method can be used on other Thermococcus strains, such as *Thermococcus onnurineus* [60]. Regardless of which method is used, β-glycosidases can be used to determine the efficiency of uptake, as both TK1761 and TK1827 genes can be used as reporter genes [93].

Shuttle Vectors *Thermococcales* can uptake foreign shuttle vectors and incorporate them into their DNA with ease. *T. kodakarensis* can uptake *E. coli* shuttle vectors and successfully be transformed [92]. Both *trpE* and the P_{gdh}-hmg_{Pf} markers were used to determine the mevinolin resistance, and successfully transformed cells were able to use the vectors for gene expression [92]. This capability extends beyond *T. kodakarensis*. Both *Pyrococcus abyssi* and *Pyrococcus furiosus* have also been successfully transformed (Table 4.3). In *P. furiosus*, the pYS3 vector—originally developed for *Sulfolobus acidocaldarius*—has been adapted to express the hmg_{Pf} to confer simvastatin resistance [114]. However, the most effective system for vector uptake is the *P. abyssi* system. Here, the archaea with a mutant *pyrE* gene are used, and the pYS3 vector has a functional copy [70]. This can create highly stable transformed vectors, with 20–30 copies per chromosome [70].

Fuel Production, Alcohol Production, and Other Biotechnologies The thermostable enzymes from *Thermococcales* have practical applications in industrial procedures. For example, the DNA polymerase from *Thermococcus litoralis*, also called Vent polymerase, has proofreading activity and is used in high-fidelity PCR applications [19]. *P. abyssi* is cultivated to extract the α-amylase enzyme, which is used in a wide range of industrial applications, such as textiles, paper, and pulp production, to break down starch into simpler products that can be used to create their respective products (Table 4.3) [98]. The α-amylase of *P. abyssi* acts as a thermostable enzyme, enabling easier and cheaper storage as compared to α-amylases of other microorganisms since the enzymes can be kept at room temperature [98].

Thermococcales are also used in a range of fuel and alcohol production. *P. furiosus* has been shown to be able to produce alcohols such as ethanol (Table 4.3) [8]. This is a high-temperature process that relies on the transformation of *P. furiosus* with alcohol dehydrogenase (AdhA) and coupling it with aldehyde ferredoxin oxidoreductase (AOR) [8]. This is important because *P. furiosus* naturally has a sugar fermentation pathway that converts glucose and other sugars into acetate [8]. Then, the introduced AOR and AdhA can be used to reduce acetate into aldehydes and then reduced further to alcohols [8]. This allows for the alcohol production at higher temperatures as compared to traditional pathways, which carries the benefit of lower cooling costs and risks of contamination [8]. Similarly, *T. kodakarensis* has been noted as a potential source of H_2 gas, which can be converted into a potent green fuel [58]. H_2 gas is produced when *T. kodakarensis* is cultivated with sugars like glucose, and in the absence of sulfur (Table 4.3). This results in membrane-bound [NiFe]-hydrogenase (Mbh) being used to break down the sugar, releasing H_2 gas [58]. By upregulating the levels of Mbh, *T. kodakarnesis* has the potential as a green fuel source [58].

Studies on *Thermococcales* histones have provided interesting evolutionary data. Research showed a deep conservation of histone variants in *Thermococcales*, which suggests that archaea and eukaryotes have independently evolved histone variants, probably due to the need for different adaptive functions in different environments.

Table 4.3 Representative model organisms of Thermococcales

Model organism	Genetic tools available	Biotechnology/ genetics	Challenges/ considerations	References
Thermococcocus kodakerniss	The strain KOD1 genome has been sequenced *TrpE* or HMG-CoA inhibition selection Can uptake *E. coli* vectors	Potential use in H_2 gas production as a green fuel. Has been demonstrated on a lab scale Uses in PCR	Low competency for cell transformation, and are obligate anaerobes. However, has fully sequenced genome	Sato et al. [95]; Fukui et al. [42]; Sato et al. [96]; Matsumi et al. [77]; Santangelo et al. [92]; Kanai [58]
Pyrococcus furiosus	Strain JCM8422 genome has been sequenced PY53 shuttle vector system with hmg_{Pff}	Alcohol and ethanol production for industrial-scale fuel production	Lacks usable reporter genes. However, has a fully sequenced genome	Waege et al. [114]; Leigh et al. [64]; Basen et al. [8]
Pyrococcus abyssi	Strain GE5 genome has been sequenced. Mutant *pyrE* gene with PY53 shuttle vector used in transformation	α-amylase production for industrial scale production of textiles, paper, pulp, etc.	Lacks usable reporter genes However, has a fully sequenced genome	Lucas et al. [70]; Cohen et al. [24]; Leigh et al. [64]; Shad et al. [98]

This highlights the evolutionary significance of chromatin organization mechanisms across domains of life [108, 117].

Thermococcales archaea serve as robust model organisms for studying life under extreme conditions. Their unique physiological traits, genetic tractability, and stable enzymes offer valuable insights into molecular adaptation, protein stability, and evolutionary biology. Ongoing research continues to uncover the potential of these microorganisms in both fundamental science and biotechnological applications [63].

4.5 Crenarchaeota

Crenarchaeota is one of the two archaeal phyla and was first described by Carl Woese [119]. Within this phylum, the order *Sulfolobales* comprises thermoacidophilic archaea primarily found in hot, acidic environments such as sulfuric hot springs and solfataric fields [65]. Members of the *Sulfolobales* have emerged as model organisms as they have unique adaptations to extreme conditions and are genetically and biochemically tractable [65]. *Sulfolobales* are mainly found in geothermal environments such as hot springs and solfatara fields. They are both thermophilic and acidophilic, with optimal growth at temperatures between 70 and 90 °C and pH values of 2–3 [65]. Their lipid monolayer membranes are unique and contain glycerol dialkyl glycerol tetraethers (GDGTs), contributing to their remarkable stability under extreme conditions [86]. *Sulfolobales* can oxidize sulfur and various reduced sulfur compounds, and some species can also utilize iron or hydrogen as electron donors. This gives them metabolic flexibility and positions them as key players in sulfur and metal biogeochemical cycles in extreme ecosystems [89].

Sulfolobales have DNA replication and repair mechanisms, some of which resemble those found in eukaryotes, a key feature that provides insights into evolutionary links between domains of life [65, 115]. Another system that made *Sulfolobales* model organisms is the CRISPR-Cas systems, which deliver adaptive immunity against viruses [115]. The thermostable enzymes that *Sulfolobales* produce, DNA polymerases, proteases, and sulfur oxygenase-reductases, are of growing interest in biotechnology for applications in PCR, bioremediation, and industrial catalysis [89]. Because of their incredible resilience and metabolic capabilities, *Sulfolobales* are studied for their roles in natural ecosystems and industrial processes such as bioleaching and high-temperature bioconversion [86].

Metabolism The *Sulfolobus* species are chemolithoautotrophs that obtain energy from the oxidation of inorganic sulfur compounds, using this energy to reduce carbon dioxide and synthesize organic compounds [69]. They utilize two primary metabolic strategies: obligate and facultative chemolithoautotrophy. In obligate chemolithoautotrophy, *Sulfolobus* oxidizes inorganic sulfur compounds like elemental sulfur and thiosulfate, generating ATP through the electron transport chain and utilizing carbon dioxide for organic compound synthesis via the Calvin–Benson–Bassham cycle [65]. In facultative chemolithoautotrophy, *Sulfolobus* can switch between aerobic and anaerobic respiration, using oxygen or sulfur compounds as electron acceptors and applying the Krebs cycle for ATP production ([28]; [69]). This metabolic adaptability allows *Sulfolobus* to thrive in extreme environments, utilizing various sulfur compounds and molecular hydrogen for energy [86].

Genomic Insights The complete genomes of several *Sulfolobus* species have been sequenced, providing profound insights into their biology (Table 4.4). For instance, the genome of *Sulfolobus acidocaldarius* DSM639 consists of a single circular chromosome with 2,225,959 base pairs and a G + C content of 36.7%, encoding 2292 proteins [21]. Similarly, *Sulfolobus solfataricus* P2 possesses a genome of 2,992,245 base pairs encoding 2977 proteins [99]. *Sulfolobus islandicus* has two strains, REY15A and HVE10/4, with genomes of 2.5 million base pairs and 2.7 million base pairs, respectively [48]. These genomic resources have facilitated the exploration of metabolic pathways, stress responses, and evolutionary relationships within the archaea.

DNA Organization, Replication, and Repair Mechanisms *Sulfolobus* species have contributed significantly to our understanding of DNA replication and repair in archaea. They possess a cell division machinery unique to all crenarchaeal orders except Thermoproteales, and it is homologous to the eukaryotic ESCRT system

[67]. Lindås et al. reported that within the products of the three-gene operon *cdv* in *Sulfolobus acidcaldarius*, the Cdv proteins CdvB and CdvC displayed homology to the eukaryotic ESCRT-III sorting complex [67]. It is a required part of the endosomal protein sorting machinery for intraluminal vesicle formation. The other product, CdvA, is a crenarchaea-specific factor and has a distant homology with proteins in the basal components of the eukaryotic cytoskeleton and nuclear envelope [67]. Lindås et al. also suggest that the analogy between the *cdv* system and the eukaryotic ESCRT-III complex has a shared system, where the primordial division machinery was retained in archaea, while the protein sorting of the system in the eukaryotic lineage evolved [67]. Exposure to DNA-damaging agents like ultraviolet (UV) irradiation induces cellular aggregation in *Sulfolobus solfataricus*, facilitating chromosomal marker exchange through homologous recombination [41]. UV radiation activates an archaeal pili system that leads to pili formation. However, Fröls et al. found that the cellular aggregation was dynamic and dependent on the dose of UV light [41]. Other stressors were tested, such as temperature shift and pH changes found in a hot spring environment, and did not induce aggregation. Unlike many archaea that utilize eukaryote-like histones for chromosome packaging, the *Sulfolobus* species employ nucleoid-associated proteins (NAPs) such as Alba, Cren7, and Sso7d [32]. These proteins are highly expressed, accounting for up to 5% of cellular protein content, and play crucial roles in genome organization and regulation [18]. Post-translational modifications, including methylation and acetylation, modulate NAP functionality, influencing gene expression and adaptation to environmental stresses [45].

Use of Antibiotics as Selection Tools The first antibiotic selectable marker for *Sulfolobus islandicus* was developed using simvastatin resistance through the overexpression of the 3-hydroxy-3-methylglutaryl coenzyme A (HMG-CoA) reductase gene (Table 4.4) [121]. HMG-CoA is an enzyme responsible for producing precursors for isoprenoid compounds, which are essential components of archaeal lipid membranes. This enzyme is inhibited by mevinolin and its derivative simvastatin, both of which target the isoprenoid biosynthetic pathway. Overexpression of HMG-CoA reductase in *S. islandicus* confers simvastatin resistance, making the antibiotic a robust selectable marker [121].

Shuttle Vectors A multicopy and non-integrative *Sulfolobus- E. coli* shuttle vector has been created for *S. acidocaldarius* and S. *solfataricus* based on the multicopy plasmid pRN1 from *S. islandicus* [10]. Another *Sulfolobus—E. coli* shuttle vector was developed for heterologous gene expression in *S. acidocaldarius* based on the pUC18 and *Sulfolobus* cryptic plasmid pRN1 (Table 4.4) [54].

Bioremediation and Other Biotechnological Applications The thermostable enzymes of the *Sulfolobus* species gained interest for industrial applications. One example is the β-galactosidase from *S. solfataricus,* which exhibits high stability and activity at elevated temperatures, making the enzyme suitable for processes involving lactose-containing fluids [83].

Table 4.4 Model organisms of crenarchaeota

Model organism	Genetic tools available	Biotechnology/ genetics	Challenges/ considerations	References
Sulfolobus solfataricus	Strain DSM639 genome has been sequenced UV radiation for homologous recombination *Sulfolobus-E. coli* shuttle vector based on multicopy plasmid pRN1 from S. *islandicus*	Use of *S. solfataricus's* β-galactosidase for its high stability and activity in high-temperature processes Enzymes are used at high temperatures for heat-resistant applications such as food stabilizer production, and conversion of organic materials into products like bioplastics and antifreeze agents	Slow growth rate Difficulty in controlling growth due to UV damage stress response	Pisani et al. [83]; Chen et al. [21]; Berkner et al. [10]; Fröls et al. [41]; Quehenberger et al. [86]; Rawat et al. [89]
Sulfolobus acidocaldarius	Strain P2 genome has been sequenced *Sulfolobus-E. coli* shuttle vector based on multicopy plasmid pRN1 from S. *islandicus* *Sulfolobus- E. coli* shuttle vector based on pUC18 and *Sulfolobus* cryptic plasmid pRN1	High temperatures allow for biofuel production, and bioremediation cleaning negative environmental factors like acidic waste Produces sulfolobicins for antimicrobial activity against archaea	Slow growth rate Harsh environments are required for growth, causing more expensive and inaccessible conditions pRN1 plasmid contains inaccessible regions for gene modification	She et al. [99]; Besse et al. [12]; Berkner et al. [10]; Hwang et al. [54]; Quehenberger et al. [86]; Rawat et al. [89]
Sulfolobus islandicus	Simvastatin as an antibiotic selection marker Strain REY15A and HVE10/4 genomes have been sequenced	Lipids help to produce heat-stable enzymes for biofuels such as bioethanol Used for industrial catalysis using enzymes to make more acid and heat-stable solutions Produces sulfolobicins for antimicrobial activity against archaea	Slow growth rate Limited selection markers	Besse et al. [12]; Guo et al. [48]; Zheng et al. [121]; Quehenberger et al. [86]; Rawat et al. [89]

Proteases, esterases, and lipases from *Sulfolobus* species have been investigated for their potential in various industrial processes due to their catalytic diversity and stability under extreme conditions [17]. For example, a thermostable extracellular lipolytic enzyme from *S. solfataricus* has been characterized, revealing its robustness and potential utility in industrial applications, such as flavor synthesis, aroma synthesis, and detergent production (Table 4.4) [22]. *Sulfolobales* are able to produce biomolecules called sulfolobicins, which have antibiotic properties (Table 4.4) [12]. They are able to withstand up to 78 °C, storage at 4 °C for 17 months, and a pH ranging from 3 to 10.7 [12]. However, more research into its mechanism and targets would be required if *Sulfolobales* are to be used as a source for an antibiotic agent.

4.6 Conclusion

The understanding of molecular, cellular, and evolutionary biology processes has advanced in recent years thanks to archaeal model organisms. These life forms are powerful tools for modern biological research. Initially, archaea were thought to inhabit only extreme environments, but are now recognized as metabolically and ecologically versatile organisms in diverse habitats, from the human microbiome to hydrothermal vents to saline lakes. Understanding how archaea work on a molecular level is like finding a missing link. It lets us compare and contrast them with bacteria and eukaryotes, revealing surprising similarities and key differences. In this chapter, we highlighted organisms belonging to the domain and why they are model organisms, as they each offer distinct advantages as model systems. Scientists have studied methanogens extensively, uncovering how they handle energy and carbon in oxygen-free environments. This research is helping us develop sustainable energy sources using biogas and methane. Halophiles are an excellent model for investigating DNA repair, gene regulation, and how organisms survive salty conditions. Thermophilic archaea and their thermostable enzymes are the models for studying protein folding and stability in extreme environments. Sulfolobales have some surprising features. Their replication machinery and CRISPR-Cas systems are similar to those found in eukaryotes. This discovery allows scientists to study early life and how genomes are controlled.

Basic scientific advancements are being made with archaeal model organisms, which also hold enormous promise for biotechnology. Stable at high temperatures, extreme pH, and salinity, their enzymes are ideal for industrial processes. DNA polymerases from *Thermococcus* and *Sulfolobus* species are already used in high-fidelity PCR. Haloarchaea produce halocins and carotenoids with antimicrobial and potential anticancer properties. Sulfolobales generate thermostable sulfolobicins and lipases that are being investigated for pharmaceutical and bioremediation purposes. Methanogens contribute to the renewable energy sector by converting CO_2 and H_2 into methane—a storable and sustainable fuel source—positioning them at the forefront of bioenergy innovation.

Furthermore, the genetic tractability of model archaea has improved dramatically. Shuttle vectors, markerless gene editing, CRISPR systems, and inducible promoters have all been developed. Archaeal genetic engineering is just as advanced as what is done with bacteria and other complex organisms. The tools are here and ready to use. Engineered archaea are being developed to perform various biotechnological tasks, including environmental remediation, biofuel manufacture, and the synthesis of valuable chemicals. The potential applications are vast and exciting.

References

1. Adams MWW, Kletzin A. Oxidoreductase-type enzymes and redox proteins involved in fermentative metabolisms of hyperthermophilic archaea. Adv Protein Chem. 1996:101–80. https://doi.org/10.1016/s0065-3233(08)60362-9.
2. Allers T. Overexpression and purification of halophilic proteins in *haloferax volcanii*. Bioeng Bugs. 2010;1:290–2. https://doi.org/10.4161/bbug.1.4.11794.
3. Allers T, Ngo H-P, Mevarech M, Lloyd RG. Development of additional selectable markers for the halophilic archaeon *Haloferax volcanii* based on the euB and trpA genes. Appl Environ Microbiol. 2004;70:943–53. https://doi.org/10.1128/aem.70.2.943-953.2004.
4. Amoozegar MA, Siroosi M, Atashgahi S, Smidt H, Ventosa A. Systematics of haloarchaea and biotechnological potential of their hydrolytic enzymes. Microbiology. 2017;163:623–45. https://doi.org/10.1099/mic.0.000463.
5. Argyle JL, Tumbula DL, Leigh JA. Neomycin resistance as a selectable marker in *Methanococcus maripaludis*. Appl Environ Microbiol. 1996;62:4233–7. https://doi.org/10.1128/aem.62.11.4233-4237.1996.
6. Atomi H, Imanaka T, Fukui T. Overview of the genetic tools in the Archaea. Front Microbiol. 2012; https://doi.org/10.3389/fmicb.2012.00337.
7. Atomi H, Reeve J. Microbe profile: *Thermococcus kodakarensis*: the model hyperthermophilic archaeon. Microbiology. 2019;165:1166–8. https://doi.org/10.1099/mic.0.000839.
8. Basen M, Schut GJ, Nguyen DM, Lipscomb GL, Benn RA, Prybol CJ, Vaccaro BJ, Poole FL, Kelly RM, Adams MW. Single gene insertion drives bioalcohol production by a thermophilic archaeon. Proc Natl Acad Sci. 2014;111:17618–23. https://doi.org/10.1073/pnas.1413789111.
9. Berg IA, Kockelkorn D, Ramos-Vera WH, Say RF, Zarzycki J, Hügler M, Alber BE, Fuchs G. Autotrophic carbon fixation in Archaea. Nat Rev Microbiol. 2010;8:447–60. https://doi.org/10.1038/nrmicro2365.
10. Berkner S, Grogan D, Albers S-V, Lipps G. Small multicopy, non-integrative shuttle vectors based on the plasmid pRN1 for *Sulfolobus acidocaldarius* and *Sulfolobus solfataricus*, model organisms of the (cren-)archaea. Nucleic Acids Res. 2007; https://doi.org/10.1093/nar/gkm449.
11. Bertram PA, Thauer RK. Thermodynamics of the formylmethanofuran dehydrogenase reaction in *Methanobacterium thermoautotrophicum*. Eur J Biochem. 1994;226:811–8. https://doi.org/10.1111/j.1432-1033.1994.t01-1-00811.x.
12. Besse A, Peduzzi J, Rebuffat S, Carré-Mlouka A. Antimicrobial peptides and proteins in the face of extremes: lessons from archaeocins. Biochimie. 2015;118:344–55. https://doi.org/10.1016/j.biochi.2015.06.004.
13. Birbir M, Eryilmaz S, Ogan A. Prevention of halophilic microbial damage on brine cured hides by extremely halophilic halocin producer strains. J Soc Leather Technol Chem. 2004;88:99–104.

14. Bitan-Banin G, Ortenberg R, Mevarech M. Development of a gene knockout system for the halophilic archaeon *Haloferax volcanii* by use of the pyrE gene. J Bacteriol. 2003;185:772–8. https://doi.org/10.1128/jb.185.3.772-778.2003.
15. Blaseio U, Pfeifer F. Transformation of *Halobacterium halobium*: development of vectors and investigation of gas vesicle synthesis. Proc Natl Acad Sci. 1990;87:6772–6. https://doi.org/10.1073/pnas.87.17.6772.
16. Buan NR. Methanogens: pushing the boundaries of biology. Emerg Top Life Sci. 2018;2:629–46. https://doi.org/10.1042/etls20180031.
17. Cabrera MÁ, Blamey JM. Biotechnological applications of archaeal enzymes from extreme environments. Biol Res. 2018; https://doi.org/10.1186/s40659-018-0186-3.
18. Cajili MK, Prieto EI. Interplay between Alba and Cren7 regulates chromatin compaction in *Sulfolobus solfataricus*. Biomolecules. 2022;12:481. https://doi.org/10.3390/biom12040481.
19. Cariello NF, Swenberg JA, Skopek TR. Fidelity of *Thermococcus litoralis* DNA polymerase (VentTM) in PCR determined by denaturing gradient gel electrophoresis. Nucleic Acids Res. 1991;19:4193–8. https://doi.org/10.1093/nar/19.15.4193.
20. Charlesworth JC, Burns BP. Untapped resources: biotechnological potential of peptides and secondary metabolites in archaea. Archaea. 2015; https://doi.org/10.1155/2015/282035.
21. Chen L, Brügger K, Skovgaard M, Redder P, She Q, Torarinsson E, Greve B, Awayez M, Zibat A, Klenk H-P, Garrett RA. The genome of *Sulfolobus acidocaldarius*, a model organism of the *Crenarchaeota*. J Bacteriol. 2005;187:4992–9. https://doi.org/10.1128/jb.187.14.4992-4999.2005.
22. Choi Y-H, Park Y-J, Yoon S-J, Lee H-B. Purification and characterization of a new inducible thermostable extracellular lipolytic enzyme from the thermoacidophilic archaeon *Sulfolobus solfataricus* P1. J Mol Catal B Enzym. 2016;124:11–9. https://doi.org/10.1016/j.molcatb.2015.11.023.
23. Clark DP, Pazdernik NJ, McGehee MR. Cloning genes for synthetic biology. Mol Biol. 2019:199–239. https://doi.org/10.1016/b978-0-12-813288-3.00007-0.
24. Cohen GN, Barbe V, Flament D, Galperin M, Heilig R, Lecompte O, Poch O, Prieur D, Quérellou J, Ripp R, Thierry J, Van der Oost J, Weissenbach J, Zivanovic Y, Forterre P. An integrated analysis of the genome of the hyperthermophilic archaeon Pyrococcus abyssi. Mol Microbiol. 2003;47:1495–512. https://doi.org/10.1046/j.1365-2958.2003.03381.x.
25. Contreras G, Thomsen J, Pfitzer M, Hafenbradl D, Kostner D, Holtmann D, Schmitz RA, Rother M, Molitor B. New perspectives for biotechnological applications of methanogens. Curr Res Biotechnol. 2022;4:468–74. https://doi.org/10.1016/j.crbiot.2022.10.001.
26. Corral P, Amoozegar MA, Ventosa A. Halophiles and their biomolecules: recent advances and future applications in biomedicine. Mar Drugs. 2019;18:33. https://doi.org/10.3390/md18010033.
27. Costa KC, Whitman WB. Model organisms to study methanogenesis, a uniquely archaeal metabolism. J Bacteriol. 2023; https://doi.org/10.1128/jb.00115-23.
28. Counts JA, Willard DJ, Kelly RM. Life in hot acid: a genome-based reassessment of the archaeal order *Sulfolobales*. Environ Microbiol. 2021;23:3568–84. https://doi.org/10.1111/1462-2920.15189.
29. Daniels L, Fuchs G, Thauer RK, Zeikus JG. Carbon monoxide oxidation by methanogenic bacteria. J Bacteriol. 1977;132:118–26. https://doi.org/10.1128/jb.132.1.118-126.1977.
30. De Lise F, Iacono R, Moracci M, Strazzulli A, Cobucci-Ponzano B. Archaea as a model system for molecular biology and biotechnology. Biomolecules. 2023;13:114. https://doi.org/10.3390/biom13010114.
31. de Macario EC, Macario AJ. Methanogenic archaea in health and disease: a novel paradigm of microbial pathogenesis. Int J Med Microbiol. 2009;299:99–108. https://doi.org/10.1016/j.ijmm.2008.06.011.
32. Ding N, Chen Y, Chu Y, Zhong C, Huang L, Zhang Z. Lysine methylation modulates the interaction of archaeal chromatin protein Cren7 with DNA. Front Microbiol. 2022; https://doi.org/10.3389/fmicb.2022.837737.

33. Dutta B, Bandopadhyay R. Biotechnological potentials of halophilic microorganisms and their impact on mankind. Beni-Suef Univ J Basic Appl Sci. 2022; https://doi.org/10.1186/s43088-022-00252-w.
34. Eckburg PB, Lepp PW, Relman DA. Archaea and their potential role in human disease. Infect Immun. 2003;71:591–6. https://doi.org/10.1128/iai.71.2.591-596.2003.
35. Elcheninov AG, Ugolkov YA, Elizarov IM, Klyukina AA, Kublanov IV, Sorokin DY. Cellulose metabolism in Halo(natrono)Archaea: a comparative genomics study. Front Microbiol. 2023; https://doi.org/10.3389/fmicb.2023.1112247.
36. Eme L, Doolittle WF. Archaea. Curr Biol. 2015;25:851–5. https://doi.org/10.1016/j.cub.2015.05.025.
37. Falb M, Müller K, Königsmaier L, Oberwinkler T, Horn P, von Gronau S, Gonzalez O, Pfeiffer F, Bornberg-Bauer E, Oesterhelt D. Metabolism of halophilic archaea. Extremophiles. 2008;12:177–96. https://doi.org/10.1007/s00792-008-0138-x.
38. Fendrihan S, Stan-Lotter H. Survival of halobacteria in fluid inclusions as a model of possible biotic survival in Martian halite. In: Teodorescu HN, Griebel HS, editors. Selected papers from EMC`04 Mars and planetary science and technology. Performantica Press; 2004. p. 9–18.
39. Fetzer S, Conrad R. Effect of redox potential on methanogenesis by *Methanosarcina barkeri*. Arch Microbiol. 1993;160:108–13. https://doi.org/10.1007/bf00288711.
40. Fiala G, Stetter KO. *Pyrococcus furiosus* sp. nov. represents a novel genus of marine heterotrophic archaebacteria growing optimally at 100°C. Arch Microbiol. 1986;145:56–61. https://doi.org/10.1007/bf00413027.
41. Fröls S, Ajon M, Wagner M, Teichmann D, Zolghadr B, Folea M, Boekema EJ, Driessen AJ, Schleper C, Albers S. UV-inducible cellular aggregation of the hyperthermophilic archaeon *Sulfolobus solfataricus* is mediated by pili formation. Mol Microbiol. 2008;70:938–52. https://doi.org/10.1111/j.1365-2958.2008.06459.x.
42. Fukui T, Atomi H, Kanai T, Matsumi R, Fujiwara S, Imanaka T. Complete genome sequence of the hyperthermophilic archaeon *Thermococcus kodakaraensis* KOD1 and comparison with *Pyrococcus* genomes. Genome Res. 2005;15:352–63. https://doi.org/10.1101/gr.3003105.
43. Gallo G, Imbimbo P, Aulitto M. The undeniable potential of thermophiles in industrial processes. Int J Mol Sci. 2024;25:7685. https://doi.org/10.3390/ijms25147685.
44. Gaudin M, Gauliard E, Schouten S, Houel-Renault L, Lenormand P, Marguet E, Forterre P. Hyperthermophilic archaea produce membrane vesicles that can transfer DNA. Environ Microbiol Rep. 2012;5:109–16. https://doi.org/10.1111/j.1758-2229.2012.00348.x.
45. Gong P, Lei P, Wang S, Zeng A, Lou H. Post-translational modifications aid archaeal survival. Biomolecules. 2020;10:584. https://doi.org/10.3390/biom10040584.
46. Grabowski B, Kelman Z. Archaeal DNA replication: eukaryal proteins in a bacterial context. Ann Rev Microbiol. 2003;57:487–516. https://doi.org/10.1146/annurev.micro.57.030502.090709.
47. Grünberger F, Schmid G, El Ahmad Z, Fenk M, Vogl K, Reichelt R, Hausner W, Urlaub H, Lenz C, Grohmann D. Uncovering the temporal dynamics and regulatory networks of thermal stress response in a hyperthermophile using transcriptomics and proteomics. mBio. 2023; https://doi.org/10.1128/mbio.02174-23.
48. Guo L, Brügger K, Liu C, Shah SA, Zheng H, Zhu Y, Wang S, Lillestøl RK, Chen L, Frank J, Prangishvili D, Paulin L, She Q, Huang L, Garrett RA. Genome analyses of Icelandic strains of *Sulfolobus islandicus*, model organisms for genetic and virus-host interaction studies. J Bacteriol. 2011;193:1672–80. https://doi.org/10.1128/jb.01487-10.
49. Haque RU, Paradisi F, Allers T. *Haloferax volcanii* for biotechnology applications: challenges, current state and perspectives. Appl Microbiol Biotechnol. 2019;104:1371–82. https://doi.org/10.1007/s00253-019-10314-2.
50. Harding T, Brown MW, Simpson AGB, Roger AJ. Osmoadaptative strategy and its molecular signature in obligately halophilic heterotrophic protists. Genome Biol Evol. 2016;8:2241–58. https://doi.org/10.1093/gbe/evw152.

51. Harrison C, Allers T. Progress and challenges in archaeal genetic manipulation. Methods Mol Biol. 2022:25–31. https://doi.org/10.1007/978-1-0716-2445-6_2.
52. Hartman AL, Norais C, Badger JH, Delmas S, Haldenby S, Madupu R, Robinson J, Khouri H, Ren Q, Lowe TM, Maupin-Furlow J, Pohlschroder M, Daniels C, Pfeiffer F, Allers T, Eisen JA. The complete genome sequence of *Haloferax volcanii* DS2, a model archaeon. PLoS One. 2010; https://doi.org/10.1371/journal.pone.0009605.
53. Hileman TH, Santangelo TJ. Genetics techniques for *Thermococcus kodakarensis*. Front Microbiol. 2012; https://doi.org/10.3389/fmicb.2012.00195.
54. Hwang S, Choi K-H, Yoon N, Cha J. Improvement of a *Sulfolobus-E. coli* shuttle vector for heterologous gene expression in *Sulfolobus acidocaldarius*. J Microbiol Biotechnol. 2015;25:196–205. https://doi.org/10.4014/jmb.1407.07043.
55. Jarrell KF, Walters AD, Bochiwal C, Borgia JM, Dickinson T, Chong JP. Major players on the microbial stage: why archaea are important. Microbiology. 2011;157:919–36. https://doi.org/10.1099/mic.0.047837-0.
56. Jia B, Liu J, Van Duyet L, Sun Y, Xuan YH, Cheong G-W. Proteome profiling of heat, oxidative, and salt stress responses in Thermococcus kodakarensis KOD1. Front Microbiol. 2015; https://doi.org/10.3389/fmicb.2015.00605.
57. Jones WJ, Whitman WB, Fields RD, Wolfe RS. Growth and plating efficiency of *Methanococci* on agar media. Appl Environ Microbiol. 1983;46:220–6. https://doi.org/10.1128/aem.46.1.220-226.1983.
58. Kanai T, Simons J-R, Tsukamoto R, Nakajima A, Omori Y, Matsuoka R, Beppu H, Imanaka T, Atomi H. Overproduction of the membrane-bound [nife]-hydrogenase in *Thermococcus kodakarensis* and its effect on hydrogen production. Front Microbiol. 2015; https://doi.org/10.3389/fmicb.2015.00847.
59. Kessler PS, Leigh JA. Genetics of nitrogen regulation in *Methanococcus maripaludis*. Genetics. 1999;152:1343–51. https://doi.org/10.1093/genetics/152.4.1343.
60. Kim YJ, Lee HS, Kim ES, Bae SS, Lim JK, Matsumi R, Lebedinsky AV, Sokolova TG, Kozhevnikova DA, Cha S-S, Kim S-J, Kwon KK, Imanaka T, Atomi H, Bonch-Osmolovskaya EA, Lee J-H, Kang SG. Formate-driven growth coupled with H_2 production. Nature. 2010;467:352–5. https://doi.org/10.1038/nature09375.
61. Kort JC, Esser D, Pham TK, Noirel J, Wright PC, Siebers B. A cool tool for hot and sour archaea: proteomics of *Sulfolobus solfataricus*. Proteomics. 2013;13:2831–50. https://doi.org/10.1002/pmic.201300088.
62. Krebs MP, Hauss T, Heyn MP, RajBhandary UL, Khorana HG. Expression of the bacterioopsin gene in *Halobacterium halobium* using a multicopy plasmid. Proc Natl Acad Sci. 1991;88:859–63. https://doi.org/10.1073/pnas.88.3.859.
63. Kumar S, Nussinov R. How do thermophilic proteins deal with heat? Cell Mol Life Sci. 2001;58:1216–33. https://doi.org/10.1007/pl00000935.
64. Leigh JA, Albers S-V, Atomi H, Allers T. Model organisms for genetics in the domain Archaea: methanogens, halophiles, *Thermococcales* and *Sulfolobales*. FEMS Microbiol Rev. 2011;35:577–608. https://doi.org/10.1111/j.1574-6976.2011.00265.x.
65. Lewis AM, Recalde A, Bräsen C, Counts JA, Nussbaum P, Bost J, Schocke L, Shen L, Willard DJ, Quax TE, Peeters E, Siebers B, Albers S-V, Kelly RM. The biology of thermoacidophilic archaea from the order *Sulfolobales*. FEMS Microbiol Rev. 2021; https://doi.org/10.1093/femsre/fuaa063.
66. Li J, Akinyemi TS, Shao N, Chen C, Dong X, Liu Y, Whitman WB. Genetic and metabolic engineering of *Methanococcus* spp. Curr Res Biotechnol. 2023;5:100115. https://doi.org/10.1016/j.crbiot.2022.11.002.
67. Lindås A-C, Karlsson EA, Lindgren MT, Ettema TJ, Bernander R. A unique cell division machinery in the archaea. Proc Natl Acad Sci. 2008;105:18942–6. https://doi.org/10.1073/pnas.0809467105.
68. Liu C, Baffoe DK, Zhan Y, Zhang M, Li Y, Zhang G. Halophile, an essential platform for bioproduction. J Microbiol Methods. 2019;166:105704. https://doi.org/10.1016/j.mimet.2019.105704.

69. Liu L-J, Jiang Z, Wang P, Qin Y-L, Xu W, Wang Y, Liu S-J, Jiang C-Y. Physiology, taxonomy, and sulfur metabolism of the *Sulfolobales*, an order of thermoacidophilic archaea. Front Microbiol. 2021; https://doi.org/10.3389/fmicb.2021.768283.
70. Lucas S, Toffin L, Zivanovic Y, Charlier D, Moussard H, Forterre P, Prieur D, Erauso G. Construction of a shuttle vector for, and spheroplast transformation of, the hyperthermophilic archaeon *Pyrococcus abyssi*. Appl Environ Microbiol. 2002;68:5528–36. https://doi.org/10.1128/aem.68.11.5528-5536.2002.
71. Maier L-K, Stachler A-E, Brendel J, Stoll B, Fischer S, Haas KA, Schwarz TS, Alkhnbashi OS, Sharma K, Urlaub H, Backofen R, Gophna U, Marchfelder A. The nuts and bolts of the *Haloferax* CRISPR-Cas system I-B. RNA Biol. 2018;16:469–80. https://doi.org/10.1080/15476286.2018.1460994.
72. Manuela TRN, Esquivel E, Wirth M, Pohlschröder. Haloferax volcanii cells lacking the flagellin FlgA2 are hypermotile Microbiology 2013;159(Pt_11):2249–58. https://doi.org/10.1099/mic.0.069617-0.
73. Manuela TS, Imam M, Pohlschröder. Haloferax volcanii flagella are required for motility but are not involved in PibD-dependent surface adhesion ABSTRACT Journal of Bacteriology 2010;192(12):3093–102. https://doi.org/10.1128/JB.00133-10.
74. Martinez-Pastor M, Tonner PD, Darnell CL, Schmid AK. Transcriptional regulation in archaea: from individual genes to global regulatory networks. Annu Rev Genet. 2017;51:143–70. https://doi.org/10.1146/annurev-genet-120116-023413.
75. Martínez-Espinosa RM. Halocins and C50 carotenoids from haloarchaea: potential natural tools against cancer. Mar Drugs. 2024; https://doi.org/10.3390/md22100448.
76. Matarredona L, Camacho M, Zafrilla B, Bonete M-J, Esclapez J. The role of stress proteins in haloarchaea and their adaptive response to environmental shifts. Biomolecules. 2020;10:1390. https://doi.org/10.3390/biom10101390.
77. Matsumi R, Manabe K, Fukui T, Atomi H, Imanaka T. Disruption of a sugar transporter gene cluster in a hyperthermophilic archaeon using a host-marker system based on antibiotic resistance. J Bacteriol. 2007;189:2683–91. https://doi.org/10.1128/jb.01692-06.
78. Morikawa M, Izawa Y, Rashid N, Hoaki T, Imanaka T. Purification and characterization of a thermostable thiol protease from a newly isolated hyperthermophilic *Pyrococcus* sp. Appl Environ Microbiol. 1994;60:4559–66. https://doi.org/10.1128/aem.60.12.4559-4566.1994.
79. Mukhopadhyay B. A reminder—peptidoglycan cell walls indeed occur in the archaeal domain, specifically in the members of *Methanobacteria* and *Methanopyri* classes. Front Microbiol. 2024; https://doi.org/10.3389/fmicb.2024.1329047.
80. Myers T, Dykstra CM. Teaching old dogs new tricks: genetic engineering methanogens. Appl Environ Microbiol. 2024; https://doi.org/10.1128/aem.02247-23.
81. Ng WV, Kennedy SP, Mahairas GG, Berquist B, Pan M, Shukla HD, Lasky SR, Baliga NS, Thorsson V, Sbrogna J, Swartzell S, Weir D, Hall J, Dahl TA, Welti R, Goo YA, Leithauser B, Keller K, Cruz R, Danson MJ, Hough DW, Maddocks DG, Jablonski PE, Krebs MP, Angevine CM, Dale H, Isenbarger TA, Peck RF, Pohlschroder M, Spudich JL, Jung K-H, Alam M, Freitas T, Hou S, Daniels CJ, Dennis PP, Omer AD, Ebhardt H, Lowe TM, Liang P, Riley M, Hood L, DasSarma S. Genome sequence of *Halobacterium* species NRC-1. Proc Natl Acad Sci. 2000;97:12176–81. https://doi.org/10.1073/pnas.190337797.
82. Peck RF, DasSarma S, Krebs MP. Homologous gene knockout in the archaeon *Halobacterium salinarum* with ura3 as a counterselectable marker. Mol Microbiol. 2000;35:667–76. https://doi.org/10.1046/j.1365-2958.2000.01739.x.
83. Pisani FM, Rella R, Raia CA, Rozzo C, Nucci R, Gambacorta A, de Rosa M, Rossi M. Thermostable β-galactosidase from the archaebacterium *Sulfolobus solfataricus* purification and properties. Eur J Biochem. 1990;187:321–8. https://doi.org/10.1111/j.1432-1033.1990.tb15308.x.
84. Pérez-Arnaiz P, Dattani A, Smith V, Allers T. *Haloferax volcanii* - a model archaeon for studying DNA replication and repair. Open Biol. 2020; https://doi.org/10.1098/rsob.200293.

85. Pohlschroder M, Schulze S, Pfeiffer F, Hong Y. *Haloferax volcanii*: a versatile model for studying Archaeal biology. J Bacteriol. 2025; https://doi.org/10.1128/jb.00062-25.
86. Quehenberger J, Shen L, Albers S-V, Siebers B, Spadiut O. *Sulfolobus*—a potential key organism in future biotechnology. Front Microbiol. 2017;8:2474. https://doi.org/10.3389/fmicb.2017.02474.
87. Quillaguamán J, Guzmán H, Van-Thuoc D, Hatti-Kaul R. Synthesis and production of polyhydroxyalkanoates by Halophiles: current potential and future prospects. Appl Microbiol Biotechnol. 2009;85:1687–96. https://doi.org/10.1007/s00253-009-2397-6.
88. Rashid N, Aslam M. An overview of 25 years of research on *Thermococcus kodakarensis*, a genetically versatile model organism for archaeal research. Folia Microbiol. 2019;65:67–78. https://doi.org/10.1007/s12223-019-00730-2.
89. Rawat M, Chauhan M, Pandey A. Extremophiles and their expanding biotechnological applications. Arch Microbiol. 2024; https://doi.org/10.1007/s00203-024-03981-x.
90. Robinson JL, Pyzyna B, Atrasz RG, Henderson CA, Morrill KL, Burd AM, DeSoucy E, Fogleman RE, Naylor JB, Steele SM, Elliott DR, Leyva KJ, Shand RF. Growth kinetics of extremely halophilic archaea (family halobacteriaceae) as revealed by arrhenius plots. J Bacteriol. 2005;187:923–9. https://doi.org/10.1128/jb.187.3.923-929.2005.
91. Rother M, Metcalf WW. Genetic technologies for archaea. Curr Opin Microbiol. 2005;8:745–51. https://doi.org/10.1016/j.mib.2005.10.010.
92. Santangelo TJ, Čuboňová L, Reeve JN. Shuttle vector expression in *Thermococcus kodakaraensis*: contributions of *cis* elements to protein synthesis in a hyperthermophilic archaeon. Appl Environ Microbiol. 2008;74:3099–104. https://doi.org/10.1128/aem.00305-08.
93. Santangelo TJ, Čuboňová L, Reeve JN. *Thermococcus kodakarensis* genetics: TK1827-encoded β-glycosidase, new positive-selection protocol, and targeted and repetitive deletion technology. Appl Environ Microbiol. 2010;76:1044–52. https://doi.org/10.1128/aem.02497-09.
94. Sarmiento FB, Leigh JA, Whitman WB. Genetic systems for hydrogenotrophic methanogens. Methods Enzymol. 2011;494:43–73. https://doi.org/10.1016/b978-0-12-385112-3.00003-2.
95. Sato T, Fukui T, Atomi H, Imanaka T. Targeted gene disruption by homologous recombination in the hyperthermophilic archaeon *Thermococcus kodakaraensis* KOD1. J Bacteriol. 2003;185:210–20. https://doi.org/10.1128/jb.185.1.210-220.2003.
96. Sato T, Fukui T, Atomi H, Imanaka T. Improved and versatile transformation system allowing multiple genetic manipulations of the hyperthermophilic archaeon *Thermococcus kodakaraensis*. Appl Environ Microbiol. 2005;71:3889–99. https://doi.org/10.1128/aem.71.7.3889-3899.2005.
97. Schult F, Le TN, Albersmeier A, Rauch B, Blumenkamp P, van der Does C, Goesmann A, Kalinowski J, Albers S-V, Siebers B. Effect of UV irradiation on *Sulfolobus acidocaldarius* and involvement of the general transcription factor TFB3 in the early UV response. Nucleic Acids Res. 2018;46:7179–92. https://doi.org/10.1093/nar/gky527.
98. Shad M, Rehman HM, Akhtar MW, Sajjad M. Structural and functional insights of starch processing α-amylase from hyperthermophilic archaeon *Pyrococcus abyssi*. Carbohydr Res. 2024; https://doi.org/10.1016/j.carres.2024.109122.
99. She Q, Singh RK, Confalonieri F, Zivanovic Y, Allard G, Awayez MJ, Chan-Weiher CC-Y, Clausen IG, Curtis BA, De Moors A, Erauso G, Fletcher C, Gordon PM, Heikamp-de Jong I, Jeffries AC, Kozera CJ, Medina N, Peng X, Thi-Ngoc HP, Redder P, Schenk ME, Theriault C, Tolstrup N, Charlebois RL, Doolittle WF, Duguet M, Gaasterland T, Garrett RA, Ragan MA, Sensen CW, Van der Oost J. The complete genome of the crenarchaeon *Sulfolobus solfataricus* P2. Proc Natl Acad Sci. 2001;98:7835–40. https://doi.org/10.1073/pnas.141222098.
100. Sieber JR, McInerney MJ, Gunsalus RP. Genomic insights into syntrophy: the paradigm for anaerobic metabolic cooperation. Ann Rev Microbiol. 2012;66:429–52. https://doi.org/10.1146/annurev-micro-090110-102844.
101. Silva-Rocha R, Pontelli MC, Furtado GP, Zaramela LS, Koide T. Development of new modular genetic tools for engineering the halophilic Archaeon *Halobacterium salinarum*. PLoS One. 2015; https://doi.org/10.1371/journal.pone.0129215.

102. Soppa J. From replication to cultivation: hot news from haloarchaea. Curr Opin Microbiol. 2005;8:737–44. https://doi.org/10.1016/j.mib.2005.10.012.
103. Sorokin DY, Kublanov IV, Yakimov MM, Rijpstra WI, Sinninghe Damsté JS. *Halanaeroarchaeum sulfurireducens* gen. nov., sp. nov., the first obligately anaerobic sulfur-respiring haloarchaeon, isolated from a hypersaline lake. Int J Syst Evol Microbiol. 2016;66:2377–81. https://doi.org/10.1099/ijsem.0.001041.
104. Sorokin DY, Messina E, Smedile F, Roman P, Damsté JS, Ciordia S, Mena MC, Ferrer M, Golyshin PN, Kublanov IV, Samarov NI, Toshchakov SV, La Cono V, Yakimov MM. Discovery of anaerobic lithoheterotrophic haloarchaea, ubiquitous in hypersaline habitats. ISME J. 2017;11:1245–60. https://doi.org/10.1038/ismej.2016.203.
105. Sorokin DY, Toshchakov SV, Kolganova TV, Kublanov IV. Halo(natrono)Archaea isolated from hypersaline lakes utilize cellulose and chitin as growth substrates. Front Microbiol. 2015; https://doi.org/10.3389/fmicb.2015.00942.
106. Stan-Lotter H, Fendrihan S. Halophilic Archaea: life with desiccation, radiation and oligotrophy over geological times. Life. 2015;5:1487–96. https://doi.org/10.3390/life5031487.
107. Stepanenko AA, Heng HH. Transient and stable vector transfection: pitfalls, off-target effects, artifacts. Mutat Res Rev Mutat Res. 2017;773:91–103. https://doi.org/10.1016/j.mrrev.2017.05.002.
108. Stevens KM, Hocher A, Warnecke T. Deep conservation of histone variants in *Thermococcales* archaea. Genome Biol Evol. 2021; https://doi.org/10.1093/gbe/evab274.
109. Sutter J-M, Tästensen J-B, Johnsen U, Soppa J, Schönheit P. Key enzymes of the semiphosphorylative Entner-Doudoroff pathway in the haloarchaeon *Haloferax volcanii*: characterization of glucose dehydrogenase, gluconate dehydratase, and 2-keto-3-deoxy-6-phosphogluconate aldolase. J Bacteriol. 2016;198:2251–62. https://doi.org/10.1128/jb.00286-16.
110. Takemasa R, Yokooji Y, Yamatsu A, Atomi H, Imanaka T. *Thermococcus kodakarensis* as a host for gene expression and protein secretion. Appl Environ Microbiol. 2011;77(7):2392–8. https://doi.org/10.1128/AEM.01005-10. Epub 2011 Jan 28
111. Talaue CO, del Rosario RC, Pfeiffer F, Mendoza ER, Oesterhelt D. Model construction and analysis of respiration in *Halobacterium salinarum*. PLoS One. 2016; https://doi.org/10.1371/journal.pone.0151839.
112. Thauer RK. The Wolfe Cycle comes full circle. Proc Natl Acad Sci. 2012;109:15084–5. https://doi.org/10.1073/pnas.1213193109.
113. Thauer RK, Kaster AK, Seedorf H, Buckel W, Hedderich R. Methanogenic archaea: ecologically relevant differences in energy conservation. Nat Rev Microbiol. 2008;6(8):579–91. https://doi.org/10.1038/nrmicro1931. Epub 2008 Jun 30
114. Waege I, Schmid G, Thumann S, Thomm M, Hausner W. Shuttle vector-based transformation system for *Pyrococcus furiosus*. Appl Environ Microbiol. 2010;76:3308–13. https://doi.org/10.1128/aem.01951-09.
115. Wagner M, van Wolferen M, Wagner A, Lassak K, Meyer BH, Reimann J, Albers S-V. Versatile genetic tool box for the crenarchaeote *Sulfolobus acidocaldarius*. Front Microbiol. 2012;3:214. https://doi.org/10.3389/fmicb.2012.00214.
116. Ward DE, Kengen SW, van der Oost J, de Vos WM. Purification and characterization of the alanine aminotransferase from the hyperthermophilic archaeon *Pyrococcus furiosus* and its role in alanine production. J Bacteriol. 2000;182:2559–66. https://doi.org/10.1128/jb.182.9.2559-2566.2000.
117. Wenck BR, Vickerman RL, Burkhart BW, Santangelo TJ. Archaeal histone-based chromatin structures regulate transcription elongation rates. Commun Biol. 2024; https://doi.org/10.1038/s42003-024-05928-w.
118. Woese CR, Fox GE. Phylogenetic structure of the prokaryotic domain: the primary kingdoms. Proc Natl Acad Sci. 1977;74:5088–90. https://doi.org/10.1073/pnas.74.11.5088.
119. Woese CR, Kandler O, Wheelis ML. Towards a natural system of organisms: proposal for the domains Archaea, Bacteria, and Eucarya. Proc Natl Acad Sci. 1990;87:4576–9. https://doi.org/10.1073/pnas.87.12.4576.

120. Yin J, Chen J-C, Wu Q, Chen G-Q. Halophiles, coming stars for industrial biotechnology. Biotechnol Adv. 2015;33:1433–42. https://doi.org/10.1016/j.biotechadv.2014.10.008.
121. Zheng T, Huang Q, Zhang C, Ni J, She Q, Shen Y. Development of a simvastatin selection marker for a hyperthermophilic acidophile, *Sulfolobus Islandicus*. Appl Environ Microbiol. 2012;78:568–74. https://doi.org/10.1128/aem.06095-11.
122. Zhong C, Wang L, Ning K. Pan-Genome Study of *Thermococcales* reveals extensive genetic diversity and genetic evidence of thermophilic adaption. Environ Microbiol. 2020;23:3599–613. https://doi.org/10.1111/1462-2920.15234.

Part II

Conventional Animal Models for Research in Health and Medicine

5 Laboratory Rat as an Animal Model for Preclinical Studies

Subramani Parasuraman, Jia Jun Chuah, and Sherly Deborah George

5.1 Introduction to the Laboratory Rat as an Animal Model

Animal experimentation, dating back over 2000 years, has significantly contributed to advancements in medicine and understanding of organism function. It, dating back to Greece, where. Aristotle (384–322 B.C.) and Galen (A.D. 129–199) were early researchers in animal studies, revealing anatomical differences through dissection. However, it wasn't until the sixteenth century that many experiments on animals began to be recorded, including William Harvey's work on the heart and blood movement in animals in 1628. In the 1800s, France became a leading centre for experimental biology and medicine. Starting around 1850, biology research accelerated, with many advancements resulting from animal experiments, but opposition to animal use emerged in the nineteenth century [53]. The selection of an animal model for biomedical research is complex due to various limitations. Key criteria include resemblance between animal species and humans, ability to reproduce diseases or pathologies, availability and size of species, and lifespan. Animal

Jia Jun Chuah and Sherly Deborah George contributed equally with all other contributors.

S. Parasuraman (✉)
Pharmacology, Toxicology and Basic Health Sciences Unit, Faculty of Pharmacy, AIMST University, Bedong, Malaysia

Centre of Excellence for Vaccine Development, Faculty of Applied Sciences, AIMST University, Bedong, Malaysia
e-mail: parasuraman@aimst.edu.my

J. J. Chuah
Pharmacology, Toxicology and Basic Health Sciences Unit, Faculty of Pharmacy, AIMST University, Bedong, Malaysia

S. D. George
Department of Physiology, Faculty of Medicine, Manipal University College Malaysia, Melaka, Malaysia

P. Kumar, V. Kothari (eds.), *Model Organisms in Biological Research*, https://doi.org/10.1007/978-981-92-0038-2_5

species like Drosophila (insects), *Danio rerio* (zebrafish), *Caenorhabditis elegans* (nematodes), Xenopus (frogs), and mammals like mice, rabbits, rats, cats, dogs, pigs, and monkeys are accepted worldwide for their phylogenetic resemblance to humans [51]. The rat (*Rattus norvegicus*) is an ideal experimental animal due to its abundance of literature, ease of handling, high fertility rate, short gestation period, low maintenance, and potential as a disease model for various human disorders. Rats constitute one of the most commonly used laboratory species. This chapter introduces rats as a species and biomedical model in research and describes the current state and a future perspectives of rats as a model for human disease.

5.2 History and Evolution of the Laboratory Rat in Research

The history of laboratory rat strains dates back over a century. Rats and humans have coexisted for several centuries. The laboratory rat commonly used in research is the domesticated form of the brown (Norway) rat, *Rattus norvegicus*, which likely originated in regions corresponding to modern-day China and Mongolia [29]. Although rats have historically been associated with disease, they have diverse roles in human society, including being a food source, companion animals, and entertainment (e.g., rat racing). By the early twentieth century, selective breeding of rats as pets was formalized: organizations such as the UK's National Mouse Club and the American Fancy Rat and Mouse Association (USA) promoted fancy rat breeding. The scientific use of rats began in the nineteenth century; the first documented experiments involving rats occurred around 1828, and by the late 1800s, laboratory rat strains were being maintained for research. The first large-scale study using laboratory rats in the United States was conducted at the Wistar Institute of Philadelphia in 1894. Shortly thereafter, breeders developed other outbred stocks; the Sprague–Dawley (SD) strain (1920s) and the Long–Evans strain (1920s, pigmented) became major workhorses [29]. These early developments laid the foundation for the rat's role as a model organism in biomedical research.

The laboratory rat was taxonomically defined as a strain of *Rattus norvegicus*. One of the most widely used strains is the Wistar rat (often an albino "Wistar" strain), which was developed specifically for research laboratories.

Table 5.1 Commonly used rat strains

Inbred strains	Usefulness as models
ACI	Congenital genitourinary anomalies, prostatic adenocarcinomas
BB/Wor	Juvenile insulin-dependent diabetes mellitus
BN (Brown Norway)	Inducible, transplantable myeloid leukaemia, hydronephrosis, bladder carcinoma
BUF (Buffalo)	Spontaneous autoimmune thyroiditis, host for transplantable Morris hepatoma
COP (Copenhagen)	Prostate adenocarcinoma
F-344 (Fischer 344)	Inbred rat model for the National Toxicology Program's Carcinogen Bioassay Program and the National Institute on Aging
LEW (Lewis)	Multiple sclerosis, various experimentally induced autoimmune diseases
LOU/C	Myeloma, production of IgG autoantibody
SHR (spontaneous hypertensive rat)	Hypertension, cardiovascular research
WF (Wistar-Furth)	Mononuclear cell leukaemia
Zucker	Obesity
Mutant strains	*Characteristics*
Brattleboro	Diabetes insipidus (autosomal recessive)
Gunn	Jaundice, kernicterus (autosomal recessive)
Nude	T cell deficient (autosomal recessive)
Obese SHR	Type 4 hyperlipoproteinemia (autosomal recessive)

The genus *Rattus* contains approximately 66 species, all of which are classified within the family Muridae and order Rodentia. For example, *Rattus rattus* (the black rat) is another well-known member of this genus, although it is more commonly found as a wild pest in human habitats [38]. Domesticated *R. norvegicus* strains have become standardized over time, facilitating reproducible experiments and comparisons across studies. A summary of the commonly used rat strains is presented in Table 5.1, and Fig. 5.1 illustrates various types of laboratory rats [59].

Fig. 5.1 Various strains of laboratory rats. (Reproduced with permission from [45] with license number 6078270884823)

5.3 The Anatomy and Physiology of Laboratory Rats

Laboratory rats are small, non-ruminant mammals with well-characterized biological parameters. Adult male rats typically weigh 300–500 g, and females weigh 250–300 g. Under standard laboratory conditions, their lifespan is approximately 2.5–3 years. Rats maintain a normal body temperature of approximately 37.5 °C and, for a 400 g rat, a basal metabolic rate of approximately 35 kcal per 24 h. Female rats reach sexual maturity at approximately 50 ± 10 days and have a gestation period of 21–23 days, with litters of approximately 8–14 pups. In cytogenetic terms, rats have a diploid chromosome number of 42 [59]. These basic life-history traits facilitate the efficient generation of experimental cohorts.

A range of physiological parameters has been established in healthy adult rats. For example, normal cardiovascular values include a mean systolic/diastolic blood pressure of approximately 116/90 mmHg and a resting heart rate of 300–500 beats per minute. Cardiac output is on the order of 50 mL/min in a 400–500 g rat. Respiratory measurements show a rate of approximately 85 breaths per minute, with a tidal volume of roughly 1.5 mL. Renal function under baseline conditions is about 5.5 mL of urine per 100 g body weight per 24 h, with typical urine pH in the range of 7.3–8.5. Comprehensive reference ranges for blood chemical analytes have also been documented. For instance, normal serum glucose levels in adult Sprague-Dawley rats are on the order of 110–115 mg/dL, and serum electrolytes such as sodium are approximately 148–150 mEq/L in healthy animals [26, 59]. These normative data (summarized in Tables 5.2, 5.3 and 5.4) provide a baseline against which experimental or pathological values in rat models can be compared [26, 59].

Table 5.2 Selected normative data for the rat[a]

Weight of adult male	300–500 g
Weight of adult female	250–300 g
Life span	2.5–3 years
Body temperature	37.5 °C
Basal metabolism rate (400-g rat)	35 kcal/24 h
Chromosome number (diploid)	42
Puberty	50 ± 10 days
Gestation	21–23 days
Litter size	8–14
Birth weight	5–6 g
Eyes open	10–12 days
Weaning	21 days
Food consumption/24 h	5 g/100 g body weight
Water consumption/24 h	8–11 mL/100 g body weight
Cardiovascular	
Arterial blood pressure	
Mean systolic	116 mmHg
Mean diastolic	90 mmHg
Heart rate	300–500 beats/min
Cardiac output	50 mL/min
Blood volume	6 mL/100 g body weight
Respiratory	
Respirations/min	85
Tidal volume	1.5 mL
Alveolar surface area (400 g rat)	7.5 m^2
Renal	
Urine volume/24 h	5.5 mL/100 g body weight
Na^+ excretion/24 h	1.63 mEq/100 g body weight
K^+ excretion/24 h	0.83 mEq/100 g body weight
Urine osmolarity	1659 mOsm/kg of H_2O
Urine pH	7.3–8.5
Urine specific gravity	1.04–1.07

[a]Reproduced with permission from [59] with license number 6012691370732

Table 5.3 Clinical chemistry reference ranges for adult rats[a]

Analyte	Units	Sprague–Dawley[b]		Fisher 344[c]	
		M	F	M	F
Serum					
Glucose	mg/dL	115 ± 16.9	111 ± 17.2	115 ± 12.5[d]	–
Urea nitrogen	mg/dL	19 ± 2.2	21 ± 3.4	15 ± 2.5[d]	–
Creatinine	mg/dL	0.70 ± 0.11	0.70 ± 0.13	–	–
Sodium	mEq/L	150 ± 3.4	148 ± 3.5	149 ± 3.0[d]	–
Potassium	mEq/L	7.00 ± 0.65	6.1 ± 0.67	4.80 ± 0.35[d]	–
Chloride	mEq/L	103.0 ± 1.90	104.0 ± 2.4	106 ± 3.0[d]	–
Calcium	mg/dL	12.0 ± 0.94	12.1 ± 0.71	10.5 ± 0.50[d]	–
Phosphorus	mg/dL	7.30 ± 1.5	5.80 ± 1.10	–	–
Magnesium		3.12 ± 0.41[e]	2.60 ± 0.21[e]	–	–
Iron	mg/dL	152 ± 70	220 ± 130 (19–21 weeks)	–	–
Total iron-binding capacity	mg/dL	368[d]	–	–	–
Alanine aminotransferase	IU/L	49 ± 24.1	69 ± 44.9	78 ± 11	49 ± 8
Aspartate aminotransferase	IU/L	95 ± 31.7	99 ± 54.5	–	–
Alkaline phosphatase	IU/L	130 ± 43.7	117 ± 41.7	49.5 ± 9.25[d]	–
Lactate dehydrogenase	IU/L	275 ± 112	–	650 ± 75	650 ± 75 (20 weeks)
Sorbitol dehydrogenase	IU/L	20 ± 5[d] (32 weeks)	–	20.0 ± 7.5[d]	–
γ-Glutamyl transpeptidase	IU/L	2.5 ± 1.25[d]	–	–	–
Creatinine kinase	IU/L	275 ± 112.5[d]	–	400 ± 50[d]	–
Protein, total	g/L	70 ± 5.0	75 ± 5	70.5 ± 4.75[d]	

(continued)

Table 5.3 (continued)

Analyte	Units	Sprague–Dawley[b]		Fisher 344[c]	
		M	F	M	F
Albumin	g/L	34 ± 2.0	40 ± 2.5	42.5 ± 3.75[d]	
Cholesterol	mg/dL	119 ± 51.3	119 ± 29.0	96.5 ± 14.25	130 ± 10.0
Triglycerides	mg/dL	266 ± 121.4	249 ± 159.7	122 ± 21.25	62.5 ± 11.25
Bilirubin	mg/dL	0.3 ± 0.16	0.4 ± 0.27	0.3 ± 0.1[d]	
Bile acids	mmol/L	40 ± 10[d] (20 weeks)	–	30 ± 10[d]	–
Uric acid	mg/dL	1.52 ± 0.30	1.25 ± 0.36	–	–
Urine					
Volume	mL/16 h	–	–	9.5 ± 4.0	9.3 ± 5.6
	mL/22 h	15.7 ± 6.7	11.0 ± 5.0		
Specific gravity	–	–	–	1.022 ± 0.007	1.017 ± 0.007
Osmolality	mOsm/kg	943 ± 32–7	995 ± 367	–	–
pH	–	7.8 ± 0.5	7.7 ± 0.5	6.0–6.5[d]	–
Chloride	mmol/L	148 ± 36	151 ± 51	–	–
Sodium	mmol/L	31 ± 11	50 ± 22	–	–
Potassium	mmol/L	121 ± 31	110 ± 45	–	–
Phosphorus	mg/dL	142 ± 34	156 ± 62	–	–
Creatinine	mg/dL	136 ± 40	116 ± 41	80 ± 28	54 ± 25
Glucose	mg/dL	–	–	9.9 ± 3.9	5.5 ± 2.3
Protein	mg/dL	–	–	98.8 ± 54.4	11.2 ± 5.5
Alkaline phosphatase	IU/L	–	–	152 ± 61	73 ± 37
Lactate dehydrogenase	IU/L	–	–	28 ± 15	16 ± 8
N-acetyl-β-glucosaminidase	IU/L	–	–	12.2 ± 7.8	5.9 ± 2.7

Aspartate aminotransferase		–	–	14.4 ± 6.5	3.6 ± 2.5
γ-Glutamyl transpeptidase	IU/L	–	–	4964 ± 780	1873 ± 215
Insulin clearance	mL/min/100 g	857 ± 178	–	–	–
Glomerular filtration rate	mL/min	275 ± 33[d]	–	–	–
Urine flow	mL/min/100 g	5.2 ± 2.0	–	–	–

[a]Values are for 12-month-old animals, unless noted otherwise, as summarized from Loeb and Quimby [99]
[b]6–18 months old
[c]58–l 12 weeks old
[d]Gender not specified
[e]Wistar strain
[Reproduced with permission from [59] with license number 6012691370732]

Table 5.4 Haematological parameters in the rat[a]

	Stock/strain					
	Crl:CD(SD)		Crl:LE		F344/DuCRL	
Parameter (units)	Male	Female	Male	Female	Male	Female
WBC (K/μL)	11 (4.0–18)	9.5 (2.4–18)	9.8 (5.6–14)	8.7 (4.9–13)	6.7 (2.1–9.6)	7.4 (1.9–12)
Neutrophils (K/μL)	3.4 (1.0–6.7)	2.6 (0.7–5.9)	4.0 (2.2–6.5)	3.2 (1.7–5.4)	2.5 (0.7–3.9)	2.4 (0.5–4.4)
Lymphocytes(K/μL)	6.5 (2.2–12)	6.1 (1.5–11.5)	5.0 (2.5–7.9)	4.6 (2.6–7.9)	3.6 (1.0–5.7)	4.4 (1.2–7.5)
Monocytes (K/μL)	0.7 (0.2–1.4)	0.6 (0.1–1.2)	0.7 (0.3–1.2)	0.7 (0.3–1.3)	0.5 (0.1–1.0)	0.5 (0.1–1.0)
Eosinophils (K/μL)	0.1 (0–0.5)	0.1 (0–0.5)	0.1 (0–0.5)	0.2 (0–0.8)	0.1 (0–0.3)	0.1 (0–0.5)
Basophils (K/μL)	0.03 (0–0.2)	0.03 (0–0.2)	0.03 (0–0.1)	0.04 (0–0.3)	0.02 (0–0.1)	0.03 (0–0.1)
Neutrophils (%)	32 (16–54)	28 (14–55)	41 (30–51)	38 (25–51)	37 (25–53)	32 (19–47)
Lymphocytes (%)	61 (38–79)	63 (37–81)	51 (41–63)	53 (38–65)	54 (40–67)	59 (42–74)
Monocytes (%)	6.3 (3–11)	6.5 (3–11)	6.6 (3–10)	7.4 (4–12)	7.6 (7.7–12)	6.8 (2.9–12)
Eosinophils (%)	1.2 (0.1–4.3)	1.4 (0.2–4.8)	1.2 (0.1–4.1)	1.6 (0.1–7.3)	0.9 (0–3.7)	1.6 (0.1–5.5)
Basophils (%)	0.3 (0–1.7)	0.3 (0–1.6)	0.3 (0–1.6)	0.4 (0–2.4)	0.4 (0–1.1)	0.4 (0–1.3)
RBC (M/μL)	7.7 (5.8–10)	7.4 (5.5–9.8)	7.6 (6.0–9.4)	7.6 (6.7–8.6)	8.3 (7.1–9.7)	8.1 (6.6–9.4)
Haemoglobin (g/dL)	17 (13–23)	16 (13–22)	16 (13–20)	16 (14–18)	16 (14–19)	16 (13–18)
Haematocrit (%)	51 (39–69)	48 (37–64)	52 (43–62)	51 (43–60)	54 (47–62)	54 (44–62)
MCV (fl)	67 (58–77)	66 (58–75)	69 (63–76)	67 (61–75)	65 (60–69)	67 (63–70)
MCH (pg)	22 (17–25)	22 (18–25)	21 (19–24)	21 (20–23)	20 (17–22)	20 (17–22)
MCHC (g/dL)	33 (27–38)	33 (27–38)	31 (29–33)	31 (29–43)	31 (27–34)	30 (26–32)
RDW (%)	16 (14–18)	15 (13–17)	16 (13–17)	14 (12–15)	15 (14–17)	13 (12–15)
Platelets (M/μL)	1.6 (0.8–2.6)	1.6 (0.8–2.4)	1.4 (0.7–1.9)	1.3 (0.6–1.7)	1.1 (0.8–1.6)	1.1 (0.7–1.5)
MPV (fl)	7.4 (5.9–10)	7.3 (5.8–9.6)	8.0 (6.7–9.2)	8.1 (5.8–9.5)	6.9 (5.8–8.2)	6.8 (5.7–8.1)

[a]Derived from vendor data (Charles River Research Models) on animals 8–10 weeks of age; mean (95% confidence interval)

[Reproduced with permission from [59] with license number 6012691370732]

5.4 Genetic Variability and Strain Selection in Rat Models

Genetic background profoundly influences experimental outcomes in rats. Laboratory rats (*Rattus norvegicus*) are widely valued for their attributes, such as high fertility, short gestation, and ease of handling [38]. Even within this species, distinct strains exhibit different physiological and behavioural characteristics. Inbred strains (produced by ≥20 generations of brother-sister matings) are genetically uniform and minimize variability, whereas outbred stocks maintain broader heterogeneity that may better reflect human populations. Careful choice of strain is therefore essential: it can reduce experimental "noise" or, when desired, model genetic diversity. For example, an inbred diabetic rat line provides a consistent metabolic phenotype for mechanistic studies, whereas an outbred stock may be preferred in early-stage drug safety studies to capture population-wide responses [78]. Genetic variability between inbred and outbred strains is depicted in Fig. 5.2.

The inbred lines, such as Fischer 344 (F344) and Lewis, and Brown Norway (BN) rats, were created by rigorous brother-sister mating to achieve fixed genotypes. The pedigree of each strain shapes its traits and research uses. For example, F344 rats (an inbred line) have a well-characterized ageing profile and are commonly used in toxicology and carcinogenicity testing, whereas Lewis and BN strains are favoured in immunology and transplantation studies. These distinct lineages underscore that "one strain does not fit all"—early history illustrates how breeders purposefully select different characteristics in different colonies. Currently, researchers have a wide palette of inbred and outbred strains to meet experimental needs. Outbred stocks, such as Wistar and SD rats, remain popular for general pharmacology and physiology; their robust growth and fertility make colony maintenance straightforward. Inbred lines (e.g. F344, Lewis, BN, and Dahl salt-sensitive/hypertensive strains) are consistent models in chronic and disease models. Critically, the strain-specific traits are dramatic. Morphological and behavioural studies have

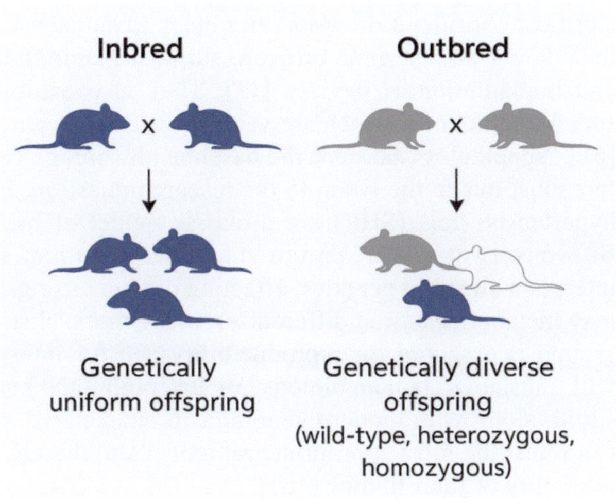

Fig. 5.2 Illustration of genetic variability: inbred (bred through brother-sister mating to fix alleles) versus outbred (maintained without sibling mating to preserve diversity) rat colonies. Appropriate strain selection balances experimental uniformity and population relevance (an image adapted from generic lab rat breeding schemes)

revealed clear differences; for instance, F344 rats are smaller with shorter long bones and lighter body weights than SD or Wistar rats [87]. Sensorimotor tests also differ by strain; Lewis and SD rats are less sensitive to mechanical stimuli than F344 rats, and even basic locomotor patterns (weight distribution on limbs) vary among strains [87]. Phenotypic variability extends to drug metabolism and physiology. For example, an in vitro comparison of hepatocytes found that SD and Wistar Han rats have broadly similar hepatic clearance rates for many drugs, yet the identities and formation rates of individual metabolites can differ significantly between strains [84]. These findings imply that a toxicologist or pharmacologist might observe the same overall clearance in the two strains, but different metabolites accumulate, which can affect toxicity or efficacy. Similarly, immune response, stress reactivity, and endocrine profiles are often strain-dependent; therefore, a model for hypertension or diabetes may require a specific genetic background.

Modern genetic engineering has diversified the rat models. Historically, rats have lagged behind mice in reverse genetics because reliable embryonic stem (ES) cells are lacking. However, the development of designer nucleases such as meganucleases, zinc-finger nucleases (ZFNs), TALENs, and CRISPR/Cas9 has revolutionized rat genetics [11]. These tools allow scientists to create precise knockouts, knockins, or point mutations directly in the rat genome with high efficiency [11, 21]. Over the past two decades, numerous genetically modified rat lines have emerged, accelerating disease modelling [11]. For example, CRISPR/Cas9 has been used to generate knock-in rats expressing Cre recombinase under the control of endogenous promoters (such as the dopamine D1 receptor or adenosine 2A receptor loci), yielding models that combine rat neurophysiology with sophisticated genetic control [62]. Other projects have "humanized" specific genes (inserting human alleles) or created null alleles for neurological disorders, immunodeficiencies, and metabolic diseases. These transgenic and CRISPR-edited strains greatly expand the translational power of rat models by aligning rat genetics more closely with human pathophysiology.

When designing translational experiments, the choice of the rat strain remains a critical consideration. Rats offer clear advantages over mice in many contexts: their larger size permits intricate surgical manipulation, multiple blood draws, and implantation of devices [11]. They also exhibit complex behaviours and social interactions, which are valuable in neuroscience and psychology research [61]. Nonetheless, because the baseline physiology can vary by strain, investigators must match the strain to the research question. For instance, spontaneously hypertensive rats (SHR) are a classic model of high blood pressure, whereas Brown Norway rats are preferred for allergic asthma studies. Sex and age further interact with strain genetics, affecting outcomes (e.g., male and female F344 rats may metabolize a drug differently). In all cases, clearly defining the strain background is essential for reproducibility and for interpreting whether the results will translate to human biology. By leveraging the known characteristics of each strain, along with modern genomics to characterize outbred stocks, researchers can select the most appropriate rat model and thereby enhance the relevance and reliability of their findings [61].

5.5 Ethical Considerations in the Use of Laboratory Rats in Research

The use of laboratory rats in research is governed by strict ethical principles. In most frameworks, animal experimentation is justified only if the anticipated benefits to human or animal health outweigh the potential harm to animals. Research involving rats (and other vertebrates) generally requires prior review and approval by an institutional animal care and use committee or equivalent body. These committees ensure that the proposed study is scientifically sound, that the number of animals is justified, and that all possible alternatives and refinements have been considered before the work begins [98]. A foundational concept in laboratory animal ethics is the "3Rs" principle [97], which seeks to minimize animal use and suffering. The 3Rs are:

- *Reduction*: The minimum number of animals necessary to achieve statistically valid results is used. This can be achieved by careful experimental design, proper control, and techniques such as power analysis to avoid under- or overuse of animals.
- *Refinement*: Modify procedures and husbandry to minimize pain, distress, and suffering. Examples include using anaesthesia and analgesia when appropriate, training personnel in humane handling, providing environmental enrichment, and defining clear humane endpoints so that animals are not subjected to unnecessary procedures.
- *Replacement*: Wherever possible, replace live animal use with alternative methods. This includes in vitro techniques (cell cultures and organoids), computational modelling, or lower organisms (invertebrates and cell lines) that can simulate the research question without using sentient vertebrates.

In practice, researchers must follow relevant legislation and guidelines. These standards have ensured the responsible use of rats while maintaining their scientific rigour. For example, many journals and funders require compliance with the ARRIVE guidelines [36], which mandate detailed reporting of experimental design, ethical review, housing conditions, and animal care in publications. Laboratory rats should be housed in clean, species-specific conditions. Overcrowding, excessive noise, lack of social contact, or inappropriate lighting can induce stress and alter physiology or behaviour. Investigators should closely monitor animals for signs of pain or distress, veterinary care must be available, and only qualified personnel should perform surgical or anaesthetic procedures. The study protocol should specify how discomfort will be minimized, including the use of anaesthesia/analgesia and euthanasia criteria [89].

Ultimately, an ethical review hinges on a cost-benefit analysis. Researchers must demonstrate that the knowledge gained is likely to be significant and that no non-animal methods can address this question. If the expected harm to animals outweighs the potential benefits, the experiment must not be conducted [89]. These ethical considerations, codified in laws, regulations, and professional guidelines, ensure that the use of laboratory rats in research is scientifically justified and conducted with the highest standards of animal welfare.

5.6 Behavioural Assessment in Rat Models

Preclinical neuroscience has long relied on rat models and standardized behavioural tests to probe motor, emotional, and cognitive functions relevant to human health. Early efforts date to the late nineteenth century, when Stewart's rotating drum (1898) recorded rat locomotion, foreshadowing modern treadmills and rotarod devices [18]. In the 1930s, Hall introduced an open-field test to quantify rodent emotionality and exploration [69]. Operant conditioning chambers (Skinner boxes) emerged in the 1930s–40s, enabling learned behaviours (lever pressing for reward) and laying the groundwork for complex cognitive assays. Together, these historical paradigms have established that rats can model diverse neurological and psychiatric processes under controlled conditions [18, 69].

5.6.1 Motor Behaviour Tests

Motor assays in rats assess general activity, coordination, and balance. A basic measure is locomotion in an open arena; rats' ambulation and exploratory patterns are recorded to detect hypo- or hyperactivity. For example, open-field distance travelled indexes spontaneous activity, while wall-hugging (thigmotaxis) is interpreted as anxiety-related behaviour [69]. More demanding tests probe coordination. The rotarod test places rats on a rotating rod, and latency to fall provides a quantitative measure of motor coordination and balance [18]. Other assays include grid/beam walking and foot-fault tasks, which count paw slips on a horizontal grid, and automated gait analysis (for example, Noldus CatWalk), which tracks paw placement and stride. Foot-fault and CatWalk analyses complement rotarod data to detect subtle neuromotor impairments [18]. These motor tests are used in models of movement disorders; for instance, lesioned rats or Parkinsonian models show reduced rotarod performance and altered gait, paralleling human motor deficits. Such assays have a high predictive validity for the motor side effects of drugs and enable translational insights (e.g., testing therapies for stroke or Parkinson's disease).

5.6.2 Emotional (Affective) Behaviour Tests

Rodent anxiety- and depression-like behaviours are typically measured using innate or stress-based paradigms. Open-field testing also yields an anxiety index: rats normally prefer a dark, enclosed periphery, and the time spent in the centre of the arena decreases with anxiety (higher *thigmotaxis*) [69]. The elevated plus maze (EPM) is a classic anxiety test with two open arms and two walled arms. Rats instinctively avoid open arms due to fear of heights and open space; therefore, increased open-arm entries or duration signifies reduced anxiety [82]. The EPM (and analogous light/dark box) is highly sensitive to anxiolytic and anxiogenic manipulations; for example, benzodiazepines reliably increase open-arm exploration [82].

Depression-like states were modelled using stress-based tasks. In the forced swim test (FST), rats are placed in inescapable water; after initial struggling, they become immobile, interpreted as "behavioural despair." Immobility time is the primary endpoint and is markedly reduced by clinically effective antidepressants [92]. Likewise, learned helplessness paradigms (uncontrollable shock) induce passivity analogous to despair. Anhedonia was assessed using sucrose preference tests (reduced intake of sweet solution) or novel-suppressed feeding. Overall, these affective tests capture components of human mood disorders; antidepressants and anxiolytics produce expected changes in rats, making such tasks useful translational screens [82, 92]. However, their construct validity is debated and must be interpreted with caution [24].

5.6.3 Cognitive Behaviour Tests

The cognitive domains in rats include learning, memory, attention, and executive function. Spatial learning and memory are most often evaluated using the Morris water maze (MWM), in which rats swim in a circular tank using distal cues to find a hidden platform. Performance (latency and path to the platform) reflects hippocampal-dependent memory. The MWM is robust, reproducible, tightly linked to hippocampal plasticity, and relatively insensitive to motivational or body mass differences [81]. The Barnes maze is a dry-land analogy. Working memory (short-term memory) is tested with the radial-arm maze: in the 8-arm maze, rats learn that only one unvisited arm contains a reward; the number of re-entries into arms already visited indexes working memory errors [93].

Nonspatial memory was assayed using spontaneous recognition tasks. In the novel object recognition (NOR) test, rats were exposed to two identical objects, and after a retention interval, one object was replaced with a novel object. Normal rats naturally explore novel objects more, indicating memory for familiar items [4]. Thus, NOR models aspects of human recognition memory and novelty preference without reinforcement.

Attention and executive function were tested using operant paradigms. The 5-Choice Serial Reaction Time Task (5-CSRTT) requires rats to detect brief light flashes in one of five holes and measures sustained visual attention, vigilance, and impulsivity. The 5-CSRTT is considered analogous to the human Continuous Performance Test. Complex tasks such as set-shifting or reversal learning probe cognitive flexibility, and intra-dimensional/extradimensional shift tasks in rats parallel the Wisconsin Card Sorting Test in humans [93]. Fully automated touchscreen platforms have recently emerged. Rats interact with visual stimuli on a screen using nose pokes, allowing cognitive tests (visual discrimination, paired associates, etc.) that are nearly identical to those used in human neuropsychological batteries. This touchscreen approach is low-stress and highly translational, as rodent performance on these tasks can be directly compared with human data [31].

Behavioural testing has evolved with technology to improve throughput and translational value. Automated video tracking and pose-estimation software now record locomotion and complex behaviours with minimal human scoring. Contemporary open-source tools use computer vision and machine learning to detect and classify behaviours with high consistency compared with manual scoring [32]. Home cage monitoring systems (e.g., IntelliCage and PhenoTyper) allow long-term continuous assessment of the natural behaviour of rats via RFID sensors and cameras. Such systems capture circadian activity rhythms, social interactions, and extended cognitive task performance with virtually no human interference [50]. The resulting large datasets have improved reproducibility (reducing between-session variability) and enabled chronic pharmacological studies that better mimic human clinical trials.

Crucially, modern behavioural assays are increasingly being designed for cross-species translation. Many rat tasks are analogous to human tests (5-CSRTT≈CPT, touchscreen tasks to computerized neuropsychological tests [31, 93]. Novel "reverse-translational" paradigms adapt human experimental protocols for rodents (e.g. probabilistic reward learning, judgment bias tasks) to target cognitive-affective processes relevant to psychiatric disorders [31]. In summary, rodent behavioural batteries, enriched by automated technology and careful validation, continue to provide a bridge between basic neuroscience and clinical conditions. The translational utility of these models lies in their ability to capture the core components of human motor, emotional, and cognitive disorders under controlled experimental conditions [81, 92].

5.7 Pharmacokinetic Testing in Rat Models

Rodent pharmacokinetic studies assess in vivo drug absorption, distribution, metabolism, and excretion, crucial for determining appropriate drug exposure, dose requirements, and other parameters for efficacious and safe studies. Pharmacokinetic studies in rats are widely utilized in drug development to comprehend the absorption, distribution, metabolism, and excretion of a compound within a living system. Sprague Dawley, Wistar, Brown Norway, F344, and Lewis rat strains are used in pharmacokinetic studies. The rodent model is also helpful in studying the drug relationships between tissue concentrations and those in serum or plasma. Pharmacokinetics in animals differ significantly from humans because rodents have higher metabolic rates and more active drug-metabolizing enzymes with faster drug elimination rates [3]. Sprague–Dawley or Wistar rats, widely used in biomedical research, may exhibit genetic differences in drug metabolism, resulting in varying pharmacokinetics within and between outbred and inbred rat strains [44]. Animal pharmacokinetic studies offer valuable insights, but human data is crucial for accurate dosing and efficacy predictions, considering factors like metabolic rate, enzyme activity, and organ function. For example, memantine's half-life in rats and mice is short (~3–4 h), while in humans it's long (60–80 h) with a C_{max}/C_{min} ratio of approximately 1.5 [6]. Hence, rodent pharmacokinetic studies offer valuable insights and further studies in humans are necessary for understanding the pharmacokinetic properties of any investigational drugs.

5.8 Rat Models in Neurological Studies

Rats have long been indispensable in neuroscience research because they offer an optimal balance of biological complexity and experimental tractability. Compared to mice, rats have larger brains and more robust behavioural repertoires [19]. These features facilitate neurosurgical procedures, neural recordings, and the study of complex behaviours. For example, rats have been trained on sophisticated cognitive and social tasks, and recent work even demonstrates pro-social "helping" behaviour driven by empathy-like processes (e.g. releasing a trapped cagemate) [10]. Rats also emit ultrasonic vocalizations (USVs) that convey affective states such as 22-kHz calls during distress. Playbacks of these calls elicit avoidance and neural responses akin to human fear processing [8], illustrating how rat models can capture ethologically relevant emotional states. This rich behavioural and affective complexity enhances the translational relevance of rat studies, as these animals can exhibit neural and psychological phenomena analogous to those in humans.

Technological advances have bolstered the utility of rat models. In the past, mice gained an edge owing to early transgenic techniques, but the gap is closing due to modern genome editing. CRISPR/Cas9 methods now allow precise genetic manipulation in rats, including targeted knockouts and knock-ins in neurons. For instance, rat lines have been created to study dopaminergic neuron function and substance abuse by editing genes in the adult brain [54]. Such tools enable the generation of disease-relevant models; transgenic rats overexpressing human amyloid or α-synuclein recapitulate Alzheimer's or Parkinson's pathology more faithfully than some mouse models. Moreover, viral vector strategies in rats permit region- or cell-type–specific expression of optogenetic or chemogenetic actuators, extending methods developed in mice into larger-brained species. In parallel, imaging and recording technologies have matured; high-field functional magnetic resonance imaging (fMRI) and functional ultrasound can map rat brain networks, and miniscope calcium imaging can monitor activity in freely behaving rats. Notably, resting-state fMRI in rats revealed large-scale functional networks similar to those in humans [46], supporting the translation of network-level findings from rodents to humans. These developments indicate that rats can be probed with invasive and non-invasive techniques that are directly analogous to human neuroimaging studies.

Rats also serve as classic models for many neurological conditions, with a focus here on their general role rather than on individual diseases. For instance, in stroke research, rat models (e.g., middle cerebral artery occlusion) reproduce many features of human infarction and have been used to test neuroprotective strategies. However, as with many preclinical studies, many interventions that are effective in rat stroke models have failed in clinical trials [28]. This underscores the translational challenge: species differences and experimental variables often limit predictive validity. However, well-designed rat studies have yielded critical insights. For example, skilled forelimb-reaching tasks in rats have demonstrated recovery trajectories after an experimental stroke that closely mirror human motor deficits [79]. Similarly, models of traumatic brain injury in rats reproduce complex pathophysiological cascades (glial responses and axonal injury) and cognitive sequelae, offering platforms to study mechanisms that would be impractical in humans. In psychiatric neuroscience, rats undergo operant and maze tasks for anxiety, depression, and impulsivity, leveraging their capacity for nuanced behaviour and social stress paradigms (such as social defeat or early life stress). Notably, rat studies of addiction and reward have been foundational; landmark work on mesolimbic dopamine and reward prediction has been performed in rats, providing a template for human addiction research. Across these domains, the physiological and genetic similarity of rats to humans allows controlled exploration of disease processes, biomarker discovery, and preclinical testing [75, 85].

Despite their value, translational limitations must be acknowledged. As noted by Chesselet [12], all animal models differ from humans in terms of their genetics, lifespan, and living conditions. For this reason, a rat is not a "mini-human," and care must be taken when extrapolating. However, when studies are carefully designed with these constraints in mind, rat models can yield critical mechanistic insight. For example, rats can be examined at multiple stages of pathology, allowing longitudinal correlation of molecular changes with behaviour, a strategy often impossible in clinical samples [12]. Furthermore, rats can be subjected to precise manipulations of genes, environment, and drugs in controlled ways. This has clarified, for instance, how specific neuronal circuits contribute to learning, memory, and affect. Electrical stimulation or lesion experiments in rats have mapped brain regions underlying behaviour (e.g., hippocampus for spatial memory and amygdala for fear conditioning). In addition, invasive recordings from the rat cortex and hippocampus have revealed network dynamics (such as place-cell sequences during navigation) that form the basis of modern computational memory models. These neurophysiological findings in rats often guide hypotheses regarding human brain function.

Overall, recent findings have underscored the translational relevance of rat neuroscience. Rat models of neurological function and dysfunction have yielded new biomarkers and therapeutic leads. For example, the use of fMRI and network analysis in transgenic rat models of Alzheimer's or autism helps identify changes that parallel patient MRI findings [46]. Studies using rat ultrasonic communication have inspired novel anxiety assays with closer human face validity [8]. Moreover, the combination of genetic tractability and system-level methods in rats may bridge the gap between cellular studies and human trials. In conclusion, rats remain a cornerstone of neurological research, and their well-characterized neurobiology, ability to perform complex behaviours, and growing toolkits for genetic and imaging methods make them uniquely suited to advancing translational neuroscience [12, 19].

5.9 Rat Models in Cardiovascular Research

Laboratory rats (*Rattus norvegicus*) have been instrumental in cardiovascular research since the early twentieth century. Their use became prominent following the establishment of the Wistar rat strain in 1894, facilitating systematic physiological and pathological studies [29]. As cardiovascular disease has become a leading cause of mortality globally, the anatomical and physiological similarities of rats to humans, especially in blood pressure regulation and cardiac structure, positioned them as an ideal model for mechanistic and therapeutic research. Commonly used rat strains in cardiovascular research with key features are summarized in Table 5.5.

One of the most notable contributions of the rat models is the field of hypertension. The SHR model, developed by Okamoto and Aoki in 1963, mimics human essential hypertension without external intervention. SHRs exhibit progressive blood pressure elevation, left ventricular hypertrophy, and end-organ damage [56]. They have been widely used to evaluate the efficacy of antihypertensive agents and study the pathophysiological changes associated with chronic hypertension. Another key model is the Dahl salt-sensitive rat, which develops hypertension in response to a high-sodium diet, mirroring salt-sensitive hypertension in humans [96].

In ischemic heart disease, the left anterior descending (LAD) coronary artery ligation technique has been established in rats as a reproducible model of myocardial infarction (MI). Following ligation, rats develop infarct-related remodelling and progressive heart failure—patterns paralleling those observed clinically in patients. Studies using this model are pivotal in the development of

Table 5.5 Commonly used rat strains in cardiovascular research

Rat strain	Key features
Wistar-Kyoto (WKY) rat	Normotensive control strain, widely used as baseline comparison for hypertensive models like spontaneous hypertensive rat (inbred strain) [95]
Spontaneously hypertensive rat (SHR)	The Spontaneously Hypertensive Rat (SHR) Colony, established in 1963, is a widely used rodent model for studying heart failure and human insulin-resistance syndromes due to its presence of essential hypertension, hyperinsulinemia, glucose intolerance, hypertriglyceridemia, and visceral obesity [73, 30]
Sprague-Dawley rat	Commonly used for general cardiovascular physiology, pharmacology, and toxicology studies
Fischer 344 (F344)	Is an inbred strain of rats and used in aging-related cardiovascular studies and cardiac toxicology research
F344BN	F344BN, a Fischer 344 × Brown Norway F1 hybrid rat strain, is used as a rodent model in aging research, examining sarcopenia, myocardial senescence, neuronal, retinal, and renal changes [83, 65]
Dahl salt-sensitive rat	Model for salt-sensitive hypertension and renal-cardiovascular pathophysiology [96]
Wistar rat	Versatile strain used in studies of vascular function, ischemia, and heart disease
Lewis rat	Used for cardiovascular inflammation models and transplant-related cardiovascular studies [41, 34]

angiotensin-converting enzyme (ACE) inhibitors such as captopril, which demonstrated reduced cardiac remodelling and mortality post-MI [63].

Beyond spontaneous and surgically induced models, genetic manipulation has advanced rat cardiovascular research. The application of CRISPR/Cas9 technology has allowed researchers to generate knockout or knock-in rat models of specific genes implicated in cardiovascular disorders. For example, rats with targeted deletions in *ApoE* or *LDLR* genes exhibit lipid abnormalities relevant to atherosclerosis research [42]. Unlike traditional mouse models, larger rats facilitate serial blood sampling and hemodynamic measurements, which make them advantageous for longitudinal studies.

The Zucker fatty rat carrying a mutation in the leptin receptor gene (*fa/fa*) represents a metabolic syndrome model with associated cardiovascular complications. Although these rats do not readily develop atherosclerosis, they are valuable for investigating the cardiac dysfunction linked to obesity, insulin resistance, and dyslipidaemia. Another strain, the JCR:LA-cp rat, develops spontaneous myocardial ischemia, arterial lesions, and insulin resistance, thus offering a more comprehensive model for studying cardiometabolic diseases [66].

From a physiological standpoint, rats have several distinct advantages. Their heart rate (300–500 bpm), blood pressure, and cardiac output values are within a manageable range for mimicking human conditions, and their larger body size allows for surgical manipulations, such as aortic banding or implantation of telemetry devices for continuous monitoring [59]. These attributes make rats especially suitable for investigating heart failure progression and pharmacological interventions.

Transgenic rats have been developed to study inherited cardiovascular disorders. For example, Ren-2 transgenic rats, which overexpress the renin gene, exhibit severe hypertension and cardiac hypertrophy [52]. Similarly, knockout models of BMPR2 and KCNK3 have been established to investigate pulmonary arterial hypertension, a condition with limited treatment options in humans [42].

In summary, rat models have significantly contributed to our understanding of cardiovascular diseases and therapeutic discovery. Classical models such as SHR and MI ligation rats remain relevant, while modern gene-edited strains continue to push the boundaries of translational research. Their continued use, balanced with ethical responsibility and technological refinement, ensures that laboratory rats remain a cornerstone of cardiovascular investigation.

5.10 Rat Models in Immunology and Infectious Disease Studies

Since the beginning of the twenty-first century, the world has witnessed multiple outbreaks of severe infectious diseases. These include the severe acute respiratory syndrome (SARS) coronavirus outbreak in 2003, the swine flu pandemic in 2009, the Middle East respiratory syndrome (MERS) coronavirus outbreak in 2012, the Zika virus epidemic in 2015, and the widespread Ebola outbreak in

West Africa from 2013 to 2016. Most recently, the COVID-19 pandemic, which began in 2019, has caused unprecedented global disruption, affecting lives and livelihoods on a massive scale [5]. Preparing for future pandemics remains a critical priority; however, attention must also be given to endemic and seasonal diseases such as influenza, a recurring and unpredictable virus from the Orthomyxoviridae family [13]. Climate change further complicates the landscape of infectious disease transmission by altering ecological conditions such as temperature, precipitation, and humidity. These changes influence disease distribution and transmission dynamics in ways that remain difficult to predict [77].

Rats are commonly used as models in immunological and infectious disease research due to their physiological and immunological similarities to humans. Infectious agents in rats can produce a range of clinical outcomes, from subclinical (asymptomatic) to overt disease. To accurately evaluate disease mechanisms and prevention strategies, studies should ideally involve large populations, such as entire colonies, rather than individual rats or small groups. While rats are valuable for testing the efficacy of therapeutic agents, it is important to recognize that antibiotics effective in humans may not always yield the same results in rats [9, 37]. Furthermore, even when laboratory rats are obtained from commercial vendors, they may harbour latent pathogens that become clinically significant under experimental stress, potentially skewing study outcomes [37].

An increasingly recognized factor influencing experimental outcomes is the rat microbiome. The gut microbiota plays a critical role in shaping immune function, modulating drug metabolism, and determining susceptibility to infectious agents. Variations in microbiome composition, affected by housing conditions, diet, and antibiotic exposure, can lead to substantial differences in immune responses and disease progression [20, 90]. As a result, standardizing or at least monitoring, the microbiome is now considered essential for ensuring reproducibility and translational relevance in preclinical infectious disease research.

Although rats are useful models for studying infectious diseases, certain limitations should be considered. Fever in rats is typically transient and not a reliable clinical indicator. Instead, signs such as weight loss, reduced food and water intake, and decreased activity are more reliable measures of disease severity. These physiological and behavioural changes should inform humane study endpoints. Additionally, elevations in serum cytokine levels can indicate disease progression and immune response severity [57]. Monitoring these physiological and immunological parameters helps ensure both ethical rigour and scientific validity in rat-based infectious disease studies.

Recent advancements in immunological reagents for rats further enhance their value in studying bacterial, viral, and fungal infections. Multiple specialized rat strains, including Lewis rats, American cotton rats, and diabetic rat strains, have been developed to support specific research needs. For instance, Wistar rats have been used successfully to evaluate the clinical efficacy and pharmacological profiles of novel antitubercular compounds for the treatment of tuberculosis [23].

5.11 Rat Models in Cancer Research

Animal models provide a more physiologically relevant environment for cancer research, allowing for the replication of tumour development under pathophysiological conditions, ensuring genomic heterogeneity in human cancers, and studying the complex tumour microenvironment, which includes immunocompetent and stromal cell populations. Because of these models, it is possible to conduct effective research on a variety of carcinogenesis-related topics that are difficult to examine in the human body [80]. In vivo studies now focus on induced and spontaneous models of disease in small animals, with more restrictions in large animals. Cancer incidence and development are influenced by external and internal factors. Induced cancer models use physical/chemical stimuli, such as light (irradiation) and genetic constructs, and are often developed using microsurgical techniques. Genetic engineering is the most efficient strategy for developing genetically programmed cancer models [58]. Experimental animal/rat models are crucial in cancer research for modelling malignant transformation and evaluating treatment effects. Current cancer rat models and cancer induction methods are summarized in Fig. 5.3.

Several rat strains are commonly utilized in cancer research, each with unique strengths and weaknesses. Sprague-Dawley, Wistar, and F344 are among the most commonly used laboratory rat strains. Comparative characteristics of common laboratory rat strains are summarized in Table 5.6. They have been selectively bred for ease of handling and high fecundity and are routinely used in carcinogenicity studies. The National Toxicology Program has recently adopted the Harlan Sprague-Dawley rat as

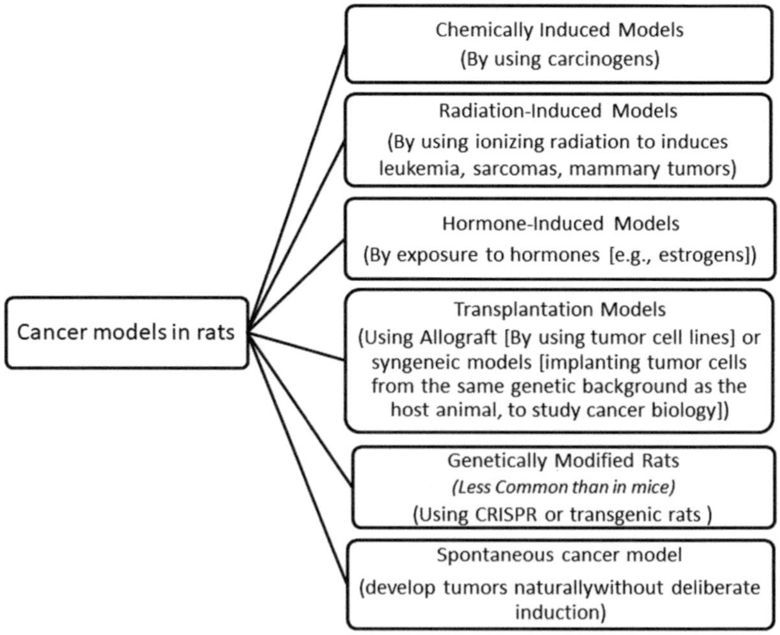

Fig. 5.3 Cancer models in rats

Table 5.6 Comparative characteristics of common laboratory rat strains

Feature	Sprague-Dawley (SD)	Wistar	Fischer 344 (F344)
Genetic type	Outbred	Outbred	Inbred
Origin/ History	SD rats developed from Wistar rats crossed with a variety of other laboratory and wild rats. SD rats developed by Sprague Dawley, Inc. in 1925 [15]	Established in 1906 at the Wistar Institute (standard outbred) [45]	Dr. Maynie Rose Curtis produced the Fischer 344 (F344) rat in 1920 from the 344th brother-sister mating of rats from the Fischer commercial breeder colony at Columbia University [47]
Body size/ morphology	Large, long-tailed albino	Medium-sized albino	Small, more uniform physique [49]
Growth rate	Fast-growing, large stature	Moderate growth	Slow growth, smaller adults
Temperament/ behaviour	Docile, calm, easy to handle	Calm but more active than SD [39]	Nervous/reactive; distinct locomotor/sensorimotor patterns [87]
Sensory/motor traits	Moderate sensitivity	Moderate	Higher mechanical sensitivity; shorter limbs [87]
Tumour incidence	High, variable spontaneous tumours [88]	Lower compared to SD/F344 [15]	High spontaneous tumour rate, common in aging studies [68]
Primary research uses	Toxicology, pharmacology, surgery, behaviour	Cardiovascular, metabolic, nutritional, pharmacology	Carcinogenicity, aging, immunology, long-term toxicology [7]
Breeding performance	High fertility; ~11 pups/ litter	Good fertility, moderate litter sizes	Lower fertility; smaller litters due to inbreeding
Limitations	Variable tumour development; substrain differences	Slightly higher follicular tumour rates in some substrains	Genetic uniformity limits external validity; anxiety/ nervousness

its standard strain. In Europe, the Han-Wistar rat is the pharmaceutical industry's preferred strain due to its lower incidence of spontaneous neoplasms [22].

The Sprague–Dawley and Wistar outbred strains are more susceptible to chemical carcinogens than other strains, such as the Marshall and August rats. Among the available strains for studying chemically induced breast carcinogenesis, the Sprague–Dawley rat is the most commonly used and is considered the most susceptible. Wistar strains, including the Wistar-Hannover, are also widely used in carcinogenicity studies [70].

5.12 Rat Models in Pain and Analgesia Research

According to the International Association for the Study of Pain, the experience of pain is described as a sensory and emotional experience associated with actual or potential tissue damage, and is undesirable. When a harmful stimulus damages a normal tissue, nociceptors which respond to such stimulus can become more sensitized, and prolonged activation may lead to increased tissue damage if not addressed promptly. In addition to nociceptor activities and nociceptive pathways, other neurophysiological processes caused by harmful stimuli are known as nociception. The inability to feel pain or respond to a painful stimulus is known as analgesia. The decrease in such sensations is known as hypoalgesia. On the other hand, hyperalgesia is the response exhibited when a stimulus causes an intensified pain response [76].

Measuring and comparing pain and analgesia between humans and animals presents a diverse range of problems mainly in animals as pain is poorly defined in them and its effects on behaviour are difficult to interpret. This is because pain involves complex sensory systems and is therefore difficult to measure. Additionally, pain is a very subjective and personal psychological experience which can be measured in humans but can be only indirectly measured in animals. Therefore, when using animals as models for analgesia studies, they are divided into animal models of nociception and antinociception [76].

Although there are many tests developed over the years to assess the nociception and pain responses in animal models, rats and mice are more commonly used in studies. In invasive biomedical studies involving animals, the use of anaesthetics and analgesics is standard. However, the use of analgesics for the behavioural responses to nociception stimulation studies is not permitted. As a result, strict ethical considerations must be applied when obtaining approval for models and protocols for studying pain in animals, with thoughtful consideration of the nature and magnitude of harmful stimulations and their duration. It is also important to consider the ability of the animal in question to tolerate such stimuli, as this may affect the evaluation of models for acute pain [76].

Several guidelines were in place by the Committee for Research and Ethical Issues of the International Association for the Study of Pain prior to conducting studies on experimental pain in conscious animals are listed as follows:

- Experiments must be carefully reviewed, and the possible outcomes clearly defined.
- The pain stimulus should, where possible, first be evaluated in humans.
- Abnormal behavioural changes in the animal from its typical behaviour should be given careful evaluation.
- For the purpose of the experiment, the minimal pain stimulus should be applied. If there are no conflicts with the objectives of the study, pain-relieving medications or techniques should be used.
- The number of animals used for the study should be kept at a minimum, and the study should be conducted in the shortest duration possible.

Although there are no problems in nociception tests that measure acute pain according to the stated guidelines, there are challenges in the application of these guidelines in chronic pain model studies. The selected animal model for the study will affect the outcome of the tests conducted; therefore, it is important to critically assess the results, with consideration of the conditions under which the tests were performed [76].

An injury to the nervous system, either by trauma, disease, or dysfunction of the endocrine system, can cause neuropathic pain, and current treatments pose a challenge. Pharmacotherapy using general pain-reducing drugs such as non-steroidal anti-inflammatory drugs (NSAIDs) and opiates has shown insufficient results in halting or reducing the pain [33].

Our current understanding of the pathophysiology of pain and the types of targets for drugs is based on studies conducted on peripheral nerve injury using rodent models. Although there are many supporting data, the translation of the obtained information into clinical applications is minimal. This may be due to the inability to differentiate how pain is processed in males versus females and the cellular response to an inflicted injury in addition to differences in pain processing in humans versus animals. However, improvements are being made in the evaluation of pain studies in animal models.

The use of rodent models treated with drugs to examine their efficacy in reducing withdrawal reactions to harmless stimuli has traditionally been investigated in the preclinical efficacy of therapeutic interventions, regardless of the type of nerve injury caused. However, there is a limited capability to predict clinical efficacy in rodents due to the recorded responses of the rodents when they are exposed to harmless stimuli reflecting the activation of their spinal reflexes instead of actual pain perception. Another challenge of using rodents in animal studies is how the male and female rodents recognise and react to pain. Female rodents do not have the need for microglia, as they lack the P2X4 receptors required for mechanical sensitivity towards pain, instead relying on the activation of the adaptive immune cells. Despite having differences in mechanical sensitivity to pain, the behavioural responses to nerve injuries due to downregulation of KCC2 and perturbation of Cl^- gradients in female rodents are identical to those of males. Over the years, differences in pain mechanisms experienced by males and females have emerged. An example is the α6 subunit, which is found on $GABA_A$ receptors, that is missing in female rodents. This highlights the importance of considering sex differences in the design of clinical trials and the development of therapies for both women and men [72].

5.13 Rat Models in Endocrine and Metabolic Disorders Studies

Metabolic disorders are a collective group of health disorders that are usually caused by either genetic or environmental factors and are not considered a single illness. Metabolic disorders can shift the body's processing and distribution of macronutrients adversely. The inability to process macronutrients such as carbohydrates, proteins, and fats properly could be caused by a recessive gene or genetic anomaly, while environmentally, it could be the type of surroundings a person grews up in, for example, poverty. Symptoms associated with metabolic disorder include weight loss or gain, seizures, jaundice, or fatigue, and these vary depending on the specific disorder. The general symptoms can be grouped into four types, which are acute symptoms, late-onset acute symptoms, progressive general symptoms, and permanent symptoms. The most common metabolic disorders affecting people globally are type 2 diabetes mellitus, obesity, and inflammatory bowel diseases [94].

Rat models have been specially bred, mostly through inbreeding with selected strains, in order to reproduce traits for metabolic syndrome studies. These rats possess characteristics that enable the identification of potential genes or mechanisms of metabolic syndrome similar to humans. For example, in obesity studies, it is difficult to translate waist circumference measurements from humans to rats, so obesity itself is often used as a defining feature. In addition, there are differences in lipid profiles between humans and rats, therefore, to define obesity as part of metabolic syndrome in rats, increased cholesterol levels and low-density lipoprotein levels are used [40].

Other rat models that are used include the obese spontaneously hypertensive rats, which exhibit a symptoms such as hypertriacylglycerolaemia, abdominal obesity, and hypertension. In this model, both male and female rats are used; however, the male rats show a more prominent increase in body weight compared to the females. Both genders are fed a standard diet but are shown to uniformly develop hyperlipidaemia. Test results have shown that there is a moderate increase in plasma cholesterol and a marked hypertriacylglycerolaemia. It is also noted that there is abnormal metabolism of the carbohydrates and proteins, with the exhibition of hyperphagia. There is also hyperinsulinism in the rats, which is correlated with either normal or slightly elevated blood glucose levels. To study hypertension-related diseases, stroke-prone rat models were developed to assess disorders such as nephropathy, cardiac hypertrophy, and atherosclerosis, which share similarities to human hypertension and stroke, which may lead to death. The metabolic disorder characteristics these rat models possess include the spontaneous development of hypertension, obesity, hyperleptinaemia, hyperlipidaemia, and hyperinsulinaemia [2].

Several rat strains are suitable for studying endocrine and metabolic disorders. Wistar rats, particularly the Spontaneously Diabetic Torii (SDT) and Goto-Kakizaki (GK) strains, are commonly used. Additionally, Zucker Diabetic Fatty (ZDF) rats, as well as Sprague-Dawley and Zucker Diabetic-Sprague-Dawley (ZDSD) rats, are frequently employed in studies of metabolic syndrome [60, 71, 91]. These genetically bred rat models provide clarity and a better understanding of the assessment of metabolic disorders and potential treatment options. They also provides further insight into as well into drug development and gene-editing technologies, which may help prevent or reduce the risk of developing metabolic diseases. In addition, the inbreeding and breeding process of these rat models demonstrate the potential for unfavourable genes to be inherited by subsequent generations, allowing metabolic disorders to be further studied.

5.14 Rat Models in Personalized Medicine

Genetic mutations cause numerous diseases, and DNA sequencing advances enable genetic diagnosis. Personalized medicine approaches require preclinical testing in cell and animal models with human target sequences. Rodent models may not always work due to species-specific genetic variation. Patient-derived cell models allow in vitro validation, while in vivo validation requires humanized models. These models guide clinical trial design, ensuring optimal dose, protein expression, and administration routes [1]. Rat models are crucial in personalized medicine, providing insights into disease mechanisms and drug response, with advances in Clustered Regularly Interspaced Short Palindromic Repeats (CRISPR) enabling development of genetically modified rat models.

5.15 Rat Models in Toxicology Studies

Rats are the most commonly used laboratory animals, especially in the field of toxicology [25]. Wistar and SD rats are commonly used in general toxicological studies. Rat strains such as Wistar, SD, F344, and Long-Evans are frequently employed in various toxicological assessments. This is because rats share a common pathways with humans when eliminating toxins from the body, and therefore, are considered among the most appropriate models for pharmacological and toxicological studies. Most drugs used in human clinical trials are first tested in rats during preclinical studies. An example of an intensively studied drug on rats is the use of tamoxifen in the treatment of 7,12-dimethylbenz[a]anthracene (DMBA)-induced rat mammary carcinoma models, and this has been translated into breast cancer therapy in humans [27].

One of the reasons why toxicity studies are conducted is to assess the safety and possible side effects of a substance. Therefore, guidelines are created for testing the safety of these potentially hazardous chemicals, which are decided by a body known as the Organisation for Economic Cooperation and Development (OECD) and the specifications of chemical testing methods are accepted internationally. These guidelines are used in the testing and assessment of hazardous chemicals by industry experts, academia, and government bodies. Before releasing a new drug into the market, it is important to conduct toxicity studies to establish its safety and efficacy as part of the development phase, and these studies should include acute, sub-acute, and chronic toxicity. It is important to note the particular drug formulation's safety and toxic characteristics so that it can be ranked and classified as per the Globally Harmonized System for classification of chemical substances and mixtures. Thus, acute oral toxicity is one of the important tests to be conducted as its data help establish the classification of hazards classification and labelling requirements and assess the risk that it may pose to the environment and human health, in addition to estimating the toxicity of said mixtures [64].

Based on OECD guidelines in acute oral toxicity studies, rodents, specifically female rats, are commonly used, and they are the only species used in detecting endocrine disruptors based on a global validation program [55]. This shows that rats play an important role in toxicity studies due to their similarity with humans. Although the laboratory rat is a suitable model for toxicology studies, upon exposure to acute intoxication, the rat will react by lowering its body core temperature using both physiological and behavioural mechanisms. This response differs from the human response when they experience acute intoxication [86]. However, on a positive note, this shows that the laboratory rat does experience responses to induced toxicity; however, it is not similar to humans, as they respond differently.

Furthermore, in toxicity testing, rats often exhibit hormetic dose–response patterns, where low doses of a toxic substance can trigger adaptive or beneficial biological effects, while higher doses produce adverse outcomes. Such biphasic responses have been observed in studies involving radiation, heavy metals, pesticides, and even pharmaceuticals [48]. For example, low doses of cadmium have been shown to upregulate antioxidant enzymes in rats, whereas high doses cause oxidative liver damage [17]. Additionally, subacute cadmium exposure at the lowest observed adverse effect level (LOAEL) has been reported to induce insulin resistance and disrupt the hormonal and metabolic functions of the liver-adipose axis in Wistar rats [67]. These findings highlight the importance of recognizing hormetic dose-response patterns for the accurate interpretation of toxicity data and the establishment of more precise safety thresholds in preclinical studies.

5.16 Rat Models in Developmental and Reproductive Toxicology Studies

In rats, there are five stages of sexual development, which are the neonatal period, infantile period, juvenile period, peripubertal period, and adolescent period [25].

The neonatal stage is characterized by Leydig cell reversion, the presence of small seminiferous tubules, and spermatogonial cells and Sertoli cells undergoing mitosis in the male rats. Furthermore, there is an absence of spermatids or spermatocytes. Upon reaching the infantile period, the sexual development in male rats starts to mature. This is shown by the Sertoli, Leydig, and spermatogonial cells undergoing proliferation instead of forming spermatocytes. Upon reaching the end of the infantile period, the blood-testis barrier is formed. The start of spermiogenesis and spermatogenesis occurs during the juvenile period. Upon the onset of puberty is the occurrence of the peripubertal period, whereby the rising gonadal hormones start to circulate, leading to sexual and reproductive maturity. By the end of puberty, it is observed that spermiogenesis is completed and the epididymis or vas deferens shows a visible presence of mature spermatozoa [25].

The main trait of the neonatal period in female rats is that oogonia and primordial follicles have undergone apoptosis. The infantile period in female rats involves development of secondary and early antral follicles with the presence of the zona pellucida. The juvenile period in female rats shows signs of enlargement of the antral follicles and the granulosa cells apoptosis. Signs of puberty upon reaching sexual maturity in female rats during the peripubertal period are marked by the observation of vaginal opening followed by irregular oestrous cycles for a few weeks before achieving full sexual maturity. The completion of puberty in female rats is shown by consistent oestrous cycles of 4- to 5 days [25].

In rats, the period for development between the neonatal and peripubertal period is between 35 and 55 days, in comparison to humans, which is between 11 and

14 years. Unlike humans, rodents reach adulthood after more than 60 days and are characterized by the maturation of the musculoskeletal system, while humans reach adulthood between 18 and 25 years. It is possible to assume that 11.8 days in rats is equivalent to one year in humans. Laboratory rats have an estimated lifespan of 2.5 to 3.5 years, based on reports in comparison to 80 years in humans. Female Wistar rats have been recorded to have higher longevity in contrast to male Wistar rats [25].

By understanding the developmental period of the rat model versus humans, a suitable timeline can be drawn when investigating pharmacology, toxicology, and drug efficacy, based on the development and age.

5.17 Applications of Rat Models in Drug Development and Translational Medicine

Rat models are essential for drug development and translational medicine because of their well-characterized biology, ease of handling, and physiological and genetic resemblance to humans. Before moving on to clinical trials, they are frequently used to evaluate the safety profile, pharmacokinetics, and pharmacodynamics of novel therapeutic agents. Rat models are also useful for assessing the effectiveness of medications in a variety of disease contexts, such as metabolic syndromes, cancer, cardiovascular disease, and neurological disorders. Because they closely resemble human pathophysiological conditions, genetically modified and disease-specific rat strains further increase the relevance of preclinical studies. These models help identify therapeutic targets and create individualized treatment plans by bridging the gap between fundamental research and clinical application.

However, most of the animal studies done are not successfully translated to humans, based on some compelling reports where 80% of the reports show no side effects or harm in animal studies, but are incapable of translating to humans. The results of the studies are influenced by the animal's body weight, age, and stage of development, as these factors can affect the metabolism of drugs, expression of genes, metabolic parameters, and other possible susceptible biological variables. The age of the animal is crucial as it should mimic the diseases in humans [25]. Another plausible explanation for the translational failures is that the viewpoint of the animal-to-human predictability concept is fundamentally flawed, or due to suboptimal experimental design and inadequate modelling. Another perspective to consider failures in translation is due to the concept of animal-to-human predictability being a misleading hypothesis, and it has not been scientifically tested thoroughly, as there are other important factors to consider, such as differences among the species and their molecular biology [43].

The physiological characteristics of rats have always been similar to those of humans, making it a favourable animal model of choice in biomedicine. It was used prior to smaller rodents such as mice due to its size, which allows for more diverse experimental procedures. For the past few decades, rats have been used as a model for studying the common diseases affecting humans, which are hypertension, aging, infectious diseases, cancer, and neurological disorders. In

1990, the first transgenic rats were developed to further mimic the diseases affecting humans [14, 16]. This milestone achievement helps to further a deeper understanding of the diseases and enables scientists to develop better drugs or treatment methods.

Examples of successful translational therapies are the treatment of heart failure, Alzheimer's disease, hypertension, and diabetes. The use of rat models in examining Alzheimer's disease led to the development of the drug Aricept, which show improvement in cognition in patients. Rat models in heart failure studies uncovered the usefulness of ACE inhibitors after the occurrence of a heart attack, which has lowered mortality rates due to heart failure by up to 19%. Additionally, advances in immunotherapies for many types of cancers, addiction studies, and stem cell therapy have used rat models [27]. These ongoing studies show how important the role of the rat model is in translational medicine and the advancements it has made in medicine.

5.18 Limitations and Challenges of Rat Models in Preclinical Research

Animal experimentation in science and medicine has proven beneficial, but concerns arise about its scientific validity and the lack of clear alternatives. From the 1920s to the 1970s, animal experiments were dominant in life sciences, especially drug development, but the landscape has evolved with the rise of complementary alternative testing methods due to challenges such as being resource-intensive, time-consuming, and often failing to accurately predict human responses. Drug discovery and development is a costly, high-risk process requiring 10–15 years and $1–2 billion for each new drug. However, 90% of candidates fail during trials and approval [74]. Around 75% of drugs from preclinical studies fail in phase II or phase III human clinical trials due to a lack of efficacy or safety. The dependence on non-human animal models is a major contributor to this failure. The biological differences between small animals and humans, such as organ structure, size, and regenerative capacity, often lead to failure in predicting drug efficacy and toxicity. Large animal models can improve predictive value, but they also introduce costs, time, and ethical considerations [35]. Limitations of rat models in preclinical research are summarized in Fig. 5.4.

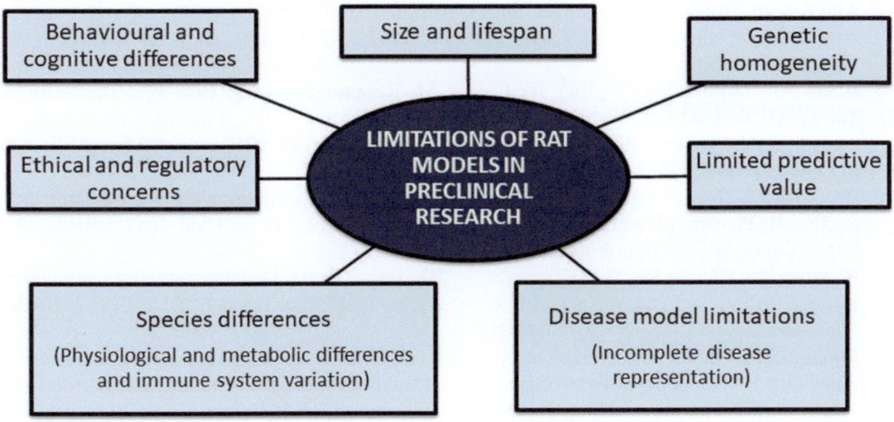

Fig. 5.4 Limitations of rat models in preclinical research

5.19 Conclusion

The laboratory rat (*Rattus norvegicus*) has been used as an animal model in biomedical research for over 150 years due to its genetic similarity to humans and complex behaviours. Rats have a relatively short lifespan, making them cost-effective for long-term studies. They are utilized in various fields such as neurology, cardiovascular research, immunology, infectious diseases, cancer, pain and analgesia, endocrine and metabolic disorders, personalized medicine, toxicology, developmental and reproductive toxicology, drug development, and translational medicine. While there are limitations such as species-specific differences and ethical concerns, advances in genetic engineering and behavioural assessment continue to enhance the precision and relevance of rat models in modern science. These developments are improving the translation of laboratory findings into clinical applications. With sustained innovation and responsible research practices, the laboratory rat remains an essential and evolving tool in the study of human health and disease.

Acknowledgement Tables 5.1, 5.2, 5.3, and 5.4 are reproduced with permission from Otto et al. [59]. Figure 5.1 is reproduced with permission from Maheshwari et al. [45].

References

1. Aartsma-Rus A, van Putten M. The use of genetically humanized animal models for personalized medicine approaches. Dis Model Mech. 2019;13(2):dmm041673. https://doi.org/10.1242/dmm.041673.
2. Aleixandre de Artiñano A, Miguel Castro M. Experimental rat models to study the metabolic syndrome. Br J Nutr. 2009;102(9):1246–53. https://doi.org/10.1017/S0007114509990729.
3. Andes D, Craig WA. Animal model pharmacokinetics and pharmacodynamics: a critical review. Int J Antimicrob Agents. 2002;19(4):261–8. https://doi.org/10.1016/s0924-8579(02)00022-5.

4. Antunes M, Biala G. The novel object recognition memory: neurobiology, test procedure, and its modifications. Cogn Process. 2012;13(2):93–110. https://doi.org/10.1007/s10339-011-0430-z.
5. Baker RE, Mahmud AS, Miller IF, Rajeev M, Rasambainarivo F, Rice BL, Takahashi S, Tatem AJ, Wagner CE, Wang L-F, Wesolowski A, Metcalf CJE. Infectious disease in an era of global change. Nat Rev Microbiol. 2022;20(4):193–205. https://doi.org/10.1038/s41579-021-00639-z.
6. Beconi MG, Howland D, Park L, Lyons K, Giuliano J, Dominguez C, Munoz-Sanjuan I, Pacifici R. Pharmacokinetics of memantine in rats and mice. PLoS Curr. 2011;3:RRN1291. https://doi.org/10.1371/currents.RRN1291.
7. Blystone CR, Elmore SA, Witt KL, Malarkey DE, Foster PMD. Toxicity and carcinogenicity of androstenedione in F344/N rats and B6C3F1 mice. Food Chem Toxicol. 2011;49(9):2116–24. https://doi.org/10.1016/j.fct.2011.05.026.
8. Bonauto SM, Greuel OM, Honeycutt JA. Playback of rat 22-kHz ultrasonic vocalizations as a translational assay of negative affective states: an analysis of evoked behavior and brain activity. Neurosci Biobehav Rev. 2023;153:105396. https://doi.org/10.1016/j.neubiorev.2023.105396.
9. Caneschi A, Bardhi A, Barbarossa A, Zaghini A. The use of antibiotics and antimicrobial resistance in veterinary medicine, a complex phenomenon: a narrative review. Antibiotics (Basel). 2023;12(3):487. https://doi.org/10.3390/antibiotics12030487.
10. Carvalheiro J, Seara-Cardoso A, Mesquita AR, de Sousa L, Oliveira P, Summavielle T, Magalhães A. Helping behavior in rats (Rattus norvegicus) when an escape alternative is present. J Comp Psychol. 2019;133(4):452–62. https://doi.org/10.1037/com0000178.
11. Chenouard V, Remy S, Tesson L, Ménoret S, Ouisse L-H, Cherifi Y, Anegon I. Advances in genome editing and application to the generation of genetically modified rat models. Front Genet. 2021;12:615491. https://doi.org/10.3389/fgene.2021.615491.
12. Chesselet M-F. A new look at animal models of neurological disorders. Neurotherapeutics. 2023;20(1):1–2. https://doi.org/10.1007/s13311-023-01366-4.
13. Choi YK. Emerging and re-emerging fatal viral diseases. Exp Mol Med. 2021;53(5):711–2. https://doi.org/10.1038/s12276-021-00608-9.
14. Christapher PV, Ganeson T, Chinni SV, Parasuraman S. Transgenic rodent models in toxicological and environmental research: future perspectives. J Pharmacol Pharmacother. 2022;13(3):258–65. https://doi.org/10.1177/0976500X221135691.
15. Clements PJM, Bolon B, McInnes E, Mukaratirwa S, Scudamore C. Animal models in toxicologic research: rodents. In: Haschek and Rousseaux's handbook of toxicologic pathology. Elsevier; 2022. p. 653–94.
16. Cozzi J, Fraichard A, Thiam K. Use of genetically modified rat models for translational medicine. Drug Discov Today. 2008;13(11–12):488–94. https://doi.org/10.1016/j.drudis.2008.03.021.
17. Das SC, Al-Naemi HA. Cadmium toxicity: oxidative stress, inflammation and tissue injury. ODEM. 2019;07(04):144–63. https://doi.org/10.4236/odem.2019.74012.
18. Deacon RMJ. Measuring motor coordination in mice. J Vis Exp. 2013;75:e2609. https://doi.org/10.3791/2609.
19. Ellenbroek B, Youn J. Rodent models in neuroscience research: is it a rat race? Dis Model Mech. 2016;9(10):1079–87. https://doi.org/10.1242/dmm.026120.
20. Ericsson AC, Franklin CL. The gut microbiome of laboratory mice: considerations and best practices for translational research. Mamm Genome. 2021;32(4):239–50. https://doi.org/10.1007/s00335-021-09863-7.
21. Fernández A, Josa S, Montoliu L. A history of genome editing in mammals. Mamm Genome. 2017;28(7–8):237–46. https://doi.org/10.1007/s00335-017-9699-2.
22. Foster JR, Frost D. The history of the rat. In: Boorman's pathology of the rat. Elsevier; 2018. p. 7–12.
23. Gaonkar S, Bharath S, Kumar N, Balasubramanian V, Shandil RK. Aerosol infection model of tuberculosis in wistar rats. Int J Microbiol. 2010;2010:426035. https://doi.org/10.1155/2010/426035.

24. Gencturk S, Unal G. Rodent tests of depression and anxiety: construct validity and translational relevance. Cogn Affect Behav Neurosci. 2024;24(2):191–224. https://doi.org/10.3758/s13415-024-01171-2.
25. Ghasemi A, Jeddi S, Kashfi K. The laboratory rat: age and body weight matter. EXCLI J. 2021;20:1431–45. https://doi.org/10.17179/excli2021-4072.
26. Harkness JE, Harkness JE, American College of Laboratory Animal Medicine, editors. Harkness and Wagner's biology and medicine of rabbits and rodents. 5th ed. Ames, Iowa, Chester: Wiley-Blackwell; American College of Laboratory Animal Medicine; 2010.
27. Hashway SA, Wilding LA. Translational potential of rats in research. In: The laboratory rat. Elsevier; 2020. p. 77–88.
28. Herson PS, Traystman RJ. Animal models of stroke: translational potential at present and in 2050. Future Neurol. 2014;9(5):541–51. https://doi.org/10.2217/fnl.14.44.
29. Hickman DL, Johnson J, Vemulapalli TH, Crisler JR, Shepherd R. Commonly used animal models. In: Principles of animal research. Elsevier; 2017. p. 117–75.
30. Holjak EJB, Savinova I, Nelson VL, Ogilvie LM, Ng AM, Edgett BA, Platt MJ, Brunt KR, Ask K, Simpson JA. An evaluation of cardiac health in the spontaneously hypertensive rat colony: implications of evolutionary driven increases in concentric hypertrophy. Am J Hypertens. 2022;35(3):264–71. https://doi.org/10.1093/ajh/hpab155.
31. Horner AE, Heath CJ, Hvoslef-Eide M, Kent BA, Kim CH, Nilsson SRO, Alsiö J, Oomen CA, Holmes A, Saksida LM, Bussey TJ. The touchscreen operant platform for testing learning and memory in rats and mice. Nat Protoc. 2013;8(10):1961–84. https://doi.org/10.1038/nprot.2013.122.
32. Isik S, Unal G. Open-source software for automated rodent behavioral analysis. Front Neurosci. 2023;17:1149027. https://doi.org/10.3389/fnins.2023.1149027.
33. Jaggi AS, Jain V, Singh N. Animal models of neuropathic pain. Fundam Clin Pharmacol. 2011;25(1):1–28. https://doi.org/10.1111/j.1472-8206.2009.00801.x.
34. Jones IKA, Orloff S, Burg JM, Haese NN, Andoh TF, Chambers A, Fei SS, Gao L, Kreklywich CN, Streblow ZJ, Enesthvedt K, Wanderer A, Baker J, Streblow DN. Blocking the IL-1 receptor reduces cardiac transplant ischemia and reperfusion injury and mitigates CMV-accelerated chronic rejection. Am J Transplant. 2021;21(1):44–59. https://doi.org/10.1111/ajt.16149.
35. Khalil AS, Jaenisch R, Mooney DJ. Engineered tissues and strategies to overcome challenges in drug development. Adv Drug Deliv Rev. 2020;158:116–39. https://doi.org/10.1016/j.addr.2020.09.012.
36. Kilkenny C, Browne W, Cuthill IC, Emerson M, Altman DG, NC3Rs Reporting Guidelines Working Group. Animal research: reporting in vivo experiments: the ARRIVE guidelines. Br J Pharmacol. 2010;160(7):1577–9. https://doi.org/10.1111/j.1476-5381.2010.00872.x.
37. Kohn DF, Barthold SW. Biology and diseases of rats. In: Laboratory animal medicine. Elsevier; 1984. p. 91–122.
38. Koolhaas JM. The laboratory rat. In: Hubrecht R, Kirkwood J, editors. The UFAW handbook on the care and management of laboratory and other research animals. 1st ed. Wiley; 2010. p. 311–26.
39. Krubaa P, Yogitha PS. Albino Wistar rats: advantages and limitations in biomedical research. SBV J Basic Clin Appl Health Sci. 2024;7(2):61–5. https://doi.org/10.4103/SBVJ.SBVJ_22_24.
40. Kwitek AE. Rat models of metabolic syndrome. Methods Mol Biol. 2019;2018:269–85. https://doi.org/10.1007/978-1-4939-9581-3_13.
41. Lai JCK, Tranfield EM, Walker DC, Dyck J, Kerjner A, Loo S, English D, Wong D, McDonald PC, Moghadasian MH, Wilson JE, McManus BM, Heart Stroke Foundation of British Columbia and Yukon, Program Project Grant Investigators. Ultrastructural evidence of early endothelial damage in coronary arteries of rat cardiac allografts. J Heart Lung Transplant. 2003;22(9):993–1004. https://doi.org/10.1016/s1053-2498(02)01163-4.
42. Lee JG, Ha CH, Yoon B, Cheong S-A, Kim G, Lee DJ, Woo D-C, Kim Y-H, Nam S-Y, Lee S-W, Sung YH, Baek I-J. Knockout rat models mimicking human atherosclerosis cre-

ated by Cpf1-mediated gene targeting. Sci Rep. 2019;9(1):2628. https://doi.org/10.1038/s41598-019-38732-2.

43. Leenaars CHC, Kouwenaar C, Stafleu FR, Bleich A, Ritskes-Hoitinga M, De Vries RBM, Meijboom FLB. Animal to human translation: a systematic scoping review of reported concordance rates. J Transl Med. 2019;17(1):223. https://doi.org/10.1186/s12967-019-1976-2.
44. Löscher W. The pharmacokinetics of antiepileptic drugs in rats: consequences for maintaining effective drug levels during prolonged drug administration in rat models of epilepsy. Epilepsia. 2007;48(7):1245–58. https://doi.org/10.1111/j.1528-1167.2007.01093.x.
45. Maheshwari R, Gupta A, Ganeshpurkar A, Chourasiya Y, Tekade M, Tekade RK. Guiding principles for human and animal research during pharmaceutical product development. In: Dosage form design parameters. Elsevier; 2018. p. 621–64.
46. Markicevic M, Savvateev I, Grimm C, Zerbi V. Emerging imaging methods to study whole-brain function in rodent models. Transl Psychiatry. 2021;11(1):457. https://doi.org/10.1038/s41398-021-01575-5.
47. Maronpot RR, Nyska A, Foreman JE, Ramot Y. The legacy of the F344 rat as a cancer bioassay model (a retrospective summary of three common F344 rat neoplasms). Crit Rev Toxicol. 2016;46(8):641–75. https://doi.org/10.1080/10408444.2016.1174669.
48. Mattson MP. Hormesis defined. Ageing Res Rev. 2008;7(1):1–7. https://doi.org/10.1016/j.arr.2007.08.007.
49. McCormick DL. Preclinical evaluation of carcinogenicity using the rodent two-year bioassay. In: A comprehensive guide to toxicology in preclinical drug development. Elsevier; 2013. p. 423–36.
50. Mingrone A, Kaffman A, Kaffman A. The promise of automated home-cage monitoring in improving translational utility of psychiatric research in rodents. Front Neurosci. 2020;14:618593. https://doi.org/10.3389/fnins.2020.618593.
51. Mukherjee P, Roy S, Ghosh D, Nandi SK. Role of animal models in biomedical research: a review. Lab Anim Res. 2022;38(1):18. https://doi.org/10.1186/s42826-022-00128-1.
52. Mullins JJ, Peters J, Ganten D. Fulminant hypertension in transgenic rats harbouring the mouse Ren-2 gene. Nature. 1990;344(6266):541–4. https://doi.org/10.1038/344541a0.
53. National Research Council, Institute for Laboratory Animal Research. Use of laboratory animals in biomedical and behavioral research. National Academies Press (US); 1988.
54. Neff EP. CRISPR improves prospects for transgenic rats. Lab Anim. 2019;48:167. https://doi.org/10.1038/s41684-019-0316-8.
55. OECD. Test no. 407: repeated dose 28-day oral toxicity study in rodents. OECD; 2008.
56. Okamoto K, Aoki K. Development of a strain of spontaneously hypertensive rats. Jpn Circ J. 1963;27:282–93. https://doi.org/10.1253/jcj.27.282.
57. Olfert ED, Godson DL. Humane endpoints for infectious disease animal models. ILAR J. 2000;41(2):99–104. https://doi.org/10.1093/ilar.41.2.99.
58. Onaciu A, Munteanu R, Munteanu VC, Gulei D, Raduly L, Feder R-I, Pirlog R, Atanasov AG, Korban SS, Irimie A, Berindan-Neagoe I. Spontaneous and induced animal models for cancer research. Diagnostics (Basel). 2020;10(9):660. https://doi.org/10.3390/diagnostics10090660.
59. Otto GM, Franklin CL, Clifford CB. Biology and diseases of rats. In: Laboratory animal medicine. Elsevier; 2015. p. 151–207.
60. Panchal SK, Brown L. Rodent models for metabolic syndrome research. J Biomed Biotechnol. 2011;2011:351982. https://doi.org/10.1155/2011/351982.
61. Parker CC, Chen H, Flagel SB, Geurts AM, Richards JB, Robinson TE, Solberg Woods LC, Palmer AA. Rats are the smart choice: rationale for a renewed focus on rats in behavioral genetics. Neuropharmacology. 2014;76(Pt B):250–8. https://doi.org/10.1016/j.neuropharm.2013.05.047.
62. Pettibone JR, Yu JY, Derman RC, Faust TW, Hughes ED, Filipiak WE, Saunders TL, Ferrario CR, Berke JD. Knock-in rat lines with Cre recombinase at the dopamine D1 and adenosine 2a receptor loci. eNeuro. 2019;6(5):ENEURO.0163-19.2019. https://doi.org/10.1523/ENEURO.0163-19.2019.

63. Pfeffer JM, Pfeffer MA, Braunwald E. Influence of chronic captopril therapy on the infarcted left ventricle of the rat. Circ Res. 1985;57(1):84–95. https://doi.org/10.1161/01.res.57.1.84.
64. Rameshwar GS, Venkatrao PU, Vithalrao NA, Pramodrao AM. OECD guidelines for acute oral toxicity studies: an overview. Int J Res Ayurveda Pharm. 2023;14(4):137–40. https://doi.org/10.7897/2277-4343.1404130.
65. Rice KM, Wu M, Blough ER. Aortic aging in the Fischer 344/NNiaHSd x Brown Norway/BiNia rat. J Pharmacol Sci. 2008;108(4):393–8. https://doi.org/10.1254/jphs.08r02cp.
66. Russell JC, Proctor SD. Small animal models of cardiovascular disease: tools for the study of the roles of metabolic syndrome, dyslipidemia, and atherosclerosis. Cardiovasc Pathol. 2006;15(6):318–30. https://doi.org/10.1016/j.carpath.2006.09.001.
67. Sarmiento-Ortega VE, Moroni-González D, Díaz A, Eduardo B, Samuel T. Oral subacute exposure to cadmium LOAEL dose induces insulin resistance and impairment of the hormonal and metabolic liver-adipose axis in Wistar rats. Biol Trace Elem Res. 2022;200(10):4370–84. https://doi.org/10.1007/s12011-021-03027-z.
68. Sass B, Rabstein LS, Madison R, Nims RM, Peters RL, Kelloff GJ. Incidence of spontaneous neoplasms in F344 rats throughout the natural life-span. J Natl Cancer Inst. 1975;54(6):1449–56. https://doi.org/10.1093/jnci/54.6.1449.
69. Seibenhener ML, Wooten MC. Use of the open field maze to measure locomotor and anxiety-like behavior in mice. J Vis Exp. 2015;96:e52434. https://doi.org/10.3791/52434.
70. Silva J, Faustino-Rocha AI, Duarte JA, Oliveira PA. Realistic aspects behind the application of the rat model of chemically-induced mammary cancer: practical guidelines to obtain the best results. Vet World. 2023;16(6):1222–30. https://doi.org/10.14202/vetworld.2023.1222-1230.
71. Singh R, Gholipourmalekabadi M, Shafikhani SH. Animal models for type 1 and type 2 diabetes: advantages and limitations. Front Endocrinol (Lausanne). 2024;15:1359685. https://doi.org/10.3389/fendo.2024.1359685.
72. Smith PA. Neuropathic pain; what we know and what we should do about it. Front Pain Res (Lausanne). 2023;4:1220034. https://doi.org/10.3389/fpain.2023.1220034.
73. Sone H, Takahashi A, Iida K, Yamada N. Disease model: hyperinsulinemia and insulin resistance part B—polygenic and other animal models. Trends Mol Med. 2001;7(8):373–6. https://doi.org/10.1016/S1471-4914(01)02072-X.
74. Sun D, Gao W, Hu H, Zhou S. Why 90% of clinical drug development fails and how to improve it? Acta Pharm Sin B. 2022;12(7):3049–62. https://doi.org/10.1016/j.apsb.2022.02.002.
75. Szpirer C. Rat models of human diseases and related phenotypes: a systematic inventory of the causative genes. J Biomed Sci. 2020;27(1):84. https://doi.org/10.1186/s12929-020-00673-8.
76. Tjølsen A, Hole K. Animal models of analgesia. In: Dickenson A, Besson J-M, editors. The pharmacology of pain. Berlin, Heidelberg: Springer Berlin Heidelberg; 1997. p. 1–20.
77. Van de Vuurst P, Escobar LE. Climate change and infectious disease: a review of evidence and research trends. Infect Dis Poverty. 2023;12(1):51. https://doi.org/10.1186/s40249-023-01102-2.
78. van Den Brandt J, Kovács P, Klöting I. Metabolic variability among disease-resistant inbred rat strains and in comparison with wild rats (Rattus norvegicus). Clin Exp Pharmacol Physiol. 2000;27(10):793–5. https://doi.org/10.1046/j.1440-1681.2000.03336.x.
79. van Lieshout ECC, Boonzaier J, Pel AJ, van Heijningen CL, Vink JJ, Visser-Meily JMA, van Tilborg GAF, Dijkhuizen RM. Translational value of skilled reaching assessment in clinical and preclinical studies on motor recovery after stroke. Neurorehabil Neural Repair. 2021;35(5):457–67. https://doi.org/10.1177/15459683211005022.
80. Vinnichuk YD, Platonov OM, Gryshchuk OO, Komisarenko SV. Experimental cancer rat models. UkrBiochemJ. 2025;97(1):5–24. https://doi.org/10.15407/ubj97.01.005.
81. Vorhees CV, Williams MT. Morris water maze: procedures for assessing spatial and related forms of learning and memory. Nat Protoc. 2006;1(2):848–58. https://doi.org/10.1038/nprot.2006.116.
82. Walf AA, Frye CA. The use of the elevated plus maze as an assay of anxiety-related behavior in rodents. Nat Protoc. 2007;2(2):322–8. https://doi.org/10.1038/nprot.2007.44.

83. Walker EM, Nillas MS, Mangiarua EI, Cansino S, Morrison RG, Perdue RR, Triest WE, Wright GL, Studeny M, Wehner P, Rice KM, Blough ER. Age-associated changes in hearts of male Fischer 344/Brown Norway F1 rats. Ann Clin Lab Sci. 2006;36(4):427–38.
84. Wang W, Teresa M, Cai J, Zhang C, Wong S, Yan Z, Khojasteh SC, Zhang D. Comparative assessment for rat strain differences in metabolic profiles of 14 drugs in Wistar Han and Sprague Dawley hepatocytes. Xenobiotica. 2021;51(1):15–23. https://doi.org/10.1080/00498254.2020.1795949.
85. Wassum KM, Ostlund SB, Maidment NT. Phasic mesolimbic dopamine signaling precedes and predicts performance of a self-initiated action sequence task. Biol Psychiatry. 2012;71(10):846–54. https://doi.org/10.1016/j.biopsych.2011.12.019.
86. Watkinson WP, Gordon CJ. Caveats regarding the use of the laboratory rat as a model for acute toxicological studies: modulation of the toxic response via physiological and behavioral mechanisms. Toxicology. 1993;81(1):15–31. https://doi.org/10.1016/0300-483x(93)90153-j.
87. Webb AA, Gowribai K, Muir GD. Fischer (F-344) rats have different morphology, sensorimotor and locomotor abilities compared to Lewis, Long-Evans, Sprague-Dawley and Wistar rats. Behav Brain Res. 2003;144(1–2):143–56. https://doi.org/10.1016/s0166-4328(03)00076-7.
88. Weber K. Differences in types and incidence of neoplasms in Wistar Han and Sprague-Dawley rats. Toxicol Pathol. 2017;45(1):64–75. https://doi.org/10.1177/0192623316672075.
89. Webster J. Animal welfare: a cool eye towards eden. 3rd ed. Wiley-Blackwell; 1995.
90. Wen L, Duffy A. Factors influencing the gut microbiota, inflammation, and type 2 diabetes. J Nutr. 2017;147(7):1468S–75S. https://doi.org/10.3945/jn.116.240754.
91. Wong SK, Chin K-Y, Suhaimi FH, Fairus A, Ima-Nirwana S. Animal models of metabolic syndrome: a review. Nutr Metab (Lond). 2016;13:65. https://doi.org/10.1186/s12986-016-0123-9.
92. Yankelevitch-Yahav R, Franko M, Huly A, Doron R. The forced swim test as a model of depressive-like behavior. J Vis Exp. 2015;97:52587. https://doi.org/10.3791/52587.
93. Yee BK, Singer P. A conceptual and practical guide to the behavioural evaluation of animal models of the symptomatology and therapy of schizophrenia. Cell Tissue Res. 2013;354(1):221–46. https://doi.org/10.1007/s00441-013-1611-0.
94. Zakir F, Mohapatra S, Farooq U, Mirza MohdA, Iqbal Z. Introduction to metabolic disorders. In: Drug delivery systems for metabolic disorders. Elsevier; 2022. p. 1–20.
95. Zhang-James Y, Middleton FA, Faraone SV. Genetic architecture of Wistar-Kyoto rat and spontaneously hypertensive rat substrains from different sources. Physiol Genomics. 2013;45(13):528–38. https://doi.org/10.1152/physiolgenomics.00002.2013.
96. Zhu Q, Hu J, Wang L, Wang W, Wang Z, Li P-L, Li N. Overexpression of MicroRNA-429 transgene into the renal medulla attenuated salt-sensitive hypertension in dahl S rats. Am J Hypertens. 2021;34(10):1071–7. https://doi.org/10.1093/ajh/hpab089.
97. The principles of humane experimental technique. Med J Australia. 1960;1(13):500. https://doi.org/10.5694/j.1326-5377.1960.tb73127.x.
98. National research council. Committee for the update of the guide for the care and use of laboratory animals. Guide for the care and use of laboratory animals. 8th ed. Washington (DC): National Academies Press (US). 2011.
99. Loeb W, Quimby F. editors. The Clinical Chemistry of Laboratory animals. 2nd ed. Taylor and Francis; Philadelphia: 1999.

The Mouse as a Cornerstone of Biomedical Research: A Biological Model for Human Health and Disease

6

Priya Mori, Hina Maniya, Mehul Chauhan, Swati Goswami, and Vijay Kumar

6.1 Introduction

The use of animal models in scientific inquiry has a long-standing history, tracing back to the fifth century B.C. However, it was during the nineteenth century that the practice of animal experimentation gained substantial momentum, particularly within the fields of medicine and biomedical sciences. Today, non-human animals are integral to research across institutions worldwide, where they serve as crucial tools for understanding human biology, pathophysiology, and the development of therapeutic interventions. The utility of animal models stems from their physiological and genetic similarities to humans. Even evolutionarily distant species, such as *Drosophila melanogaster* (fruit fly), *Danio rerio* (zebrafish), and *Caenorhabditis elegans* (nematode), have proven valuable in biomedical research due to shared biological processes and conserved gene functions. These organisms have enabled researchers to investigate fundamental mechanisms underlying disease and development [28, 57]. Among all model organisms, the mouse (*Mus musculus*) stands out

P. Mori · M. Chauhan · S. Goswami
Postbiotics and Foodomics Lab, Department of Microbiology, School of Science, RK University, Rajkot, Gujarat, India

H. Maniya
Postbiotics and Foodomics Lab, Department of Microbiology, School of Science, RK University, Rajkot, Gujarat, India

Department of Microbiology, Krishna School of Science, Dr. Kiran and Pallavi Patel Global University (KPGU), Vadodara, Gujarat, India

V. Kumar (✉)
Postbiotics and Foodomics Lab, Department of Microbiology, School of Science, RK University, Rajkot, Gujarat, India

Department of Microbiology and MLT, Bhagwan Mahavir College of Basic and Applied Sciences, Bhagwan Mahavir University, Surat, Gujarat, India

P. Kumar, V. Kothari (eds.), *Model Organisms in Biological Research*, https://doi.org/10.1007/978-981-92-0038-2_6

as the most widely used in biomedical research. Advances in genetic engineering and biotechnology have dramatically enhanced the mouse's utility as a model system. Technologies such as CRISPR/Cas9 gene editing, conditional knockouts, and humanized mouse models have opened new avenues for disease modeling and therapeutic testing. This acceleration in tool development has enabled more precise investigations into human disease mechanisms and potential treatments [12].

The foundational rationale for animal experimentation lies in the expectation that such research will ultimately benefit human health. The anatomical and physiological similarities between humans and other vertebrates support this approach, as many organs and biological systems are highly conserved. Consequently, certain animals develop diseases resembling human conditions, including diabetes, cancer, cardiovascular disorders, and neurological illnesses [31]. Animal models are indispensable in assessing the safety and efficacy of novel drugs and therapeutic strategies. They offer valuable insights into the potential adverse effects of investigational drugs, such as hepatotoxicity, teratogenicity, reproductive toxicity, and carcinogenicity [88].

Despite their scientific value, the use of mice in research is subject to strict ethical and regulatory oversight. In countries where applicable, Institutional Animal Ethics Committees (IAECs) or equivalent review bodies assess proposed mouse studies to ensure both scientific validity and humane treatment. These assessments are guided by national and international regulations rooted in the principles of the 3Rs—Reduction, Refinement, and Replacement originally proposed by [89, 101]. The 3Rs aim to minimize the number of animals used, refine procedures to reduce pain or distress, and encourage the use of alternative methods where feasible [47].

In recent years, a fourth "R"—Responsibility, has been introduced by ethicists and researchers, emphasizing the moral obligation of scientists not only to the animals involved but also to the integrity and societal impact of their research [101]. As such, ethical review processes now often operate under the 4Rs framework, underscoring a comprehensive commitment to humane, responsible, and scientifically rigorous animal research [62].

6.2 Why Mice? The Advantages of Using Mice in Research

Mice are among the most widely used animal models in biomedical research due to a combination of practical, genetic, and economic advantages. Their small size makes them easy to house and handle, reducing space requirements and facilitating large-scale studies [10]. Moreover, mice have a short generation time of 8–10 weeks, accelerating breeding and enabling the rapid study of heritable traits and multigenerational effects. Importantly, mice share approximately 95–98% of their genes with humans, and many of their physiological systems, such as the immune, nervous, and cardiovascular systems, are highly conserved, making them ideal for modeling human diseases [102]. The mouse genome is not only well-characterized but also fully sequenced, enabling precise genetic manipulation. This has led to the development of numerous genetically modified mouse strains, including transgenic, knockout, and knock-in models that allow researchers to study specific gene functions and replicate human disease conditions at the molecular level. Mice are also relatively cost-effective compared to larger animals, which makes them suitable for high-throughput screening and long-term experiments [1].

When compared to other model organisms such as rats, zebrafish, or non-human primates, mice often strike the best balance between genetic relevance and practicality. While rats are sometimes preferred for behavioral and cardiovascular studies due to their size and complex behavior, they lack the same breadth of genetic tools available for mice [5]. Non-human primates offer close genetic and physiological parallels to humans but raise greater ethical concerns, require more resources, and are subject to stricter regulations. Therefore, for many areas of biomedical research, especially those involving genetics, immunology, oncology, and metabolic diseases, mice remain the preferred model organism [54].

6.3 Types of Mouse Models (Fig. 6.1)

6.3.1 Inbred Strains

Inbred mouse strains are commonly employed in biomedical research (Fig. 6.1) because of their genetic homogeneity, which is established by sibling mating for 20 or more generations [78]. Such homozygosity at all loci, with the exception of

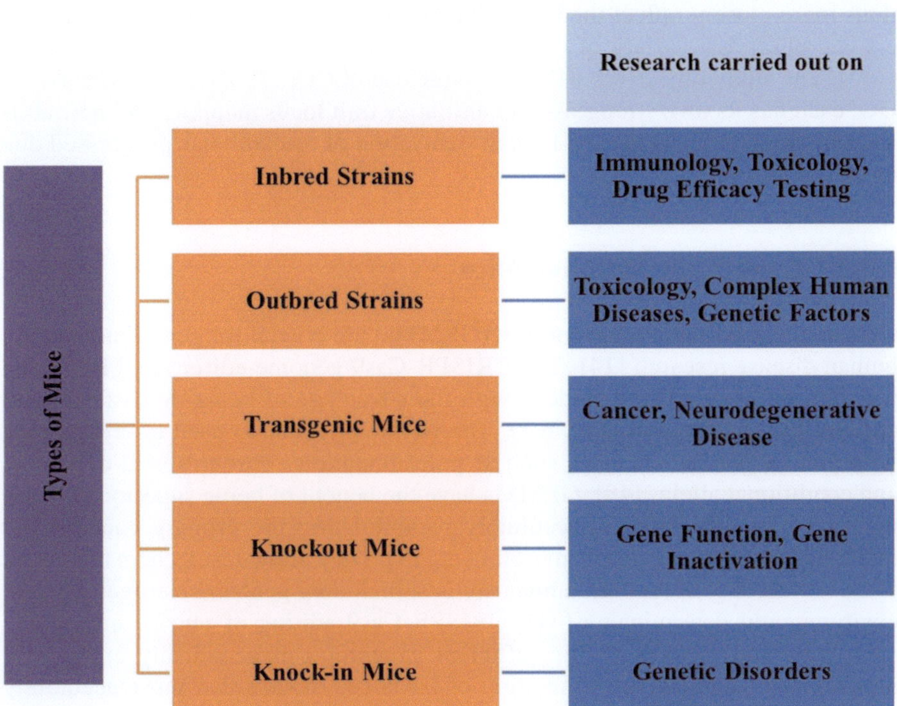

Fig. 6.1 Types of various mouse with its research applicability

sex differences, leads to predictable traits and homogeneous responses to experimental procedures [23]. The purity of the genetics in inbred strains can be detected through microsatellite markers and multiplex PCR [78]. Inbred strains offer advantages for studying complex genetic traits and age-related diseases, as demonstrated in a large-scale aging study involving 28 inbred strains. Their genetic diversity within strains and genomic similarity to humans make them useful in genome-wide association studies to determine candidate genes for disease pathogenesis [99]. Researchers should, however, keep in mind strain-specific differences and the influence of genetic background when conducting studies and interpreting outcomes [23].

6.3.2 Outbred Strains

To mimic human genetic variation in research, outbred mouse models like the Diversity Outbred (DO) and Collaborative Cross (CC) have been developed [92, 93]. Such models are more appropriate for drug development and toxicity research (Fig. 6.1) since they can reproduce a wider array of genetic variance compared to traditional inbred strains. CC mice are various inbred strains that share the same founders [92], while DO mice are genetically distinct individuals that arise by randomly mating pairs of eight founder strains. Since these populations more closely reflect the heterogeneity of the human population, researchers can observe a variety of responses to drugs and environmental challenges [107]. The study objectives will decide the use of CC, CC-RIX (F1 hybrids), or DO; each has its own strengths in quantitative trait locus mapping and heritability estimation [55]. Mouse-to-human translation of research has progressed due to these models.

6.3.3 Genetically Modified Mice

Genetically modified mouse models (GEMMs) are crucial for gene function and human disease research [14]. The CRISPR-Cas9 genome-editing tool has made it easier to develop these models, with the advantage of being much faster and cheaper than in the past (Fig. 6.1). The technology can be used to create mice with precise genetic changes, such as point mutations, chromosomal deletions, and conditional alleles [48]. GEMMs have the benefit of being superior to transplant and in vitro models, accurately recapitulating the primary cancers and maintaining the tumor microenvironment [14]. They allow researchers to control those variables confounding human studies, including genetic heterogeneity and environmental determinants [44]. CRISPR-Cas9 editing of murine embryonic stem cells and zygotes enables efficient generation of genetically engineered mouse lines, enabling the generation of advanced models that more accurately mimic human diseases [44, 68].

6.3.4 Transgenic Mice

Transgenic mice are produced by inserting foreign genes into their genomes using different techniques. The conventional method uses pronuclear microinjection, wherein DNA or gene-editing reagents are microinjected into the male pronucleus of a one-cell embryo [61]. This technique, established by the advances in reproductive science, genetics, and micromanipulation (Fig. 6.1), has been commonly employed for more than 35 years [91]. Other techniques are lentiviral transduction of spermatozoa with subsequent in vitro fertilization and embryo transfer, providing a less complex protocol for producing transgenic mice [15]. The TARGATT™ system was developed more recently using integrase-mediated targeted transgenesis via pronuclear microinjection. This method enables site-specific single-copy transgene insertion with high efficiency and accurate transmission over generations, evading random insertion and position effects problems inherent with conventional techniques [18].

6.3.5 Knockout Mice

Knockout mice are useful tools for investigating gene function in vivo by disabling specific genes (Fig. 6.1). The Cre/loxP system is widely utilized, where loxP sites are placed around target exons, which are subsequently removed when Cre enzyme is expressed [4]. Gene inactivation can be done through several methods, such as the introduction of artificial DNA sequences to stop transcription and translation [41]. Constitutive and conditional knockout mice have been generated to explore intricate biological processes, including those of progranulin in frontotemporal lobar degeneration [8]. Recently, the CRISPR/Cas9 system has also been used as a potent tool for producing knockout mouse embryonic stem cell lines. The process includes transfecting cells with paired CRISPR sgRNAs-Cas9 plasmids, leading to targeted gene knockout [103]. These methods enable scientists to investigate the impact of gene inactivation in diverse tissues and organs of the body.

6.3.6 Knock-In Mice

Knock-in mice, essential for the investigation of gene function, can be effectively produced by using CRISPR/Cas9-based genome editing methods (Fig. 6.1). These techniques enable accurate insertion or alteration of particular gene sequences. Easi-CRISPR, with the use of long single-stranded DNA donors, provides high efficiency (30–60%) for the generation of knock-in and conditional knockout models [74]. The two-cell homologous recombination (2C-HR)-CRISPR approach, in which CRISPR reagents are delivered at the two-cell stage of the embryo, increases efficiency (up to 95%) for large insertions [45]. CRISPR-READI is a combination of AAV-delivered HDR donor and Cas9/sgRNA RNP electroporation, allowing high-throughput production of diverse knock-in models [17]. These newer methods are highly superior to the traditional ones and enable the production of reporter mice, conditional knockouts, and gene replacements [33]. These methods have greatly transformed the field, rendering the production of intricate mouse models easier and more efficient.

Table 6.1 outlines different types of mouse models commonly used in biomedical research, detailing their specific advantages, relevance to human biology and disease, and key applications. This table serves to emphasize the critical role of each model in advancing our understanding of disease mechanisms and evaluating therapeutic strategies, supported by relevant literature references.

Table 6.1 Description of various mouse models with their application, advantages, and relevance to diseases

Mouse model type	Advantages in research	Relevance to human biology/disease	Key applications	References
Inbred strains	Genetic uniformity: Minimizes variation in experimental outcomes, enhancing statistical power and replicability. Enables consistent reactions to treatments or stimuli Stable phenotype: Reliable and stable physical and physiological traits Baseline for genetic manipulation: offers a consistent background to examine the impact of introduced genetic alterations	Although not an exact replication of human genetic variation, certain inbred strains can simulate some features of human physiology or disease with a specific genetic etiology Useful for studying fundamental biological processes without the confounding effect of genetic variation	Immunology studies (e.g., transplantation research) Toxicology studies require consistent responses Identifying specific gene functions Drug efficacy testing on a standardized genetic background	[96, 100]

(continued)

Table 6.1 (continued)

Mouse model type	Advantages in research	Relevance to human biology/ disease	Key applications	References
Outbred strains	Genetic diversity: More closely approximates human population genetic heterogeneity Expanded response spectrum: May detect a spectrum of responses to test conditions, with the possibility of identifying subpopulations with differences in susceptibility or resistance Mimicking human variability: Used for investigating multigenic, environmentally influenced diseases, as well as human populations	Better for modeling diseases with variable penetrance and expressivity seen in humans Useful for studies in toxicology and pharmacology to forecast the spectrum of responses of humans to drugs or exposures to the environment	Toxicology and pharmacology research aimed at exploring population-level effects Modeling complex human diseases with an important role of genetic background Determining genetic factors affecting susceptibility or resistance to diseases	[19, 35]
Transgenic mice	Gain-of-function studies: Enables examination of the outcomes of overexpression of a gene or expression of a novel gene Disease modeling: Can be designed to produce genes causing human diseases, offering models in which to investigate disease mechanisms and test therapy Reporter gene studies: Introduction of genes that produce detectable signals (e.g., fluorescent proteins) to monitor cells or gene expression	Extensively employed to simulate several human diseases through the introduction of disease-inducing genes or analysis of the implications of certain protein overexpression Can provide information about the function of particular genes in development and physiology	Cancer research (researching oncogenes) Neurodegenerative disease research (e.g., expressing mutant amyloid precursor protein in the Alzheimer's models) Developmental biology (examining the function of individual genes during development) Drug target validation	[21, 59]

(continued)

Table 6.1 (continued)

Mouse model type	Advantages in research	Relevance to human biology/disease	Key applications	References
Knockout mice	Loss-of-function studies: Permits the examination of the function of a given gene by watching what happens when it is not present Essential gene identification: Assists in identifying whether a gene is required for development, survival, or certain physiological processes Disease modeling: Knocking out genes that normally suppress disease can create models for studying those diseases	Crucial for understanding the function of individual genes in various biological pathways and disease processes Widely applied to simulate human diseases resulting from loss or malfunctioning of specific genes (e.g., cystic fibrosis)	Studying gene function and its role in specific pathways Modeling gene inactivation-caused genetic disorders Drug target discovery and validation Understanding developmental processes	[52, 64]
Knock-in mice	Accurate gene modification: Allows for investigation of the influence of individual mutations (e.g., point mutations, small deletions/insertions) on gene function and phenotype Humanization studies: Genes in mice can be substituted with human equivalents to provide more appropriate models of human diseases or drug screens Reporter knock-in: Reporter gene insertion into defined loci to investigate the regulation of endogenous gene expression	Extremely useful in modeling human genetic diseases due to individual mutations Enables the investigation of small alterations in protein function and their influence on disease Useful for preclinical testing of drugs targeting specific mutant proteins	Modeling genetic disorders with known mutations (e.g., Huntington's disease) Investigating the functional effects of individual amino acid substitutions in proteins Creating humanized mouse models for drug development Examining endogenous gene regulation	[34]

6.4 Applications of Mice in Biomedical Research

Mice serve as indispensable models for understanding a wide spectrum of human diseases due to their genetic, physiological, and anatomical similarities to humans, along with their well-characterized genome and availability of genetically modified strains. Their use spans numerous disease categories, enabling researchers to explore pathogenesis, evaluate potential therapies, and uncover molecular mechanisms of disease progression, as shown in Fig. 6.2.

6.4.1 Cancer

Mouse models have significantly advanced our understanding of cancer biology. Mice have been critical in identifying how cancer starts, spreads, and how it can be treated. Scientists can alter specific genes in mice to mimic human cancers and then test how tumors react to new drugs. Genetically engineered mouse models (GEMMs), such as transgenic and knockout strains, allow precise manipulation of oncogenes and tumor suppressor genes. Models like the MMTV-PyMT mouse have been instrumental in breast cancer research, while Apc^Min/+ mice are widely used

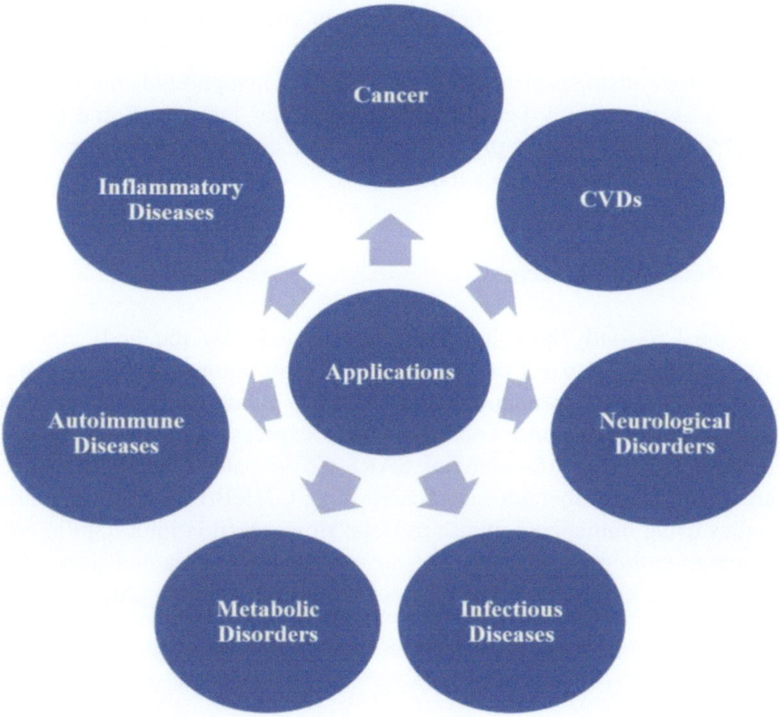

Fig. 6.2 Various applications of mice in biological research

to study colorectal cancer. Patient-derived xenograft (PDX) models, where human tumor tissues are implanted in immunodeficient mice, are also used to test drug responses and tumor behavior in a more clinically relevant setting [16, 98].

6.4.2 Cardiovascular Diseases

Mice are widely used to model atherosclerosis, hypertension, myocardial infarction, and heart failure. Mice are used to study heart attacks, high blood pressure, and clogged arteries. These models have shown how lifestyle, diet, and genes affect heart health. For example, ApoE−/-and LDLR−/− mice are essential for studying lipid metabolism and atherosclerosis. Surgical models like coronary artery ligation are used to study myocardial infarction and cardiac remodeling [40, 42].

6.4.3 Neurological Disorders

Mouse models help scientists understand diseases of the brain and nervous system, including memory loss, tremors, and seizures. Mouse models are pivotal in investigating neurodegenerative diseases such as Alzheimer's, Parkinson's, and Huntington's disease. For Alzheimer's, APP/PS1 and 5xFAD mice exhibit amyloid plaque formation and memory deficits. Parkinson's disease research often uses mice overexpressing α-synuclein or treated with MPTP to induce dopaminergic neuron loss. Models for autism spectrum disorder (ASD), schizophrenia, and epilepsy also exist and are used to study behavioral and molecular features. Some neurotropic viruses show different disease progression in mice compared to humans, affecting the translational accuracy of these models [73].

6.4.4 Infectious Diseases

Mice are used to study how infections spread and how the body fights them. Scientists also use mice to test vaccines and treatments before using them in humans. Immunocompetent and immunodeficient mice are widely used to study bacterial, viral, and parasitic infections. Transgenic mice expressing human receptors (e.g., hACE2 mice) were crucial for studying SARS-CoV-2 pathogenesis [20, 22, 46, 76]. Mice are also used in tuberculosis research (e.g., C3HeB/FeJ strain), malaria, and HIV studies using humanized mouse models. However, it is important to note that certain human pathogens, such as *Neisseria gonorrhoeae*, cannot infect standard laboratory mice, limiting the scope of preclinical investigations [70].

6.4.5 Metabolic Disorders

Mouse models are extensively used to study obesity, insulin resistance, and diabetes mellitus. They are used to study how the body controls sugar and fat, and what goes wrong in diabetes and obesity. The ob/ob and db/db mice are classic models for type 2 diabetes, while NOD mice develop autoimmune type 1 diabetes. These models provide insights into pancreatic beta-cell function, glucose metabolism, and therapeutic interventions. While murine models have been used to study type 2 diabetes, species-specific metabolic differences can influence disease progression and therapeutic response [63, 69, 106].

6.4.6 Autoimmune and Inflammatory Diseases

Mice have been instrumental in unraveling the mechanisms of diseases like rheumatoid arthritis, multiple sclerosis, and lupus. These models help researchers understand why the immune system sometimes attacks the body's own tissues, causing chronic illness. The collagen-induced arthritis (CIA) model is used for rheumatoid arthritis, while experimental autoimmune encephalomyelitis (EAE) mimics aspects of multiple sclerosis. These models help identify immunological pathways and test immunotherapies [65, 67].

Limitations of Mouse Models in Host–Pathogen Studies
While mice have been invaluable for infectious disease research, they are not universally susceptible to all human pathogens. Several microbes, such as *Neisseria gonorrhoeae*, *Mycobacterium leprae*, and human-specific strains of *Plasmodium falciparum*, do not establish natural infections in mice due to species-specific receptor incompatibilities or immune system differences. In some cases, pathogens can infect mice but fail to produce symptoms analogous to those seen in humans, as with certain influenza virus strains or *Helicobacter pylori*-induced gastritis. These discrepancies can limit the translational relevance of murine infection models and necessitate the use of humanized mice or alternative systems to better mimic human disease [9, 94].

Table 6.2 presents a comprehensive overview of how mouse models have contributed to the understanding and treatment of various targeted diseases. It highlights the specific therapies or medications employed, their intended applications, the role of mouse models in the research process, and the broader scientific and clinical impacts supported by relevant references.

Table 6.2 Role of mouse models in targeted disease research, therapeutic applications, and scientific impact

Targeted disease	Treatment/ medication/therapy	Used for	Mouse model contribution	Impact	References
Cancer therapies	Trastuzumab (Herceptin)	HER2-positive breast cancer	Genetically engineered mice that overexpress the HER2/neu gene were used to study breast tumor development and to test monoclonal antibodies targeting HER2	Trastuzumab became one of the first targeted cancer therapies and has significantly improved survival in breast cancer patients	[32]
	Immune checkpoint inhibitors (e.g., anti-PD-1/PD-L1 drugs)	Melanoma, lung, kidney, and bladder cancers	Mice engineered to express human immune checkpoint molecules helped test drugs that re-activate the immune system against tumors	Revolutionized cancer treatment with longer-lasting responses and fewer side effects than traditional chemotherapy	[6]
Vaccines	COVID-19 Vaccines (e.g., Pfizer, Moderna)	SARS-CoV-2 infection	Humanized ACE2 transgenic mice were vital for early testing of mRNA vaccines and understanding how the virus infects cells	Enabled rapid and safe vaccine development during the global pandemic	[20, 22, 46, 76]
	HPV Vaccine (Gardasil)	Preventing cervical and other HPV-related cancers	Mouse studies were essential to identify viral proteins that could stimulate protective immunity, leading to vaccine development	Dramatically reduced rates of HPV infection and cervical cancer	[2]
Diabetes treatments	Insulin Therapy and Drug Development (e.g., Metformin, GLP-1 agonists)	Type 1 and Type 2 diabetes	NOD mice mimicked autoimmune destruction of insulin-producing cells in type 1 diabetes, while ob/ob and db/db mice modeled obesity and type 2 diabetes	Led to a better understanding of disease mechanisms and the development of life-saving drugs that control blood sugar and reduce complications	[72]

Cardiovascular disease	Statins	Lowering cholesterol and preventing heart attacks	ApoE knockout mice were used to understand how cholesterol builds up in arteries and how statins can prevent this	Statins are now among the most widely prescribed drugs worldwide	[77]
Neurological disorders	Alzheimer's Drug Research	Investigating treatments for memory loss and cognitive decline	5xFAD and APP/PS1 mice are engineered to develop amyloid plaques and other brain changes seen in Alzheimer's patients	Helped screen numerous drugs, paving the way for clinical trials (e.g., aducanumab)	[30]
Autoimmune diseases	Multiple Sclerosis (MS) Drugs (e.g., Natalizumab, Fingolimod)	Reducing relapse and nerve damage in MS	AE mice allowed researchers to test immunomodulatory drugs that reduce inflammation in the nervous system	Several FDA-approved treatments for MS have been developed from mouse-based research	[97]

6.5 Ethical Considerations and Regulations

6.5.1 The Concept of the 3Rs (Replacement, Reduction, Refinement) and Their Importance in Ethical Animal Research

The 3Rs-Replacement, Reduction, and Refinement are fundamental principles in ethical animal research. These guidelines aim to minimize animal suffering while ensuring scientific rigor, forming the basis of regulations and institutional policies worldwide, as shown in Fig. 6.3.

6.5.1.1 Replacement: Reducing Dependence on Animal Models

Replacement encourages the use of non-animal methods whenever possible. Advances in in vitro cell culture systems, organ-on-a-chip technology, and in silico computer modeling offer alternative approaches that can replicate biological processes without using live animals. For example, human induced pluripotent stem cells (iPSCs) allow researchers to study disease mechanisms and drug responses with greater relevance to human biology. Although these technologies continue to evolve, some complex physiological processes, such as immune responses and systemic drug metabolism, still require animal models [24, 26].

6.5.1.2 Reduction: Minimizing the Number of Animals Used

Reduction aims to decrease the number of animals used in experiments by improving study design, statistical analysis, and data sharing. Strategies such as power calculations, refined imaging techniques, and longitudinal studies allow researchers to obtain meaningful data from fewer animals. Additionally, initiatives like the ARRIVE (Animal Research: Reporting of In Vivo Experiments) guidelines ensure that studies are designed and reported transparently, reducing unnecessary duplication. Collaborative research efforts and open-access databases further support this principle by enabling scientists to build on existing data rather than conducting redundant experiments [66].

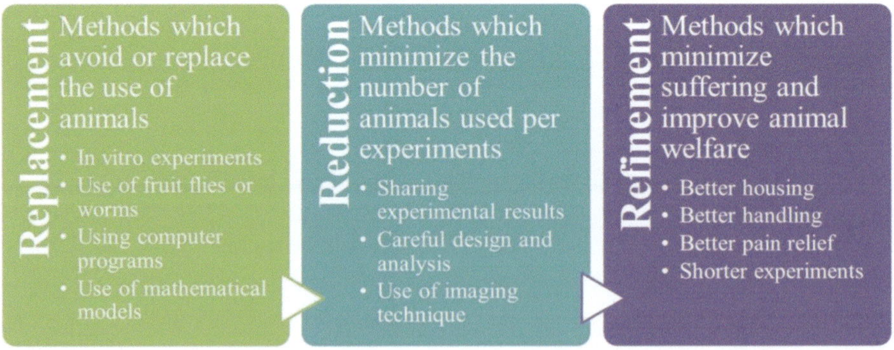

Fig. 6.3 Concept of the 3Rs in the experimental mouse model

6.5.1.3 Refinement: Enhancing Animal Welfare

Refinement focuses on improving experimental procedures to minimize stress, pain, and suffering. This includes the use of better anesthesia, analgesia, and humane euthanasia methods. Environmental enrichment, such as social housing, nesting materials, and exercise opportunities, helps maintain natural behaviors, reducing physiological stress that could otherwise affect experimental outcomes. Organizations like the National Research Council (NRC) and the Association for Assessment and Accreditation of Laboratory Animal Care (AAALAC) provide guidelines for proper housing, handling, and veterinary care, ensuring high welfare standards in research facilities [13].

6.5.2 Implementation of the 3Rs and Future Perspectives

Many countries have incorporated the 3Rs into legal frameworks. For instance, the EU Directive 2010/63/EU enforces strict ethical regulations, while institutions like NC3Rs (UK) fund research into alternative models. Despite progress, some areas of biomedical research, such as neuroscience, immunology, and infectious diseases, still require whole-animal models to capture complex interactions within living systems. However, continuous advancements in 3D bioprinting, artificial intelligence-driven modeling, and human-on-a-chip technologies offer promising alternatives that may further reduce reliance on animal testing in the future [43].

6.5.3 The Role of Institutional Animal Care and Use Committees (IACUCs) or Equivalent Ethics Review Boards in Ensuring the Humane Treatment of Research Animals

Ethical and regulatory oversight ensures that the use of mice in research adheres to high standards of animal welfare. In countries where applicable, Institutional Animal Ethics Committees (IAECs) in India, IACUCs in the United States, or equivalent review bodies evaluate proposed mouse studies for compliance with ethical guidelines, legal frameworks, and the principles of the 3Rs-Reduction, Refinement, and Replacement, originally proposed by Russell and Burch (1959). Before initiating a study, researchers must justify the choice of mice as the model organism and outline measures to minimize pain and distress. These committees also ensure that murine-specific housing, enrichment, handling, and veterinary care meet welfare standards. Periodic inspections and protocol reviews help maintain compliance, and mouse-specific training for research staff reinforces best practices. By integrating ethical considerations with scientific objectives, these oversight mechanisms safeguard the humane treatment of mice while enabling high-quality, reproducible research [75].

6.5.4 The Regulatory Frameworks and Guidelines That Govern Animal Research in India

In India, the use of animals in scientific research is governed by a well-defined legal and ethical framework to ensure humane treatment and responsible experimentation. The Prevention of Cruelty to Animals Act (PCA), 1960, serves as the foundation for these regulations, with oversight provided by various committees and regulatory bodies. The Committee for the Purpose of Control and Supervision of Experiments on Animals (CPCSEA), under the Ministry of Environment, Forest and Climate Change (MoEFCC), is the primary authority overseeing animal research. All institutions conducting research, breeding, or education involving animals must be registered with CPCSEA and adhere to its guidelines. CPCSEA mandates the establishment of Institutional Animal Ethics Committees (IAECs) within research institutions, ensuring that all animal experiments comply with ethical principles, including the 3Rs (Replacement, Reduction, and Refinement) [56].

The Breeding of and Experiments on Animals (Control and Supervision) Rules, 1998, further regulate the care and use of animals in research, requiring institutions to maintain proper housing, nutrition, and health monitoring. Experiments on large animals, such as dogs and monkeys, require special approval at the national level. IAECs play a crucial role in evaluating experimental protocols, monitoring animal welfare, and ensuring compliance with national guidelines. Additionally, other organizations such as the Indian National Science Academy (INSA), the Medical Council of India (MCI), and the Pharmacy Council of India (PCI) have established ethical guidelines encouraging reduced animal use in biomedical and pharmaceutical research [85].

India has also made significant progress in promoting alternatives to animal testing. The ban on cosmetic animal testing (2014) and the prohibition of importing animal-tested cosmetics (2015) highlight the country's commitment to ethical research. Emerging alternatives, such as in vitro models, computational simulations, and human-based studies, are being encouraged to reduce reliance on animal experimentation. Despite these advancements, animal models remain necessary in certain research areas, such as studying complex physiological responses, drug development, and disease modeling. However, with continued advancements in organoids, 3D cell cultures, and artificial intelligence-driven research, India is moving towards a future where animal use in research will be minimized while maintaining scientific progress [36, 51, 81].

6.5.5 Public Concerns About Animal Research and the Importance of Transparency and Open Communication

Public concerns about animal research primarily revolve around ethical considerations, animal welfare, and the necessity of using live animals in scientific studies. Many advocacy groups and individuals question the moral justification of subjecting animals to experimentation, particularly when alternative methods are being developed. Transparency and open communication play a crucial role in addressing these concerns and fostering public trust [84].

Institutions and researchers must actively engage with the public by providing clear, evidence-based explanations of why animal research is sometimes necessary, what measures are taken to minimize animal suffering, and how such research contributes to medical and scientific advancements. Public outreach efforts, such as educational programs, open-access research publications, and collaborations with ethics committees, can help bridge the gap between scientific institutions and society. Moreover, regulatory bodies and research institutions should establish and adhere to strict guidelines that ensure humane treatment, ethical justification, and continuous efforts to reduce, refine, and replace animal research where possible [83].

The implementation of transparency measures, such as public reporting on animal use statistics, ethical review processes, and institutional compliance with international standards, can further alleviate public concerns. Additionally, fostering a culture of open dialogue through public forums, media engagement, and collaboration with animal welfare organizations can help address misconceptions and promote informed discussions on the ethical implications of animal research. By embracing transparency, the scientific community can enhance public confidence in the necessity, integrity, and ethical commitment of biomedical research [39, 80].

6.6 Housing and Husbandry: Ensuring Animal Welfare

6.6.1 The Importance of Providing Appropriate Housing, Environmental Enrichment, and Care for Research Mice

Ensuring proper housing and husbandry for research mice is fundamental to their welfare and the integrity of scientific data. Mice are nocturnal, highly social animals that exhibit behaviors such as burrowing, climbing, foraging, and nesting. Housing environments should facilitate these natural behaviors to promote well-being. Social housing is crucial, as mice thrive in stable groups that prevent stress and abnormal behaviors arising from isolation. Environmental conditions, including appropriate lighting, sound levels, temperature, and humidity, also play a significant role [82]. Light levels should be low to protect light-sensitive strains, and ultrasonic noise from sources like dripping taps and computer monitors should be minimized to prevent stress [90]. The ideal room temperature for mice is between 20 °C and 24 °C (68–75 °F), with relative humidity maintained at 45–65%, but providing nesting materials allows them to regulate their microenvironment effectively [37].

Environmental enrichment is essential for allowing mice to express species-specific behaviors and reducing stress. Providing nesting materials, such as shredded paper or cotton squares, helps them build nests, which are crucial for thermoregulation and comfort. Hiding structures like igloos or tunnels offer a sense of security, while foraging opportunities, such as puzzle feeders, encourage natural food-seeking behaviors. Additionally, running wheels and climbing structures provide physical activity, supporting overall health [95]. Consistent and humane husbandry practices further contribute to maintaining animal welfare and ensuring reliable research outcomes. Non-aversive handling techniques, such as tunnel handling or cupping, reduce stress and improve the human-animal relationship. Regular health monitoring helps identify signs of illness or distress, enabling timely intervention. Proper sanitation and cage maintenance prevent disease outbreaks, though over-cleaning should be avoided to preserve scent markings essential for social interactions [71]. Training mice to cooperate with procedures using gentle techniques and food reinforcements can also minimize stress during experimental manipulations. By prioritizing appropriate housing, enrichment, and humane husbandry, researchers can improve the well-being of laboratory mice while ensuring scientifically reliable and ethically sound research outcomes [87].

6.6.2 Environmental Factors Can Influence Research Outcomes and the Need for Standardized Conditions

Environmental factors such as light exposure, noise pollution, and social interactions significantly influence the physiological and behavioral responses of laboratory mice, thereby impacting research outcomes. Disruptions to the light-dark cycle can affect glucose metabolism, immune responses, and cognitive function in mice. Similarly, noise pollution, including ultrasonic sounds imperceptible to humans, can elevate stress hormones and alter behavior, influencing experimental results in neuroscience and behavioral studies [11].

Social interactions among mice are also critical for maintaining stable physiological and behavioral responses. Mice housed in isolation often exhibit increased anxiety-like behaviors, altered immune function, and changes in brain neurochemistry, which can compromise study outcomes. Conversely, overcrowding can lead to social stress, aggression, and increased corticosterone levels, further complicating data interpretation [29].

Given these factors, strict adherence to standardized environmental conditions is essential for ensuring consistency in experimental results. Researchers must carefully regulate temperature, humidity, lighting, noise levels, and social interactions to minimize confounding variables. Institutions and regulatory bodies, such as the National Research Council (NRC) and the Association for Assessment and Accreditation of Laboratory Animal Care (AAALAC), provide guidelines to maintain optimal environmental conditions for laboratory mice. Standardized protocols, along with continuous monitoring and refinement of husbandry practices, not only enhance data reliability but also improve animal welfare, reinforcing the ethical principles of biomedical research [38].

The background microbiome of laboratory mice is another critical but often overlooked factor influencing experimental outcomes. Most laboratory mice are conventionally raised and harbor complex microbial communities that vary between facilities, suppliers, and even rooms within the same institution [27, 50]. These microbial differences can significantly alter immune development, metabolic function, and susceptibility to disease, thereby affecting the reproducibility and interpretation of results. For example, variations in gut microbiota composition have been shown to influence responses to infection, cancer progression, and the efficacy of immunotherapies [25]. In contrast, gnotobiotic mice reared under germ-free conditions or colonized with defined microbial consortia offer greater control over microbiome variables, but are less commonly used due to higher maintenance costs and specialized housing requirements. Recognizing and, where feasible, standardizing or documenting microbiome composition is essential to improve experimental reliability and cross-study comparability.

6.6.3 The Importance of Minimizing Stress and Pain in Research Animals

Minimizing stress and pain in research animals is a fundamental ethical obligation and a critical factor in ensuring the reliability of experimental results. Stress and pain can significantly alter an animal's physiological and behavioral responses, leading to skewed data that may compromise the validity of scientific findings. Stress-induced changes in hormonal levels, immune responses, and metabolic processes can confound research outcomes, making it difficult to distinguish between genuine experimental effects and stress-related artifacts. Moreover, persistent pain or distress can impact appetite, sleep patterns, and social interactions, further affecting an animal's overall health and behavior. This is particularly concerning in studies related to neuroscience, pharmacology, and behavioral research, where even slight variations in stress levels can lead to misinterpretation of results [60].

To mitigate these issues, researchers and regulatory bodies emphasize the principle of the "Three Rs"—Replacement, Reduction, and Refinement. Refinement involves the continuous improvement of animal housing, handling, and experimental procedures to minimize pain and distress. This includes providing appropriate environmental enrichment, such as nesting materials, social housing to reduce isolation stress, and using proper anaesthesia and analgesia for procedures that might cause pain. Furthermore, handling techniques such as tunnel handling for mice, rather than tail handling, have been shown to reduce stress and improve the reliability of behavioral studies [79].

6.7 Alternatives to Mouse Research and Future Directions

6.7.1 Ongoing Efforts to Reduce Reliance on Mice in Research

Efforts to reduce the use of mice in biomedical research have gained momentum through the development of refined alternative methods tailored to areas traditionally reliant on murine models. In toxicology, high-throughput in vitro assays using murine hepatocytes and liver organoids have replaced certain acute toxicity tests, significantly lowering mouse numbers [7, 105]. In neurobiology, primary neuronal cultures derived from mouse embryonic stem cells are increasingly used to evaluate neurotoxicity and synaptic function, replacing some live-animal behavioral assays [53]. Similarly, intestinal organoids generated from mouse tissue now allow researchers to investigate gut physiology, host–pathogen interactions, and drug absorption without requiring large numbers of live mice [105].

6.7.1.1 In Vitro Murine-Derived Models

Murine-derived in vitro systems, such as three-dimensional brain organoids and heart-on-a-chip devices seeded with mouse cardiomyocytes, are emerging as powerful tools for modeling organ-specific physiology and disease [7, 105]. These approaches provide data that closely correlate with in vivo murine responses, enabling preliminary screening before confirmation in a reduced set of live animals. In immunology, ex vivo assays using murine splenocytes or lymph node cells are replacing certain vaccine potency tests, cutting down the total number of mice required while still preserving immunological relevance [53].

6.7.1.2 In Silico Models for Murine Research

Computational modeling tailored to murine physiology is increasingly applied in pharmacokinetics, oncology, and immunology. By integrating genetic, metabolic, and phenotypic datasets from mice, these models can predict disease progression, drug efficacy, and toxicity [86]. Such predictions allow researchers to refine experimental design and limit in vivo validation to the minimum number of animals needed for statistical robustness.

6.7.1.3 Targeted Databases for Alternative Methods in Mouse Research

Resources such as the Non-Animal Technologies (NAT) Database now include sections devoted to alternatives applicable to murine studies, such as validated in vitro immunotoxicity assays using murine cells [3, 58]. By consolidating these alternatives, the database supports researchers in selecting replacement or reduction strategies while maintaining experimental validity.

6.7.2 Areas Where the Mouse Remains the Model of First Choice

Despite the emergence of alternatives, the mouse continues to be the primary model in several critical research domains. In neurodegenerative disease research, transgenic models like the APP/PS1 mouse for Alzheimer's disease remain irreplaceable for studying complex neuronal networks and cognitive decline [49]. In immunology, genetically defined inbred strains provide unparalleled reproducibility for infection and autoimmune disease studies [53]. Oncology heavily relies on genetically engineered mouse models (GEMMs) and patient-derived xenografts (PDXs) to replicate tumor biology in a whole-organism context [104].

6.7.3 Limitations of Current Alternatives in Mouse-Focused Research

While organoid cultures, murine-derived cell assays, and computational models have reduced live mouse usage, they cannot replicate whole-body interactions, systemic immune responses, or complex behaviors [53]. For example, in vitro assays fail to capture the intricate interplay between organ systems, and in silico predictions are limited by the quality and completeness of available datasets [86]. These limitations reinforce the need for carefully justified use of live mice in studies that require integrated physiological responses.

6.7.4 Future Directions in Refining the Use of Mice

Technological advancements are enabling more precise and ethical use of mice in research. Genome-editing tools such as CRISPR/Cas9 have streamlined the generation of targeted mouse models, allowing fewer animals to yield more informative results [53]. Non-invasive imaging technologies and longitudinal study designs enable repeated measurements in the same animal, reducing overall animal numbers [104]. Coupled with growing availability of murine-derived organoids, tissue chips, and computational simulations, these innovations are expected to refine experimental design and align with the Reduction and Refinement principles of the 3Rs [7, 105].

6.8 Conclusion

Mice have undeniably become one of the most valuable and versatile model organisms in biomedical research. Their genetic, physiological, and anatomical similarities to humans, combined with a well-characterized genome and the availability of advanced genetic tools, have enabled groundbreaking discoveries across numerous disciplines, from cancer and cardiovascular disease to infectious, neurological, and metabolic disorders. This chapter has highlighted the diverse types of mouse models, including inbred, outbred, transgenic, knockout, and knock-in strains, and their respective advantages in simulating human diseases, elucidating gene function, and evaluating novel therapies. Mouse models have not only contributed to the foundational understanding of disease mechanisms but have also been instrumental in translating preclinical findings into real-world medical advances, including the development of life-saving drugs, targeted cancer therapies, and vaccines. The ethical framework surrounding animal research, particularly the principles of the 3Rs (Replacement, Reduction, and Refinement), has continued to evolve alongside scientific progress, guiding researchers to design humane, responsible, and high-quality experiments. While the development of alternative approaches such as organ-on-a-chip, in vitro assays, and computer simulations represents an exciting frontier in biomedical science, these technologies are not yet sufficient to fully replace animal models in many complex research areas. Therefore, a balanced integration of animal and non-animal methods, informed by ethical responsibility and scientific rigor, remains essential. As we look to the future, continued innovations in genome editing, artificial intelligence, and personalized medicine will further enhance the utility of mouse models. At the same time, the global scientific community must remain committed to refining ethical standards and promoting transparency in research. Through this synergy of scientific innovation and ethical responsibility, mouse models will continue to play a pivotal role in advancing human health and medical science.

References

1. Adhikari S, Hackett R. Travelling through time with Disease Models & Mechanisms. Dis Model Mech. 2025;18:dmm052259. https://doi.org/10.1242/dmm.052259.
2. Ahn J, Peng S, Hung C, Roden RBS, Best SR. Prophylactic immunization with human papillomavirus vaccines induces oral immunity in mice. Laryngoscope. 2018;128(1):16–20. https://doi.org/10.1002/lary.26772.
3. Akkermans A, Chapsal JM, Coccia EM, Depraetere H, Dierick JF, Duangkhae P, Goel S, Halder M, Hendrikseni C, Levis R, Pinyosukhee K, Pullirschl D, Sanyal G, Shin L, Sitrin R, Smith D, Stickings P, Terao E, Uhlrich S, Viviani L, Webster J. Animal testing for vaccines: implementing replacement, reduction and refinement: challenges and priorities. Biologicals. 2020;68:92–107. https://doi.org/10.1016/j.biologicals.2020.07.010.
4. Antonson P, Humire P, Gustafsson JÅ. Estrogen receptor-α knockout mice. In: Eyster KM, editor. Estrogen receptors, Methods in molecular biology, vol. 1366. New York: Humana Press; 2016. p. 425–30. https://doi.org/10.1007/978-1-4939-3127-9_33.
5. Barbee RW, Turner PV. Incorporating laboratory animal science into responsible biomedical research. ILAR J. 2019;60(1):9–16. https://doi.org/10.1093/ilar/ilz017.
6. Barham W, Hsu M, Liu X, Harrington SM, Hirdler JB, Gicobi JK, Zhu X, Zeng H, Pavelko KD, Yan Y, Mansfield AS, Dong H. A novel humanized PD-1/PD-L1 mouse model permits direct comparison of antitumor immunity generated by FDA-approved PD-1 and PD- L1 inhibitors. Immuno Horizons. 2023;7(1):125–39. https://doi.org/10.4049/immunohorizons.2200054.
7. Bhatia SN, Ingber DE. Microfluidic organs-on-chips. Nat Biotechnol. 2014;32(8):760–72. https://doi.org/10.1038/nbt.2989.
8. Boddaert J, Wils H, Kumar-Singh S. Methods to investigate the molecular basis of progranulin actions on brain and behavior in vivo using knockout mice. In: Bateman A, Bennett H, Cheung S, editors. Progranulin, Methods in molecular biology, vol. 1806. New York: Humana Press; 2018. p. 233–53. https://doi.org/10.1007/978-1-4939-8559-3_16.
9. Brehm MA, Jouvet N, Greiner DL, Shultz LD (2013) Humanized mice for the study of infectious diseases. Curr Opin Immunol. 2013;25(4):428–35. https://doi.org/10.1016/j.coi.2013.05.012.
10. Bryda EC. The mighty mouse: the impact of rodents on advances in biomedical research. Mo Med. 2013;110(3):207–11. PMCID: PMC6179510
11. Cacioppo JT, Hawkley LC, Norman GJ, Berntson GG. Social isolation. Ann N Y Acad Sci. 2011;1231(1):17–22. https://doi.org/10.1111/j.1749-6632.2011.06028.x.
12. Canales CP, Walz K. The mouse, a model organism for biomedical research. In: Cellular and animal models in human genomics research. Academic; 2019. p. 119–40. https://doi.org/10.1016/B978-0-12-816573-7.00006-7.
13. Caron A, Lelong C, Pascual MH, Benning V. Miniaturized blood sampling techniques to benefit reduction in mice and refinement in nonhuman primates: applications to bioanalysis in toxicity studies with antibody–drug conjugates. J Am Assoc Lab Anim Sci. 2015;54(2):145–52.

14. Castle KD, Chen M, Wisdom AJ, Kirsch DG. Genetically engineered mouse models for studying radiation biology. Transl Cancer Res. 2017;6(Suppl 5):S900–13. https://doi.org/10.21037/tcr.2017.06.19.
15. Chandrashekran A, Casimir C, Dibb N, Readhead C, Winston R. Generating transgenic mice by lentiviral transduction of spermatozoa followed by in vitro fertilization and embryo transfer. In: Federico M, editor. Lentiviral vectors and exosomes as gene and Protein delivery tools, Methods in molecular biology, vol. 1448. New York: Humana Press; 2016. p. 95–106. https://doi.org/10.1007/978-1-4939-3753-0_8.
16. Chen J, Liao S, Xiao Z, Pan Q, Wang X, Shen K, Wang S, Yang L, Guo F, Liu H, Pan Q. The development and improvement of immunodeficient mice and humanized immune system mouse models. Front Immunol. 2022;13:1007579. https://doi.org/10.3389/fimmu.2022.1007579.
17. Chen S, Sun S, Moonen D, Lee C, Lee AYF, Schaffer DV, He L. CRISPR-READI: efficient generation of knockin mice by CRISPR RNP electroporation and AAV donor infection. Cell Rep. 2019;27(13):3780–9. https://doi.org/10.1016/j.celrep.2019.05.103.
18. Chen-Tsai RY. Integrase-mediated targeted transgenics through pronuclear microinjection. In: Larson M, editor. Transgenic mouse, Methods in molecular biology, vol. 2066. New York: Humana; 2020. p. 35–46. https://doi.org/10.1007/978-1-4939-9837-1_3.
19. Chia R, Achilli F, Festing MF, Fisher EM. The origins and uses of mouse outbred stocks. Nat Genet. 2005;37(11):1181–6. https://doi.org/10.1038/ng1665.
20. Clever S, Volz A. Mouse models in COVID-19 research: analyzing the adaptive immune response. Med Microbiol Immunol. 2023;212(2):165–83. https://doi.org/10.1007/s00430-022-00735-8.
21. Crispo M, Meikle MN, Delerue F, Benavides F. Genetically altered rodents. In: Rodent quality control: genes and bugs: monitoring health and genetics of laboratory animals. Cham: Springer International Publishing; 2024. p. 29–51. https://doi.org/10.1007/978-3-031-59103-7_2.
22. Dinnon KH III, Leist SR, Schäfer A, Edwards CE, Martinez DR, Montgomery SA, West A, Yount BL Jr, Hou YJ, Adams LE, Gully KL, Brown AJ, Huang E, Bryant MD, Choong IC, Glenn JS, Gralinski LE, Sheahan TP, Baric RS. A mouse-adapted model of SARS-CoV-2 to test COVID-19 countermeasures. Nature. 2020;586(7830):560–6. https://doi.org/10.1038/s41586-020-2708-8.
23. Doolittle ML, Maynard RD, Ackert-Bicknell CL. Inbred mouse strains in the study of bone disease. In: Zaidi M, editor. Encyclopedia of bone biology. 1st ed. Elsevier Inc; 2020. p. 150–9.
24. Eggel M, Würbel H. Internal consistency and compatibility of the 3Rs and 3Vs principles for project evaluation of animal research. Lab Anim. 2021;55(3):233–43. https://doi.org/10.1177/0023677220968583.
25. Ericsson AC, Franklin CL. Manipulating the gut microbiota: methods and challenges. ILAR J. 2015;56(2):205–17. https://doi.org/10.1093/ilar/ilv021.
26. Fenwick N, Griffin G, Gauthier C. The welfare of animals used in science: how the "Three Rs" ethic guides improvements. Can Vet J. 2009;50(5):523. PMCID: PMC2671878.
27. Franklin CL, Ericsson AC. Microbiota and reproducibility of rodent models. Lab Anim. 2017;46(4):114–22. https://doi.org/10.1038/laban.1222.
28. Fernandes MR, Pedroso AR. Animal experimentation: a look into ethics, welfare and alternative methods. Rev Assoc Med Bras. 2017;63:923–8. https://doi.org/10.1590/1806-9282.63.11.923.
29. Fone KC, Porkess MV. Behavioural and neurochemical effects of post-weaning social isolation in rodents-relevance to developmental neuropsychiatric disorders. Neurosci Biobehav Rev. 2008;32(6):1087–102. https://doi.org/10.1016/j.neubiorev.2008.03.003.
30. Forner S, Kawauchi S, Balderrama-Gutierrez G, Kramár EA, Matheos DP, Phan J, Javonillo DI, Tran KM, Hingco E, da Cunha C, Rezaie N, Alcantara JA, Baglietto-Vargas D, Jansen C, Neumann J, Wood MA, MacGregor GR, Mortazavi A, Tenner AJ, LaFerla FM, Green

KN. Systematic phenotyping and characterization of the 5xFAD mouse model of Alzheimer's disease. Sci Data. 2021;8(1):270. https://doi.org/10.1038/s41597-021-01054-y.
31. Franco NH. Animal experiments in biomedical research: a historical perspective. Animals. 2013;3(1):238–73. https://doi.org/10.3390/ani3010238.
32. Fry EA, Taneja P, Inoue K. Clinical applications of mouse models for breast cancer engaging HER2/neu. Integr Cancer Sci Ther. 2016;3(5):593–603. https://doi.org/10.15761/ICST.1000210.
33. Fujii W. Generation of knock-in mouse by genome editing. In: Hatada I, editor. Genome editing in animals, Methods in molecular biology, vol. 1630. New York: Humana Press; 2017. p. 91–100. https://doi.org/10.1007/978-1-4939-7128-2_8.
34. Fujii W. Generation of knock-in mouse by genome editing. In: Genome editing in animals: methods and protocols. New York: Springer US; 2023. p. 99–109. https://doi.org/10.1007/978-1-4939-7128-2_8.
35. Fuochi S, Rigamonti M, Iannello F, Raspa M, Scavizzi F, de Girolamo P, D'Angelo L. Phenotyping spontaneous locomotor activity in inbred and outbred mouse strains by using Digital Ventilated Cages. Lab Anim. 2021;50(8):215–23. https://doi.org/10.1038/s41684-021-00793-0.
36. Ganguly P. Organoids and 3D in vitro models as a platform for precision medicine (PM): an update. Organ. 2024;3(3):165–73. https://doi.org/10.3390/organoids3030011.
37. Gaskill BN, Rohr SA, Pajor EA, Lucas JR, Garner JP. Some like it hot: mouse temperature preferences in laboratory housing. Appl Anim Behav Sci. 2009;116(2–4):279–85. https://doi.org/10.1016/j.applanim.2008.10.002.
38. Gelfo F, Mandolesi L, Serra L, Sorrentino G, Caltagirone C. The neuroprotective effects of experience on cognitive functions: evidence from animal studies on the neurobiological bases of brain reserve. Neuroscience. 2018;370:218–35. https://doi.org/10.1016/j.neuroscience.2017.07.065.
39. Genzel L, Froudarakis E, Rapti G. Openness and transparency in animal research: why and how. Eur J Neurosci. 2024;60(11):6866–73. https://doi.org/10.1111/ejn.16597.
40. Getz GS, Reardon CA. Animal models of atherosclerosis. Arterioscler Thromb Vasc Biol. 2012;32(5):1104–15. https://doi.org/10.1161/ATVBAHA.111.237693.
41. Gianazza E, Miller I, Guerrini U, Palazzolo L, Laurenzi T, Parravicini C, Eberini I. What if? Mouse proteomics after gene inactivation. J Proteome. 2019;199:102–22. https://doi.org/10.1016/j.jprot.2019.03.008.
42. Golforoush P, Yellon DM, Davidson SM. Mouse models of atherosclerosis and their suitability for the study of myocardial infarction. Basic Res Cardiol. 2020;115(6):73. https://doi.org/10.1007/s00395-020-00829-5.
43. Grimm H, Biller-Andorno N, Buch T, Dahlhoff M, Davies G, Cederroth CR, Maissen O, Lukas W, Passini E, Törnqvist E, Olsson IAS, Sandström J. Advancing the 3Rs: innovation, implementation, ethics and society. Front Vet Sci. 2023;10:1185706. https://doi.org/10.3389/fvets.2023.1185706.
44. Gruzdev A, Scott GJ, Hagler TB, Ray MK. CRISPR/Cas9-assisted genome editing in murine embryonic stem cells. In: Allen I, editor. Mouse models of innate immunity, Methods in molecular biology, vol. 1960. New York: Humana Press; 2019. p. 1–21. https://doi.org/10.1007/978-1-4939-9167-9_1.
45. Gu B, Posfai E, Rossant J. Efficient generation of targeted large insertions by microinjection into two-cell-stage mouse embryos. Nat Biotechnol. 2018;36(7):632–7. https://doi.org/10.1038/nbt.4166.
46. Gu H, Chen Q, Yang G, He L, Fan H, Deng YQ, Wang Y, Teng Y, Zhao Z, Cui Y, Li Y, Li XF, Li J, Zhang NN, Yang X, Chen S, Guo Y, Zhao G, Wang X, Luo DY, Wang H, Yang X, Li Y, Han G, He Y, Zhou X, Geng S, Sheng X, Jiang S, Sun S, Qin CF, Zhou Y. Adaptation of SARS-CoV-2 in BALB/c mice for testing vaccine efficacy. Science. 2020;369(6511):1603–7. https://doi.org/10.1126/science.abc4730.
47. Hansen LA. Institution animal care and use committees need greater ethical diversity. J Med Ethics. 2013;39(3):188–90. https://doi.org/10.1136/medethics-2012-100982.

48. Henao-Mejia J, Williams A, Rongvaux A, Stein J, Hughes C, Flavell RA. Generation of genetically modified mice using the CRISPR–Cas9 genome-editing system. Cold Spring Harb Protoc. 2016;2016(2):pdb-prot090704. https://doi.org/10.1101/pdb.prot090704.
49. Homberg JR, Adan RAH, Alenina N, Asiminas A, Bader M, Beckers T, Begg DP, Blokland A, Burger ME, van Dijk G, Eisel ULM, Elgersma Y, Englitz B, Fernandez-Ruiz A, Fitzsimons CP, van Dam AM, Gass P, Grandjean J, Havekes R, Henckens MJA, Herden C, Hut RA, Jarrett W, Jeffrey K, Jezova D, Kalsbeek A, Kamermans M, Kas MJ, Kasri NN, Kiliaan AJ, Kolk SM, Korosi A, Korte SM, Kozicz T, Kushner SA, Leech K, Lesch KP, Lesscher H, Lucassen PJ, Luthi A, Ma L, Mallien AS, Meerlo P, Mejias JF, Meye FJ, Mitchell AS, Mul JD. The continued need for animals to advance brain research. Neuron. 2021;109(15):2374–9. https://doi.org/10.1016/j.neuron.2021.04.004.
50. Hufeldt MR, Nielsen DS, Vogensen FK, Midtvedt T, Hansen AK. Variation in the gut microbiota of laboratory mice is related to both genetic and environmental factors. Comp Med. 2010;60(5):336–47. PMID: 21262117; PMCID: PMC2958200
51. Husain A, Meenakshi DU, Ahmad A, Shrivastava N, Khan SA. A review on alternative methods to experimental animals in biological testing: recent advancement and current strategies. J Pharm Bioallied Sci. 2023;15(4):165–71. https://doi.org/10.4103/jpbs.jpbs_130_23.
52. Jiao SY, Yang YH, Chen SR. Molecular genetics of infertility: loss-of-function mutations in humans and corresponding knockout/mutated mice. Hum Reprod Update. 2021;27(1):154–89. https://doi.org/10.1093/humupd/dmaa034.
53. Justice MJ, Dhillon P. Using the mouse to model human disease: increasing validity and reproducibility. Dis Model Mech. 2016;9(2):101–3. https://doi.org/10.1242/dmm.024547.
54. Kato S, Sugawara H, Furuta T, Abe O, Akai H. CT lymphangiography of the thoracic duct in mice: direct mesenteric versus popliteal lymph node puncture. Eur Radiol Exp. 2025;9(1):22. https://doi.org/10.1186/s41747-025-00568-z.
55. Keele GR. Which mouse multiparental population is right for your study? The collaborative cross inbred strains, their F1 hybrids, or the diversity outbred population. G3: Genes Genomes Genet. 2023;13(4):jkad027. https://doi.org/10.1093/g3journal/jkad027.
56. Khan MN. Balancing act: navigating regulations and ethical considerations in animal experiments-intrinsic from international, Indian laws and Hinduism perspectives. Indian J Health Sci Biomed Res Kleu. 2024;17(3):185–91. https://doi.org/10.4103/kleuhsj.kleuhsj_31_24.
57. Kiani AK, Pheby D, Henehan G, Brown R, Sieving P, Sykora P, Marks R, Falsini B, Capodicasa N, Miertus S, Lorusso L. Ethical considerations regarding animal experimentation. J Prev Med Hyg. 2022;63(2 Suppl 3):E255. https://doi.org/10.15167/2421-4248/jpmh2022.63.2S3.2768.
58. Kinter LB, DeHaven R, Johnson DK, DeGeorge JJ. A brief history of use of animals in biomedical research and perspective on non-animal alternatives. ILAR J. 2021;62(1–2):7–16. https://doi.org/10.1093/ilar/ilab020.
59. Kumari M, Faizaan A, Shalini S, Kumar A. Transgenic animals in biotechnology. In: Introduction to pharmaceutical biotechnology, volume 1 (Second edition), Basic techniques and concepts. Bristol: IOP Publishing; 2024. p. 5–1.
60. Langford DJ, Bailey AL, Chanda ML, Clarke SE, Drummond TE, Echols S, Glick S, Ingrao J, Klassen-Ross T, LaCroix-Fralish ML, Matsumiya L, Sorge RE, Sotocinal SG, Tabaka JM, Wong D, van den Maagdenberg AMJM, Ferrari MD, Craig KD, Mogil JS. Coding of facial expressions of pain in the laboratory mouse. Nat Methods. 2010;7(6):447–9. https://doi.org/10.1038/nmeth.1455.
61. Larson MA. Pronuclear microinjection of one-cell embryos. In: Larson M, editor. Transgenic mouse, Methods in molecular biology, vol. 2066. New York: Humana; 2020. p. 27–33. https://doi.org/10.1007/978-1-4939-9837-1_2.
62. Lee KH, Lee DW, Kang BC. The 'R'principles in laboratory animal experiments. Lab Anim Res. 2020;36:1–3. https://doi.org/10.1186/s42826-020-00078-6.
63. Li M, Ding L, Cao L, Zhang Z, Li X, Li Z, Xia Q, Yin K, Song S, Wang Z, Du H, Zhao D, Li X, Wang Z. Natural products targeting AMPK signaling pathway therapy, diabetes

mellitus, and its complications. Front Pharmacol. 2025;16:1534634. https://doi.org/10.3389/fphar.2025.1534634.

64. Limaye A, Cho K, Hall B, Khillan JS, Kulkarni AB. Genotyping protocols for genetically engineered mice. Curr Protoc. 2023;3(11):e929. https://doi.org/10.1002/cpz1.929.
65. Lin X, Liu Y, Ma L, Ma X, Shen L, Ma X, Chen Z, Chen H, Li D, Su Z, Chen X. Constipation induced gut microbiota dysbiosis exacerbates experimental autoimmune encephalomyelitis in C57BL/6 mice. J Transl Med. 2021;19:1–16. https://doi.org/10.1186/s12967-021-02995-z.
66. Lindsjö J, Fahlman Å, Törnqvist E. Animal welfare from mouse to moose—implementing the principles of the 3Rs in wildlife research. J Wildl Dis. 2016;52(2S):S65–77. https://doi.org/10.7589/52.2S.S65.
67. Liu X, Yang L, Su Z, Ma X, Liu Y, Ma L, Ma X, Ma M, Liu X, Zhang K, Chen X. Acupoint catgut embedding alleviates experimental autoimmune encephalomyelitis by modulating neuroinflammation and potentially inhibiting glia activation through JNK and ERK pathways. Front Neurosci. 2025;18:1520092. https://doi.org/10.3389/fnins.2024.1520092.
68. Low BE, Kutny PM, Wiles MV. Simple, efficient CRISPR-Cas9-mediated gene editing in mice: strategies and methods. In: Proetzel G, Wiles M, editors. Mouse models for drug discovery, Methods in molecular biology, vol. 1438. New York: Humana Press; 2016. p. 19–53. https://doi.org/10.1007/978-1-4939-3661-8_2.
69. Luo L, Zuo Y, Dai L. Metabolic rewiring and inter-organ crosstalk in diabetic HFpEF. Cardiovasc Diabetol. 2025;24(1):155. https://doi.org/10.1186/s12933-025-02707-7.
70. Ma M, Powell DA, Weyand NJ, Rhodes KA, Rendón MA, Frelinger JA, So M. A natural mouse model for Neisseria colonization. Infect Immun. 2018;86(5):e00839–17. https://doi.org/10.1128/IAI.00839-17.
71. Makowska IJ, Franks B, El-Hinn C, Jorgensen T, Weary DM. Standard laboratory housing for mice restricts their ability to segregate space into clean and dirty areas. Sci Rep. 2019;9(1):6179. https://doi.org/10.1038/s41598-019-42512-3.
72. Mathisen AF, Vacaru AM, Unger L, Lamba EM, Mardare OAM, Daian LM, Ghila L, Vacaru AN, Chera S. Molecular profiling of NOD mouse islets reveals a novel regulator of insulitis onset. Sci Rep. 2024;14(1):14669. https://doi.org/10.1038/s41598-024-65454-x.
73. McKean NE, Handley RR, Snell RG. A review of the current mammalian models of Alzheimer's disease and challenges that need to be overcome. Int J Mol Sci. 2021;22(23):13168. https://doi.org/10.3390/ijms222313168.
74. Miura H, Quadros RM, Gurumurthy CB, Ohtsuka M. Easi-CRISPR for creating knock-in and conditional knockout mouse models using long ssDNA donors. Nat Protoc. 2018;13(1):195–215. https://doi.org/10.1038/nprot.2017.153.
75. Mohan S, Huneke R. The role of IACUCs in responsible animal research. ILAR J. 2019;60(1):43–9. https://doi.org/10.1093/ilar/ilz005.
76. Muruato A, Davis-Gardner ME, Vanderheiden A, Lokugamage K, Langsjoen RM, Schindewolf C, Plante JA, Weaver SC, Debbink K, Plante KS, Routh AL, Crocquet PA, Walker D, Suthar MS, Shi PY, Xie X, Menachery VD. Mouse-adapted SARS-CoV-2 protects animals from lethal SARS-CoV challenge. PLoS Biol. 2021;19(11):e3001284. https://doi.org/10.1371/journal.pbio.3001284.
77. Nakamura K, Sasaki T, Cheng XW, Iguchi A, Sato K, Kuzuya M. Statin prevents plaque disruption in apoE-knockout mouse model through pleiotropic effect on acute inflammation. Atherosclerosis. 2009;206(2):355–61. https://doi.org/10.1016/j.atherosclerosis.2009.02.014.
78. Naseem A, Ahire S, Ingle A. Multiplexing: an efficient way of genetic monitoring of laboratory mice using microsatellite markers. Agric Sci Dig. 2022;42(3):358–64. https://doi.org/10.18805/ag.D-5470.
79. Olsson IAS, Sherwin CM. Behaviour of laboratory mice in different housing conditions when allowed to self-administer an anxiolytic. Lab Anim. 2006;40(4):392–9. https://doi.org/10.1258/002367706778476389.
80. Ormandy EH, Weary DM, Cvek K, Fisher M, Herrmann K, Hobson-West P, McDonald M, Milsom W, Rose M, Rowan A, Zurlo J, von Keyserlingk MAG. Animal research, account-

ability, openness and public engagement: report from an international expert forum. Animals. 2019;9(9):622. https://doi.org/10.3390/ani9090622.

81. Park G, Rim YA, Sohn Y, Nam Y, Ju JH. Replacing animal testing with stem cell-organoids: advantages and limitations. Stem Cell Rev Rep. 2024;20(6):1375–86. https://doi.org/10.1007/s12015-024-10723-5.
82. Peirson SN, Brown LA, Pothecary CA, Benson LA, Fisk AS. Light and the laboratory mouse. J Neurosci Methods. 2018;300:26–36. https://doi.org/10.1016/j.jneumeth.2017.04.007.
83. Pietrzykowski T. Ethical review of animal research and the standards of procedural justice: a European perspective. J Bioeth Inq. 2021;18(3):525–34. https://doi.org/10.1007/s11673-021-10111-5.
84. Pound P, Blaug R. Transparency and public involvement in animal research. Altern Lab Anm. 2016;44(2):167–73. https://doi.org/10.1177/026119291604400210.
85. Qadri SS, Ramachandra SG. Laws, regulations, and guidelines governing research animal care and use in India. In: Laboratory animals. Academic; 2018. p. 237–61. https://doi.org/10.1016/B978-0-12-849880-4.00008-8.
86. Raies AB, Bajic VB. In silico toxicology: computational methods for the prediction of chemical toxicity. WIREs Comput Mol Sci. 2016;6(2):147–72. https://doi.org/10.1002/wcms.1240.
87. Ratuski AS, Weary DM. Environmental enrichment for rats and mice housed in laboratories: a metareview. Animals. 2022;12(4):414. https://doi.org/10.3390/ani12040414.
88. Regenberg A, Mathews DJ, Blass DM, Bok H, Coyle JT, Duggan P, Faden R, Finkel J, Gearhart JD, Hillis A, Hoke A. The role of animal models in evaluating reasonable safety and efficacy for human trials of cell-based interventions for neurologic conditions. J Cereb Blood Flow Metab. 2009;29(1):1–9. https://doi.org/10.1038/jcbfm.2008.98.
89. Russell WMS, Burch RL. The principles of humane experimental technique. Med J Aust. 1959;1(13):500. https://doi.org/10.5694/j.1326-5377.1960.tb73127.x.
90. Sales GD, Wilson KJ, Spencer KEV, Milligan SR. Environmental ultrasound in laboratories and animal houses: a possible cause for concern in the welfare and use of laboratory animals. Lab Anim. 1988;22(4):369–75. https://doi.org/10.1258/002367788780746188.
91. Saunders TL. The history of transgenesis. In: Larson M, editor. Transgenic mouse, Methods in molecular biology, vol. 2066. New York: Humana; 2020. p. 1–26. https://doi.org/10.1007/978-1-4939-9837-1_1.
92. Schmidt S. Capturing genetic diversity: the power of the CC and DO mouse models. Environ Health Perspect. 2018;126(1):014003. https://doi.org/10.1289/EHP2385.
93. Schmidt CW. Diversity outbred: a new generation of mouse model. Environ Health Perspect. 2015;123(3):A64–7. https://doi.org/10.1289/ehp.123-A64.
94. Shultz LD, Brehm MA, Garcia-Martinez JV, Greiner DL. Humanized mice for immune system investigation: progress, promise and challenges. Nat Rev Immunol. 2012;12(11):786–98. https://doi.org/10.1038/nri3311.
95. Slater AM, Cao L. A protocol for housing mice in an enriched environment. J Vis Exp. 2015;100:52874. https://doi.org/10.3791/52874.
96. Standley A, Xie J, Lau AW, Grote L, Gifford AJ. Working with miraculous mice: Mus musculus as a model organism. Curr Protoc. 2024;4(10):e70021. https://doi.org/10.1002/cpz1.70021.
97. Steinman L. The discovery of natalizumab, a potent therapeutic for multiple sclerosis. J Cell Biol. 2012;199(3):413–6. https://doi.org/10.1083/jcb.201207175.
98. Stripecke R, Münz C, Schuringa JJ, Bissig KD, Soper B, Meeham T, Yao LC, Di Santo JP, Brehm M, Rodriguez E, Wege AK, Bonnet D, Guionaud S, Howard KE, Kitchen S, Klein F, Saeb-Parsy K, Sam J, Sharma AD, Trumpp A, Trusolino L, Bult C, Shultz L. Innovations, challenges, and minimal information for standardization of humanized mice. EMBO Mol Med. 2020;12(7):e8662. https://doi.org/10.15252/emmm.201708662.
99. Sundberg JP, Berndt A, Sundberg BA, Silva KA, Kennedy V, Smith RS, Schofield PN. Approaches to investigating complex genetic traits in a large-scale inbred mouse aging study. Vet Pathol. 2016;53(2):456–67. https://doi.org/10.1177/0300985815612556.

100. Tam WY, Cheung KK. Phenotypic characteristics of commonly used inbred mouse strains. J Mol Med. 2020;98(9):1215–34. https://doi.org/10.1007/s00109-020-01953-4.
101. Tannenbaum J, Bennett BT. Russell and Burch's 3Rs then and now: the need for clarity in definition and purpose. J Am Assoc Lab Anim Sci. 2015;54:120–32.
102. Uhl EW, Warner NJ. Mouse models as predictors of human responses: evolutionary medicine. Curr Pathobiol Rep. 2015;3:219–23. https://doi.org/10.1007/s40139-015-0086-y.
103. Wettstein R, Bodak M, Ciaudo C. Generation of a knockout mouse embryonic stem cell line using a paired CRISPR/Cas9 genome engineering tool. In: Turksen K, editor. Embryonic stem cell protocols, Methods in molecular biology, vol. 1341. New York: Humana Press; 2015. https://doi.org/10.1007/7651_2015_213.
104. Workman P, Aboagye EO, Balkwill F, Balmain A, Bruder G, Chaplin DJ, Eccles SA. Guidelines for the welfare and use of animals in cancer research. Br J Cancer. 2010;102(11):1555–77. https://doi.org/10.1038/sj.bjc.6605642.
105. Yin X, Mead BE, Safaee H, Langer R, Karp JM, Levy O. Engineering stem cell organoids. Cell Stem Cell. 2016;18(1):25–38. https://doi.org/10.1016/j.stem.2015.12.005.
106. Zhang X, Chen Y, Qian J, Si Y, Wang C, Wang J, He Q, Ma J. Animal models in the study of diabetic erectile dysfunction: mechanisms and applications. Front Endocrinol. 2025;16:1512360. https://doi.org/10.3389/fendo.2025.1512360.
107. Zuberi A, Lutz C. Mouse models for drug discovery. Can new tools and technology improve translational power? ILAR J. 2017;57(2):178–85. https://doi.org/10.1093/ilar/ilw021.

Humanized Animal Models in Biological Research

7

Stanzin Nurboo, Sakshi Yadav, Sonam Ranwa, and Aamir Nazir

Abbreviations

BAC	Bacterial artificial chromosome
BRG	C.Cg-$Rag2^{tm1Fwa}Il2rg^{tm1Sug}$
CAR-T	Chimeric antigen receptor T-cell therapy
CDSCO	Central Drugs Standard Control Organization
CDX	Cell-line derived xenografts
CT-RMGT	Chromosome transfer- and recombinase-mediated genomic transfer
DENV	Dengue virus
DMEs	Drug metabolizing enzymes
EMA	European Medicines Agency
EPA	Environmental Protection Act
ESC	Embryonic stem cell
FMT	Fecal microbiota
GEAC	Genetic Engineering Appraisal Committee
HAC	Human artificial chromosomes
hCF	Human chromosome fragments
HIS	Human immune system
HSP	Hereditary spastic paraplegia
MAC	Mouse artificial chromosomes
MMCT	Micro-cell mediated chromosome transfer
MPS	Microphysiological systems
NK	Natural killer cells
NOD	Non-obese diabetic

S. Nurboo · S. Yadav · S. Ranwa · A. Nazir (✉)
Division of Toxicology and Experimental Medicine, CSIR-Central Drug Research Institute, Lucknow, India
e-mail: anazir@cdri.res.in

P. Kumar, V. Kothari (eds.), *Model Organisms in Biological Research*,
https://doi.org/10.1007/978-981-92-0038-2_7

NOG	NODShi.Cg-*Prkdc*^{scid}*Il2rg*^{tm1Sug}
NSG	NOD.Cg-*Prkdc*^{scid} *Il2rg*^{tm1Wjl}
PD-1	Programmed cell death protein 1
PDX	Patient-derived xenografts
RCGM	Review Committee on Genetic Manipulation
RMGR	Recombination-mediated genomic replacement
RRGS	$hSIRP\alpha^{+}$ $Rag1^{-/-}$ $Il2rg^{-/Y}$
SCID	Severe combined immunodeficiency
S-RMCE	Sequential recombinase-mediated cassette exchange
TALEN	Transcription activator-like effector nucleases
YAC	Yeast artificial chromosomes
ZFN	Zinc-finger nucleases

7.1 Introduction: What Are Humanized Animal Models?

Animal models provide significant insights into studying advanced therapies, disease pathologies, and treatment strategies, and their employment in biomedical research is ever-increasing. The presence of a wide variety of animal models in the research field offers a significant advantage in studying the complex human biological system with more specificity, allowing researchers to choose a fitting system for particular diseases, interventions, or biological processes. This also enables us to take an alternative route in studying disease biology, drug efficacy, therapeutic interventions, and system biology, avoiding the ethical and logistical restrictions associated with human tissues and samples. However, there are substantial differences between animal models and humans at cellular, molecular, physiological, and systemic levels, which result in complexities in translating the outcomes generated from the preclinical models to clinical studies. Employing larger animals that are phylogenetically closer to humans, such as primates, can significantly reduce the gap to some extent, but primates are high-cost, require long cultivation time, and are difficult to maintain in large quantities [56].

Mice have been extensively used to generate humanized models, driven by the emergence of numerous immunodeficient strains that facilitate the efficient engraftment of various human immune system (HIS) components, patient-derived xenografts (PDX), and cell-line derived xenografts (CDX) [16]. Additionally, other model organisms have also been used for various engraftment purposes in studying drug efficacy and disease treatment.

Humanized animal models have garnered significant attention over the past two decades and have the potential to overcome these inconsistencies. These are the new generation of models that mimic disease pathology with higher validity and efficacy, thus providing a valuable tool for studying human diseases and drug development. The purpose is to develop a model with an appropriate system that carries human genes, immune system components, human cells, tissues, and even organs, which enables researchers to comprehend the disease progression and drug treatment more accurately and with greater relevance by simulating human physiology. A recent report showing a cross-species difference between the inhibitory activity of immune checkpoint inhibitor PD-1 (programmed cell death protein 1) in mouse and human revealed that humanized mice carrying human PD-1 showed reduced anti-tumor T-cell activity as compared to the wild type [59]. The infamous first-in-human trial for TGN1412 (superagonistic anti-CD28 antibody) resulted in massive cytokine storm and multiorgan failure in six individuals, which the conventional models used in preclinical testing failed to predict [90]. This highlights the significance of humanization, as later studies revealed that human peripheral blood mononuclear cell-reconstituted NOD-RAG1$^{-/-}$A$\beta^{-/-}$HLADQ $^{(tg+ \text{ or } tg-)}$ IL-2Rγc$^{-/-}$ mice resulted in adverse events and even mortality due to robust human cytokine release, thereby recapitulating the adverse effects in humans. In addition, humanization is an innovative tool in modern biomedical research that enhances the translational outcomes generated from preclinical studies to clinical trials with greater success. Here, we provide a detailed overview of the current progress in the generation of humanized animal models and their applications in various fields of biomedical research, improving the translational efficacy and clinical outcomes. Although the latest advances and emerging technologies have enabled the development of humanized models with great translation potential, the quest for a model that simulates all the aspects of human biology is yet to be resolved.

7.2 Humanized Animal Models Used in Biomedical Research

Humanization of animal models is a powerful technique that bridges the physiological gap between conventional animal models and humans. Several animal models have been humanized to mimic human physiology, allowing us to address a range of concerns regarding translational outcomes with greater accuracy and efficacy. While the majority are vertebrates, due to their high genetic homology with humans, invertebrate models have also been used for humanization, as they offer some distinct advantages in studying molecular pathways, neurodegenerative diseases, and high-throughput screening. However, invertebrate models lack adaptive immunity and complex organ systems, which limits their usage for humanization. The following sections will discuss various humanized animal models used, their applications, and key advances in biomedical research.

7.2.1 Vertebrate Models

Many vertebrate models, such as rodents (mice and rats), pigs, zebrafish, and non-human primates, are being employed for humanization and each model holds a unique attribute that makes it an appropriate model for various humanization purposes. One of the extensively investigated and widely utilized species for humanization is the mouse model, which shares a high genetic similarity with humans and has a well-characterized immune system. A detailed overview of the models commonly employed is given below, and further details are listed in Table 7.1.

7.2.1.1 Humanized Mouse Model

Humanized mouse models are generated by introducing functional human cells, tissues, or components of the immune system. The primary requirement for generating a humanized model is to develop highly immunodeficient mice, ensuring successful engraftment with a low rate of rejection. The first step in establishing a humanized mouse requires a mutant nude mouse that is T-lymphocyte-deficient due to the absence of the *Foxn1* gene. A mutation in the gene coding for the catalytic domain of the DNA-dependent protein kinase (*Prkdc*) resulted in a severe combined immunodeficiency (SCID) phenotype in CB17 mice, which lack both B and T lymphocytes. This allows the engraftment of various human cells or tissues, including hematopoietic stem cells (Hu-DRC-SCID), liver hematopoietic cells, thymus, and lymph nodes from fetuses [61], peripheral blood mononuclear cells (hu-PBL-SCID) [67], thereby recapitulating parts of the human immune system. Apart from that, these SCID mice could also be engrafted with other human cells and tissues, including pancreatic islet cells and skin cells [9]. However, the engraftment rate remained low, as these mice retained natural killer (NK) cells and other innate immune cells, and also exhibited leakiness as they developed T and B cells with age. Crossing SCID mice with non-obese diabetic (NOD) mice resulted in the generation of NOD/*scid* mice, which negates the impediments caused by NK cells and innate immunity. The most important breakthrough in the generation of humanized mice was the development of immunodeficient NOD/SCID rg^{null}($Ilrg^{-/-}$) mice [41]. The model is generated by knocking out the gene coding for the common cytokine-receptor γ chain of the interleukin-2 receptor (IL-2R) [39], which results in the complete loss of NK cells as well as a deficiency in mature B and T lymphocytes. This led to the development of various immunodeficient strains, which became a gold standard for humanization with a high rate of successful engraftment and are referred to as "the new generation of humanized mice." This includes NOD.Cg-$Prkdc^{scid}$ $Il2rg^{tm1Wjl}$ (NSG) mice, NODShi.Cg-$Prkdc^{scid}Il2rg^{tm1Sug}$ (NOG) mice, and C.Cg-$Rag2^{tm1Fwa}Il2rg^{tm1Sug}$ (BRG) mice. Additionally, signal regulatory protein α (SIRP α), a transmembrane protein primarily expressed in monocytes and macrophages, shows polymorphism in the NOD strain and interacts with CD47 [82]. This results in the prevention of xenograft rejection mediated by macrophages, as NOD mice SIRPα also shows binding affinity towards human CD47. A next-generation THX humanized mouse was generated by grafting non-γ-irradiated, genetically myeloablated Kit^{W-41J} mutant immunodeficient pups with human cord blood $CD34^+$ cells, followed by hormonal conditioning, resulting in the generation of a mature, class-switched human antibody response mimicking the human immune development [15]. A recent report also showed the transplantation of Cas9-edited and genetically modified human $CD34^+$ cells in mice in order to recapitulate the functional changes that may occur in the human immune system upon targeted gene knockout [70]. Additional studies using humanized mice are listed in Table 7.1.

Table 7.1 Brief overview of the humanized animal models used in biomedical research

Model	Species	Humanized system/ organ	Strain/model	Application	Ref.
Vertebrate models	Mouse	Immune system components	BALB/c-*Rag1*$^{-/-}$*Ɣc*$^{-/-}$BALB/c-*Rag2*$^{-/-}$*γc*$^{-/-}$	Humanized model for Zika virus infection	[78]
			NOD.Cg-*Prkdc*^{scid} *Il2rg*$^{tm1Wjl/SzJ}$	Humanized model for tumor xenotransplantation	[42]
				(CAR) T-cell therapy for AML	[53]
				Humanized patient-derived xenograft model for evaluation of lung cancer immune response	[63]
			NOD/SCID/IL2r^{null}	Model for HIV infection	[17]
			NOD.Cg*Kit*^{W41J} *Tyr*$^{+}$ *Prkdc*^{scid} *Il2rg*^{tm1Wjl}/ThomJ	Humanized patient-derived xenograft model for evaluation of ovarian cancer	[81]
		Liver	NOD.*Fah*$^{-/-}$ *Rag2*$^{-/-}$ *Il2rg*$^{-/-}$ (FRGN)	Humanized model for non-alcoholic steatohepatitis.	[55]
			TK- NOD/Shi-scid/IL-2Rγ (TK-NOG)	Model for drug oxidation activity in humanized liver	[86]
			FRG, TK-NOG, uPA-SCID	Humanized liver for metabolic disorder study	[54]
		Skin	NOD-*scid IL2rγ*^{null} (NSG)	Model to study human cutaneous memory T cells	[46]
		iPSC-derived neuronal progenitor cells (NPCs)	P0 *Rag1*$^{-/-}$	Generation of a humanized neuronal chimeric mouse model by engrafting hiPSC-derived NPCs	[13]
	Rats	Liver	*Fah*$^{-/-}$*Rag2*$^{-/-}$*IL2rg*$^{-/-}$ (FRG)	Humanized liver model to study drug testing	[99]
		hESC-derived dopaminergic neurons	Female Sprague–Dawley rats	Humanized α-synuclein rat model to study α-synuclein pathology	[35]
	Pigs	Kidney	Porcine	Xenotransplantation of humanized kidney	[1]
	Zebrafish	Oligodendrocytes	*GPR17*-humanized zebrafish	Humanized zebrafish to evaluate pro-remyelinating drugs	[30]
Invertebrate models	*Drosophila*	DNA replication machinery	Orc6-based *drosophila*	Evaluation of the Orc6 gene in MGS	[3]
	C. elegans	Membrane trafficking	*vps-45 (tm246)*	Studying pathogenic mutations in VPS45 associated with SCN	[28]

(CAR)T chimeric antigen receptor T-cell therapy, *AML* acute myeloid leukemia, *HIV* human immunodeficiency virus, *iPSC* induced pluripotent stem cell, *hESC* human embryonic stem cell, *MGS* Meier–Gorlin syndrome, *SCN* severe congenital neutropenia

7.2.1.2 Humanized Rat Model

Rats exhibit higher functional similarity with humans compared to mice and are considered to be a preferred rodent model for studying various physiological, pharmacological, and toxicological parameters. It is also a preferred species for modelling metabolic diseases and cognitive functions, and its humanization offers a significant advantage due to its larger body size and longer lifespan, thus providing a higher resolution of tissues and organs and a longer experimental window. Similar to the immunodeficient mice, immunodeficient rats have been generated through various genetic mutations, including the *Foxn1* gene [71], to develop T cell-deficient nude rats, *Rag1/2* [69] to prevent radiosensitivity, *Il2rg* [58] to remove NK cells, along with *Prkdc,* to generate the SCID phenotype. With the advancement in CRISPR/Cas9 technology, a transgenic rat carrying human *SIRPα* (*hSIRPα*) is intercrossed with SG ($Prkdc^{-/-}Il2r\gamma^{-/-}$) rats to generate $hSIRP\alpha^{+}Prkdc^{-/-}Il2r\gamma^{-/-}$ (NSGL) rats. The rate of successful engraftment in NSGL rats is relatively higher with various human cancer cells and embryonic stem cells that can be efficiently engrafted. NSGL rats also support the engraftment of human HSCs, which allows us to recapitulate the human immune system in rats (HIS-rats) [96]. In addition, other immunodeficient strains are also developed, such as $Rag1^{-/-}$ $Rag2^{-/-}$ $Il2rg^{-/Y}$ (SD-RG) rats that support human lung cancer xenografts [33], and $hSIRP\alpha^{+}$ $Rag1^{-/-}$ $Il2rg^{-/Y}$ (RRGS) rats that show complement levels similar to humans and are humanized using $hCD34^{+}$ cells [62]. Thus, humanized rats are a vital tool for human stem cell research and for studying the safety and efficacy of therapeutic agents.

7.2.1.3 Humanized Pig Model

Pigs are considered to be a significant model to study human biology as they share various attributes with humans in terms of anatomy, physiology, immunology, and genome. Based on this, one of the first immunodeficient pig models has been generated for humanization by manipulating *RAG1*, *RAG2*, *ARTEMIS*, or *IL2RG* to develop SCID pigs [8]. Cancer cell lines have been transplanted in these SCID pigs, and tumors were developed at the site of injection with a low rate of rejection. Human IPSCs have also been engrafted in SCID pigs with biallelic *RAG2* mutations, and they developed teratomas representing the three germ layers [51]. Advances in the CRISPR-Cas9 technology-mediated genome editing have enabled cross-species transplantation (xenotransplantation) of tissues and organs more efficiently because of the physiological similarity between pigs and humans. Various reports suggest the xenotransplantation of genetically engineered (GE) pig counterparts, such as pig corneas, heart, kidney, liver, and pancreatic islets in human patients [92]. Developing human-pig chimeras is an emerging field, and it holds significant potential in transplantation. GE pigs that lack a particular cell type lineage are transplanted with hIPSCs to rescue the mutated embryo by generating the human counterpart, thereby developing a human-pig chimera [27]. These novel techniques hold significant potential in preclinical research to study drug development, toxicological studies and disease biology.

7.2.1.4 Humanized Zebrafish Model

Zebrafish are a powerful biological model owing to their small size, low cost of maintenance, rapid generation time, large number of progenies, transparency, and genetic homology to humans. Therefore, it is employed for studying various aspects, including genetic mutations, neurological, endocrine, and cardiovascular disorders, drug efficacy, and toxicity. Although humanized mice are considered the gold standard for engraftment purposes, zebrafish offer distinct advantages, which led to the development of various immunodeficient models for humanization. This includes mutation in *Rag1, Rag2*, or both *Prkdc* and *Il2r*, and are mostly used for studying the transplantation of human tumors. Also, one of the first humanized zebrafish that expresses human hematopoietic-specific cytokines was developed to promote the survival and differentiation of human HSCs and progenitor cells, ultimately resulting in the enhancement of primary human leukemia cell proliferation [75]. Furthermore, due to its high regenerative capacity and ease of genetic manipulation, zebrafish are a perfect alternative model to recapitulate a human disease model, with a rejection-free environment, enabling multiple drug tests with low-cost maintenance and a faster reproduction rate.

7.2.1.5 Other Humanized Models

Several other vertebrate models are often used for humanization, including rabbits, Syrian hamsters and nonhuman primates for various therapeutic purposes. Immunodeficient rabbits have been developed by employing CRISPR-Cas9 technology to manipulate RAG1, RAG2, IL2RG, FOXN1, and IgH, thereby generating a model for humanization [25, 79]. Humanized rabbit models have been widely used in the production of monoclonal antibodies [89]. Similarly, hamsters have also been employed for humanization strategies. Transplantation of the human pancreatic cancer cells in *IL2RG* gene knockout Syrian hamsters resulted in better recapitulation of the metastasis development as compared to immunodeficient mice [64].

7.2.2 Invertebrate Models

Invertebrate models serve as a potential alternative to vertebrate models in biomedical research due to their simplicity in anatomy and physiology, ease of genetic manipulation, and large-scale maintenance. The high maintenance cost and ethical concerns associated with using vertebrate models make invertebrate models a better candidate for carrying out large-scale drug testing, toxicity studies, and drug efficacy evaluations. Moreover, the significant degree of evolutionarily conserved similarity at the molecular and genetic level enables researchers to study numerous biological processes relevant to human diseases. Landmark discoveries of some basic biological principles, including RNA interference and apoptosis in *Caenorhabditis elegans,* and valuable insights about the innate immunity and regulation of organ growth in *Drosophila melanogaster,* further strengthen the value of these model organisms in biomedical research.

The evolutionarily conserved sequence similarity between humans and invertebrate models such as *C. elegans* (~60% homology) and *Drosophila* (~65% homology) has enabled researchers to explore numerous biological pathways and diseases relevant to humans. Despite the shared similarities, invertebrate models fail to simulate the human system due to subtle differences in their gene expression profiles and other regulatory elements, which result in only a rough approximation of human diseases. In addition to this, traditional transgenic models have also been extensively used in studying various human diseases. However, these are not exclusive humanized models as they do not carry a human-specific gene inserted by replacing the invertebrate counterpart. While there is very little evidence regarding the complete humanization of these invertebrate models, attempts have been made to create various disease models by replacing a specific genome sequence with that of the human counterpart by using advanced genome editing technologies such as CRISPR-Cas9 systems. For instance, a mutation in the *GNAO1* gene, which encodes for a G protein, Gαo, results in a severe neurodevelopmental disorder encompassing infantile seizure, movement disorder, and cognitive dysfunction. Replacing the *Drosophila* Gαo with the human ortholog resulted in proper expression of the humanized protein by fully recapitulating the functions of the protein, including normal development, lifespan, and memory formation [77]. Similarly, hereditary spastic paraplegia (HSP), a group of rare degenerative neurological disorders, is characterized by stiffness in the lower limb and progressive weakness. Human kinesin light chain 4 (*KLC4*), having high sequence homology with *C. elegans klc-2,* is associated with autosomal dominant HSP. A humanized *C. elegans* model is generated by replacing the *klc-2* gene with h*KLC4,* retaining many of the functions of *klc-2* in the worms [29]. A systemic humanization attempt was also made in yeast by replacing 414 essential genes with their human ortholog to determine the functional conservation of these genes, resulting in humanization of nearly 47% of the genes. Also, some genes that share similar biological processes (e.g., sterol biosynthesis) are highly replaceable, while others are not (e.g., DNA replication initiation) [47]. A brief overview of the different animal models used for humanization by recapitulating human systems or organs for various purposes is listed in Table 7.1.

7.3 Types of Humanizations in Animal Models

Humanization of animal models is an invaluable tool for recapitulation of human biology, physiology, and disease response, thereby increasing the translational efficacy in biomedical research. Animal models are humanized by employing various strategies that involve the development of transgenic strains, the introduction of human cells, tissues, or organs, the replacement of the gut microbiota with the human counterparts, or the development of a model to study drug metabolism and toxicity (Fig. 7.1).

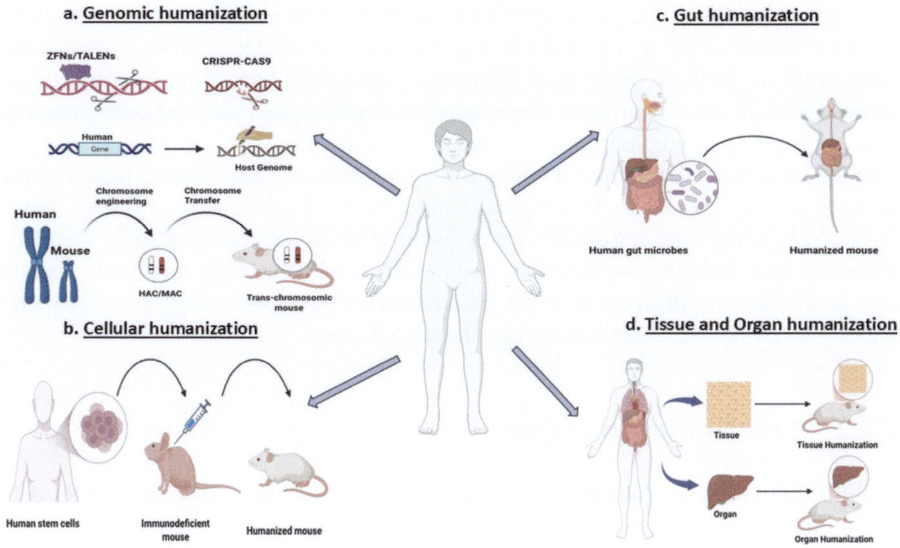

Fig. 7.1 Types of humanizations in animal models

7.3.1 Genomic Humanization

Genomic humanization of animal models involves the introduction of a specific region of the human genome into the host genome, thereby simulating the expression profile of humans. Multiple approaches have been developed to create different kinds of genomically humanized animals, contributing immensely to the biomedical research field. One of the early techniques employed to develop humanized animals is the transgenic insertion of yeast artificial chromosomes (YACs) and bacterial artificial chromosomes (BACs), which integrate at random chromosomal positions [22]. However, this method involves the pronuclear injection of DNA, which can disrupt the host genome by silencing gene expression and causing uncontrolled copy number [19]. Advances in genome editing technology led to the establishment of various knock-in and knock-out strains using zinc-finger nucleases (ZFNs), transcription activator-like effector nucleases (TALENs), and CRISPR-Cas9 systems, resulting in the efficient generation of genomically humanized animal models [44]. Another method involves the target integration of the human DNA sequence into the corresponding region of embryonic stem cells (ESCs) of the mouse genome by homologous recombination and recombination-mediated genomic replacement (RMGR) for precise humanization [88]. Sequential recombinase-mediated cassette exchange (S-RMCE) technology allows for the production of mice inserted with large Mb-sized human genomic regions [52]. This is further enhanced by the generation of transchromosomic (Tc) animals by inserting human chromosome fragments (hCFs) using micro-cell mediated chromosome transfer (MMCT) into mouse ESCs. Some of these hCFs are heritable, which led to the generation of novel heritable Tc strains [84], and some carry chromosome 21 to develop a Down syndrome model [43]. However, there are some drawbacks in using hCFs to develop Tc models, including inefficient germline transmission and partial trisomy for orthologous host genes. To overcome these, various methods have been developed, including human artificial chromosomes (HACs), mouse artificial chromosomes (MACs), and chromosome transfer- and recombinase-mediated genomic transfer (CT-RMGT). To further enhance the efficacy of humanized animal models, the integration of chromosome transfer and genome editing proved to be of great value. Large-sized genomic cloning using HACs/MACs and chromosomal transfer using MMCT are powerful tools allowing Mb-sized gene insertion without any harm to the host. Therefore, it exhibits invaluable significance in recapitulating the human physiology, improving the understanding of disease biology, and developing effective therapy.

7.3.2 Cellular Humanization

Cellular humanization involves the xenotransplantation of a specific human cell population or components of the cell into the host animal, thereby creating an environment that closely resembles the human physiological environment. The transplanted cell carries a complete set of genetic information and has an advantage over genomic humanization in terms of studying physiological processes that involve multiple genomic interactions. Cellular humanization is widely implicated in studying immunology, oncology, and regenerative medicine. Rodents, especially mice, have been extensively explored in this aspect, resulting in the generation of various immunodeficient mouse models including NOD, NSG, NOG, and BRG strains [39, 85]. Using such models, human innate immune cells and adaptive immunity can be generated by transplanting the human hematopoietic stem cells (HSCs), thereby reconstituting the functional immune system [49]. To avoid rejection and graft versus host disease (GVHD) in HSCs, a novel approach involving the xenotransplantation and humanization of iPSCs has gained significant attention in recent times. This includes the generation of iPSC-derived HSCs to develop humanized immune systems, iPSC-derived dopaminergic neurons to treat Parkinson's disease models in nonhuman primates [31], and iPSC-derived lung progenitor cells to further enhance the development of stem cell-based lung therapeutics [91]. Studying cancer cells by directly transplanting patient-derived xenografts (PDX) in immunodeficient models is another important aspect of cellular humanization that helps in the characterization of these tumors, as they retain their histological and genetic features, thereby providing a platform for translational cancer research [34]. However, there are still concerns to be addressed pertaining to engraftment efficiency, survival of xenografts and compatibility within the host environment.

7.3.3 Gut Microbiome Humanization

The method of humanizing the gut microbiome in model organisms involves modifying or replacing the entire gut microbiota or introducing a selected subset of human-derived microbial communities in order to recapitulate the human microbiome. Mice are considered the preferred model for humanizing the gut microbiome. This involves the generation of germ-free (GF) mice, characterized by the absence of the complete microbiome, thereby serving as a blank microbial background that is a significant tool in investigating the biology of the human gut microbiome without the interference of the host microbiome. The humanization technique of GF mice involves the transplantation of fecal microbiota (FMT) into the gut, thereby colonizing the host gut with human microbiota and simulating the gut microbiome physiology of humans. This allows us to study host-microbiome interaction, immune system development, and dysbiosis-related diseases. Colonization of specific bacterial strains instead of whole FMT in GF mice also revealed their potential role in developing immunity in the host [60]. However, the efficacy of colonization varies from 44% to 77% in GF mice, and a comparative study significantly revealed a higher colonization efficacy in rats than in mice [93]. Moreover, other alternative options include pig and non-human primate models that share high GM similarity with humans. Humanization of the gut microbiome turned out to be a powerful tool to study and recapitulate multiple human disorders, including diabetes, obesity, gut inflammation, and to identify the role of the gut-brain axis in neurodevelopment and neurodegeneration [36].

7.3.4 Tissue and Organ Humanization

Tissue and organ humanization provides a comprehensive understanding of human physiology for studying drug responses, disease biology, and organ transplantation. Immunodeficient mice have been robustly used in this regard, and various human tissues or organs are either transplanted or generated in the host system to study the efficacy of transplantation, drug metabolism, toxicity, and regenerative medicine. Recapitulating the features of human drug metabolism and modeling the infection of the human hepatitis virus draws significant attention, leading to the generation of various humanized liver mice. Laboratory animals and humans have different drug response pathways, enzyme activity patterns, and disposition mechanisms, leading to translational failure from preclinical outcomes to clinical studies. In addition to immunodeficient strains, engraftment of human hepatocytes in mice involves the elimination of murine hepatocytes by expressing urokinase plasminogen activator (*uPA*) [72] or fumaryl acetoacetate hydrolase (*fah)* knockout ($Fah^{-/-}$) [4] that induces death of murine hepatocytes, thus providing an exclusive niche to human hepatocytes. Cytochrome P450 plays a crucial role in drug metabolism and both human and mice expresses different subtypes of the enzyme. Engraftment of human hepatocytes in *uPA/scid* mice significantly recapitulated the human counterpart by expressing P450 subtypes similar to the donor liver [37]. Additionally, the route of drug disposition pattern followed by humanized mice reflects exactly that of humans, suggesting the efficacy of the engraftment [95]. Apart from that, humanized liver mice are also implicated in studying infectious diseases involving hepatitis viruses. Successful infectious models have been generated using $Fah^{-/-}Rag2^{-/-}IL\text{-}2rg^{-/-}$ mice strains to develop Hepatitis B virus (HBV) models [5] and Hepatitis C virus (HCV) models [6].

Various other humanized models have also been developed, including humanized skin mice for modelling infectious diseases and allograft rejection [74], humanized islet mice for modelling diabetic models and islet transplantation [10], humanized nervous system mice for modelling glial cell development and differentiation [32] and humanized intestine models to study metabolic disorders and gut microbiome [20]. However, concerns remain in terms of efficacy, transplant rejection, ethical and regulatory issues, which need to be resolved. With advances in stem cell research technology, the focus is now shifting towards the development of human organoids, as it has the potential to revolutionize biomedical research by providing faster outcomes, easy accessibility and handling, and a more accurate representation of the human biological system than animal models, thereby bridging the gap between conventional models and humans [45].

7.4 Application of Humanized Models in Translational Biomedical Research

7.4.1 Immune Oncology

Recently, the use of humanized models to study cancer has skyrocketed by compromising the immune system of the animals, which makes them immunodeficient and allows engrafting with human cancer cell lines or patient-derived xenografts. Such as, engrafting multiple myeloma cell precursors in MISTRG6 mice to develop the full spectrum of plasma cell dyscrasia without change in genetic diversity [21]. Humanized mouse models are highly exploited to create specific strains to study different cancers by using the CRISPR-Cas9 system. These models are used to understand the complexity of immune systems towards the development, progression, and treatment of cancer. Different therapies like CAR T cell therapy, immune checkpoint blockade therapy, and antisense therapy have been developed since. The rising trend in personalized cancer therapy using HIS-mice as mouse avatars widens the treatment options by minimizing the risks of side effects.

Immune-oncology models are reproduced through engrafting cancer cell lines or patient-derived xenografts with matched or unmatched HSCs or PBMCs to study the interactions between the immune system and tumorigenesis. For instance, HSC-reconstituted MISTRG mice engrafted with the human melanoma cell line Me290 featured human-like immune responses to melanoma infiltration. Human-VEGF inhibitor treatment fully diminished the tumor growth [76]. Hepatocellular carcinoma-PDX engrafted HIS-NSG when treated with pembrolizumab (PD-1 inhibitor) or ipilimumab (CTLA-4 inhibitor) reactivated CD8+ T cells and diminished tumor growth [100].

Fully autologous humanized immune-oncology models are complex but highly effective for studying cancer development. Introduction of a mutant NPM1 gene into HSCs to develop leukemia by lentivirus engraftment into NSG mice. Since only about 20% of HSCs were transduced, some developed into leukemia, while the rest formed a matching human immune system. Similarly, mice engrafted with CD34+ fetal liver cells modified with MLL-AF9 developed mixed-lineage leukemia [94]. In another study, [40] used the same HSCs to generate both B cell leukemia and human T cells, which were then engineered into CD19-specific CAR T cells. These CAR T cells were tested on mice with autologous leukemia, allowing researchers to study therapy without interference from mismatched immune responses.

Humanized models are invaluable in predicting clinical efficacy, characterizing pharmacodynamics, and refining therapeutic strategies for both investigational and approved immunotherapies. They play a critical role in oncology drug development, enabling mechanistic evaluation of immune checkpoint inhibitors. Tiragolumab, a humanized anti-TIGIT antibody (immune checkpoint inhibitor) being tested in combination with atezolizumab for PD-L1–high metastatic non-small-cell lung cancer (NSCLC), has received FDA Breakthrough Therapy designation as of January 2021 (NCT04294810). This therapeutic rationale originates from the complementary roles of TIGIT and PD-1/PD-L1 pathways in mediating immune exhaustion and tumor immune evasion, thereby making dual-checkpoint blockade a biologically compelling strategy.

In addition, dostarlimab (JEMPERLI®), approved by the FDA in 2021, is a humanized IgG4 kappa monoclonal antibody directed against PD-1, indicated for mismatch repair-deficient (dMMR) endometrial cancer and other dMMR solid tumors. It binds PD-1 with high affinity, preventing interaction with its ligands (PD-L1 and PD-L2) and thereby restoring cytotoxic T-cell function. Dostarlimab was derived from a mouse antibody and humanized via CDR grafting and affinity maturation (NCT06333314).

7.4.2 Infectious Diseases

Human diseases like AIDS, malaria, filariasis, and dengue hemorrhagic fever are caused by organisms that do not infect mice or other lab animals, which restricts the use of these animals to study these diseases. So, to study human-tropic infectious diseases, humanized mouse models were developed. These mice are introduced with human cells or tissues to mimic parts of the human body. Because of this change, they can be infected by these human-specific disease-causing organisms. That means researchers can study how the diseases work and test possible vaccines in these specially altered mice.

Due to the presence of HIV host cell types, like CD4+ T cells and macrophages, HIS-mice can be used to develop HIV infection without the need for virus adaptation and allow the study [57]. This aids the development of various HIV therapies, such as HIV vaccine development, anti-HIV CAR-T immunotherapy, tyrosine kinase inhibitors, neutralizing antibodies, and gene therapy [48]. Recently, humanized DRAGA mice were used to study cellular immunotherapy for HIV by treating HIV-specific CAR/CXCR5 T cells in an antiretroviral therapy, which revealed complete suppression of viral replication, suggesting that HIV infection was treatment-responsive in the DRAGA mice [73].

HIS-NSG mice, when infected with dengue virus (DENV), show similar symptoms to humans, such as DENV viremia, thrombocytopenia, liver damage, and the production of anti-DENV IgM antibodies. It is also used to study DENV-related disease mechanisms. When depleting human NK cells in IL-15 and Flt3-L-injected HIS-NSG mice, the human NK cells protected the mice from severe DENV infection and helped reduce liver pathology [18].

Due to the specificity of *Plasmodium falciparum* to humans, mouse models to study malaria were restricted to murine *Plasmodium* strains. To study human immune response to malaria infection, IL-15 and Flt3-L injected HIS-NSG mice were used, where human NK cells destroyed the infected erythrocytes and released IFN-γ in response to *P. falciparum*. To study the liver stage of *P. falciparum*, DRAGA mice with CD34+ fetal liver cells are developed. Recently developed MISTRGFah mice with humanized liver, immune system, and circulating human red blood cells are used to study all the stages of malaria infection [80].

Infections with hepatitis B, C, and D viruses can be developed in HEP-mice, which are engrafted with human hepatocytes. HIL-mice with dual humanization of the immune system and liver are developed by engrafting human fetal liver cells or human bone marrow MSCs into immunodeficient mice. HIL-FRGS mice were studied for liver cirrhosis by infecting them with the chronic hepatitis B virus [97]. Similarly, HIL-NSG mice were studied for HBV-induced liver inflammation and the role of CD206+ macrophages. Treatment with anti-GM-CSF shows a reduction in CD206+ macrophage infiltration, resulting in a decrease in liver fibrosis [83]. Reducing human T cells or macrophages or IFNα-2a administration in HIL-NSG mice decreases the severity of HCV diseases. These dual-humanized mouse models provide a valuable platform for studying immune-targeted treatments. In a recent study, immunodeficient NSG mice were used as recipients of human peripheral blood mononuclear cells (hPBMCs), in order to assess the immunogenicity of chimeric HBV immunogen produced in CRISPR/Cas9-edited *N. benthamiana*, showing activation of antigen-specific B cells [11].

Pritelivir (development codes AIC316 and BAY 57–1293) is a novel, direct-acting antiviral agent under advanced clinical development for the treatment of herpes simplex virus (HSV) infections, particularly important in immunocompromised patients. It acts through a distinct mechanism by inhibiting the viral helicase–primase complex, a critical enzyme system required for HSV DNA replication. Pritelivir is being developed by AiCuris Anti-Infective Cures AG, and has progressed to Phase III clinical trials evaluating its safety and efficacy in patients with acyclovir-resistant mucocutaneous HSV infections.

Reflecting its clinical promise and unmet medical need, the U.S. Food and Drug Administration (FDA) granted pritelivir Fast Track designation in 2017 and subsequently awarded Breakthrough Therapy designation in 2020. These regulatory designations underscore both the innovative nature of its antiviral mechanism and its potential to address limitations of existing HSV therapies.

7.4.3 Autoimmune Diseases

Autoimmunity arises due to the intricate interactions between the immune system and self-antigens. Apart from genetic and environmental risk factors, viral infections or a new immunotherapy can trigger autoimmune diseases. HIS-mice provide some features of the autoimmunity model, but it still lacks the complexity of the human immune system. Different humanized mouse models are developed to understand many aspects of autoimmune diseases.

Immunodeficient mice were used as rheumatoid arthritis models by transferring PBMCs from affected patients. NOG mice with injected CD4+ memory T and B cells sustained the production of rheumatoid factors [38]. HSC-engrafted NOD-$Prkdc^{SCID}$ mice are used to induce arthritis by intra-articular injection of Chlamydia trachomatis [12]. Similarly, HIS-NSG mice are used to study acute inflammatory arthritis through intra-articular injection of complete Freud's adjuvant and demonstrated local joint erosion with no signs of systemic joint erosion [66].

Early attempts to model type 1 diabetes (T1D) used NRG-Akita mice, but these developed hyperglycemia without a human immune system. Transferring T1D patient immune cells (HLA-DR4 diabetic antigen-pulsated CD4+ T cells or PBMCs) into DRAG mice caused some islet damage, but did not lead to full diabetes [87]. A more accurate model came from using HLA-DQ8-transgenic HIS-NSG mice, which were immunized with HLA-DQ8-restricted insulin B-chain peptides. This led to CD3+ T-cell infiltration, islet destruction, and hyperglycemia in 60% of the mice.

Systemic lupus erythematosus (SLE) is caused by pathogenic B cells and the production of autoantibodies. BRG and NSG mice are engrafted with PBMCs from SLE patients to develop features like anti-dsDNA and deposition of C3 immune complexes in the kidney [2, 65]. However, the limitation includes the fact that disease reproduction varies among the donors and the requirement of a large amount of PBMCs. Pristane-induced SLE in HIS-NSG mice produced lupus nephritis, pulmonary serositis, and human antinuclear antibodies. The applications of various humanized animal models used in biomedical research are listed in Table 7.2.

Table 7.2 Applications of humanized models in translational biomedical research

Model organism	Strain	Humanized Version	Applications	Key features	Reference
Immunodeficient Mice	NSG (NOD-scid IL2Rγnull), NOG	Engrafted with human tissues, HSCs	Modeling human immune system, cancer research, infectious diseases, transplantation	Functional immune system absent, allowing acceptance of human cells and tissues	[14]
Transgenic Mice	C57BL/6-Tg(APP, PS1)	Express human genes	Alzheimer's disease, Huntington's disease, and Parkinson's disease	Carry foreign genetic material to model human disease mechanisms	[98]
Rowett Nude Rats	Immunodeficient strain	Humanized liver rat models	Xenotransplantation, organ transplantation, immune response studies	Lack functional immune system, enabling engraftment of human tissues	[24]
Pig	Göttingen minipig, Yucatan	Humanized pig models for organ transplantation	Cardiovascular research, diabetes, surgical training, and regenerative medicine	Close organ size and physiology to humans	[50]
Non-Human Primates	Rhesus macaque, Cynomolgus monkey	Humanized immune system models	Preclinical testing of immunotherapies, infectious disease research, and transplantation	High genetic similarity to humans	[23]

7.5 Role of Humanized Models in Studying Regulatory Toxicology

Humanized models have been extensively employed in studying the safety, efficacy, and toxicity of various drug candidates for the treatment of human diseases. While traditional animal models such as rodents, *Drosophila*, zebrafish, and *C. elegans* provide significant preclinical outcomes; however, these animal models differ greatly from humans in terms of drug response and metabolism due to the presence of a different set of drug metabolizing enzymes (DMEs) such as cytochrome P450 family [68], which often leads to toxicity as well as translational failure during clinical trials. Humanized animal models play a crucial role in bridging the gap by mimicking human physiology, thereby streamlining the translational process with greater efficiency. Several mouse models have been generated through P450 humanization [7], providing valuable insights into drug metabolism and toxicity in a simulated human environment. In addition, human liver chimeric mouse models, including the alb-uPA mouse (urokinase-type plasminogen activator expressed under the albumin promoter), the fah−/−/Il2rg−/−/Rag2−/− mouse, and the TK-NOG mouse, have been used to study drug regulatory toxicology, including human toxicity, profiling of human drug metabolism, drug-drug interaction, pharmacokinetics, and pharmacogenetic studies, and other DME studies. However, some challenges need to be overcome, including variability in chimerism, reminiscence of murine liver that expresses mouse DMEs, as well as strain-specific limitations. Emerging technologies further increase the understanding of regulatory studies by integrating novel approaches like organoid culture, microphysiological systems (MPS), and computational model generation, enhancing safety and clinical efficacy.

7.6 Regulatory Status and Ethical Considerations

Experimentation on animal models is the basis of scientific studies, including biomedical research, disease modeling, drug development, and toxicity studies by providing countless outcomes and human-relevant response predictions. However, studies in animal models often fail to replicate the particular outcome in human patients due to differences in the physiology, anatomy, metabolism, and drug response. A report in 2015 suggests that more than 50% of the data generated from preclinical research is not reproducible in the USA alone in the past two decades, costing approximately USD 28 billion per year [26]. Moreover, potential therapeutic translation to phase I trials decreased from 10.4% between 2003 & 2007 to 6% to 7% between 2011 and 2017, raising serious concerns in the scientific community. To address this and many other concerns related to animal usage, various regulatory bodies and agencies from different countries have been established to regulate and develop novel approaches to lessen the burden on animal models and to adopt an alternative route. The major biosafety regulatory framework for humanized models in India is based on the Rules of 1989, under the Environment (Protection) Act

(EPA), 1986, which is established under "Manufacture, use, import, export and storage of hazardous microorganisms/genetically engineered organisms or cells, Rules 1989." Multiple regulatory bodies have been established to ensure the proper usage of humanization techniques for scientific purposes, focusing on the clinical outcomes and avoiding any kind of exploitation of animal models. This includes the Review Committee on Genetic Manipulation (RCGM) under the Department of Biotechnology (DBT) and the Genetic Engineering Appraisal Committee (GEAC) under the Ministry of Environment and Forests (MoEF), as well as the Central Drugs Standard Control Organization (CDSCO). The IAEC (Institutional Animal Ethics Committee) and IBSC (Institutional Biosafety Committee) are the institutional review committees that regulate the ethical and biosafety concerns associated with research involving humanized animal models. The Food and Drug Administration (FDA), USA, has approved new approach methods (NAMs) as one such strategy in an attempt to reduce the usage of animals, shifting the focus towards new emerging concepts. The enactment of the FDA Modernization Act 2.0 in December 2022 allows the usage of alternatives like in vitro and in silico models, by removing the legal requirement of preclinical drug testing in animal models. Similarly, the European Medicines Agency (EMA) published revised guidelines to address first-in-human (FIH) and early-phase clinical trials (CTs).

Humanization of animal models also faces severe ethical issues due to the involvement of stem cell transplantation (iPSCs), human-animal chimeras, and lifelong immunosuppression in animal models, among many others. The 3Rs (replacement, reduction, and refinement) concept puts a great emphasis on the limits of animal usage and on developing alternative strategies and new generation models.

7.7 Limitations and Future Directions

Humanized models are an invaluable tool, boosting the efficacy of translational research and providing a platform to assess drug development, response, safety, and toxicity in a simulated human environment. It greatly reduces the burden of using animals in large numbers without any clinical outcomes. However, multiple challenges remain a concern in generating humanized models. Species difference often leads to incomplete humanization due to the presence of different sets of regulators and DMEs, which results in variation in the drug response due to the interference of host metabolic pathways that lead to replication failure during clinical trials. Also, novel methods would allow proper maturation and development of human T-cells in HIS-mice, which requires HLA matching between the human HSCs and mouse epithelial cells. Human drug response involves multiple organs working in sync to produce desirable effects. Lack of humanized multi-organ models leads to prediction failure, which can be overcome by adopting advanced technologies like multi-organ on-chip models. Apart from that, there are ethical constraints on the availability of tissues, organs, or iPSCs and the generation of chimeric models. In addition, regulatory gaps due to a lack of standardized protocols for humanization and high expenses on the model development pose major challenges.

3D organoids derived from pluripotent stem cells (PSCs) or adult stem cells have shown great potential in complementing the existing model systems as well as better recapitulation of the in vivo systems as compared to the 2D cell culture system or other in vitro settings. This provides a more physiologically relevant human setting with faster and robust outcomes, enabling patient specificity, in-depth biological studies with a wide variety of applications, including drug discovery, disease study, and cell therapy. However, this model system is still under development, and much of the current excitement stems from the potential it holds to revolutionize disease research, ultimately reducing the burden on the existing animal models.

7.8 Conclusion

In conclusion, humanized animal models are a substantial tool in understanding the mechanism of disease biology and human physiology. It has provided invaluable insights into the development and assessment of various human-based immunotherapies. Mice have been extensively explored in this regard for humanization; however, larger models, including rats and pigs, offer many significant advantages in terms of studying human physiology, disease biology, drug development, and toxicity, owing to their larger anatomy and similarities with humans. Increases in the understanding of xenograft rejection and advances in gene editing technology have further enabled the development of more advanced humanized models, replacing the conventional model systems, thereby providing more accurate and faster results. Future studies focus on the development of complex multi-organ models to accurately mimic human physiology, in studying drug response and metabolism. Generation of human animal chimeras, stem cell-cultured 3D organoids, and organ-on-chip technologies are an emerging tool that further enhances the understanding of human biology and bridges the gap between animal data and human health outcomes.

References

1. Anand RP, Layer JV, Heja D, Hirose T, Lassiter G, Firl DJ, et al. Design and testing of a humanized porcine donor for xenotransplantation. Nature. 2023;622(7982):393–401. https://doi.org/10.1038/s41586-023-06594-4.
2. Andrade D, Redecha PB, Vukelic M, Qing X, Perino G, Salmon JE, et al. Engraftment of peripheral blood mononuclear cells from systemic lupus erythematosus and antiphospholipid syndrome patient donors into BALB-RAG-2 $-/-$IL-2R$\gamma-/-$ mice: a promising model for studying human disease. Arthritis Rheum. 2011;63(9):2764–73. https://doi.org/10.1002/art.30424.
3. Article H. Humanized Drosophila Model of the Meier-Gorlin. 2020;216(December):995–1007.
4. Azuma H, Paulk N, Ranade A, Dorrell C, Al-Dhalimy M, Ellis E, et al. Robust expansion of human hepatocytes in $Fah^{-/-}/Rag2^{-/-}/Il2rg^{-/-}$ mice. Nat Biotechnol. 2007;25(8):903–10. https://doi.org/10.1038/nbt1326.
5. Bility MT, Cheng L, Zhang Z, Luan Y, Li F, Chi L, et al. Hepatitis B virus infection and immunopathogenesis in a humanized mouse model: induction of human-specific liver fibrosis and

M2-like macrophages. PLoS Pathog. 2014;10(3):e1004032. https://doi.org/10.1371/journal.ppat.1004032.
6. Bissig KD, Wieland SF, Tran P, Isogawa M, Le TT, Chisari FV, et al. Human liver chimeric mice provide a model for hepatitis B and C virus infection and treatment. J Clin Invest. 2010;120(3):924–30. https://doi.org/10.1172/JCI40094.
7. Bissig KD, Han W, Barzi M, Kovalchuk N, Ding L, Fan X, et al. P450-humanized and human liver chimeric mouse models for studying xenobiotic metabolism and toxicity. Drug Metab Dispos. 2018;46(11):1734–44. https://doi.org/10.1124/dmd.118.083303.
8. Boettcher AN, Loving CL, Cunnick JE, Tuggle CK. Development of severe combined immunodeficient (SCID) pig models for translational cancer modeling: future insights on how humanized SCID pigs can improve preclinical cancer research. Frontiers in Oncology. 2018;8(NOV):1–14. https://doi.org/10.3389/fonc.2018.00559.
9. Bosma MJ, Carroll AM. The scid mouse mutant: definition, characterization, and potential uses. Annu Rev Immunol. 1991;9(1):323–50. https://doi.org/10.1146/annurev.iy.09.040191.001543.
10. Brehm MA, Bortell R, DiIorio P, Leif J, Laning J, Cuthbert A, Yang C, et al. Human immune system development and rejection of human islet allografts in spontaneously diabetic NOD-Rag1null IL2rγnull Ins2Akita mice. Diabetes. 2010;59(9):2265–70. https://doi.org/10.2337/db10-0323.
11. Caras I, Ionescu I, Pantazica A, Eerde V, Steen H, Heldal I, Haugslien S, Tucureanu C, et al. Humanized mouse model reveals the immunogenicity of Hepatitis B virus vaccine candidates produced in CRISPR / Cas9-edited Nicotiana benthamiana. 2025:1–13. https://doi.org/10.3389/fimmu.2025.1479689.
12. Chang NH, Inman RD, Dick JE, Wither JE. Bone marrow-derived human hematopoietic stem cells engraft NOD/SCID mice and traffic appropriately to an inflammatory stimulus in the joint. J Rheumatol. 2010;37(3):496–502. https://doi.org/10.3899/jrheum.090317.
13. Chen C, Kim WY, Jiang P. Humanized neuronal chimeric mouse brain generated by neonatally engrafted human iPSC-derived primitive neural progenitor cells. JCI Insight. 2016;1(19):1–12. https://doi.org/10.1172/jci.insight.88632.
14. Chen J, Liao S, Xiao Z, Pan Q, Wang X, Shen K, Wang S, et al. The development and improvement of immunodeficient mice and humanized immune system mouse models. Front Immunol. 2022;13:1–13. https://doi.org/10.3389/fimmu.2022.1007579.
15. Chupp DP, Rivera CE, Zhou Y, Xu Y, Ramsey PS, Xu Z, Zan H, et al. A humanized mouse that mounts mature class-switched, hypermutated and neutralizing antibody responses. Nat Immunol. 2024;25(8):1489–506. https://doi.org/10.1038/s41590-024-01880-3.
16. Chuprin J, Buettner H, Seedhom MO, Greiner DL, Keck JG, Ishikawa F, Shultz LD, Brehm MA. Humanized mouse models for immuno-oncology research. Nat Rev Clin Oncol. 2023;20(3):192–206. https://doi.org/10.1038/s41571-022-00721-2.
17. Colas C, Volodina O, Béland K, Pham TNQ, Li Y, Dallaire F, et al. Generation of functional human T cell development in NOD/SCID/IL2rγnull humanized mice without using fetal tissue: application as a model of HIV infection and persistence. Stem Cell Rep. 2023;18(2):597–612. https://doi.org/10.1016/j.stemcr.2023.01.003.
18. Costa VV, Ye W, Qingfeng C, Teixeira MM, Preiser P, Ooi EE, Chena J. Dengue Virus-Infected Dendritic Cells but Not Monocytes , Activate Natural Killer Cells through a Contact-Dependent. MBio. 2017;8(4):pii: e00741-17.
19. Costantini F, Radice G, Lee JJ, Perry W, Son HJ. Insertional mutations in transgenic mice. Prog Nucleic Acid Res Mol Biol. 1989;36(C):159–69. https://doi.org/10.1016/S0079-6603(08)60169-5.
20. Daharsh L, Lohani SC, Ramer-Tait AE, Li Q. Characterization of double humanized BLT-mice with stable engraftment of a human gut bacterial microbiome. Front Microbiomes. 2024;3(July):1–16. https://doi.org/10.3389/frmbi.2024.1404353.
21. Das R, Strowig T, Verma R, Koduru S, Hafemann A, Hopf S, Kocoglu MH, et al. Microenvironment-dependent growth of preneoplastic and malignant plasma cells in humanized mice. Nat Med. 2016;22(11):1351–7. https://doi.org/10.1038/nm.4202.

22. Devoy A, Bunton-Stasyshyn RKA, Tybulewicz VLJ, Smith AJH, Fisher EMC. Genomically humanized mice: technologies and promises. Nat Rev Genet. 2012;13(1):14–20. https://doi.org/10.1038/nrg3116.
23. Deycmar S, Gomes B, Charo J, Ceppi M, Cline JM. Spontaneous, naturally occurring cancers in non-human primates as a translational model for cancer immunotherapy. J Immunother Cancer. 2023;11(1):e005514. https://doi.org/10.1136/jitc-2022-005514.
24. Festing MF, Lovell D, Sparrow S, MAY D, Connors TA. An athymic nude mutation in the rat. Nature. 1978;274(5669):365–6. https://doi.org/10.1038/274365a0.
25. Flisikowska T, Thorey IS, Offner S, Ros F, Lifke V, Zeitler B, et al. Efficient immunoglobulin gene disruption and targeted replacement in rabbit using zinc finger nucleases. PLoS One. 2011;6(6):e21045. https://doi.org/10.1371/journal.pone.0021045.
26. Freedman LP, Cockburn IM, Simcoe TS. The economics of reproducibility in preclinical research. PLoS Biol. 2015;13(6):1–9. https://doi.org/10.1371/journal.pbio.1002165.
27. Garry DJ, Garry MG. Interspecies chimeras and the generation of humanized organs. Circ Res. 2019;124(1):23–5. https://doi.org/10.1161/CIRCRESAHA.118.314189.
28. Gengyo-ando K, Tateyama M, Mitani S, Ando H, Nakai J. A humanized Caenorhabditis elegans model for studying pathogenic mutations in VPS45, a protein essential for membrane trafficking, associated with severe congenital neutropenia. Micropubl Biol. 2023.
29. Gümüşderelioğlu S, Resch L, Brock T, Luxton GWG, Cope H, Tan QKG, et al. A humanized Caenorhabditis elegans model of hereditary spastic paraplegia-associated variants in KLC4. DMM Dis Models Mech. 2023;16(8):dmm050076. https://doi.org/10.1242/dmm.050076.
30. Häberlein F, Mingardo E, Merten N, Schulze Köhling NK, Reinoß P, Simon K, et al. Humanized zebrafish as a tractable tool for in vivo evaluation of pro-myelinating drugs. Cell Chem Biol. 2022;29(10):1541–1555.e7. https://doi.org/10.1016/j.chembiol.2022.08.007.
31. Hallett PJ, Deleidi M, Astradsson A, Smith GA, Cooper O, Osborn TM, et al. Successful function of autologous iPSC-derived dopamine neurons following transplantation in a non-human primate model of Parkinson's disease. Cell Stem Cell. 2015;16(3):269–74. https://doi.org/10.1016/j.stem.2015.01.018.
32. Han X, Chen M, Wang F, Windrem M, Wang S, Shanz S, Xu Q, Oberheim NA, et al. Forebrain engraftment by human glial progenitor cells enhances synaptic plasticity and learning in adult mice. Cell Stem Cell. 2013;12(3):342–53. https://doi.org/10.1016/j.stem.2012.12.015.
33. He D, Zhang J, Wu W, Yi N, He W, Lu P, Li B, et al. A novel immunodeficient rat model supports human lung cancer xenografts. FASEB J. 2019;33(1):140–50. https://doi.org/10.1096/fj.201800102RR.
34. Hidalgo M, Amant F, Biankin AV, Budinská E, Byrne AT, Caldas C, et al. Patient-derived xenograft models: an emerging platform for translational cancer research. Cancer Discov. 2014;4(9):998–1013. https://doi.org/10.1158/2159-8290.CD-14-0001.
35. Hoban DB, Shrigley S, Mattsson B, Breger LS, Jarl U, Cardoso T, et al. Impact of α-synuclein pathology on transplanted hESC-derived dopaminergic neurons in a humanized α-synuclein rat model of PD. Proc Natl Acad Sci USA. 2020;117(26):15209–20. https://doi.org/10.1073/pnas.2001305117.
36. Hsiao EY, McBride SW, Hsien S, Sharon G, Hyde ER, McCue T, et al. Microbiota modulate behavioral and physiological abnormalities associated with neurodevelopmental disorders. Cell. 2013;155(7):1451–63. https://doi.org/10.1016/j.cell.2013.11.024.
37. Hu Y, Wu M, Nishimura T, Zheng M, Peltz G. Human pharmacogenetic analysis in chimeric mice with "humanized livers.". Pharmacogenet Genomics. 2013;23(2):78–83. https://doi.org/10.1097/FPC.0b013e32835cb2c7.
38. Ishikawa Y, Usui T, Shiomi A, Shimizu M, Murakami K, Mimori T. Functional engraftment of human peripheral T and B cells and sustained production of autoantibodies in NOD/LtSzscid/IL-2Rγ−/− mice. Eur J Immunol. 2014;44(11):3453–63. https://doi.org/10.1002/eji.201444729.
39. Ito M, Hiramatsu H, Kobayashi K, Suzue K, Kawahata M, Hioki K, Ueyama Y, Koyanagi Y, Sugamura K, et al. Nod/scid/⁄⁄⁄. Bone. 2002;100(9):3175–82. https://doi.org/10.1182/blood-2001-12-0207.Supported.

40. Jin CH, Xia J, Rafiq S, Huang X, Hu Z, Zhou X, Brentjens RJ, et al. Modeling anti-CD19 CAR T cell therapy in humanized mice with human immunity and autologous leukemia. EBioMedicine. 2019;39:173–81. https://doi.org/10.1016/j.ebiom.2018.12.013.
41. Katano I, Ito R, Kamisako T, Eto T, Ogura T. Mice for generation of humanized mice. 2014;63(January):321–30.
42. Kaur K, Jewett A. Differences in tumor growth and differentiation in NSG and humanized-BLT mice; analysis of human vs. humanized-BLT-derived NK expansion and functions. Cancers. 2023;15(1):112. https://doi.org/10.3390/cancers15010112.
43. Kazuki Y, Gao FJ, Yamakawa M, Hirabayashi M, Kazuki K, Kajitani N, et al. A transchromosomic rat model with human chromosome 21 shows robust Down syndrome features. Am J Hum Genet. 2022;109(2):328–44. https://doi.org/10.1016/j.ajhg.2021.12.015.
44. Kim S, Kim D, Cho SW, Kim J, Kim JS. Highly efficient RNA-guided genome editing in human cells via delivery of purified Cas9 ribonucleoproteins. Genome Res. 2014;24(6):1012–9. https://doi.org/10.1101/gr.171322.113.
45. Kim J, Koo BK, Knoblich JA. Human organoids: model systems for human biology and medicine. Nat Rev Mol Cell Biol. 2020;21(10):571–84. https://doi.org/10.1038/s41580-020-0259-3.
46. Klicznik MM, Benedetti A, Gail LM, Varkhande SR, Holly R, Laimer M, Stoecklinger A, et al. A novel humanized mouse model to study the function of human cutaneous memory T cells in vivo in human skin. Sci Rep. 2020;10(1):1–15. https://doi.org/10.1038/s41598-020-67430-7.
47. Konermann S. 乳鼠心肌提取 HHS public access. J Clin Sleep Med. 2016;12(2):263–6. https://doi.org/10.1126/science.aaa0769.Systematic.
48. Kuhlmann AS, Haworth KG, Barber-Axthelm IM, Ironside C, Giese MA, Peterson CW, et al. Long-term persistence of anti-HIV broadly neutralizing antibody-secreting hematopoietic cells in humanized mice. Mol Ther. 2019;27(1):164–77. https://doi.org/10.1016/j.ymthe.2018.09.017.
49. Lan P, Tonomura N, Shimizu A, Wang S, Yang YG. Reconstitution of a functional human immune system in immunodeficient mice through combined human fetal thymus/liver and CD34+ cell transplantation. Blood. 2006;108(2):487–92. https://doi.org/10.1182/blood-2005-11-4388.
50. Larsen MO, Rolin B. Use of the Göttingen minipig as a model of diabetes, with special focus on type 1 diabetes research. ILAR J. 2004;45(3):303–13. https://doi.org/10.1093/ilar.45.3.303.
51. Lee K, Kwon DN, Ezashi T, Choi YJ, Park C, Ericsson AC, Brown AN, et al. Engraftment of human iPS cells and allogeneic porcine cells into pigs with inactivated RAG2 and accompanying severe combined immunodeficiency. Proc Natl Acad Sci USA. 2014a;111(20):7260–5. https://doi.org/10.1073/pnas.1406376111.
52. Lee EC, Liang Q, Ali H, Bayliss L, Beasley A, Bloomfield-Gerdes T, et al. Complete humanization of the mouse immunoglobulin loci enables efficient therapeutic antibody discovery. Nat Biotechnol. 2014b;32(4):356–63. https://doi.org/10.1038/nbt.2825.
53. Lichtman E, Du H, Shou P, Song F, Suzuki K, Li G, et al. HHS Public Access. 2021;27(11):3141–53. https://doi.org/10.1158/1078-0432.CCR-20-2540.Preclinical.
54. Luo Y, Lu H, Peng D, Ruan X, Eugene Chen Y, Guo Y. Liver-humanized mice: a translational strategy to study metabolic disorders. J Cell Physiol. 2022;237(1):489–506. https://doi.org/10.1002/jcp.30610.
55. Ma J, Tan X, Kwon Y, Delgado ER, Zarnegar A, DeFrances MC, Duncan AW, et al. A novel humanized model of NASH and its treatment with META4, A potent agonist of MET. Cmgh. 2022;13(2):565–82. https://doi.org/10.1016/j.jcmgh.2021.10.007.
56. Ma M, Ge JY, Nie YZ, Li YM, Zheng YW. Developing humanized animal models with transplantable human iPSC-derived cells. Front Biosci (Landmark Ed). 2024;29(1):34. https://doi.org/10.31083/j.fbl2901034.
57. Marsden MD. Benefits and limitations of humanized mice in HIV persistence studies. Retrovirology. 2020;17(1):1–6. https://doi.org/10.1186/s12977-020-00516-2.
58. Mashimo T, Takizawa A, Voigt B, Yoshimi K, Hiai H, Kuramoto T, et al. Generation of knockout rats with X-linked severe combined immunodeficiency (X-SCID) using zinc-finger nucleases. PLoS One. 2010;5(1):e8870. https://doi.org/10.1371/journal.pone.0008870.

59. Masubuchi T, Chen L, Marcel N, Wen GA, Caron C, Zhang J, et al. Functional differences between rodent and human PD-1 linked to evolutionary divergence. Sci Immunol. 2025;10(103):1–39. https://doi.org/10.1126/sciimmunol.ads6295.
60. Mazmanian SK, Cui HL, Tzianabos AO, Kasper DL. An immunomodulatory molecule of symbiotic bacteria directs maturation of the host immune system. Cell. 2005;122(1):107–18. https://doi.org/10.1016/j.cell.2005.05.007.
61. McCune JM, Namikawa R, Kaneshima H, Shultz LD, Lieberman M, Weissman IL. The SCID-hu mouse: murine model for the analysis of human hematolymphoid differentiation and function. Science. 1988;241(4873):1632–9. https://doi.org/10.1126/science.2971269.
62. Ménoret S, Renart-Depontieu F, Martin G, Thiam K, Anegon I. Efficient generation of human immune system rats using human CD34+ cells. Stem Cell Rep. 2024;19:1255–63. https://doi.org/10.1016/j.stemcr.2024.07.005.
63. Meraz IM, Majidi M, Meng F, Shao RP, Ha MJ, Neri S, et al. An improved patient-derived xenograft humanized mouse model for evaluation of lung cancer immune responses. Cancer Immunol Res. 2019;7(8):1267–79. https://doi.org/10.1158/2326-6066.CIR-18-0874.
64. Miao JX, Wang JY, Li HZ, Guo HR, Chard Dunmall LS, Zhang ZX, et al. Promising xenograft animal model recapitulating the features of human pancreatic cancer. World J Gastroenterol. 2020;26(32):4802–16. https://doi.org/10.3748/WJG.V26.I32.4802.
65. Mihaylova N, Chipinski P, Bradyanova S, Velikova T, Ivanova-Todorova E, Chausheva S, et al. Suppression of autoreactive T and B lymphocytes by anti-annexin A1 antibody in a humanized NSG murine model of systemic lupus erythematosus. Clin Exp Immunol. 2020;199(3):278–93. https://doi.org/10.1111/cei.13399.
66. Misharin AV, Haines IK, Rose S, Gierut AK, Hotchkiss RS, Perlman H. Development of a new humanized mouse model to study acute inflammatory arthritis. J Transl Med. 2012;10(1):1–9. https://doi.org/10.1186/1479-5876-10-190.
67. Mosier, D. (1988). *Transfer of functional human immune system to mice* (pp. 256–259).
68. Nelson DR, Zeldin DC, Hoffman SMG, Maltais LJ, Wain HM, Nebert DW. Fp140101 1..18. Pharmacogenetics. 2003;450(1):1–18. https://doi.org/10.1097/01.fpc.0000054151.92680.31.
69. Noto FK, Adjan-Steffey V, Tong M, Ravichandran K, Zhang W, Arey A, et al. Sprague Dawley RAG2-null rats created from engineered spermatogonial stem cells are immunodeficient and permissive to human xenografts. Mol Cancer Ther. 2018;17(11):2481–9. https://doi.org/10.1158/1535-7163.MCT-18-0156.
70. Pal P, Gao S, Gao H, Qin X, Cella M, Wang Q, et al. Establishment of a reverse genetics system for studying human immune functions in mice. Sci Adv. 2025;11(28):eadu1561. https://doi.org/10.1126/sciadv.adu1561.
71. Panzer SE, Wilson NA, Verhoven BM, Xiang D, Rubinstein CD, Redfield RR, et al. Complete B cell deficiency reduces allograft inflammation and intragraft macrophages in a rat kidney transplant model. Transplantation. 2018;102(3):396–405. https://doi.org/10.1097/TP.0000000000002010.
72. Perkins J, Kayi MARKA. A modified urokinase plasminogen regeneration without activator I n d u c e s. Hum Gene Ther. 1995;1037(August):1029–37. Downloaded from online . liebertpub.com by Uc Davis Libraries University of California Davis on 01/04/15. For personal use only.
73. Pumtang-On P, Thron LK, Goodarzi N, Davey BC, Sevcik EN, Coleman-Fuller N, et al. Humanized DRAGA mice are a valuable model to study novel immunotherapies for HIV-1. The Journal of Immunology. 2025;214(11):3017–27. https://doi.org/10.1093/jimmun/vkaf185.
74. Racki WJ, Covassin L, Brehm M, Pino S, Ignotz R, Dunn R, et al. NOD-scid IL2rγnull mouse model of human skin transplantation and allograft rejection. Transplantation. 2010;89(5):527–36. https://doi.org/10.1097/TP.0b013e3181c90242.
75. Rajan V, Melong N, Wong WH, King B, Spencer Tong R, Mahajan N, et al. Humanized zebrafish enhance human hematopoietic stem cell survival and promote acute myeloid leukemia clonal diversity. Haematologica. 2020;105(10):2391–9. https://doi.org/10.3324/haematol.2019.223040.

76. Rongvaux A, Willinger T, Martinek J, Strowig T, Gearty SV, Teichmann LL, et al. Development and function of human innate immune cells in a humanized mouse model. Nat Biotechnol. 2014;32(4):364–72. https://doi.org/10.1038/nbt.2858.
77. Savitsky M, Solis GP, Kryuchkov M, Katanaev VL. Humanization of drosophila gαo to model gnao1 paediatric encephalopathies. Biomedicine. 2020;8(10):1–16. https://doi.org/10.3390/biomedicines8100395.
78. Schmitt K, Charlins P, Veselinovic M, Kinner-Bibeau L, Hu S, Curlin J, Remling-Mulder L, et al. Zika viral infection and neutralizing human antibody response in a BLT humanized mouse model. Virology. 2018;515(January):235–42. https://doi.org/10.1016/j.virol.2017.12.026.
79. Song J, Yang D, Ruan J, Zhang J, Chen YE, Xu J. Production of immunodeficient rabbits by multiplex embryo transfer and multiplex gene targeting. Sci Rep. 2017;7(1):1–9. https://doi.org/10.1038/s41598-017-12201-0.
80. Song Y, Shan L, Gbyli R, Liu W, Strowig T, Patel A, et al. Combined liver–cytokine humanization comes to the rescue of circulating human red blood cells. Science. 2021;371(6533):1019–25. https://doi.org/10.1126/science.abe2485.
81. Steinkamp MP, Lagutina I, Brayer KJ, Schultz F, Burke D, Pankratz VS, et al. Humanized patient-derived xenograft models of disseminated ovarian cancer recapitulate key aspects of the tumor immune environment within the peritoneal cavity. Cancer Res Commun. 2023;3(2):309–24. https://doi.org/10.1158/2767-9764.CRC-22-0300.
82. Takenaka K, Prasolava TK, Wang JCY, Mortin-Toth SM, Khalouei S, Gan OI, Dick JE, Danska JS. Polymorphism in Sirpa modulates engraftment of human hematopoietic stem cells. Nat Immunol. 2007;8(12):1313–23. https://doi.org/10.1038/ni1527.
83. Tan-Garcia A, Lai F, Sheng Yeong JP, Irac SE, Ng PY, Msallam R, Tatt Lim JC, et al. Liver fibrosis and CD206+ macrophage accumulation are suppressed by anti-GM-CSF therapy. JHEP Reports. 2020;2(1):100062. https://doi.org/10.1016/j.jhepr.2019.11.006.
84. Tomizuka K, Yoshida H, Uejima H, Kugoh H, Sato K, Ohguma A, et al. Functional expression and germline transmission of a human chromosome fragment in chimaeric mice. Nat Genet. 1997;16(2):133–43. https://doi.org/10.1038/ng0697-133.
85. Traggiai E, Chicha L, Mazzucchelli L, Bronz L, Piffaretti JC, Lanzavecchia A, et al. Development of a human adaptive immune system in cord blood cell-transplanted mice. Science. 2004;304(5667):104–7. https://doi.org/10.1126/science.1093933.
86. Uehara S, Yoneda N, Higuchi Y, Yamazaki H, Suemizu H. Cytochrome P450-dependent drug oxidation activities and their expression levels in liver microsomes of chimeric TK-NOG mice with humanized livers. Drug Metab Pharmacokinet. 2022;44:100454. https://doi.org/10.1016/j.dmpk.2022.100454.
87. Viehmann Milam AA, Maher SE, Gibson JA, Lebastchi J, Wen L, Ruddle NH, et al. A humanized mouse model of autoimmune insulitis. Diabetes. 2014;63(5):1712–24. https://doi.org/10.2337/db13-1141.
88. Wallace HAC, Marques-Kranc F, Richardson M, Luna-Crespo F, Sharpe JA, Hughes J, et al. Manipulating the mouse genome to engineer precise functional syntenic replacements with human sequence. Cell. 2007;128(1):197–209. https://doi.org/10.1016/j.cell.2006.11.044.
89. Weber J, Peng H, Rader C. From rabbit antibody repertoires to rabbit monoclonal antibodies. Exp Mol Med. 2017;49(3):e305–12. https://doi.org/10.1038/emm.2017.23.
90. Weißmüller S, Kronhart S, Kreuz D, Schnierle B, Kalinke U, Kirberg J, et al. TGN1412 induces lymphopenia and human cytokine release in a humanized mouse model. PLoS One. 2016;11(3):1–19. https://doi.org/10.1371/journal.pone.0149093.
91. Wetsel RA, Wang D, Calame DG. Therapeutic potential of lung epithelial progenitor cells derived from embryonic and induced pluripotent stem cells. Annu Rev Med. 2011;62:95–105. https://doi.org/10.1146/annurev-med-052009-172110.
92. Wolf E, Kemter E, Klymiuk N, Reichart B. Genetically modified pigs as donors of cells, tissues, and organs for xenotransplantation. Anim Front. 2019;9(3):13–20. https://doi.org/10.1093/af/vfz014.
93. Wos-Oxley ML, Bleich A, Oxley APA, Kahl S, Janus LM, Smoczek A, et al. Comparative evaluation of establishing a human gut microbial community within rodent models. Gut Microbes. 2012;3(3):1–16. https://doi.org/10.4161/gmic.19934.

94. Xia J, Hu Z, Yoshihara S, Li Y, Jin CH, Tan S, Li W, Chen Q, Sykes M, et al. Modeling human leukemia immunotherapy in humanized mice. EBioMedicine. 2016;10:101–8. https://doi.org/10.1016/j.ebiom.2016.06.028.
95. Xu D, Wu M, Nishimura S, Nishimura T, Michie SA, Zheng M, et al. Chimeric TK-NOG mice: a predictive model for cholestatic human liver toxicity. J Pharmacol Exp Ther. 2015;352(2):274–80. https://doi.org/10.1124/jpet.114.220798.
96. Yang X, Zhou J, He J, Liu J, Wang H, Liu Y, et al. An immune system-modified rat model for human stem cell transplantation research. Stem Cell Rep. 2018;11(2):514–21. https://doi.org/10.1016/j.stemcr.2018.06.004.
97. Yuan L, Jiang J, Liu X, Zhang Y, Zhang L, Xin J, et al. HBV infection-induced liver cirrhosis development in dual-humanised mice with human bone mesenchymal stem cell transplantation. Gut. 2019;68(11):2044–56. https://doi.org/10.1136/gutjnl-2018-316091.
98. Zhang JX, Lai YH, Mi PY, Dai XL, Zhang R, Zhang ZJ, et al. Rescue of cognitive deficits in APP/PS1 mice by accelerating the aggregation of β-amyloid peptide. Alzheimer's Research and Therapy. 2019;11(1):1–16. https://doi.org/10.1186/s13195-019-0560-6.
99. Zhang L, Ge JY, Zheng YW, Sun Z, Wang C, Peng Z, et al. Survival-assured liver injury preconditioning (SALIC) enables robust expansion of human hepatocytes in fah−/−Rag2−/−IL2rg−/− rats. Adv Sci. 2021;8(19):1–14. https://doi.org/10.1002/advs.202101188.
100. Zhao Y, Shuen TWH, Toh TB, Chan XY, Liu M, Tan SY, et al. Development of a new patient-derived xenograft humanised mouse model to study human-specific tumour microenvironment and immunotherapy. Gut. 2018;67(10):1845–54. https://doi.org/10.1136/gutjnl-2017-315201.

Part III

Alternative Model Organisms for Research in Health and Medicine

Working with Worms: Protocols and Some Practical Considerations for a *C. elegans* Laboratory

8

Nidhi Thakkar, Gemini Gajera, Hirva Shah, Adarshi Bangiawala, Dhyey Nagariya, Pranshu Dixit, and Vijay Kothari

Model organisms have remained the mainstay of biological research for a long. With increasing ethical concerns with respect to the use of higher animals (primates, pigs, rats, mice, etc.), researchers have found the alternative model organisms like the fruit fly (*Drosophila melanogaster*), zebrafish (*Danio rerio*), and the nematode worm *Caenorhabditis elegans* quite useful for research in various fields of biology. Since the basic biology of the worm lifecycle and details of its life in the lab are already available, researchers can now directly focus on designing experiments tailored to their specific scientific questions [20]. Wild type and transgenic strains of *C. elegans* have found widespread applications in labs working in areas of microbiology [43], diabetes [4], neurobiology [44], wound healing [5], aging and longevity [55], stress biology [49], developmental biology [52], toxicity [48], and anthelmintic [47] studies, etc. The community of worm researchers is growing globally, and many new protocols are being reported for working with *C. elegans*. While platforms like WormBook and WormAtlas serve as excellent resources (Table 8.1) for beginners as well as experienced worm researchers in general, each type of assay has its own set of intricacies. Our lab has been using *C. elegans* for microbiological and non-microbiological assays for over a decade, and in this chapter, we are providing some protocols as well as certain additional experimental insights we have gained while learning and optimizing those protocols.

N. Thakkar · G. Gajera · H. Shah · A. Bangiawala · D. Nagariya · P. Dixit · V. Kothari (✉)
Institute of Science, Nirma University, Ahmedabad, Gujarat, India
e-mail: vijay.kothari@nirmauni.ac.in

P. Kumar, V. Kothari (eds.), *Model Organisms in Biological Research*,
https://doi.org/10.1007/978-981-92-0038-2_8

Table 8.1 Resources for *C. elegans* researchers

Resource	Website	Description
WormBase	http://www.wormbase.org/	An international consortium of biologists and computer scientists providing accessible information about the genetics, genomics and biology of *C. elegans* and related nematodes
WormBook	http://www.wormbook.org/	A comprehensive resource covering various aspects of *C. elegans* and other nematodes. It includes WormMethods, a collection of experimental protocols; WormHistory, featuring personal perspectives on the development of the field; and the Worm Breeder's Gazette, an informal newsletter
WormAtlas	http://www.wormatlas.org/	An NIH-funded centralized resource for the structural anatomy of *C. elegans* and other nematodes
Caenorhabditis Genetics Center	https://www.cbs.umn.edu/research/resources/cgc	Source of stocks of wild-type and mutant nematode strains
Open Worm Science	http://www.openworm.org/	Aimed at building the first comprehensive computational model of *C. elegans*

8.1 General Protocols

Though setting up a basic *C. elegans* lab may not be very expensive, execution of advanced protocols with this worm is dependent on the availability of specialized equipment. Requirements for a functional worm laboratory are listed in Table 8.2. In the following text, wherever 'worm activity' is mentioned, that refers to quantification of worm motility through an automated worm tracker (Phylumtech ARENA).

8.1.1 Maintenance of *C. elegans*

8.1.1.1 Harvesting Worms from an Old Plate

1. Adult worms, maintained on *E. coli* OP50 (used as food for *Caenorhabditis elegans*) seeded NGM agar plate (~maximum 7 days old; 90 mm petri plate) in incubator, are collected from the old plate using ~10 mL sterile M9 buffer in 15 mL sterile centrifuge tube ('V' bottom), and are allowed to settle down for 5–10 min at ambient temp (not more than 25 °C).

8.1.1.2 Washing

1. Supernatant is removed and pellet is reconstituted with 5–10 mL sterile M9 buffer.
2. The above-mentioned step is repeated for 5–6 times (till the supernatant appears clear) and then ~300 μl of sterile M9 buffer is added to the worms settled at the bottom.

Table 8.2 Requirements for setting up a *Caenorhabditis elegans* laboratory

Sr. No.	Item	Description
Basic requirements		
1	Temperature-controlled room	An air-conditioned room is preferred, particular in countries where ambient temp exceed 25 °C
2	Incubator with cooling capacity	Wild type strain to be maintained at 20–22 °C; some transgenic strains may have still lower temp requirement (e.g. 15 °C for the strain CL4176)
3	Biosafety cabinet	While *C. elegans* itself is a non-pathogenic organism, labs employing it as a host for pathogenic microorganisms will need a biosafety cabinet
4	Microscopes	Stereomicroscope/dissecting microscope for empirical observation Compound binocular microscope with a halogen light source (yellow light) for worm counting
5	Worm pick tools	Platinum wire loop mounted on glass pipettes for individual worm transfer
6	Nematode Growth Medium (NGM) (3 g/L NaCl, 1 M $CaCl_2$, 1 M $MgSO_4$, 2.5 g/L peptone, 5 mg/mL cholesterol, 1 M phosphate buffer of pH 6, 17 g/L agar-agar)	While preparing NGM, $MgSO_4$, $CaCl_2$, and phosphate buffer are autoclaved separately, and then added into the sterile mixture of the rest of the contents. Cholesterol is dissolved in ethanol, and filter-sterilized NGM agar is used in worm maintenance, as well as in any assays requiring solid media
7	Petri dishes and multiwell plates	Petri dishes for various sizes e.g. 90 mm, 50 mm, 35 mm for different assay formats Multiwell plates: While 6-well and 24-well plates are okay for routine experiments, 96-well and 384-well plates are more suitable for high throughput screening in the worm platform Choice of plates is defined often by the population size of the worms, along with other factors like incubator space, objective of the assay, etc.
8	*Escherichia coli* OP50 (uracil auxotroph), and appropriate growth media (e.g., nutrient broth) for it	Live *E. coli* OP50 may be obtained from CGC, USA. Lyophilized E. coli OP50 may be purchased from commercial suppliers (e.g. LabTIE B.V., the Netherlands) *E. coli* wild-type K12 strain, K12-derived strain HT115 (D3) can also be used as an alternative of *E. coli* OP50 [60] While non-pathogenic *E. coli* is used as a standard lab food for *C. elegans*, theoretically, any non-pathogenic bacteria can be used to feed these worms. Three probiotic strains were shown to support worm survival at par with the worm's regular lab food *E. coli* OP50 [37]

(continued)

Table 8.2 (continued)

Sr. No.	Item	Description
9	M9 buffer (3 g/L KH_2PO_4; 6 g/L Na_2HPO_4; 5 g/L NaCl); S-buffer (129 ml 0.05 M K_2HPO_4, 871 ml 0.05 M KH_2PO_4, 5.85 g NaCl)	For washing and transferring worms While M9 is the most commonly used buffer/medium, few other options, like K medium and S buffer, are also available
10	Bleaching reagents	Worm synchronization (sodium hypochlorite + NaOH)
11	Refrigerator	For storing worm-containing plates
12	Centrifuge (with temp control)	For pelleting worms or bacteria
Advanced requirements		
1	Deep freezer (−80 °C)	For long-term storage of worm stocks (glycerol stocks)
2	Fluorescence microscope	For working with GFP/RFP-labelled transgenic strains
3	PCR machine (thermal cycler)	For genotyping or expression analysis
4	Microinjector setup	For creating transgenic strains
5	Imaging system attached to microscope	To capture images for publications or time-lapse studies
6	Worm tracker	For behavioural analysis and motility tracking (e.g., WMicrotracker by Phylumtech S.A. Argentina; WormLab® by MBF Bioscience, USA, etc.)
7	Worm synchronizer	For obtaining age-synchronized populations of *C. elegans*
8	Confocal microscope	For imaging the worm body at different depths

3. The liquid droplets containing worms are placed on the peripheral surface of one side of the NGM plate pre-seeded with E. coli OP50 culture.

   [Preparing pre-seed plates: At the centre of the NGM agar plate, add 100–150 μL of freeze-dried *E. coli* OP50 suspension (OD_{625} = 1) prepared in sterile distilled water. With the help of a pipette, and by tilting plate in a circular fashion, make the bacterial suspension evenly spread over a greater area in the centre of the Petri plate.]
4. The plate is incubated at 22 °C till gravid adults are visible (usually after 3–4 days) on the plate, upon microscopic examination. Such plates containing adult worms should be used within 7 days, or else they should be transferred to a refrigerator.

   [A note on 'chunking': As an alternative method of worm transfer, "chunking" of agar piece containing worms from an old to a new plate can also be done. Wherein, using a sterile scalpel, a ~1 cm^2 chunk of agar from the old NGM plate containing *C. elegans* is placed in an inverted (agar side down) position onto a fresh NGM plate pre-seeded with *E. coli* OP50. This agar chunk is allowed to remain on the fresh plate for approximately 30 min, enabling the worms to migrate onto the new plate. After that, the agar chunk is carefully removed from the fresh plate, and the latter is incubated at 22 °C until further use.]

8.1.2 Synchronization

1. Gravid adults are collected into a 15 mL centrifuge tube from an old plate using M9 buffer and are allowed to settle down.
2. Washing is done as described in Sect. 8.1.1.2.
3. Bleaching solution, equal amount of 1 N NaOH and 4% sodium hypochlorite (~500 μL each, adjusted according to worm density), is added to the worm pellet, and vortexed for ~1 min to facilitate the release of eggs from within the worm bodies. The released eggs should be visible microscopically in 15 mL centrifuge tube when placed in a tilted position under a microscope (4X).
4. Eggs are then collected by centrifuging at ~1800–2000 rpm for 3 min. The supernatant is removed and the pellet containing eggs is mixed with 15 ml of sterile M9 buffer and centrifuged at ~1800–2000 rpm for 3 min.
5. This procedure of adding fresh M9 buffer, followed by centrifugation, and removal of supernatant is repeated 5–6 times (depending on the pellet size) to completely remove the bleaching solution.
6. After the final round of centrifugation, the supernatant is discarded, and the egg-pellet is reconstituted with ~100 μL of antibiotic solution (kanamycin or streptomycin at 10 mg/mL; a combination of antibiotics may also be tried) to protect them from bacterial contamination.
7. Drops of this egg suspension are then pipetted onto the periphery (along half of the diameter of the plate) of a pre-seeded NGM agar plate, containing *E. coli* OP50 in the centre, and the plate is then incubated at 22 °C. During incubation, eggs will hatch, and the progeny worms will start migrating towards the bacterial food in the centre of the plate.
8. Usually, after around 30 h of incubation, L3-L4 stage worms will be largely present near the bacterial food in the centre. One or two mL of sterile M9 buffer is then added in the centre of the plate, while keeping the plate tilted, so that the M9 liquid flow is directed towards the side of the plate, opposite to where previously eggs were loaded.
9. Majority of the worms thus collected will be in the same phase of their life cycle (L3-L4), and hence this population can be considered as synchronized, and can be used further for various assays.

8.1.3 How to Get Gnotobiotic Worms?

If *C. elegans* used in an assay is fed with live *E. coli* OP50, this monoculture microbiota in the worm intestine may have an influence on worm survival, fertility, and its response to various experimental conditions. Use of gnotobiotic worms, which do not harbour any microbial flora in their intestine, can safeguard against the possible interference of gut biota. If worm intestine is already having presence of non-pathogenic bacteria, it was fed on, they may not allow proper colonization by a pathogenic bacteria when worms are used as a model host for any such pathogen. Similarly, while evaluating the efficacy of any probiotic product in a worm model, their gnotobiotic status is a prerequisite. We recently demonstrated prophylactic efficacy of a *Panchagavya* formulation against subsequent pathogen-challenge to the gnotobiotic worms [8]. In this case, since the test formulation itself has a rich metagenomic profile, use of gnotobiotic worms becomes necessary for evaluation of such a product's efficacy. Though gnotobiotic worms are in a somewhat 'hungry' state, and that may also have its own impact on the outcome of an assay and interpretation of results; 'non-gnotobiotic state' perhaps may be a greater confounding factor. In one of our recent studies, while there was no significant difference in worm activity between gnotobiotic and non-gnotobiotic worms, the non-gnotobiotic worms responded more actively to the root extract of *Withania somnifera* (Fig. 8.1). Hence, careful consideration should be made while making a choice between gnotobiotic or non-gnotobiotic worms for a particular assay. The protocol for obtaining gnotobiotic worms is described below.

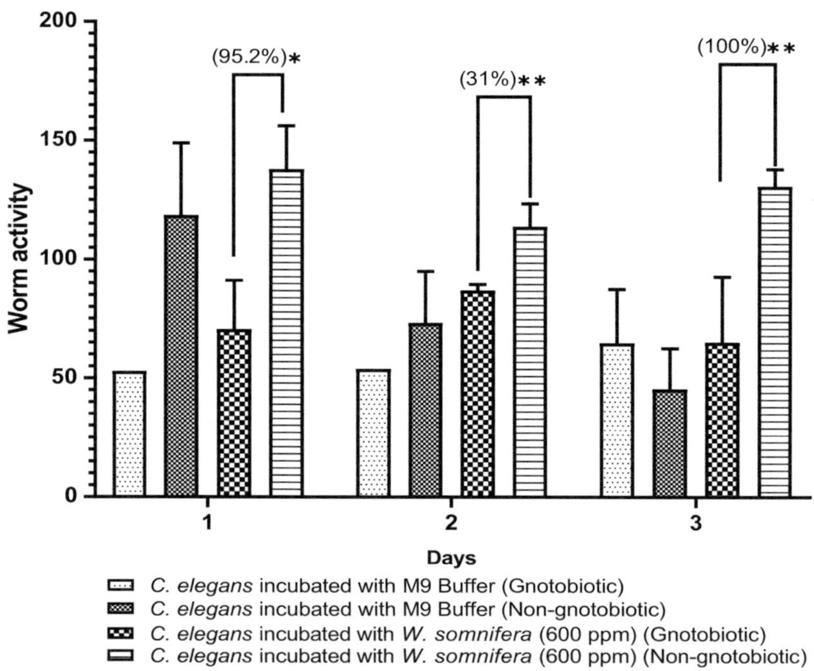

Fig. 8.1 *E. coli* OP50-fed worms respond to *Withania somnifera* root extract more positively than their gnotobiotic counterparts. No significant difference between the activity of control populations (not treated with extract) of gnotobiotic and non-gnotobiotic worms was observed. However, upon treatment with *W. somnifera* root extract, non-gnotobiotic worms exhibited a significantly higher activity, as compared to gnotobiotic worms. This assay was performed in a 24-well plate, wherein each well contained approximately 100 worms in 1 mL of M9 buffer. The worm tracker used for capturing worm activity was set at 23 °C, and the acquisition lapse time was kept at 15 min

1. Synchronized worms are collected from NGM agar plates in a sterile centrifuge tube using 1–2 mL M9 buffer per plate, and subsequently washed twice with M9 buffer to remove the residual *E. coli* OP50.
2. Further, these L3-L4 stage synchronized worms are centrifuged at ~1000 rpm for 2 min.
3. Supernatant is discarded carefully, and the worm pellet is reconstituted with 200 μL M9 buffer, to be transferred onto newly prepared NGM agar plates (free from any bacterial food).
4. These plates are then incubated at 22 °C for 48 h. This duration, in our experience, is the minimum required for the worms to digest all the bacterial food within their gut and attain the gnotobiotic state. These gnotobiotic worms can then be collected for further use.
5. Their gnotobiotic character can be confirmed by crushing these worms and streaking the resulting suspension on a sterile nutrient agar plate. Following incubation, this plate should not develop any bacterial colonies if the worms crushed were gnotobiotic.

8.1.4 Cryopreservation

Cryopreservation of *C. elegans* is essential for long-term storage of valuable strains without the need for continuous sub-culturing, which can lead to genetic drift or contamination over time. It allows researchers to preserve genetic lines in a stable state, ensuring consistency and reproducibility in experiments. Additionally, it serves as a backup in case of culture loss. The protocol for the same follows.

1. A freshly starved population of *C. elegans* at the L1-L2 larval stage is collected using M9 buffer, transferred into sterile centrifuge tubes, and washed twice with sterile S buffer, allowing the worms to settle at the bottom after each wash.
2. Discard the supernatant carefully from the tube and reconstitute the worm pellet in equal volume (~500 μL of each component) of S buffer and Freezing solution (S buffer +30% glycerine).
3. Dispense the worm suspensions in the cryovials, ensuring appropriate labelling.
4. Place the cryovials in the Cryo 1 °C Cooler freezing container (e.g., Tarson 525000) to facilitate controlled freezing and store them in a −80 °C freezer.
5. Rejuvenation of worms stored at −80 °C is done by thawing the frozen vials for 15–20 min at room temperature until the ice is completely converted to a liquid state. (Since the recovery rate may not be very high, it is advisable to keep the number of originally cryopreserved worms as high as possible.)
6. Transfer the thawed suspension onto the freshly prepared NGM plate pre-seeded with *E. coli* OP50. In our experience, worms preserved this way can be recovered safely for up to 6 months. Periodic survival tests can be done to ensure worm safety.

8.2 Worm as a Screening Platform for Various Biological Activities

C. elegans can be used as a useful pre-clinical screening platform for detecting various types of biological activities in a library of natural products or synthetic compounds. During these screening exercises, the effect of test products on the worm population is measured by quantifying different lifespan and healthspan parameters such as worm motility, metabolic activity, fertility, etc. Protocols for quantifying these important parameters are presented in the following sub-sections. Most of these assays can be done in solid as well as liquid media, and with different population sizes, but the choices must be made carefully.

8.2.1 Assay for Anthelmintic Activity

Since the burden of nematode infections in humans, animals, and plants is considerable, there is a clear need for dedicated efforts for the discovery of novel anthelmintic compounds. Despite not being a parasitic nematode, *C. elegans* has been considered a useful platform for anthelmintic and nematicide discovery [30]. While the protocol [9] described in the next paragraph is based on liquid media, assay for evaluation of anthelmintic activity can also be done in solid media too. To evaluate the influence of the physical state of the assay medium on worm susceptibility to anthelmintics, we conducted a comparative analysis. While the initial baseline worm activity was not found to be different between those on NGM agar or in M9 buffer, the magnitude of initial shock (within the first hour of toxin challenge) was higher in liquid media. However, over a four-day period, worm activity on solid media was statistically similar to that in liquid media (Fig. 8.2). It should be noted that solid media (devoid of anthelmintic agent) supports progeny formation from the third day onward, which is not the case with liquid media. Though progeny formation was not observed in the presence of anthelmintic compounds, to avoid any possibility of interference from progenies (which may occur in the presence of a weak anthelmintic agent, or in the presence of sub-lethal doses of a potent anthelmintic agent too), it is advisable to perform this assay in liquid media.

1. Gnotobiotic worms (obtained as described in Sect. 8.1.3) are picked from the NGM agar plate, and are distributed into different wells of a 24-well plate (flat bottom; surface non-treated) containing M9 buffer. Ten (hermaphrodite) worms are delivered per well. This is followed by the addition of the required volume of test compound. The total volume in each well is kept at 1 mL.
2. Three replicates for each concentration of test compounds are set. Control wells contain worms in M9 buffer (with no test compound). Benzimidazole (1000 ppm) and ivermectin (1 ppm) are used as positive controls. Since these compounds are dissolved in dimethylsulfoxide (DMSO), an appropriate vehicle control (i.e., worms in M9 buffer+0.5%v/v DMSO) is also set.

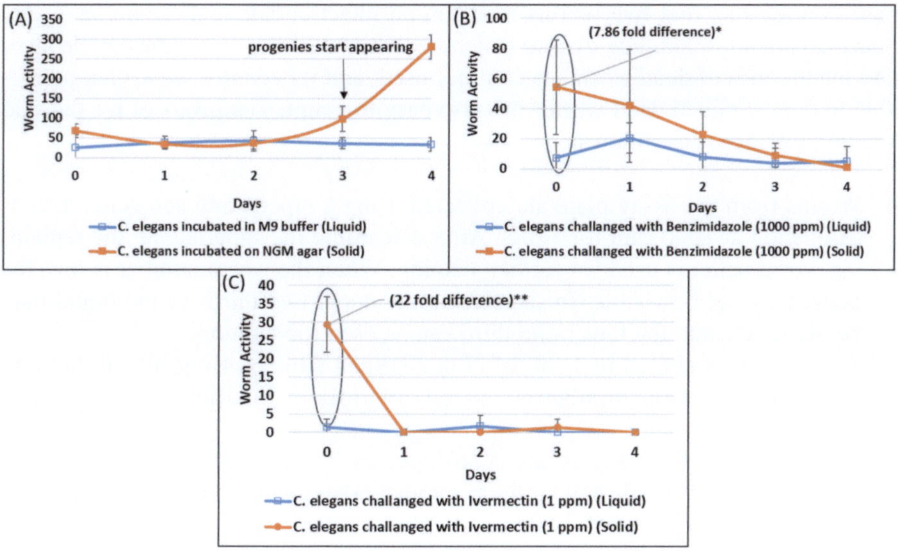

Fig. 8.2 Magnitude of anthelmintic activity of ivermectin and benzimidazole remains identical in liquid and solid media. The population size was 50 ± 10 gnotobiotic worms per well. Each well contained 1 mL of media. (**a**) *Control worm population.* Worm activity till the second day, in solid (36.34 ± 15.3) or liquid (42.67 ± 26.1) media, remained statistically the same. Higher activity in worms on solid media, day three onward, is attributable to progenies. (**b**) *Effect of ivermectin on worm population.* Within the first hour of ivermectin (1 ppm)-challenge, worm activity in liquid media (1.33 ± 2.30) was much lower than that in solid media (29.33 ± 7.50). However, such a difference was not observed with lower (0.5 ppm) ivermectin concentration, for which the worm activity values in liquid and solid media were 4.6 ± 6.4 and 20 ± 10.58 (not plotted). (**c**) *Effect of benzimidazole on worm population.* Within first hour of benzimidazole challenge, worm activity in liquid media (7.33 ± 1.15) was much lower than that in solid media (57.66 ± 26.16). However, this difference disappeared thereafter

3. These 24-well plates are incubated at 22 °C for 5-days, and a live-dead count is performed under a microscope (4X) on a daily basis.
4. Non-moving straight worms are considered dead. Plates are tapped to confirm the absence of movement in apparently dead worms. On the last day of the experiment, when plates could be opened, dead-looking worms were touched gently with a straight wire to further confirm the lack of response. Live-dead staining, described in the next section, can also be performed for further confirmation.
5. Plot worm survival as a function of time.

8.2.2 Live-Dead Staining

In anthelmintic research or toxicity studies, often worms die at LD_{50} and LD_{100} doses, and to complement the microscopic observation of live-dead count, live-dead staining is useful. This staining method can distinguish live worms from dead ones based on the fact that live worms maintain an intact cuticle and membranes, and hence do not allow staining dyes to penetrate; while the integrity of the cuticle layer and membranes of dead worms is compromised, and it becomes somewhat permeable to dyes. Stained dead worms are also easy to count. The protocol for the said staining [39, 56] is as follows.

1. Worms from the assay plate are collected using a pipette and subjected to centrifugation at 1000 rpm for 5 min. After discarding the supernatant, the remaining worm pellet is used for further staining. When the worm number is low, the pellet may not be visible. In such cases, the upper two-thirds of the liquid may be discarded, and the lower one-third can be used for staining.
2. Worms are incubated in 1 mL of either Trypan Blue (0.4 mg/mL in distilled water) for 20 min, or neutral red (20 mg/mL in distilled water) for 2 h at 22 °C.
3. After incubation, the worms are centrifuged at 700 rpm for 1 min, followed by two washes with sterile M9 buffer to remove excess dye. The stained worms are gently resuspended in M9 buffer, and a drop of the suspension is placed onto a microscope slide for observation under a light microscope (4X).
4. Dead worms will appear red or blue, depending on the dye used, while live worms will remain unstained. Images and videos of worms stained through this protocol can be seen respectively in Gajera et al. [9] and at: 10.17605/OSF.IO/S9BMZ.

8.2.3 Lifespan and Healthspan Assay

This assay is quite useful in the field of aging and longevity research. While many nutraceutical formulations are claimed to impart longevity, preliminary validation of their claims can be tried in the *C. elegans* model. Traditionally, such assays rely on a simple live-dead count; however, longevity achieved at the cost of health makes little sense. Hence, we recommend simultaneous quantification of lifespan and worm activity (motility) in such assays, which can be achieved through automated worm tracker machines. Though the use of automated worm trackers may look simple, rapid, and less labour-intensive, various factors influencing the performance of such machines need to be optimized before relying on data generated by them.

Besides considering lifespan and healthspan simultaneously, another important consideration in such assays is whether to use anti-fertility agents like FuDR (5-fluoro-2′-deoxyuridine). While FuDR can avoid interference from progenies during such assays, and its use is quite common in worm labs globally, its effect should not always be assumed to be neutral [31]. We recently reported how the use of FuDR-pre-treated worms masked the positive effect of *W. somnifera* root extract on worm population, i.e., false-negative results may arise owing to the influence of FuDR [53]. The protocol for simultaneous assessment of lifespan and healthspan is described below.

1. Synchronized (L3-L4 stage) gnotobiotic worms are incubated at 22 ± 1 °C with different concentrations of the test formulation in 24-well plates. Control wells will contain worms incubated in the absence of the test formulation.
2. Each well contains 1 mL total volume (995 μL M9 buffer +5 μL test formulation). Appropriate vehicle control containing (0.5%v/v DMSO + worms) is also included in the experiment, if the test formulation is dissolved in DMSO.
3. Two different 24-well plates may be set. One containing 10 worms per well, and to be observed through manual microscopic count. Another containing ≥25 worms per well, to be read through an automated motility tracker. These plates are to be monitored over a 12-day period (or till almost all worms die in the control wells). Before that, an important endpoint should be the day when the control population experiences ~50% death.
4. Plot the worm survival or activity as a function of time for the full duration of the assay, using appropriate tools, e.g., GraphPad Prism or Microsoft Excel®.

8.2.4 Metabolic Activity

For capturing the metabolic activity of living cells, the MTT assay is the most commonly used method. However, this is not suitable for *C. elegans* (we have tried to optimize the MTT assay for this worm in our lab, but no linear correlation between the number of live worms and the amount of dye metabolized was found). Another widely used method for capturing metabolic activity is the Alamar Blue® assay, which is widely employed to quantify viability or metabolic activity of microbial cells [42] or eukaryotic cell lines [13]. However, its use with *C. elegans* is not yet that common. When the Alamar Blue dye is added to the wells containing living cells or organisms, the metabolically active cells reduce the blue dye (resazurin) to a pink or red coloured product (resorufin), and the intensity of this colour is proportional to the viability of the biological entity. Various parameters like concentration of the dye, incubation period, and population size of the biological entity (or cell number), etc. are factors that can influence the outcome of this assay. Data showing the correlation between worm number and assay readout (OD_{570}) is presented in Fig. 8.3. Based on these results, the following protocol was finalized.

1. Synchronized worms are incubated at 22 ± 1 °C in 24-well plates, wherein each well contains approximately 100 worms in M9 buffer (900 μl), and one hundred μL of Alamar Blue® (Thermo Fisher), making the total volume 1 mL.
2. Appropriate abiotic controls (containing the dye and other media components, but no worms) are also included in the assay.
3. Sufficient number of replicates should be included in the assay, as the amount of dye reduced may be quantified at multiple end-points in time. The first endpoint may be the time when visible colour change can be detected, and that usually will take >48 h. Content from one or more wells can be used at each time point for reading OD.

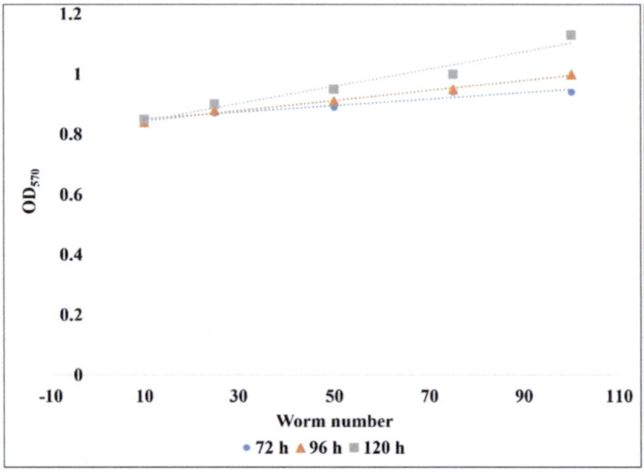

Worm number	Correlation coefficient (r) between OD_{570} values for all time-points and worm number
10	0
25	0.18
50	1
75	0.98
100	0.98

Incubation Time (h)	72	96	120
Correlation coefficient (r) between worm number and OD_{570} for each time-point	0.96	0.99	0.97

Fig. 8.3 Worm population's metabolic response to the Alamar Blue dye as a function of incubation length and worm number. From the correlation (left table) between OD_{570} values and worm number (10–100) for three different incubation lengths, incubation of worms with the dye for any time period between 72 and 120 h seems acceptable, with 96 h end-point being apparently the best. The correlation (right table) between OD_{570} (for all three incubation period end-points) and worm number was acceptable for worm population size 50–100. Population size <50 may be avoided for this assay

4. To quantify the amount of dye reduced, full content from each assay well is transferred into a separate plastic vial (1.5 mL), followed by centrifugation (13,600 g at 25 °C) for 10 min.
5. Then the supernatant is read at 570 nm. This wavelength reads the intensity of the pink colour of the reduced product.
6. If the effect of any product is being checked on worm metabolic activity, OD_{570} between control and experimental wells can be compared to draw a conclusion.

8.2.5 Screening for Anti-stress Activity

C. elegans is widely used as a model for anti-stress research, too. Different types of stressors (e.g., mitotoxins, neurotoxins, oxidizing agents, heat, etc.), at sub-lethal doses, can be used to induce stress in this worm. While various transgenic strains of *C. elegans* are available for specific purposes, the wild-type N2 strain remains the most commonly used strain, in which a large variety of assays can be done by varying the type of stressor. Some examples of stressors are shown in Fig. 8.4, along with brief notes on their relevance to worm experiments.

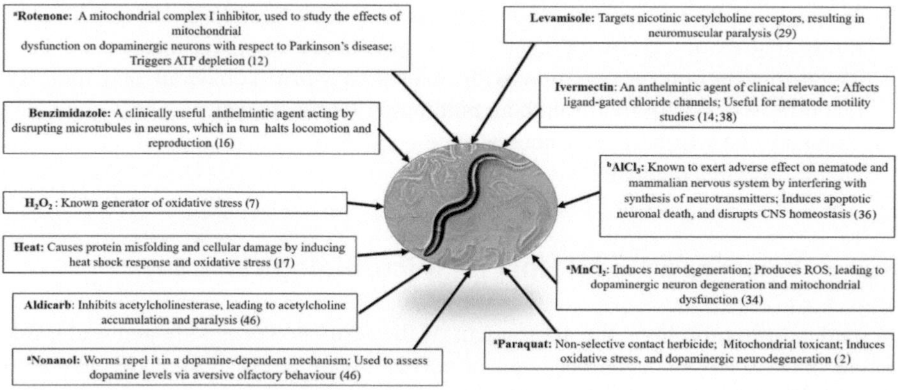

Fig. 8.4 Different stressors useful for inducing a variety of stress in *C. elegans*. H_2O_2: Hydrogen peroxide, $AlCl_3$: Aluminium Chloride, $MnCl_2$: Manganese Chloride, CNS: Central Nervous System, ROS: Reactive Oxygen species; [a]Induces symptoms relevant to Parkinson's disease; [b]Induces symptoms relevant to Alzheimer's disease

A simple protocol for assessing anti-stress (or toxicity-alleviating) potential of any test formulation is described below. While the worms can be challenged with toxins in the presence or absence of the test formulation, we suggest that the worms pre-treated with the test formulation should subsequently be challenged with toxins, thus avoiding the simultaneous presence of toxin and the test formulation in the assay well. This guards against complications in result interpretation owing to possible interaction of test formulation with the stressor/toxin.

1. Gnotobiotic worms are separated into two groups, one for the control and one for the experimental. The experimental worm population is incubated in 5 mL liquid media (contained in conical flask of 25 mL volume) containing the test formulation for 48 h at 22 °C, while the control worm group is kept in liquid media devoid of the test formulation.
2. Intermittent shaking should be provided throughout incubation for the sake of homogeneity, and this will also prevent worms from remaining settled at the bottom.
3. After incubation, transfer the content from each flask into sterile centrifuge tubes (15 mL) and rotate at 1000 rpm for 2 min. The resulting pellet should be mixed with fresh sterile M9 buffer (5 mL), and again centrifuged. Repeat this washing step to ensure complete removal of the residual test formulation from the experimental worm population. (Though control worm population is not treated with any test formulation, this washing step is done with them too, to make them as 'equivalent' as possible to the experimental population.)
4. Worms obtained at the end of the above processing are distributed in 24-well plates. Approximately 100 worms are added per well. Add sub-lethal concentrations of an appropriate stressor chemical, and incubate the plates at 22 °C for a sufficient number of days.

5. While a single positive control compound cannot be recommended for different kinds of stressors, cortisol (7.5–15 mM) can be useful in many cases [59].
6. Make microscopic observation with respect to worm behaviour and motility. Quantify worm activity through an automated worm-tracker.
7. Compare the worm activity between the control and worm populations to draw an appropriate conclusion.

8.3 Worm as a Model Host for Microbiological Assays

In medical microbiology, the assessment of the virulence of a pathogen under different experimental conditions is an important requirement. Virulence cannot be assessed in the absence of a host, and *C. elegans* has served well as a model host for many human-pathogenic bacteria and fungi (Table 8.3). Owing to its susceptibility to many human pathogens, *C. elegans* is an excellent platform for screening large libraries of natural products [35] as well as synthetic compounds [10] for preliminary in vivo anti-pathogenic activity. The anti-pathogenic activity of any bioactive molecule or extract may be owing to its antagonistic effect on the bacterial virulence and/or agonistic effect on host immunity. To get the relevant mechanistic insights, we are describing different microbiological assays in Sects. 8.3.1, 8.3.2, 8.3.3, 8.3.4, 8.3.5 and 8.3.6. Various types of controls to be included in these assays and certain other commonly applicable points are explained below.

1. Sterility Control: Sterile M9 buffer containing neither bacteria nor worms.
2. Survival Control: M9 buffer containing worms (no bacteria added).

Table 8.3 List of human–pathogenic microorganisms for which *C. elegans* can be used as a model host

Bacteria	Fungi
Gram-positive	*Aspergillus niger* [24]
Clostridium difficile [18]	
Staphylococcus aureus [28]	*Candida albicans* [33]
Streptococcus pyogenes [21]	
Gram-negative	*Candida auris* [15]
Chromobacterium violaceum [23]	
Escherichia coli [51]	
Klebsiella pneumoniae [58]	*Cryptococcus gattii* [25]
Proteus mirabilis [41]	
Pseudomonas aeruginosa [27]	
Salmonella enterica serovar Typhi [32]	*Cryptococcus neoformans* [1]
Serratia marcescens [40]	
Shigella sonnei [57]	
Vibrio cholerae [45]	
Vibrio parahaemolyticus [6]	

3. Toxicity Control: Worms in M9 buffer supplemented with test formulation.
4. Infection Control: Worms in M9 buffer +100 μL of the bacterial culture broth (OD_{764} = 1.50 ± 0.05). These wells should not contain any test compound.
5. Vehicle Control: For anti-pathogenic assay, vehicle control will contain (worms + solvent used to dissolve the test compound + bacteria); while for anti-infective assay, it will comprise (worms + solvent-pre-treated pathogen). This control becomes relevant when test formulations are dissolved in a solvent other than water.
6. Positive Control: Worms + bacteria + an antibiotic to which the test pathogen is susceptible. The concentration of the antibiotic used here must be potent enough to exert anti-virulence or bactericidal effect, as the case may be.
7. Optical density of bacterial suspension for pigmented and non-pigmented bacteria is to be measured respectively at 764 nm and 625 nm [22].
8. Test compounds should be used at non-growth-inhibitory concentrations if their anti-virulence activity is being assessed. If they inhibit the growth partially, then equalization of OD between the control and experimental bacterial populations must be done before using them to challenge the worm population.

8.3.1 Anti-pathogenic Assay

Worms are challenged with the pathogen in the absence or presence of the test formulation. This assay allows the test formulation to exert its possible effect simultaneously on the host (immunomodulatory effect) as well as the pathogen (anti-pathogenic effect).

1. Test pathogen is grown at a temperature and in a medium appropriate for supporting its virulence. For example, Pseudomonas broth for *P. aeruginosa*; glycerol-supplemented nutrient broth for *S. marcescens*. Inoculum used for growing pathogens here should be prepared from its colonies suspended in normal saline (0.85%w/v NaCl), whose OD_{625} is adjusted between 0.08 and 0.10 to achieve equivalence to McFarland turbidity standard 0.5. This inoculum is added into the growth media at 10%v/v.

   [Note: Some pathogens will be required to be grown on agar for up to 48 h, before being used for inoculum preparation. Lesser incubation length will yield smaller colonies in which certain traits, e.g. pigmentation, are not fully expressed, and inoculum prepared using them may fail to exert sufficient virulence in subsequent in vivo assays.]
2. Ten worms of the L3-L4 stage contained in 900 μL of M9 buffer in each well of a 24-well plate are challenged with 100 μL of the test pathogen (OD_{764}:1.50 ± 0.05), in the absence or presence of test compound, wherein neither the bacterium nor the worms are pre-exposed to test compound. Volume of M9 buffer should be adjusted in such a way that the total volume per well remains 1 mL.
3. Incubation is done at 22 °C for 5 days with live-dead microscopic count once a day.

4. Better survival in experimental wells should indicate an anti-pathogenic and/or immunomodulatory effect of the test formulation, as the compound is present during the infection process and can act on both the pathogen and the host simultaneously. Improved survival may result from inhibition of bacterial virulence, direct antibacterial activity, or enhancement of the host's immune response.
5. If any test compound confers protection on the worm population in the face of pathogen challenge in this assay, to differentiate the effect of the test compound on the worm from that on the pathogen, assays described in Sects. 8.3.2 and 8.3.3 may be done.

8.3.2 Anti-infective Assay

This assay is designed to detect whether our test formulation can interfere with host-pathogen attachment, i.e., initiation of the infection process. Here, the pathogenic bacteria grown in the presence of the test formulation are washed and subsequently allowed to attack the worms (having no previous exposure to the test formulation).

1. Test pathogen is grown at a temperature and in a medium appropriate for supporting its virulence. For example, Pseudomonas broth for *P. aeruginosa*; glycerol-supplemented nutrient broth for *S. marcescens*. Medium supplemented with or without the test compound is to be used for the growth of experimental and control bacterial populations, respectively. Inoculum used for growing pathogens here should be prepared from its colonies suspended in normal saline (0.85%w/v NaCl), whose OD_{625} is adjusted between 0.08 and 0.10 to achieve equivalence to McFarland turbidity standard 0.5. This inoculum is added into the growth media at 10%v/v.
2. Post-incubation, 100 μL of the culture broth (OD_{764}:1.50 ± 0.05) is added to 900 μL of M9 buffer containing 10 worms (L3-L4 stage) in each well of a 24-well plate (surface non-treated; HiMedia). This plate is incubated at 22 °C for 5 days, and worm survival is monitored through a live-dead microscopic count, once a day. A higher number of worms may be used, and in that case, worm activity can be monitored using a worm tracker machine.
3. Plot worm survival as a function of time for control and experimental populations.
4. Better survival in experimental wells should indicate an anti-infective effect of the test formulation, suggesting that bacterial virulence or host-pathogen adhesion is compromised due to prior exposure of the pathogen to the test compound.

8.3.3 Prophylactic Assay

This assay is designed to detect the immunomodulatory activity in a test product, wherein gnotobiotic worms pre-exposed to the test product are subsequently challenged with the pathogen (having no previous exposure to the test product) in an assay medium devoid of the test product.

1. Gnotobiotic worms are incubated in M9 buffer supplemented with test compound/formulation, at 22 °C, in 1.5 ml vials for 72–96 h. The duration of this incubation may need to be optimized for each test formulation separately.
2. Following incubation, the worms are washed twice with M9 buffer and then challenged with the test pathogen (not previously exposed to the test compound) in 24-well plates. For this purpose, 100 μL of the culture broth (OD = 1.50 ± 0.05) of the bacterial pathogens grown in an appropriate medium is to be mixed with the rest of the assay content containing an appropriate number of worms.
3. Worm survival is monitored over a 5-day period with live-dead count once a day under a microscope. Worm motility may also optionally be quantified through an automated worm tracker.
4. Better survival in experimental wells should indicate a prophylactic effect of the test compound, suggesting that prior exposure to this compound enhances the worm's resistance to subsequent pathogen challenge. This effect may result from modulation of host defenses rather than direct antimicrobial action, as the pathogen used in this assay has not been exposed to the test compound.

8.3.4 Post-infection Assay

This assay somewhat mimics the real-life situation, wherein the patient approaches the physician when he/she has already been infected, and symptoms are visible. In this assay, for the first few hours, the pathogen is allowed to attack the worm host in the absence of the test formulation, and once the infection is established, the test formulation is added (before initiation of death in the worm population) to observe whether it can confer any beneficial effect on the already infected host population, combating the pathogen.

1. Ten worms contained in M9 buffer are first challenged with the pathogen (100 μL of culture suspension; OD = 1.50 ± 0.05) grown in appropriate culture media, and the pathogen is then allowed for 3–6 h to establish infection, before adding the test compound into the well. Duration for allowing the pathogen to establish infection before the entry of the pathogen in the assay system should be decided carefully, depending on the speed of the pathogen for killing the host population.
2. Survival and motility of worms are observed over a 5-day period, through live-dead count under a microscope and a worm tracker, respectively.
3. Better survival in experimental wells should indicate therapeutic efficacy of the test formulation, demonstrating its ability to aid the host in combating an already established infection. This may result from direct antimicrobial action, virulence attenuation, and/or host immunomodulation.

8.3.5 Post-antibiotic Effect (PAE)

This assay investigates whether bacterial progeny, derived from parent cells exposed to an antibiotic, exhibit any attenuation in their virulence compared to the parent cells. This phenomenon is of clinical relevance in real-life situations, wherein inside the body of a patient already on antibiotic treatment, the available concentration of the antibiotic will keep on decreasing with time, and soon approach sub-inhibitory levels. Though these low concentrations may allow the pathogenic bacteria to replicate inside the host, the resulting progenies might not exhibit virulence at par to the parent cells, owing to the possible post-antibiotic effect [50]. The phenomenon of bacterial progeny exhibiting reduced growth and/or virulence after previously being exposed to an antimicrobial agent can be an important one with respect to deciding the outcome of a host-pathogen interaction. While such an effect with conventional growth-inhibitory antibiotics has been reported as post-antibiotic effect, a similar effect on bacterial virulence exerted by certain plant extracts can more specifically be labelled as Post-Extract Effect (PEE; 10.32388/359873).

1. Pathogen is grown in an appropriate medium supplemented with the test compound (preferably with the lowest effective concentration). Control will be the pathogen grown in the same medium with no test compound.
2. Cells from the above tubes are collected through centrifugation, and the cell pellet is washed with normal saline to remove any traces of the test compound. Cells thus obtained will be used as an inoculum (OD_{625}:0.08–0.10) for a fresh medium containing no test compound. Following incubation, 100 μL of this culture broth (OD = 1.50 ± 0.05) will be used to challenge the worm population in a 24-well plate, as described in previous sections.
3. Worm survival and/or worm activity is quantified over a five-day period, and the virulence of the pathogen cells derived from the antibiotic-exposed parent is compared with that of control pathogen cells.
4. Reduced killing of the worms in experimental wells should be taken as an indication of the presence of PAE. In case of positive confirmation of the PAE, this assay can be continued with the next generation of pathogen originally derived from antibiotic-exposed parents, till the time of disappearance of the PAE.

8.3.6 Resistance Development Assay

When a bacterial population is repeatedly exposed to a particular antimicrobial agent, it can experience strong selection pressure, potentially leading to the development of resistant phenotypes. The ability of bacteria to adapt and acquire resistance after continuous exposure to an effective antimicrobial agent is a crucial consideration in determining the long-term viability of an antimicrobial agent in the market. The following protocol can be used to simulate a repeated exposure scenario and assess resistance development in the target pathogen.

1. To simulate the selection pressure exerted by strong antimicrobial agents, the test pathogen is grown in a medium supplemented with the test compound/extract (preferably at the highest effective concentration, so as to exert high selection pressure in favour of resistance development).

2. Growth of the pathogen thus obtained is to be used for next subculturing in liquid (or on solid media) supplemented with the test compound, and such subculturing event may be continued till 10 (or more) times.
3. The pathogen cells (100 μL of the culture broth; OD = 1.50 ± 0.05) obtained at the end of the tenth such subculturing will then be used to challenge the worm population. The control pathogen population will have no previous exposure to the test compound.
4. Monitor the worm survival and/or worm activity over a five-day period. If worm survival between the control and experimental wells is at par, that will indicate the absence of any resistance to the test compound in the pathogenic bacteria. However, if higher worm survival in observed in the experimental wells, that will indicate the development of partial or full resistance developed by the pathogen.
5. Positive control in this assay should be an antibiotic against which the pathogen is already known to be capable of developing resistance.

8.4 Important Parameters Affecting Worm Assays

While the different assays described above are useful, their execution requires certain precautions with respect to various parameters/factors that can influence the final outcome of the assay and subsequent interpretation. Some of these parameters are described below, with supporting experimental data.

8.4.1 Host to Pathogen Ratio

The same number of bacterial pathogens may exert varying lethality towards different sizes of host populations. Adding too dense a bacterial inoculum can kill the worm population too fast without providing any chance to the test formulation to show its effect, and hence raising a probability of false negative results. On the contrary, an inoculum too low in density or volume may show the test formulation to be more potent than it actually is, increasing the probability of false positive results.

8.4.2 Growth Media Used for Culturing Pathogens Prior to In Vivo Assay

Since bacterial virulence is influenced by the media used to grow it, the choice of media must be made carefully. The chosen media must allow secretion of virulence factors by the pathogen. For example, in the case of *Pseudomonas aeruginosa*, production of one of its virulence factors, pyocyanin, is supported better when grown in Pseudomonas broth than in general-purpose media like nutrient broth. (https://airdrive.eventsair.com/eventsairwesteuprod/production-bsac-public/b32b2125d3c1454ea596eaeb62af4b27).

8.4.3 Physiological State of the Worm Population

C. elegans serves as a robust model organism for studying physiological and toxicological responses to potential drugs under investigation, due to its well-defined developmental stages, genetic tractability, and ease of maintenance. Since each developmental stage may have specific physiological traits, selection of the synchronized worms in the most appropriate larval stage for a particular assay is of paramount importance. On the other hand, for certain assays, mixed-stage (non-synchronized) worm population may be the right choice, e.g. while determining whether a newly discovered anthelmintic agent can be effective against different life cycle stages of the nematodes. These parameters, namely, the developmental stage, physiological state, and population density, can significantly influence the outcome and reproducibility of the experiments. Since motility is an important behavioural trait in *C. elegans*, and a reliable indicator of its health [3], we did a comparison of the worm activity recorded through an automated worm tracker, between synchronized (L3-L4) and mixed-state worm populations, across various population densities (Table 8.4). Activity in worm populations ranging from 1 to 100 individuals was quantified using a worm tracker machine, which is a behavioural tracking system that enables quantitative analysis of locomotion and movement patterns. Our observations revealed that the linearity between worm activity and population size is compromised, both for synchronized as well as mixed-stage worms, when population size is small (e.g., 5–10). That means in experiments, where it is expected for the worm activity to go down over time (e.g., while assessing the killing effect of any pathogen or toxin), it is advisable to keep the starting worm population high.

Furthermore, the correlation between population size and worm activity maintained its linearity to an acceptable level till fifth day in both population types, establishing the five-day period as a reasonable operational window for behavioural and assays.

8.4.4 Choosing an Appropriate End-Point

In an assay running over multiple days, there may be more than one end-points, and a particular day's end-point may be more informative than others. Selection of the most relevant end-point is crucial. An appropriately defined end-point ensures consistency across replicates and makes a valid comparison between different studies possible. Observations made at a premature or delayed end-point in time can lead to misinterpretation of data, mask true effects or introduce unwarranted confounding variations. The selection of end-point should be informed by the kinetics of the biological response being investigated. For example, during an anti-pathogenic assay, in case of a pathogen killing all host worms in less than a day, a 24-hour end-point is valid, but additional end-points in time, say 72 h and 120 h, can provide information on whether the anti-pathogenic effect is sustained over a longer time period. Having multiple end-point measurements in a large screening exercise can help distinguish the weak leads from stronger ones. An example of this scenario can

Table 8.4 Influence of population size and physiological state on motility captured by an automated worm tracker

	Synchronized worms						
	Worm activity						
Worm number	**Day 0**	**Day 1**	**Day 2**	**Day 3**	**Day 4**	**Day 5**	**'r' value**[b]
1	5 ± 6.32	4.22 ± 10.51	1.56 ± 2.40	2.89 ± 5.93	4.89 ± 11.68	1 ± 1.50	0.52
5	11.9 ± 15.29	7 ± 6.73	9.89 ± 9.18	2.56 ± 4.98	6.22 ± 8.57	4.44 ± 7.40	0.73
10	21.56 ± 17.20	15.11 ± 12.86	14.56 ± 16.51	5.44 ± 5.68	10.11 ± 10.59	5.44 ± 6.98	0.89
25	19.33 ± 19.11	24.33 ± 12.33	20.33 ± 19.49	14.78 ± 15.25	16.89 ± 16.20	4 ± 4.90	0.80
50	41.56 ± 29.22	41.11 ± 21.53	39.33 ± 28.22	29.11 ± 19.97	26.33 ± 25.22	27.22 ± 21.18	0.93
75	75.33 ± 43.44	70.33 ± 26.76	65.11 ± 26.85	70.33 ± 23.28	63 ± 27.35	55 ± 31.29	0.89
100	88.44 ± 32.36	70 ± 13.44	73.33 ± 17.94	69.67 ± 25.76	65.22 ± 20.37	40.22 ± 22.47	0.88
'r'value[a]	0.98	0.98	0.99	0.97	0.97	0.91	
	Mixed state worms[c]						
	Worm activity						
Worm number	**Day 0**	**Day 1**	**Day 2**	**Day 3**	**Day 4**	**Day 5**	**'r' value**[b]
1	1 ± 2.00	0	0	6.44 ± 12.84	1.33 ± 4.00	0.67 ± 2.00	0.19
5	12.67 ± 13.07	12.67 ± 11.78	8.11 ± 11.79	9.67 ± 10.65	5.33 ± 6.98	5.11 ± 11.19	0.92
10	16.11 ± 10.15	19.33 ± 15.61	17 ± 19.85	7.33 ± 7.98	9.67 ± 10.37	7 ± 12.57	0.83
25	35.78 ± 20.03	44.22 ± 19.12	39.67 ± 27.70	34.11 ± 20.02	20.11 ± 17.03	21.33 ± 16.36	0.81
50	61 ± 15.31	56.11 ± 9.93	61.44 ± 24.11	50.89 ± 14.09	38.22 ± 17.28	34.78 ± 23.49	0.90
75	95.67 ± 34.95	102.78 ± 31.38	88.56 ± 43.68	82.67 ± 35.17	82.22 ± 44.62	57 ± 35.40	0.88
100	86.83 ± 18.20	97.67 ± 27.20	93.17 ± 44.43	72 ± 22.20	66.33 ± 32.46	50.33 ± 32.34	0.88
'r' value[a]	0.96	0.96	0.98	0.93	0.91	0.95	

r: correlation coefficient; Day zero readings are taken 1 h post addition of worms in the assay well containing 1 mL of M9 buffer

[a]Correlation between population size and worm activity on each particular day

[b]Correlation between worm activity and population size across a six-day period

[c]In case of mixed-state worms, interference from progenies canot be avoided. As some of the worm generate offsprings, overall population size increases in those wells

be viewed in Gajera et al. [10]. Similarly, in toxicity assays, time points corresponding to death/paralysis of 50% and 90–100% of the worm population can be important [9]. In lifespan assays, a time-point when 50% of the worms are live in the control population can be an important one [53].

8.4.5 Choosing an Appropriate Positive Control

An appropriately selected positive control guards against false-negative results. However, selection of the most appropriate compound as a positive control at a correct concentration is not always that straightforward. For example, while screening a library of synthetic compounds or natural products for their possible anti-Alzheimer or anti-Parkinson activity in worm model, a satisfactory positive control cannot be found, as no therapeutic molecule with convincing efficacy against these disease conditions is still available.

While conventional bactericidal antibiotics are the first choice as a positive control in an antibacterial assay, choosing the correct antibiotic and using it at the correct concentration is important. While antibiotics in current clinical use offer a good choice as positive control in anti-pathogenic and post-infection assays, finding an antibiotic suitable as positive control in prophylactic and anti-infective assays can be quite challenging, because while all antibiotics may not confer prophylactic benefit on the host in the face of subsequent pathogen attack, most of them at sub-lethal levels also do not exhibit anti-virulence property. While antibiogram of the test pathogen generated through conventional disc diffusion assay can prove a useful indicator of which antibiotics should be used as positive control, and the broth dilution assay for determining IC50 and minimum inhibitory concentration (MIC) can inform us about the appropriate concentration to be used, always the in vitro antibacterial effect of an antibiotic may not correlate with its in vivo anti-pathogenic efficacy [54].

8.4.6 Toxicity Control

Whether the main objective of a study is to assess the safety of the test compounds, it is always recommended to include a toxicity control in the assay, preferably for each test concentration. For example, while screening various test substances for possible antibacterial effect, almost all the attention is on investigating the test substance-pathogen interaction, but it is extremely important not to forget to ask whether the test substance has any negative effect on worm survival. Certain compounds can exert antibacterial as well as anthelmintic activity simultaneously. If a toxicity control (i.e., worms + test compound) is not included in the assay, we may get false negative results with respect to the anti-pathogenic activity of compounds being screened. In such cases, either we may completely miss the detection of anti-pathogenic activity or underrate the actual magnitude of activity. One of our recent publications [10] presents some such examples. Inclusion of such toxicity controls also permits simultaneous discovery of anthelmintic activity in compounds being investigated, with a hope for antibacterial activity! An example of this kind of serendipitous finding is Gajera and Kothari [11], wherein all the identified anthelmintic hits were originally procured with an intention of antibacterial screening.

Though we are naming the control described in this section as "toxicity control," it can also help detect if there is any positive effect of the test compound on the worms, as that may also be a confounding factor.

8.4.7 Concentration of the Vehicle Solvent (DMSO)

C. elegans is increasingly being used in pre-clinical research for screening libraries of natural products and synthetic compounds for various biological activities. Many of these test compounds are not water-soluble, and need to be dissolved in other solvents for assay purposes. Among the alternative solvents, DMSO is perhaps the most common one. When test substances are dissolved in DMSO not only is it necessary to include an appropriate vehicle control in the assay (to ensure that DMSO is not interfering) [19], but also it is desired to maintain the DMSO concentration in the assay medium at the lowest possible level for two reasons. One, DMSO at higher concentrations may impact worm survival (Fig. 8.5a) and activity (Fig. 8.5b) negatively, but even at non-lethal concentrations, it may potentiate the activity of test compounds.

Knowing DMSO's effect on bacterial virulence assumes high importance, particularly while evaluating anti-pathogenic activity of different natural products or synthetic compounds dissolved in it. In our experience, DMSO up to 3%v/v did not affect the virulence of *Chromobacterium violaceum* or methicillin-resistant

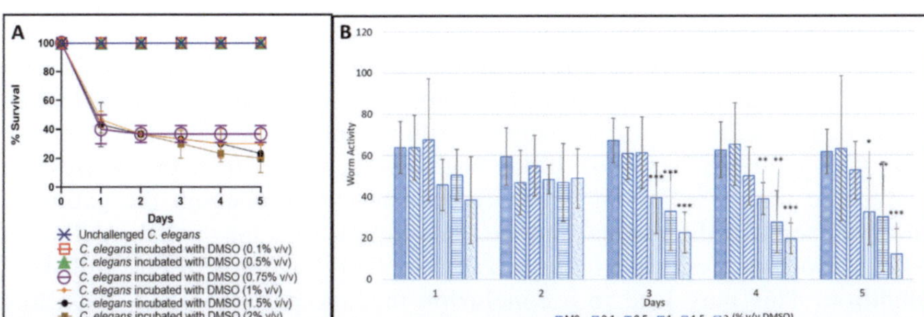

Fig. 8.5 DMSO impacts worm survival and activity negatively in a dose-dependent manner. (**a**) Ten worms (L3-L4) were present per well containing 1 mL of M9 buffer, and survival was monitored through a microscopic live-dead count. DMSO at 0.75–2%v/v caused a reduction in worm survival by 40–46.66%*** as per the first day, and 20–36.66%*** as per the fifth day end-point. (**b**) Fifty worms (L3-L4) were present per well, containing 1 mL of M9 buffer, and worm activity was quantified through an automated worm tracker. These assays were done in 24-well plates. DMSO was found to be safe for worms in both assays till 0.5%v/v. *$p \leq 0.05$, ** $p \leq 0.01$, ***$p \leq 0.001$

Staphylococcus aureus (MRSA), but at ≥2.5%v/v it attenuated *P. aeruginosa's* virulence marginally. DMSO at 1–2%v/v affected growth and pigmentation, but not virulence in the latter [26]. Whether it can potentiate the effect of any test compound dissolved in it, remains to be investigated.

8.4.8 Selection of Appropriate Assay Format and Method of Recording Output

A protocol can be considered robust when it yields identical results despite minor variations in the assay format. We assessed the performance of the anti-pathogenic protocol described in this chapter in two different assay formats, i.e., the same assay performed in a 24-well plate as well as a 96-well plate. While the worm number per well was maintained identically (10 worms per well) in both formats, assay volume differed (1 mL for the 24-well plate, and 0.2 mL for the 96-well plate), and the worm density too. Despite these variations, the conclusion drawn from both assay formats regarding the anti-pathogenic efficacy of the antibiotic ofloxacin was the same (Fig. 8.6a).

However, in another such assay performed to check whether exposing the worms to heat in dishes made of different materials (glass versus plastic) generates different results, we found the difference in heat susceptibility of worms kept in different material dishes to be quite high. Heat susceptibility of the transgenic *C. elegans* strain CL4176 could be observed in the worm population kept in a glass dish, clearly in the form of complete paralysis, following a 3-h heat shock at 35 °C. But similar heat-shock to worms kept in a plastic dish could only slow their movement, but none of the worms were completely paralyzed. This observation highlights the notable impact of the plate material on the observed phenotype (Fig. 8.6b).

In the third assay of this series (Fig. 8.6c), we checked whether relying on motility tracking solely as a readout during an assay to monitor pathogen-induced worm killing is feasible. We found that while this serves well when the pathogen is able to kill all worms within a short time period, it does not when the pathogen is slow-killing. In the latter case, when the worms are challenged with a relatively less virulent bacterial pathogen, the pathogen kills only some of the worms, and the remaining worms may use the pathogen cells as food, and register higher activity (motility) owing to their "well-fed" condition. This may lead to a conclusion that the pathogen had no killing effect on the worms, unless and until a parallel microscopic count is also run to confirm the results of the automated motility tracker. It shows that the choice of the method used for quantifying the assay output has to be made with great care, and wherever possible, confirmation through an alternative method is recommended.

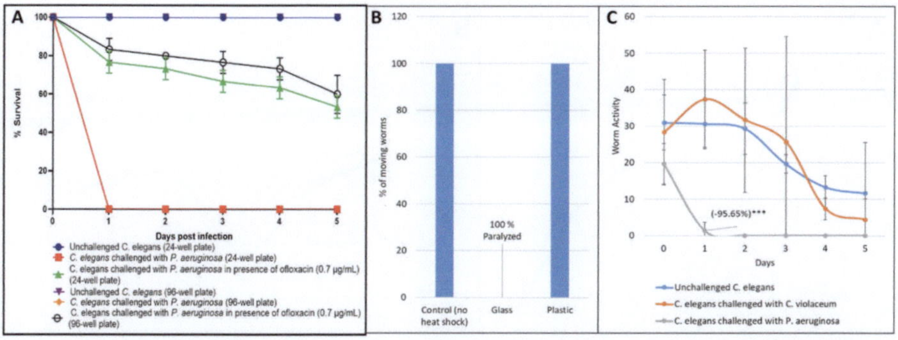

Fig. 8.6 (**a**) Anti-pathogenic assay in two different formats (24-well plate and 96-well plate) gives identical results. Worm survival was monitored manually through a microscope. As per the fifth day endpoint, survival values in the ofloxacin wells in 24-well (53.33 ± 5.77%***) and 96-well (60 ± 10%***) plates were statistically at par ($p = 0.37$). Progeny worms were apparent in all wells containing Ofloxacin, day three onward, irrespective of the assay format. (**b**) Heat susceptibility of worms is influenced by the type of container used to house worms during heat shock. Though heat-induced clumping was observed in both plastic and glass plates, worms in the latter experienced a complete lack of movement as opposed to the partial paralysis in worms put in a plastic plate. This assay was performed in small dishes containing 10 worms on NGM agar. (**c**) Automated capturing of worm activity during an anti-pathogenic assay is compatible with fast-killing bacteria, but not with slow-killing bacteria. *P. aeruginosa* kills all worms within the first 24 h, and hence worm activity quickly approaches zero. However, in case of *C. violaceum*, though some worms are killed day three onward, the lines for worm activity pertaining to control and experimental wells run almost parallel ($p = 0.73$), since the live worms in experimental wells register higher motility. *C. violaceum* could kill 86 ± 5.77% worms by the fifth day

8.4.9 Oxygen Availability and Incubation Medium

Though during most assays, it is okay to incubate worms in ambient air conditions, in some assays lasting longer periods, parafilm (Merck) may be applied on the boundaries of multiwall plates to prevent evaporation. This may also be done if any of the assay components is volatile. Parafilm may affect air exchange (i.e., oxygen availability), and that in turn may have an influence on worm activity. Even while using *C. elegans* as a model host for anaerobic bacterial pathogens, incubation under oxygen-limited conditions may be warranted. In these cases, determining the effect of oxygen availability, or confirming the absence of any such effect, becomes important. In our experience, worm activity remains unaffected over a five-day period, irrespective of whether the multiwall places containing worms were wrapped with parafilm (Fig. 8.7a). Two layers of parafilm were applied in this experiment.

Though M9 buffer is a widely used medium in which worms are kept for various assays, if some of the test compounds are not soluble in it, then the assay may need to be done in water. In our experience, 50 worms kept in either sterile distilled water or M9 buffer in 24-well plates exhibited identical ($p = 0.44$) motility (Fig. 8.7b).

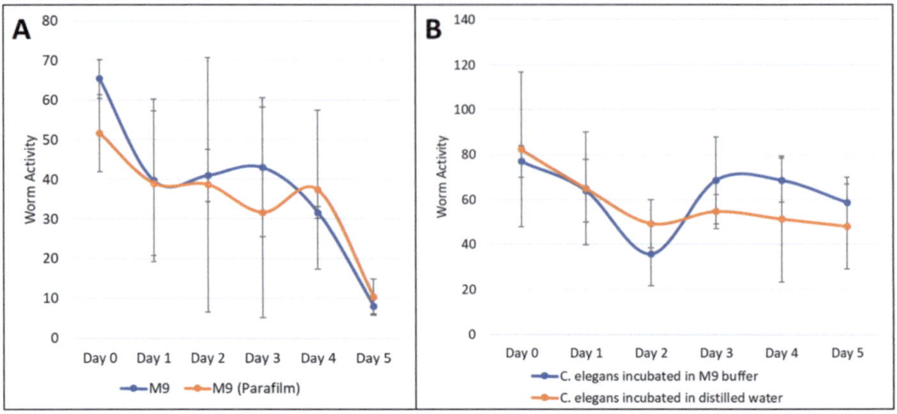

Fig. 8.7 (**a**) Wrapping the multi-well plates with parafilm had no significant impact on worm activity. (**b**) Worms incubated in M9 buffer or distilled water did not differ in terms of their activity over a five-day period

Final Comments

The community of worm researchers globally has benefited by sharing the protocols among themselves. Making protocols openly available allows other labs to test them and use them with or without modifications. This helps in validating the robustness of an experimental protocol. We hope the protocols presented in this chapter, along with additional experimental tips listed in Box 8.1, will be helpful to worm researchers globally. While *C. elegans* is an extremely useful model organism in biological research, its procurement from authentic sources is not as easy as for microorganisms. The way different countries have their own microbial culture collection centres, such worm repositories do not exist in most countries. Transport of this worm among labs poses challenges owing to its short lifespan, sensitivity to temperatures >25 °C, absence of hardy stages (like bacterial spores) in its life cycle, etc. Transgenic strains and automated worm trackers are quite costly for an average resource-constrained lab. We hope, in the long run, many countries will have their own worm repositories housing wild-type as well as transgenic *C. elegans* strains, which will reduce the need for international transfer, and the associated longer transit times and higher costs. Despite the increasing popularity of this worm in the global research community, from our personal experience, at times, we have found a certain lack of open-mindedness among some members of grant review panels while evaluating the grant proposals involving alternative model organisms like *C. elegans*. Perhaps, few of them are still reluctant in deviating from the long-established mouse, rat, or pig models, despite the possibility of reducing the sacrifice of higher animals, if alternative models like worm, fly, or zebrafish are used at an earlier stage of research. We hope for a more open-minded approach from review panels of various funding agencies in the future, particularly, for example, while handling proposals dealing with the validation of certain traditional medicine formulations, already having a long history of safe usage by humans.

Box 8.1: Some Useful Tips for Working with *C. elegans*

1. During microscopic observation, do not assume a non-moving worm as dead. Confirm its non-responsiveness by shaking the plate or touching it gently with a wire. All dead worm bodies will not be always straightened.
2. While performing the live-dead count, do pay special attention to worms who are not dead, but appear moving abnormally or paralyzed, as this can have an important bearing on your interpretation of the data/observation.
3. For manual microscopic count-based assays, keep worm population size limited to a maximum to 20, and the 24-well format is most suitable for this in our experience. For higher population size, assay should be shifted to automated worm-tracking formats. Manual counting in 96-wells may be more challenging if the wells are deep, and there is a higher probability of missing few wells. Six-wells plates are also not preferred for manual microscopic count, as the area to be screened under the microscope will increase.
4. Consider carefully whether your assay will yield most relevant results with gnotobiotic worms, or well-fed worms.
5. For microbiological assays, avoid use of any antibiotics during worm maintenance, as some of them may impart prophylactic benefit to worms during subsequent pathogen challenge.
6. While clumping of worms, which is a sign of them being under stress, is detected better on solid media; motility quantification is done better in liquid media.
7. Unless and until necessary, avoid use of anti-fertility agents like FUdR, as that may prove to be a confounding factor while interpreting the results.
8. In our opinion, measurement of lifespan alone in worms is less useful, than measuring it along with healthspan.
9. Automated worm tracking systems do minimize the probability of human error, and are less labour-intensive, but they should not be viewed as an alternative to manual microscopic count, as the qualitative differences in worm motility are better observed through direct visual observation. Assays based on automated tracking, also requires higher number of worms.
10. *C. elegans* being sensitive to temperatures above 25 °C, it becomes difficult to work with this worms in hot climate, even if the lab is air-conditioned. Some transgenic strains required even lower temperatures (less than 20 °C). Many resource-constrained labs do not have fully air-conditioned areas, and their incubators struggle to maintain required temperature. Hence many of the experiments with *C. elegans* may fail during peak summer seasons. For labs in warmer parts of the globe, it is advised to plan most of their worm assays outside the peak summer.

Acknowledgement Authors thank Nirma University for infrastructural support. NT acknowledges support from Nirma University, and Gujarat Government's SHODH scheme for doctoral fellowship. We thank Xavier's Research Foundation, Ahmedabad for providing the CL4176 strain of *C. elegans*. Vaidehee Popat, Shraddha Makvana, and Feni Mehta are thanked for their assistance during optimization of Alamar Blue assay.

Conflict of Interest Authors declare no conflict of interest. Names of any manufacturers of any scientific product mentioned in this chapter are just for example purpose, and not to endorse them.

References

1. Ahamefule CS, Ezeuduji BC, Ogbonna JC, Moneke AN, Ike AC, Jin C, Wang B, Fang W. *Caenorhabditis elegans* as an infection model for pathogenic mold and dimorphic fungi: applications and challenges. Front Cell Infect Microbiol. 2021;11:751947. https://doi.org/10.3389/fcimb.2021.751947.
2. Bora S, Vardhan GS, Deka N, Khataniar L, Gogoi D, Baruah A. Paraquat exposure over generation affects lifespan and reproduction through mitochondrial disruption in *C. elegans*. Toxicology. 2021;447:152632. https://doi.org/10.1016/j.tox.2020.152632.
3. Buckingham SD, Partridge FA, Sattelle DB. Automated, high-throughput, motility analysis in *Caenorhabditis elegans* and parasitic nematodes: applications in the search for new anthelmintics. Int J Parasitol Drugs Drug Resist. 2014;4(3):226–32. https://doi.org/10.1016/j.ijpddr.2014.10.004.
4. de Torre MP, Vizmanos JL, Cavero RY, Calvo MI. A novel *in vivo* method Using *Caenorhabditis elegans* to evaluate α-glucosidase inhibition by natural products for type 2 diabetes treatment. Pharmaceuticals. 2024;17(12):1685.
5. Demenj M, Žabčić M, Vukomanović M, Ilić-Tomić T, Milivojević D, Tomić S, Živanović D, Babić Radić MM. Design of the Multi-Bioactive Graphene-Oxide/gelatin/alginate scaffolds as dual ECM-mimetic and specific wound healing phase-target therapeutic concept for advanced wound healing. Pharmaceutics. 2025;17(1):89. https://doi.org/10.3390/pharmaceutics17010089.
6. Faleye OS, Lee JH, Lee J. Selected flavonoids exhibit antibiofilm and antibacterial effects against vibrio by disrupting membrane integrity, virulence and metabolic activities. Biofilms. 2023;6:100165. https://doi.org/10.1016/j.bioflm.2023.100165.
7. Fu X, Tang Y, Dickinson BC, Chang CJ, Chang Z. An oxidative fluctuation hypothesis of aging generated by imaging H_2O_2 levels in live *Caenorhabditis elegans* with altered lifespans. Biochem Biophys Res Commun. 2015;458(4):896–900. https://doi.org/10.1016/j.bbrc.2015.02.055.
8. Gajera G, Funde S, Palep H, Kothari V. Duration of fermentation affects microbiome composition and biological activity of an Indian traditional formulation–Panchagavya. J Ayurveda Integr Med. 2024;15(2):100880. https://doi.org/10.1016/j.jaim.2023.100880.
9. Gajera G, Godse C, DeSouza A, Mehta D, Kothari V. Potent anthelmintic activity of a colloidal nano-silver formulation (Silversol®) against the model worm *Caenorhabditis elegans*. BMC Res Notes. 2023a;16(1):130. https://doi.org/10.1186/s13104-023-06392-1.
10. Gajera G, Henriksen N, Cox B, Kothari V. Identification of anti-pathogenic activity among in silico predicted small-molecule inhibitors of *Pseudomonas aeruginosa* LasR or nitric oxide reductase (NOR). Drug Target Insights. 2023b;17:101. https://doi.org/10.33393/dti.2023.2638.
11. Gajera G, Kothari V. Identification of anthelmintic activity in a small chemical library through random screening using the worm *Caenorhabditis elegans* as a model helminth. bioRxiv. 2024; https://doi.org/10.1101/2024.08.02.606330.
12. Gonzalez-Hunt CP, Luz AL, Ryde IT, Turner EA, Ilkayeva OR, Bhatt DP, Hirschey MD, Meyer JN. Multiple metabolic changes mediate the response of *Caenorhabditis elegans* to

the complex I inhibitor rotenone. Toxicology. 2021;447:152630. https://doi.org/10.1016/j.tox.2020.152630.
13. Hamid R, Rotshteyn Y, Rabadi L, et al. Comparison of alamar blue and MTT assays for high through-put screening. Toxicol In Vitro. 2004;18(5):703–10. https://doi.org/10.1016/j.tiv.2004.03.012.
14. Hernando G, Bouzat C. *Caenorhabditis elegans* neuromuscular junction: GABA receptors and ivermectin action. PLoS One. 2014;9(4):e95072. https://doi.org/10.1371/journal.pone.0095072.
15. Hernando-Ortiz A, Eraso E, Jauregizar N, de Groot PW, Quindós G, Mateo E. Efficacy of the combination of amphotericin B and echinocandins against Candida auris in vitro and in the *Caenorhabditis elegans* host model. Microbiol Spectrum. 2024;12(1):e02086–23. https://doi.org/10.1128/spectrum.02086-23.
16. Holden-Dye L, Walker R. Anthelmintic drugs and nematocides: studies in *Caenorhabditis elegans*. In: WormBook: the online review of C. elegans biology; 2014. p. 1–29. https://doi.org/10.1895/wormbook.1.143.2.
17. Hosono R, Kuno S, Midsukami M. Temperature-sensitive mutations causing reversible paralysis in *Caenorhabditis elegans*. J Exp Zool. 1985;235(3):409–21. https://doi.org/10.1002/jez.1402350313.
18. Hu C, Garey KW. Nonmammalian models to study Clostridioides difficile infection; a systematic review. Anaerobe. 2023;79:102694. https://doi.org/10.1016/j.anaerobe.2023.102694.
19. Hughes S, van Dop M, Kolsters N, van de Klashorst D, Pogosova A, Rijs AM. Using a Caenorhabditis elegans Parkinson's disease model to assess disease progression and therapy efficiency. Pharmaceuticals. 2022;15(5):512. https://doi.org/10.3390/ph15050512.
20. Hulme SE, Whitesides GM. Chemistry and the worm: *Caenorhabditis elegans* as a platform for integrating chemical and biological research. Angew Chem Int Ed. 2011;50(21):4774–807.
21. Jansen WT, Bolm M, Balling R, Chhatwal GS, Schnabel R. Hydrogen peroxide-mediated killing of *Caenorhabditis elegans* by streptococcus pyogenes. Infect Immun. 2002;70(9):5202–7. https://doi.org/10.1128/iai.70.9.5202-5207.2002.
22. Joshi C, Kothari V, Patel P. Importance of selecting appropriate wavelength, while quantifying growth and production of quorum sensing regulated pigments in bacteria. Recent Pat Biotechnol. 2016;10(2):145–52.
23. Kanekar S, Devasya RP. Growth-phase specific regulation of cvi I/R based quorum sensing associated virulence factors in Chromobacterium violaceum by linalool, a monoterpenoid. World J Microbiol Biotechnol. 2022;38(2):23. https://doi.org/10.1007/s11274-021-03208-x.
24. Kassam R, Kamil D, Rao U. *Caenorhabditis elegans* as bait for isolating promising biocontrol fungi against Meloidogyne incognita from soils across India. Indian Phytopathol. 2021;74:739–52. https://doi.org/10.1007/s42360-021-00354-y.
25. Kitisin T, Muangkaew W, Sukphopetch P. *Caenorhabditis elegans* DAF-16 regulates lifespan and immune responses to Cryptococcus neoformans and Cryptococcus gattii infections. BMC Microbiol. 2022;22(1):162. https://doi.org/10.1186/s12866-022-02579-x.
26. Kothari V, Thakkar N, Gajera G. Investigation on various parameters influencing virulence of bacterial pathogens towards the model host *Caenorhabditis elegans*, and major mode of worm killing by different pathogens. Antimicrobial Chemotherapy Conference (ACC 2025), Feb 4–5, 2025. Organized by Global Antibiotic Research & Development Partnership (GARDP) and British Society for Antimicrobial Chemotherapy (BSAC); 2025. https://airdrive.eventsair.com/eventsairwesteuprod/production-bsac-public/b32b2125d3c1454ea596eaeb62af4b27
27. Kumar L, Brenner N, Brice J, Klein-Seetharaman J, Sarkar SK. Cephalosporins interfere with quorum sensing and improve the ability of *Caenorhabditis elegans* to survive Pseudomonas aeruginosa infection. Front Microbiol. 2021;12:598498. https://doi.org/10.3389/fmicb.2021.598498.
28. Mangu JC, Stylianou M, Olsson PE, Jass J. Per-and polyfluoroalkyl substances enhance Staphylococcus aureus pathogenicity and impair host immune response. Environ Pollut. 2022;314:120294. https://doi.org/10.1016/j.envpol.2022.120294.

29. Martin RJ, Robertson AP, Buxton SK, Beech RN, Charvet CL, Neveu C. Levamisole receptors: a second awakening. Trends Parasitol. 2012;28(7):289–96. https://doi.org/10.1016/j.pt.2012.04.003.
30. Mathew MD, Mathew ND, Miller A, Simpson M, Au V, Garland S, Gestin M, Edgley ML, Flibotte S, Balgi A, Chiang J. Using *C. elegans* forward and reverse genetics to identify new compounds with anthelmintic activity. PLoS Negl Trop Dis. 2016;10(10):e0005058.
31. McIntyre G, Wright J, Wong HT, Lamendella R, Chan J. Effects of FUdR on gene expression in the *C. elegans* bacterial diet OP50. BMC Res Notes. 2021;14(1):207. https://doi.org/10.1186/s13104-021-05624-6.
32. Mir DA, Balasubramaniam B, VenkataKrishna LM, Muthubharathi BC, Balamurugan K. Proteomic analysis of *Caenorhabditis elegans* against Salmonella Typhi toxic proteins. Genes Immun. 2021;22(2):75–92. https://doi.org/10.1038/s41435-021-00132-w.
33. Mokoena NZ, Steyn H, Hugo A, Dix-Peek T, Dickens C, Gcilitshana OM, Sebolai O, Albertyn J, Pohl CH. Eicosapentaenoic acid influences the pathogenesis of Candida albicans in *Caenorhabditis elegans* via inhibition of hyphal formation and stimulation of the host immune response. Med Microbiol Immunol. 2023;212(5):349–68. https://doi.org/10.1007/s00430-023-00777-6.
34. Moline B. Characterization of manganese-induced Neurodegenration in *C. elegans* treated with winterberry leaf extract. Honors College; 2023.
35. Osorio-Paz I, Valle-Jiménez X, Brunauer R, Alavez S. Vanillic acid improves stress resistance and substantially extends life span in *Caenorhabditis elegans*. J Gerontol Ser A. 2023;78(7):1100–7. https://doi.org/10.1093/gerona/glad086.
36. Page KE, White KN, McCrohan CR, Killilea DW, Lithgow GJ. Aluminium exposure disrupts elemental homeostasis in *Caenorhabditis elegans*. Metallomics. 2012;4(5):512–22. https://doi.org/10.1039/c2mt00146b.
37. Patel P, Joshi C, Kothari V. Supernatants of the probiotic bacterial cultures at sub-MIC levels attenuate virulence of pathogenic bacteria towards the model host *Caenorhabditis elegans*. Infect Disord Drug Targets (Formerly Current Drug Targets-Infectious Disorders). 2020;20(6):867–77. https://doi.org/10.2174/1871526519666191211141827.
38. Patil VM, Verma S, Masand N. Prospective mode of action of Ivermectin: SARS-CoV-2. Eur J Med Chem Rep. 2022;4:100018. https://doi.org/10.1016/j.ejmcr.2021.100018.
39. Phiri AM, De Pomerai DI, Buttle DJ, Behnke JM. A novel assay for the detection of anthelmintic activity mediated by cuticular damage to nematodes: validation on *Caenorhabditis elegans* exposed to cysteine proteinases. Parasitology. 2017;144(5):583–93.
40. Pradel E, Zhang Y, Pujol N, Matsuyama T, Bargmann CI, Ewbank JJ. Detection and avoidance of a natural product from the pathogenic bacterium Serratia marcescens by *Caenorhabditis elegans*. Proc Natl Acad Sci. 2007;104(7):2295–300. https://doi.org/10.1073/pnas.0610281104.
41. Prithika U, Balamurugan K. Evaluation of changes in *Caenorhabditis elegans* immune response during bacterial infection: a single nematode approach. Microbes Infect. 2021;23(9–10):104846. https://doi.org/10.1016/j.micinf.2021.104846.
42. Rampersad SN. Multiple applications of Alamar Blue as an indicator of metabolic function and cellular health in cell viability bioassays. Sensors. 2012;12(9):12347–60. https://doi.org/10.3390/s120912347.
43. Rossi SG, Fisher ML. The microbiome of *C. elegans* enhances growth of S. Aureus during infection. Access Microbiol. 2024; https://doi.org/10.1099/acmi.0.000802.v1.
44. Roussos A, Kitopoulou K, Borbolis F, Palikaras K. *Caenorhabditis elegans* as a model system to study human neurodegenerative disorders. Biomolecules. 2023;13(3):478.
45. Sahu SN, Lewis J, Patel I, Bozdag S, Lee JH, LeClerc JE, Cinar HN. Genomic analysis of immune response against Vibrio cholerae hemolysin in Caenorhabditis elegans. PLoS One. 2012;7(5):e38200. https://doi.org/10.1371/journal.pone.0038200.
46. Sammi SR, Jameson LE, Conrow KD, Leung MC, Cannon JR. *Caenorhabditis elegans* neurotoxicity testing: novel applications in the adverse outcome pathway framework. Front Toxicol. 2022;4:826488. https://doi.org/10.3389/ftox.2022.826488.

47. Shahlehi S, Kasah R, Petalcorin MI, Najim MM. Phytochemical analysis and anthelmintic activity of Andrographis paniculata, Azadirachta indica, and Litsea elliptica leaves extracts against *Caenorhabditis elegans*. J Wildl Biodivers. 2024;8(3):365–88. https://doi.org/10.5281/zenodo.12580790.
48. Silva AC, Viçozzi GP, Farina M, Ávila DS. *Caenorhabditis elegans* as a model for evaluating the toxicology of Inorganic nanoparticles. J Appl Toxicol. 2024; https://doi.org/10.1002/jat.4704.
49. Somuah-Asante S, Sakamoto K. Stress buffering and longevity effects of amber extract on *Caenorhabditis elegans* (*C. elegans*). Molecules. 2022;27(12):3858. https://doi.org/10.3390/molecules27123858.
50. Srimani JK, Huang S, Lopatkin AJ, You L. Drug detoxification dynamics explain the postantibiotic effect. Mol Syst Biol. 2017;13(10):948. https://doi.org/10.15252/msb.20177723.
51. Suryawinata N, Yokosawa R, Tan KH, Lai AL, Sone R, Mori I, Noma K. Dietary E. coli promotes age-dependent chemotaxis decline in *C. elegans*. Sci Rep. 2024;14(1):5529. https://doi.org/10.1038/s41598-024-52272-4.
52. Tessa O, Steven Z. *C. elegans* as a model to study mitochondrial biology and disease. Semin Cell Dev Biol. 2024;154(Part A):48–58. https://doi.org/10.1016/j.semcdb.2023.04.006. ISSN 1084-9521.
53. Thakkar N, Gajera G, Mehta D, Nair S, Kothari V. Withania somnifera root extract (LongeFera™) confers beneficial effects on health and lifespan of the model worm *Caenorhabditis elegans*. Drug Target Insights. 2025a;19:18. https://doi.org/10.33393/dti.2025.3368.
54. Thakkar N, Gajera G, Kothari V. Correlation of *in vitro* antibiogram with *in vivo* anti-pathogenic efficacy of different antibiotics against three gram-negative bacterial pathogens in the *Caenorhabditis elegans* infection model. In: Proceedings of the 4th International Electronic Conference on Antibiotics, 21–23 May. Basel: MDPI; 2025b. https://sciforum.net/paper/view/22968.
55. Wang Z, Arnold JC. Cannabinoids and healthy ageing: the potential for extending healthspan and lifespan in preclinical models with an emphasis on *Caenorhabditis elegans*. GeroScience. 2024;46(6):5643–61. https://doi.org/10.1007/s11357-024-01162-8.
56. Wijaya CS, Meng X, Yang Q, Xu S. Protocol to induce wounding and measure membrane repair in *Caenorhabditis elegans* epidermis. STAR Protoc. 2020;1(3):100175.
57. Wong BC, Tan HS. Shigella sonnei and Shigella flexneri infection in *Caenorhabditis elegans* led to species-specific regulatory responses in the host and pathogen. Microbial Genomics. 2025;11(1):001339. https://doi.org/10.1099/mgen.0.001339.
58. Yang WH, Chen PH, Chang HH, Kwok HL, Stern A, Soo PC, Chen JH, Yang HC. Impaired immune response and barrier function in GSPD-1-deficient *C. elegans* infected with Klebsiella pneumoniae. Curr Res Microbial Sci. 2023;4:100181. https://doi.org/10.1016/j.crmicr.2023.100181.
59. Yasuda K, Kubo Y, Murata H, Sakamoto K. Cortisol promotes stress tolerance via DAF-16 in *Caenorhabditis elegans*. Biochem Biophys Rep. 2021;26:100961. https://doi.org/10.1016/j.bbrep.2021.100961.
60. Zečić A, Dhondt I, Braeckman BP. The nutritional requirements of *Caenorhabditis elegans*. Genes Nutr. 2019;14:1–3. https://doi.org/10.1186/s12263-019-0637-7.

Caenorhabditis elegans as a Model Organism for Proteinopathies

9

Priyanka Vimal, Deepanshu Verma, Prasun Kumar Bhunia, and Prasad Kasturi

9.1 Introduction

Proteinopathies are a collection of diseases caused by the misfolding and aggregation of specific proteins, leading to cellular toxicity and degeneration. These conditions are central to many age-related neurodegenerative disorders that are age-dependent, such as Alzheimer's, Parkinson's, Frontotemporal Dementia with Parkinsonism (FTDP), Huntington's, and amyotrophic lateral sclerosis (ALS). In all these conditions, specific proteins like Amyloid beta (Aβ), Tau, TDP-43, *α-synuclein,* or polyglutamine (polyQ) expanded proteins misfold and aggregate, disrupting various cellular processes, including proteostasis, mitochondrial function, synaptic signalling, and ultimately contributing to cell death. The pathological process usually occurs with age, when protein quality control (PQC) systems fail and disturb cellular protein homeostasis (proteostasis). As a result, misfolded proteins accumulate to form pathological aggregates that cause proteotoxicity. These aggregates either accumulate in intracellular or extracellular regions and can spread through cells and tissues in a prion-like manner, contributing to disease progression. Understanding the mechanisms of these age-related proteinopathies is necessary for therapeutic development.

P. Vimal · D. Verma · P. K. Bhunia · P. Kasturi (✉)
School of Biosciences and Bioengineering, Indian Institute of Technology Mandi, Kamand, Himachal Pradesh, India
e-mail: prasadkasturi@iitmandi.ac.in

P. Kumar, V. Kothari (eds.), *Model Organisms in Biological Research*, https://doi.org/10.1007/978-981-92-0038-2_9

As current research efforts focus on improving proteostasis during ageing, this chapter offers a comprehensive overview of PQC systems, ageing pathways, and molecular mechanisms that drive proteinopathies. A central focus of this chapter is the use of the model organism, *Caenorhabditis elegans* (*C. elegans*), to study the molecular basis of proteinopathies. Due to its genetic tractability, short lifespan, and conserved biological processes, *C. elegans* has become a valuable and ideal model system for studying ageing and age-related neurodegenerative diseases. We will describe available transgenic *C. elegans* strains expressing various human disease-related proteins that can recapitulate key features of protein aggregation and toxicity. Overall, this chapter illustrates how *C. elegans* models contribute to our understanding of proteinopathies.

9.2 Proteostasis or PQC

Proteins are the building blocks of cells and serve several essential functions, such as structural support, biochemical reactions, and signalling. Proteins are biological polymeric structures made up of 20 different amino acids. They need to maintain proper structural morphology or a 3D-folded structure for functionality. If they fail to attain a properly folded structure, they become misfolded and result in their aggregation. The protein quality control (PQC) systems monitor protein synthesis, folding, and degradation to maintain protein homeostasis (or proteostasis) [4]. These three processes (synthesis, folding, and degradation) are also considered as a proteostasis network (PN) that facilitates the timely production of correctly folded proteins in required quantities. It also removes excess and misfolded proteins, either through autophagy or proteasomal degradation. Molecular chaperones are key components of the PN, promoting correct protein folding and conformational stability. Collectively, these systems function to prevent the accumulation of toxic misfolded proteins.

The proteomes of eukaryotic cells exhibit considerable complexity. For example, in human cells, there are over 10,000 distinct proteins, with proteome composition differing across cell types and tissues. Properly balanced proteostasis is essential for cellular proteome integrity and health. Decline or imbalance of proteostasis is linked to ageing and age-related degenerative disorders [38]. To maintain proteome balance, the actions of three nodes of the PN (protein synthesis, folding, and degradation) must be coordinated (Fig. 9.1).

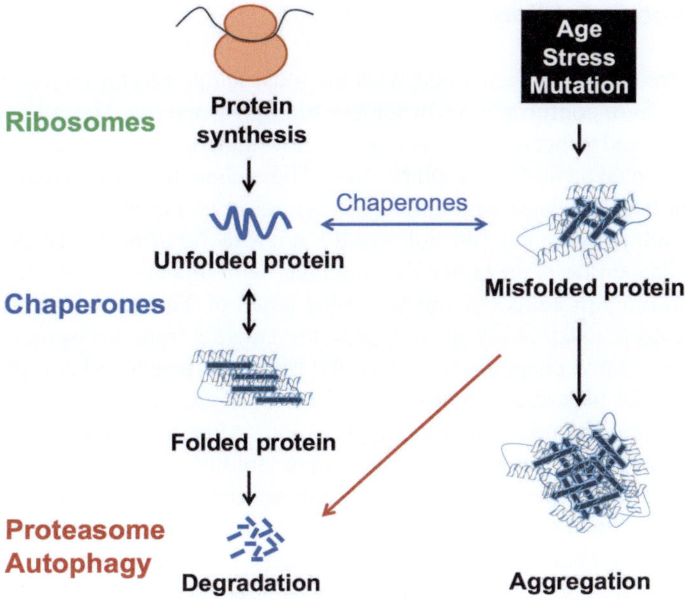

Fig. 9.1 Proteostasis network (PN) consisting of three components that include protein synthesis (ribosomes), protein folding (chaperones), and protein degradation (proteasome and autophagy)

9.2.1 Protein Synthesis

Cells make new proteins by gene expression (transcription) and protein synthesis (translation). Protein synthesis occurs in two steps: transcription and translation. In transcription, RNA polymerase and transcription factors produce mRNA, which carries genetic information for coding proteins. Translation begins on the ribosome, where mRNA is decoded into a polypeptide chain, which undergoes folding and modifications to become a functional protein. Proteins exhibit significant variability in quantity or abundance, ranging from less than 50 molecules per cell for certain proteins (such as transcription factors) to over ten million copies (such as histones or ribosomal proteins). The regulation of protein abundance is essential for basic cellular functions and enabling the stoichiometric assembly of macromolecular complexes, such as respiratory chain complexes, ribosomes, and proteasomes. Thus, protein synthesis is a crucial cellular process and the first step in the PQC. However, with age, gene expression and translation undergo significant changes, leading to reduced synthesis and errors in protein synthesis, resulting in faulty proteins [97]. The production of faulty proteins burdens proteostasis, contributing to age-related diseases, including neurodegeneration. This highlights the importance of protein synthesis as a quality-control mechanism for maintaining cellular function and overall health.

9.2.2 Protein Folding

The majority of freshly generated proteins must adopt certain three-dimensional (3D) structures or conformations to achieve biological activity. The folding process, formerly believed to occur spontaneously, is now understood to need a special class of proteins known as molecular chaperones. These chaperons are required for many proteins, particularly those with intricate structures or numerous domains, to achieve effective folding within a physiologically relevant timeframe. By definition, a molecular chaperone is an entity that interacts with and facilitates the folding or assembly of other proteins without becoming a part of them [37]. The protein folding begins co-translationally as polypeptides emerge from ribosomes, aided by ribosome-associated chaperones such as the ribosome-associated complex (RAC). Chaperones like heat shock proteins (HSPs) and chaperonins like TRiC assist in folding and prevent misfolding or aggregation. After translation, cytosolic and organellar chaperones help maintain proper folding, especially under stress. However, conditions like stress, ageing, and translational errors can lead to misfolding. The HSPs identify their substrates by recognizing the exposed hydrophobic amino acids and unstructured polypeptide chains, which are common characteristics of non-native states. Chaperones inhibit aggregation and facilitate folding in both ATP-dependent and independent fashion [37]. When correct folding fails, chaperones facilitate the degradation of defective proteins.

In some instances, in vivo folding occurs co-translationally, meaning it begins before the completion of protein production while the nascent chain is still linked to the ribosome. Conversely, many proteins fold completely in the cytoplasm after getting out of the ribosome, while others undertake folding in the endoplasmic reticulum (ER) or mitochondria, after their translocation across membranes. Molecular chaperones first identify damaged proteins and assist in their refolding and repair. If the damage is excessive or under settings detrimental to protein repair, compromised proteins are designated for destruction.

9.2.3 Protein Degradation

Maintaining a functional cellular proteome requires continuous protein turnover primarily through regulated degradation pathways. Protein degradation is a crucial and regulated mechanism for modulating the functional proteins' abundance. Protein breakdown or degradation is not just elimination or waste of cellular resources; it is a recycling process, since the amino acids derived from the damaged protein are repurposed for the synthesis of new proteins. Effective protein breakdown is crucial for cellular survival under circumstances that cause significant cellular damage. Damaged, misfolded, or excess proteins are eliminated via the ubiquitin-proteasome system (UPS) or autophagy-lysosomal degradation [21, 26]. Additional proteases, including intra- and extracellular types such as

AAA+ proteases, also contribute to protein degradation [95]. These mechanisms preserve cellular homeostasis, especially under stress, and are vital for ageing and disease regulation.

The UPS is the principal pathway for degrading short-lived and defective proteins. It involves two steps: ubiquitination and proteasomal degradation. Ubiquitination is a type of posttranslational modification in which a ubiquitin protein is attached to target proteins. This process involves three steps: activation (by E1), conjugation (by E2), and ligation (by E3 ligases). Polyubiquitin chains can be extended by E4 ligases. Deubiquitinating enzymes (DUBs) reverse this process. The main functional proteosome (26S) is composed of a catalytic core (20S) and a regulatory complex (19S). It recognizes the ubiquitinated target proteins and degrades in an ATP-dependent manner [6]. Autophagy complements UPS by targeting larger cellular structures, including protein aggregates and damaged organelles. Autophagosomes, double-membraned vesicles formed near the ER, sequester cargo for lysosomal degradation. This pathway is crucial for metabolic balance and is often impaired during ageing, contributing to age-related diseases. Collectively, UPS, autophagy, and protease systems coordinate protein quality control and are integral to cellular stress responses, ageing, and neurodegeneration.

9.2.4 Protein Aggregation

Protein folding is a finely balanced process; most native conformations are only marginally stable. Misfolded proteins can arise due to conformational stressors such as heat, oxidative damage, or ageing. Protein misfolding refers to structurally abnormal proteins that lack physiological function, often forming non-functional or toxic aggregates. Many disorders, including hereditary conditions, stem from the accumulation of misfolded proteins that escape degradation and form intracellular or extracellular aggregates. There are two types of protein aggregates: amorphous and amyloid fibrils. In the amyloid fibrils, parallel or anti-parallel β-strands are oriented perpendicular to the longitudinal axis of the fibril (cross-β structure), displaying unique staining capabilities. While amyloids are predominantly associated with pathological aggregation, amorphous aggregates are limited to diseases (α-crystallin aggregation in cataract). Despite their differing morphologies, both amyloid fibrils and amorphous aggregates can disrupt cellular function by interacting abnormally with membranes or macromolecules, including other proteins and RNA [20]. Misfolding and aggregation underlie many diseases commonly referred to as proteinopathies, including neurodegenerative disorders like Alzheimer's, Parkinson's, and Huntington's diseases [119]. Protein aggregation is not limited to the brain and can affect peripheral tissues, contributing to various systemic diseases like cataract, type 2 diabetes, cancer, and atherosclerosis.

9.2.5 Proteostasis Decline: A Hallmark of Ageing

The proteostasis capacity decreases with age due to compromised protein synthesis and degradation, resulting in the accumulation of misfolded proteins and their aggregation. The decline in proteostasis is considered a primary hallmark of ageing and a main risk factor for proteinopathies [70]. Age-dependent decrease in proteostasis capacity affects key PQC systems, including the cytosolic stress response (also known as the heat shock response), the organellar stress response (also known as the unfolded protein response), and proteasomal degradation. As a result, uncontrolled protein abundances, misfolding, and aggregation increase and negatively impact the organism's lifespan. These age-associated changes in proteostasis are evolutionarily conserved, highlighting proteostasis decline as a central feature in neurodegenerative and other age-related protein aggregation diseases.

9.3 Ageing and Age-Related Diseases

As individuals age, different types of genetic damage, primarily in the form of genetic mutations, impact an individual's fitness over time. This diminished fitness lowers the ability to combat diseases and heightens the vulnerability to develop various age-related conditions such as neurodegeneration, cardiovascular disease, cancer, diabetes, and musculoskeletal disorders [49]. The ageing population, when affected by these age-related diseases, elevates the socioeconomic burden on society [25]. A myriad of studies on ageing have identified several hallmarks of ageing that can be classified into primary, integrative, and antagonistic categories [70]. Among these distinguishing features, loss of proteostasis has attracted significant attention from research groups. It is generally understood as a disruption in proteostasis, often leading to an unbalanced proteome comprising uncontrolled abundance, misfolded, and aggregated proteins [38]. Compromised proteostasis and consequent protein misfolding/aggregation result in several diseases commonly known as proteinopathies. Signalling pathways that are involved in PQC or proteostasis also regulate ageing and are conserved across species, including worms, flies, mice, and humans. The nematode worm *C. elegans* serves as an exemplary model for investigating ageing and its associated pathways, which include the insulin/IGF-1-like signalling (IIS) pathway, dietary restriction, TOR signalling, mitochondrial respiration, and endoplasmic reticulum (ER) protein quality control [60]. These ageing-related signalling pathways are intricately connected to neurodegenerative proteinopathies in multicellular organisms.

The initial pathway identified to affect ageing in animals is the IIS pathway, which is extensively conserved throughout animals [5]. Numerous results indicate that the reduction of the IIS pathway in yeast, *Drosophila*, mice, and *C. elegans* extends lifespan [50]. It also plays a role in stress control and modulates many age-related genes and protein homeostasis during ageing. Activation of the IIS pathway results in the inactivation of a downstream transcription factor (FOXO/*daf-16*). The DAF-16 transcription factor regulates metabolism, longevity, and stress resistance by controlling its target gene expression [64]. The signalling cascade indicates that mutations in the IIS receptor (*daf-2)* in *C. elegans* exhibit increased longevity, whereas mutations in the *daf-16* demonstrate reduced lifespan [51]. In addition, HSF-1 and Nrf/SKN-1 transcription factors also function downstream of the IIS pathway to regulate stress response and ageing [12, 61]. Extensive research work on these mutant worms revealed that the IIS pathway regulates ageing and proteostasis [127], suggesting that the model organisms offer a better way to study ageing and age-associated diseases. This chapter will comprehensively provide information on *C. elegans* models for proteinopathies and their advantages.

9.4 *C. elegans* as a Model Organism

Caenorhabditis elegans is a little nematode that has garnered significant interest in recent decades. It has emerged as a pivotal model organism for investigating several diseases, particularly those associated with ageing [113]. It has surpassed other model organisms owing to the benefits it provides to the scientific community. The utilization of *C. elegans* in ageing research has substantially enhanced our comprehension of ageing, as well as the complex signalling networks and biological processes associated with it. It possesses numerous advantages compared to other model organisms, rendering it appropriate for ageing studies.

C. elegans is a little, non-parasitic roundworm residing in soil, with adults measuring 1 mm in length. It primarily consumes bacteria (*E. coli* OP50) and can be cultivated on an axenic diet. It can be cryopreserved at −80 °C or − 196 °C, facilitating the maintenance of extensive cultures and long-term storage, and can be revived whenever required. Culturing and maintaining *C. elegans* is far easier and more affordable than that of other model organisms [15]. Its lifespan is approximately 20 days when cultured at 20 ° C. Its complete genome has been sequenced, leading to the identification of numerous essential genes implicated in various signalling pathways by forward and reverse genetics. It possesses a transparent body that enables observation of its cells and various tissues at distinct developmental stages. This attribute renders it an excellent model for microscopy-based research. Fluorescent proteins conjugated to a protein of interest can be utilized to investigate diverse developmental processes, assess numerous mutants and their functions, and characterize protein interactions and aggregation [74].

It is a hermaphrodite that self-fertilizes, with males occurring rarely at 0.1–0.2% frequency. The hermaphrodite possesses a reproductive cycle lasting 2.5–4 days, during which a single worm can produce up to 300 eggs. The population is predominantly isogenic, and the adult stage can be attained roughly 48 hours post-hatching. It progresses through four larval stages (L1 to L4) post-hatching before maturing into an adult, or it may enter the Dauer larval stage after L2 to endure unfavourable conditions. This represents a developmentally stunted dispersal phase. It can re-enter the L4 stage and resume normal development when conditions become optimal. The worms possess a constant number of somatic cells, specifically 959; hence, research teams have delineated the developmental trajectory of each cell from fertilization to adulthood, thus elucidating the ancestry of every cell [103]. Utilizing electron micrographs, the morphology of each cell, including 302 neurons, has been recreated. It has significant genetic similarities with humans (about 60–80%), as well as some signalling pathways (Fig. 9.2). One of the notable advantages of working with *C. elegans* is its very participatory and collaborative scientific community.

A primary benefit of utilizing *C. elegans* is the simplicity of genetic modification. Through the application of basic genetic crossings and gene knockdown techniques, we can produce diverse mutant strains. Microinjection can be utilized to introduce foreign DNA into the gonads of adolescent adult organisms, leading to the creation of transgenic worms. RNA interference (RNAi) can be used to silence genes in worms by merely feeding them bacteria that express a specific gene's double-stranded RNA. The CRISPR–Cas9 technology enables precise genome editing to produce desired mutants. We can conduct studies necessitating tissue-specific expression of specific genes to comprehend their overall or tissue-specific functions. This benefit has also resulted in the creation of numerous disease models, including those for neurodegenerative disorders. This chapter will examine the diverse disease models created and their role in elucidating the molecular functions and mechanisms of various proteinopathies associated with ageing.

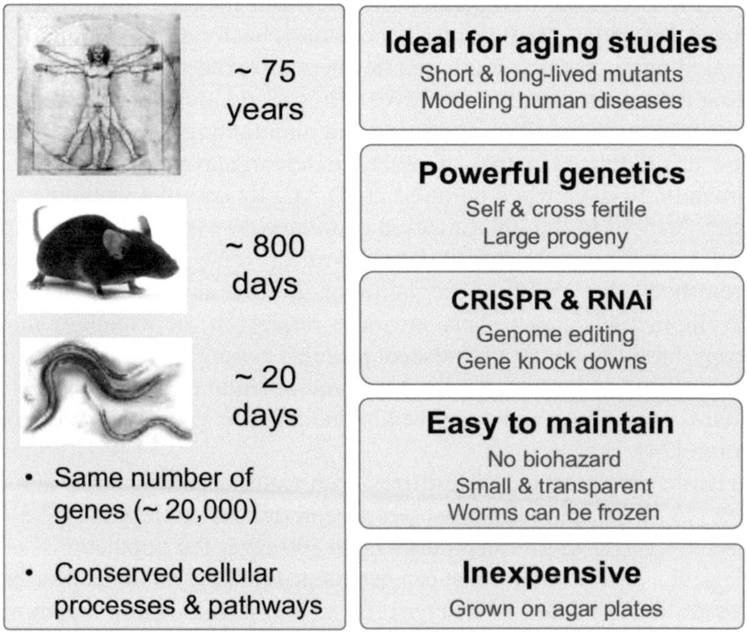

Fig. 9.2 Advantages of *C. elegans* as a model organism

9.4.1 Genetic Strategies for Disease Modelling in *C. elegans*

The nematode *C. elegans* serves as a powerful model organism for investigating the molecular basis of human diseases. Researchers employ several strategies to manipulate specific genetic features within *C. elegans*. One approach involves altering the endogenous homolog of a disease-related gene through sequence modification or regulation of gene expression. Alternatively, transgenic techniques can be applied to introduce and express the human version of the disease-associated gene. For genes conserved between *C. elegans* and humans, diverse methodologies can be implemented. These include gene deletion or suppression to reduce expression levels, overexpression of target genes, or introducing disease-associated mutations into the *C. elegans* homolog. Additionally, chimeric constructs can be developed by combining the *C. elegans* homolog with both wild-type and mutant domains of the corresponding human protein. Such constructs offer valuable insights into protein functionality and mechanisms of disease. To directly study the effects of human genes, transgenic *C. elegans* models can be designed to express either the wild-type or mutant human protein. These proteins can be localized to specific tissues such as muscles or neurons, or expressed throughout the organism. Furthermore, single-copy knock-ins of wild-type human genes can replace the *C. elegans* homolog, enabling precise functional analysis of human disease genes in vivo. Upon introducing these genetic manipulations, researchers can analyse a wide range of phenotypes to better understand gene function and disease pathology. Developmental phenotypes such as defects in embryogenesis, larval arrest, or abnormal growth patterns may emerge. Behavioural phenotypes, including impaired locomotion, disrupted chemotaxis, or altered sensory responses, can offer valuable clues for systematic phenotyping and drug repurposing. Cellular and molecular phenotypes, such as protein aggregation, mitochondrial dysfunction, or dysregulated signalling pathways, provide mechanistic insights into disease. Tissue-specific phenotypes, such as muscle degeneration or neuronal loss, are particularly useful for modelling human diseases that impact these systems [18].

9.5 *C. elegans* Model for Frontotemporal Dementia

Frontotemporal dementia with Parkinsonism (FTDP) is a rare hereditary neurological disorder, predominantly inherited through an autosomal dominant pattern. Approximately 40% of cases have a family history of dementia or movement disorders, underscoring its strong genetic basis. FTDP is characterized by dysfunction in the frontal and temporal lobes, driven by the abnormal accumulation of tau proteins, which are critical for stabilizing microtubules. This tau accumulation disrupts neuronal function, leading to hallmark symptoms such as behavioural changes, speech impairments, and motor difficulties resembling parkinsonism [121]. FTDP manifests between the ages of 40 and 60, in contrast to Alzheimer's disease, which commonly presents after age 65. This early onset, combined with diagnostic challenges, underscores FTDP's significance in tauopathy and neurodegeneration research.

The disease manifests in two forms: familial and sporadic, with familial cases accounting for 1–5% of instances. Familial cases often involve early onset and are driven by mutations in the *MAPT* gene that encodes tau protein [102, 110].

9.5.1 FTDP Types

Given its diverse clinical presentations, FTDP is classified into two primary subtypes: behavioural variant (bvFTDP) and language variant (lvFTDP). These subtypes arise from differences in genetic mutations and affected brain regions. The bvFTDP accounts for more than 50% of FTDP cases, primarily affecting the right frontal and temporal lobes. Degeneration in these regions leads to profound changes in personality, behaviour, and interpersonal conduct. In the early stages, bvFTDP is characterized by compulsive behaviours, apathy or lack of motivation, and alterations in eating habits. Despite these behavioural changes, memory and cognitive abilities remain relatively preserved in the initial phases, making bvFTDP distinct from other dementias like Alzheimer's disease. The lvFTDP is marked by a progressive decline in language abilities. It is further divided into two distinct subtypes: semantic dementia (SD) and progressive non-fluent aphasia (PNFA). Semantic dementia is a slowly progressive disorder typically affecting individuals aged 60 and older. It constitutes 20–25% of FTDP cases and is characterized by degeneration in the temporal lobes. The phenotypes include i) difficulties in understanding words and their meanings, ii) memory impairments, iii) prosopagnosia (difficulty recognizing faces), and iv) higher prevalence of depression. The progressive non-fluent aphasia represents the second most common form of FTDP, accounting for approximately 25% of cases. It is associated with left frontal lobe atrophy and generally manifests later in life, around the age of 63. Key characteristics of PNFA include: i) slow, non-fluent speech. ii) difficulty forming grammatically correct sentences, and iii) paraphasic errors (word substitutions or distortions).

9.5.2 Tau Protein

Tau protein, a critical microtubule stabilizer, is encoded by the *MAPT* gene (microtubule-associated protein tau). The tau protein is composed of four domains: the N-terminal domain, proline-rich domain, MT-binding domain, and C-terminal domain. There are 16 exons in the *MAPT* gene that can generate six isoforms of tau (by alternative splicing of exons 2, 3, and 10). These splicing events create structural variations in the N-terminal and microtubule-binding (MT-binding) domains, which are essential for tau's biological functions [104]. Among these, the N-terminal domain and MT-binding domain are critical for microtubule stabilization and intracellular transport. While the exact role of the N-terminal domain remains unclear, it is thought to contribute to microtubule stabilization and interactions with other cellular components. Alternative splicing of exons 2 and 3 generates the 0 N, 1 N, and 2 N tau isoforms, which differ in their functional implications based on whether one, both, or neither exon is included [68]. The MT-binding domain, indispensable for stabilizing microtubules and maintaining cytoskeletal integrity, exists in two isoforms determined by the inclusion or exclusion of exon 10, and the number of repeats determines its binding affinity for the tubules:

- 3-repeat (3R) isoform: Lacks exon 10 and contains three repeat regions in the MT-binding domain.
- 4-repeat (4R) isoform: Includes exon 10, which enhances microtubule-binding affinity but increases susceptibility to aggregation and toxicity.

Several mutations in the *MAPT* gene are linked to tauopathies that significantly influence tau function and aggregation (Box 9.1).

Box 9.1 Mutations and Phosphorylation in Tau Protein

Missense Mutations

Missense mutations alter tau's amino acid sequence, disrupting microtubule assembly and destabilizing microtubules [42]. Consequently, unbound tau accumulates in the cytosol of neurons. These mutations often affect phosphorylation sites and disrupt the normal splicing of *MAPT* mRNA, further impairing tau's functionality. Notably, missense mutations in the exon 10 region increase 4R tau expression, either by destabilizing the stem-loop structure or altering splicing enhancers and silencers. These mutations also promote post-translational modifications, such as lysine acetylation, which inhibit tau degradation, enhance aggregation, and amplify neurotoxicity.

Splicing Mutations

Tau protein predominantly localizes to axonal regions of neurons, where it stabilizes microtubules and supports intracellular transport of vesicles and organelles. This axonal enrichment is due to tau's stronger microtubule-binding affinity in axons compared to dendrites. Tau localization and function are isoform-dependent, influenced by alternative splicing patterns. Mutations near exon 10 of the *MAPT* gene, particularly in intronic regions, disrupt normal splicing and lead to an overproduction of the 4R tau isoform. An imbalance between 3R and 4R tau isoforms causes mislocalization of 4R tau to dendrites, where tau is typically absent. This mislocalization results in synaptic dysfunction, microtubule instability, and impaired transport, contributing to neuronal dysfunction and the development of tauopathies such as FTDP-17 [102].

(continued)

Box 9.1 (continued)

Hyperphosphorylation Sites

Tau's interaction with microtubules is negatively regulated by phosphorylation, which reduces its microtubule-binding affinity. Tau isoforms contain approximately 80 phosphorylation sites, primarily at serine, threonine, and tyrosine residues. Hyperphosphorylation is a hallmark of tauopathies, including FTDP-17, and is often exacerbated by missense mutations. Hyperphosphorylated tau detaches from microtubules, aggregates into neurofibrillary tangles, and impairs axonal transport and neuronal signalling. Additionally, other post-translational modifications, such as lysine acetylation, hinder tau degradation and clearance, further amplifying its pathological effects [118].

9.5.3 PTL-1 Models

The *C. elegans* genome encodes a protein that is analogous to tau (a member of the microtubule-associated protein (MAP) family). The gene was named as *ptl-1* for *p*rotein with *t*au-*l*ike repeats 1. This protein is characterized as a basic, thermostable molecule containing multiple phosphorylation sites that align with tau's functional properties. Sequence similarity between the PTL-1 protein and tau is predominantly concentrated in the C-terminal region, with less homology observed in the N-terminal region [34]. The *ptl-1* gene makes PTL-1A and PTL-1B, two isoforms through alternative splicing. These isoforms exhibit significant parallels to mammalian tau, particularly in their structural and functional features. Despite their differences in microtubule assembly proficiency, the isoforms share fundamental traits, such as amino acid composition, charge distribution, hydrophobicity, size and flexibility. The expression of *ptl-1* is dynamically regulated during development. In embryonic stages, it is predominantly expressed in epidermal cells and select neuronal populations. This expression pattern shifts during postembryonic development, becoming more localized to mechanosensory neurons. Such spatial and temporal regulation highlights the gene's role in facilitating specialized cellular and neuronal functions across developmental phases [34]. PTL-1 plays critical roles in processes such as egg hatching and touch responsiveness. It is also implicated in neuronal integrity maintenance and lifespan regulation. Loss-of-function of *ptl-1* induces neuronal ageing signatures, suggesting that PTL-1 within individual neurons to preserve neuronal health [19]. Studies on mutant *ptl-1* worms revealed disturbed movement of motor proteins and their cargo proteins, providing evidence of PTL-1's function in coordinating neuronal transport and motor activities. Together, these findings underscore PTL-1's multifunctional contributions to neuronal health, structural stability, and intracellular transport.

9.5.4 Tauopathy Models

Transgenic *C. elegans* strains expressing human tau associated with FTDP (*hTau-FTDP-17*) have been developed to study the underlying disease mechanisms at the molecular level. These models replicate key pathological features of FTDP, enabling detailed studies on tau aggregation, neurodegeneration, and the role of tau mutations. Expression of either wild-type tau or the *hTau-FTDP-17* mutant in *C. elegans* neurons leads to a progressively worsening uncoordinated phenotype as the organisms age. This neurodegenerative phenotype and the insoluble tau accumulation are typical of tauopathies. The first model was generated by overexpression of the *h1N4R-MAPT* gene under the *aex-3* promoter for pan-neuronal expression. Three transgenic strains were generated: one expressing the 1N4R wild-type *MAPT* and two expressing the P301L and V337M mutant variants of *FTDP-17*. These mutant worms displayed a severe uncoordinated phenotype compared to the wild-type tau line, with the severity of symptoms progressively increasing with age. Among these models, the V337M mutant displayed the most aggressive phenotype, characterized by the accumulation of tau aggregates and reduced survival, emphasizing its role in a dysfunctional and progressively deteriorating nervous system. It has been shown by ultrastructural studies that abnormal tau protein inclusions form in the axoplasm, which further degenerate axons. These findings reinforce the connection between tau mutations, motor dysfunction, and their broader implications for tauopathies and neurodegenerative diseases [53]. In a different worm model, various tau variants were expressed under the *mec-7* promoter in the touch neurons. The wild-type construct included two isoforms, both lacking N-terminal inserts. Transgenic lines expressing wild-type tau exhibited only mild sensitivity to touch responses compared to wild-type non-transgenic worms. In contrast, mutant transgenic lines demonstrated significantly reduced touch responses, which progressively deteriorated with age. Among the mutants, tau R406 lines displayed an earlier onset of functional decline compared to the mutant tau P301L lines, highlighting distinct levels of impairment caused by these tau variants. The results get worsened with age, showing compromised microtubule parallel by complete loss of touch response.

To study tau aggregation, the Fatourous group developed two distinct transgenic models, pro-aggregation and anti-aggregation lines [30]. The pro-aggregation model, generated by the deletion of K280, promotes tau aggregation significantly. This line shows severe phenotypes, including pronounced tau accumulation, uncoordinated movement in adult worms, and defects in axonal integrity and pre-synaptic structures. The anti-aggregation model, on the other hand, involves the introduction of mutations I277P and I308P, which inhibit tau aggregation. While this line produces lower levels of aggregated tau, it exhibits only mild phenotypes in comparison to the pro-aggregation model [54]. The differences between these two models provide critical insights into how specific tau mutations influence aggregation propensity and neurodegeneration. Building on the discussion of mutations primarily affecting the repeat region of tau, a novel mutation, A152T, has been identified, distinguishing itself by residing outside the repeat region in the N-terminal domain. This mutation is notable for its minimal impact on tau aggregation, offering valuable insights into tau's diverse roles beyond microtubule binding and stabilization. It highlights tau's participation in other cellular processes, broadening our understanding of its functional versatility [118]. To further investigate the implications of this mutation, transgenic lines were developed by pan-neuronal expression of 2N4R tau (wildtype and mutant A152T). The organisms expressing the A152T mutant exhibited an earlier onset of paralysis compared to those expressing wild-type tau. This phenotype was linked to morphological defects in mechanosensory neurons observed in the mutant lines. Furthermore, these animals demonstrated reduced lifespans, emphasizing the detrimental impact of the A152T mutation on neuronal health and organismal longevity [88]. Available strains are shown in Fig. 9.3 and listed in Table 9.1.

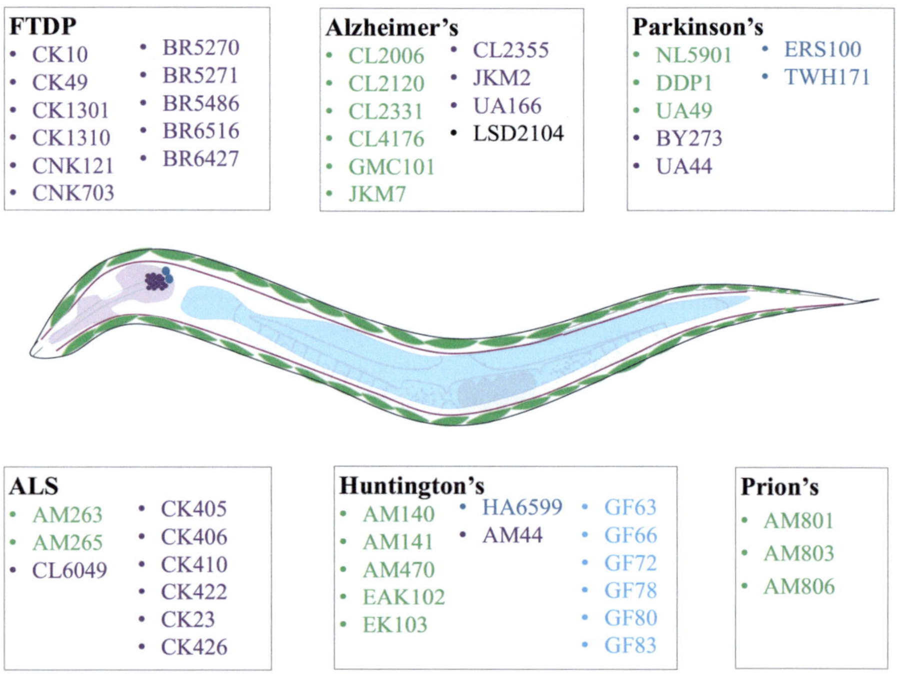

Fig. 9.3 *C. elegans* disease models and strains

Table 9.1 List of *C. elegans* disease models and available strains

Disease	Protein	Strain	Genotype	Expression	Phenotype	Availability on CGC	References
FTDP	Tau	CK10, CK49, CK1301, CK1310	aex-3p::4R1N hTau (WT, V337M, P301L) + myo-2p::gfp	Constitutive pan-neuronal	Loss of coordination, accumulation of insoluble tau fibrils and shortened lifespan	No	Kraemer et al. [53]
		NA	mec-7p::tau WT (0N4R, 0N3R)	Mechanosensory neurons	Gradual and mild decline in touch sensitivity, minimal detectable tau buildup	No	Miyasaka et al. [75]
		NA	mec-7p::tau (P301L, R406W)	Mechanosensory neurons	Deficits in touch sensitivity, substantial tau aggregation in the PLM tail neuron	No	Miyasaka et al. [75]
		BR5270, BR5485, BR5944, BR5706	*Pro-aggregation lines:* rab-3p::F3ΔK280 + myo-2p::mCherry	Constitutive pan-neuronal	Movement difficulties at a younger age, rapid formation of aggregates and disruption in presynaptic transmission	Yes	Fatouros et al. [30]
		BR5271, BR5486, BR6516, BR6427	*Anti-aggregation lines:* rab-3p::F3ΔK280 (I277P) (I308P) + myo-2p::mCherry	Constitutive pan-neuronal	No noticeable movement difficulties; minor impact on nervous system development	Yes	Fatouros et al. [30]
		CNK121	mec-4p::MAPT/Tau F3ΔK281::mCherry	Touch receptor neurons	Prominent neurotoxicity and mechanosensory deficits	No	Sandhof et al. [93]

(continued)

Table 9.1 (continued)

Disease	Protein	Strain	Genotype	Expression	Phenotype	Availability on CGC	References
Alzheimer's	Amyloid-beta	CL2006, CL2120	unc-54p::Aβ1–42 (WT) or dimer Aβ1–42	Constitutive muscle	Amyloid deposits and paralysis	Yes	Link [66]
		GMC101	CL354 (unc-54p::DA-Aβ1–42) + CL26 (mtl-2p::gfp)	Constitutive muscle	Paralysis after 48 hrs of temperature shift to 25 °C	Yes	McColl et al. [73]
		CL2331	myo-3p::GFP::A-beta (3–42)			Yes	Link et al. [67]
		CL2241, CL2355	smg-1 (cc546); snb-1p:: Aβ1–42 + mtl-2p::gfp	Inducible pan-neuronal	Impaired chemotactic response and accumulation of amyloid deposits	Yes	Xu et al. [122]
		CL4176	smg-1(cc546); myo-3p:: Aβ1–42	Inducible body-wall muscle	Swift onset of paralysis, oxidative stress occurring prior to Aβ fibril development.	Yes	Drake et al. [27]
		UA166	eat-4p::ssAβ1–42(N-term) + eat-4p::gfp + myo-2p::mCherry	Glutamatergic neurons	Glutamatergic loss of neurons, age-dependent manner	No	Treusch et al. [106]
		JKM2	rgef-1p::Signal peptide- A-beta (1–42)::hsp-3(IRES)::wrmScarlet-A-beta (1–42)	Pan-neuronal	Age-dependent progressive increase in aggregation in neurons.	No	Gallrein et al. [32]

		JKM7	*myo-3*p::Signalpeptide-A-beta(1–42)::*hsp-3* (IRES)::wrmScarlet-A-beta(1–42)	Body wall muscle	Age-dependent progressive increase in aggregation in the body muscle.	No	Gallrein et al. [32]
		LSD2104	*hsp16.2p*::ssSel1::FLAG::super folder GFP::spacer:: human Amyloid beta1–42::let-858 3'UTR; pRF4 *rol-6 (su1006)*	Heat shock induces the secretion and localization of sfGFP::Aβ to different tissues over time	Aβ aggregation toxicity	Yes	Jongsma et al. [46]

(continued)

Table 9.1 (continued)

Disease	Protein	Strain	Genotype	Expression	Phenotype	Availability on CGC	References
Parkinson's	Alpha-synuclein	BY273, UA18, UA31, UA44	dat-1p::α-syn (WT, A53T) + dat-1p::gfp	Dopaminergic neurons	A substantial reduction in dopaminergic neurons and noticeable dendritic damage.	No	Starr et al. [101]
		NL5901	unc-54p::α-syn::yfp + unc-119(+)	Constitutive muscle	Inclusion formation with ageing.	Yes	Van Ham et al. [109]
			aex-3p:: α-syn (WT, A53T) + aex-3p::GFP/dat-1p::GFP	Constitutive pan-neuronal	Motility defects.	No	
		DDP1	unc-54p::alpha-synuclein::CFP + unc-54p::alpha-synuclein::YFP(Venus)	Constitutive muscle	slower pharyngeal pumping rates, reduced movement and lifespan.	Yes	Bodhicharla et al. [14]
		UA49	unc-54p::α-syn::gfp	Constitutive muscle	Misfolding and aggregation of alpha-synuclein.	No	Hamamichi et al. [35]
			unc-51p::α-syn (WT, A53T, A30P) + unc-51::eGFP	Constitutive pan-neuronal	No motor impairment and developmental abnormalities.	No	Kuwahara et al. [56]
			mec-7p::α-syn (WT, A53T)	Mechanosensory neurons	Mild to moderate deficits in sensory touch sensitivity.	No	Kuwahara et al. [56]
			dat-1p::α-syn (A30P, A53T, A56P) + dat-1p::mCherry	Dopaminergic neurons	A56P exhibit substantial dopamine (DA) depletion, whereas A30P and A53T show a reduction in DA levels. Impairments in DA- dependent behaviour.	No	Karpinar et al. [48]

			unc-51p::α-syn (S129A, S129D) + unc-51p::eGFP	Constitutive pan-neuronal	Early-onset severe motor impairment, restricted growth, and disruptions in synaptic function.	No	Kuwahara et al. [57]
		ERS100	dat-1p::mCherry + dat-1p:: hSNCA::Venus	Dopaminergic neurons	Degenerative signs in dopaminergic neurons	Yes	Vozdek et al. [115]
		TWH179	dat-1p::alpha-synuclein(A30P) + ges-1p:: DsRed			Yes	
			snb-1p::lrrk2 (WT, G2019S, R1441C, KD, R1441C/ KD) + mec-4p::gfp	Constitutive pan- neuronal	G201S mutants heightened DA neuron vulnerability and promoted degeneration	No	Saha et al. [92]
		UA118	dat-1p::lrrk2(G2019S) + dat-1p::gfp	Dopaminergic neurons	Dopamine depletion, neuron degeneration, behavioural deficit and locomotor dysfunction	No	Liu et al. [69]
		SGC722, SGC851, SGC856, SGC862	dat-1p::lrrk2 (WT, R1441C, G2019S, K1347A) + dat-1p:: gfp	Dopaminergic neurons	Neuron degeneration, Behavioural and motor dysfunction worsen	No	Yao et al. [124]
		SG900, SGC910	dat-1p::gfp, dat1p::lrrk2(R1441C/ A2016T)	Dopaminergic neurons	Double mutants exhibit a comparable phenotype to R1441C and G2019S LRRK2, Dopaminergic impairment and progressive neurodegeneration	No	Yao et al. [124]

(continued)

Table 9.1 (continued)

Disease	Protein	Strain	Genotype	Expression	Phenotype	Availability on CGC	References
ALS	FUS		rgef-1p::fus (WT, R514G, R521G, R522G, R524S, P525L)	Constitutive pan-neuronal	Accumulation of cytoplasmic FUS aggregates, shorter lifespan, P525L, FUS513, FUS501 associated with paralysis	No	Murakami et al. [80]
	TDP-43	CK405, CK406, CK410, CK422, CK423, CK426	snb-1p::tdp-43 (WT, G290A, A315T, M337V) + snb-1p:: gfp	Constitutive pan- neuronal	Accumulation of truncated aggregates leads to motor neuron degeneration	No	Liachko et al. [65]
		CL6049	snb-1p::hTDP-43/3′ long UTR + mtl-2p::GFP	Pan-neuronal	Uncoordinated and abnormal motor neuron synapse	yes	Ash et al. [2]
		CL1681, CL1682, CL2609	unc-25p::snb-1p::gfp + snb-1p::tdp-43; regf-1p:: DsRed2; unc122p::rfp	Constitutive pan- neuronal	Disorganized motor function accompanied by irregular motor neuron synapses	No	Ash et al. [2]
	SOD-1		hsp-16.2p::sod-1 (WT, A4V, G37R, G93A) + myo-3p::sod-1 (WA4V):: gfp	Heat-shock inducible muscles	Formation of aggregates in response to oxidative stress	No	Oeda et al. [83]
			snb-1p::sod-1(WT, G85R):: yfp	Constitutive pan- neuronal	Soluble oligomers and insoluble aggregates show profound motor impairments	No	Wang et al. [117]

			sng-1p::sod-1(WT, A4V, G37R, G93C)::gfp	Constitutive pan- neuronal	Mutant homodimers exhibit heightened aggregate formation relative to heterodimers, though G85R heterodimers demonstrate greater toxicity	No	Witan et al. [120]
		AM263, AM265	unc-54p::sod-1(WT, G85R, G93A, G127 ins TGGG stop)::yfp	Constitutive muscles	Mutants exhibit structurally diverse aggregates, distinct biophysical characteristics, and mild movement impairments	Yes	Gidalevitz et al. [33]
			unc-25p::sod-1(G93A)::gfp	GABAergic motor neurons	Gradually worsening impairment in motor function, formation of aggregates and disruptions in axonal navigation	No	Le et al. [62]

(continued)

Table 9.1 (continued)

Disease	Protein	Strain	Genotype	Expression	Phenotype	Availability on CGC	References
Huntington's	Huntingtin		osm-10p::gfp + osm-10::htt (Q2, 23, 95, 150)	Chemosensory neurons	Expression of Q150 induces ASH degeneration	No	Faber et al. [29]
		HA659	osm-10p::gfp + osm-10p:: htt (150)	Chemosensory neurons	Identification of poly Q enhancers	Yes	
		ID1, ID24	mec-3p::htt57 (Q19, 88, 128)::gfp; mec-3p::htt57 (Q19, 88, 128)::cfp+ mec-3p::yfp	Mechanosensory neurons	Desensitized to posterior touch, Mec phenotypes, substantial aggregate sediments and PLM axon deformities	No	Parker et al. [87]
		EAK102, EAK103	unc-54p::htt513 (Q15, Q128)::yfp	Constitutive muscles	Length-driven accumulation and toxicity, Q128 animals show motor malfunction	Yes	Lee et al. [63]

	PolyQ		unc-54p::polyQ (19, 82):: gfp/yfp/cfp	Constitutive muscles	Length-dependent aggregation, developmental delays and impaired motility	No	Satyal et al. [94]
		AM140, AM141	unc-54p::polyQ (0, 19, 29, 35, 37, 40, 44, 64, 82):: yfp	Constitutive muscles	Aggregation threshold at Q35–40, motility impairments	Yes	Morley et al. [78]
		AM78, AM80, AM81, AM83, AM85, AM87; AM303, AM305, AM308, AM313, AM322, AM324	gef-1p::polyQ (0, 19, 35, 40, 67, 86)::cfp/yfp	Constitutive pan- neuronal	Length-dependent polyQ aggregation, dysfunctional neurons	No	Brignull et al. [16]
		GF63, GF66, GF72, GF78, GF80, GF83	vha-6p::polyQ::YFP	Constitutive intestinal	Aggregation threshold at Q44	Yes	Mohri-Shiomi and Garsin [77]

(continued)

Table 9.1 (continued)

Disease	Protein	Strain	Genotype	Expression	Phenotype	Availability on CGC	References
Prion			ric-19p::prp + ric-19::gfp	Constitutive pan- neuronal	Elevated PrP changes morphology, motor dysfunction, reduced lifespan	No	Park and Li [86]
			mec-7p::prp (WT, PG13) + str-1::gfp; mec-7::gfp	Mechanosensory neurons	Mutants exhibit a gradual decline in tail-touch sensitivity, without neuronal degeneration	No	Biza et al. [10]
		AM801, AM803, AM806	unc-54p::sup35(RΔ2–5, NM, R2E2)::yfp	Constitutive muscle	Profound mitochondrial dysfunction, larval arrest, delayed development and impaired proteostasis	No	Nussbaum-Krammer et al. [82]

Abbreviations: *4R1N hTau* human tau isoform with 4 microtubule binding repeats and 1 N terminal insert, *0N4R* no N-terminal Insert, 4 microtubule binding repeats, *0N3R* no N-terminal Insert, 3 microtubule binding repeats, *F3ΔK280* mutant form of tau repeat domain, *F3ΔK281* mutant form of tau repeat domain, *DA* dutch (E22Q) + Artic (E22G) mutation, *ssAβ1–42* signal sequence fused to the N-terminus of Aβ1–42, *hSNCA* human α-synuclein, *RΔ2–5* sup35 prion domain with deletion of repeats 2–5, *NM* N + M domain of Sup35, *R2E2* synthetic variant with 2 extra oligopeptide repeats, *Q* glutamine

9.6 *C. elegans* Model for Alzheimer's Disease

Alzheimer's disease (AD) is the sixth leading cause of death in the United States and is estimated to affect 35 million people worldwide. With mild impaired learning and memory, this neurodegenerative disorder presents substantial challenges to healthcare systems. With the disease progression, patients experience neuropsychiatric and behavioural changes, ultimately leading to underperform routine activities. AD exists in two main forms, familial AD (early-onset) and sporadic AD (late-onset). The early onset accounts for ~5% of cases and is largely driven by genetic mutations in genes encoding for amyloid precursor protein (APP) and presenilins (PSEN1 and PSEN2). The sporadic form generally manifests in older adults and does not follow a genetic inheritance pattern. Despite differences in their origins, both forms of ADs are characterized by hallmark pathological features such as amyloid plaques (Aβ42 aggregates in the extracellular region) and neurofibrillary tangles (hyperphosphorylated tau protein aggregates in the intracellular region) [52]. These pathological changes result in neuronal dysfunction, chronic inflammation, and eventual cell death, contributing to the progressive decline of cognition observed in patients with AD.

9.6.1 Amyloid Precursor Protein (APP)

It belongs to the APP-related protein family, which includes orthologs such as APL-1 in *C. elegans* and APPL (Amyloid Precursor Protein-Like) in *Drosophila*, making it highly conserved across species. In humans, the APP gene locus is on chromosome 21, comprises 18 exons, and produces eight isoforms through alternative splicing. Structurally, APP is composed of three regions: (1) extracellular domain (important for intercellular interactions), (2) amyloid-beta (Aβ) domain (prone to cleavage, releasing toxic peptides), and (3) cytoplasmic region (critical for intracellular signalling and processing). APP's multi-domain structure allows it to function in neuronal growth, repair, and synaptic communication. However, its improper cleavage drives the formation of Aβ peptides, which are central to Alzheimer's pathology.

9.6.2 The Link Between Aβ and Tau Aggregation

While Aβ accumulation plays a pivotal role in AD pathology, tau aggregation is a hallmark feature of a broader group of disorders (tauopathies), including Pick's disease and frontotemporal dementia. However, in AD, these two pathologies are closely interconnected. The Aβ accumulation precedes tau pathology. This initiates a cascade of events that includes induction of tau hyperphosphorylation; as a result, tau loses its ability to stabilize microtubules, leading to cytoskeletal collapse and aggregation into neurofibrillary tangles (NFTs). These NFTs, in conjunction with Aβ-induced oxidative stress and inflammation, exacerbate neuronal toxicity [128].

Although the precise molecular mechanisms linking Aβ toxicity to tau aggregation remain unresolved, their interaction is widely considered to synergize in accelerating the progression of AD. Experimental models highlight that Aβ acts as a "trigger," while tau pathology acts as a "driver" of neurodegeneration. Given this interdependence, this discussion primarily emphasizes APP and Aβ, as they represent the starting point for plaque formation and aggregation. Understanding APP mutations, processing, and their contributions to Aβ production is key to unravelling early-stage mechanisms of Alzheimer's disease.

Unlike humans, *C. elegans* does not naturally produce amyloid-beta (Aβ) peptides due to a lack of the corresponding amino acid sequence and the β-secretases required for amyloid precursor protein (APP) processing. To overcome this limitation, Christopher Link generated a transgene with human Aβ under the *unc-54* promoter to express it in *C. elegans* muscle cells [66]. The initial transgenic strains (CL2006 and GMC101) were specifically designed for muscle cell-specific expression. The expression of Aβ in the worm's muscle cells caused paralysis and premature death. However, later studies revealed that the CL2006 strain does not generate the full-length Aβ1–42 peptide but instead produces a truncated Aβ3–42 variant. To address this issue, the GMC101 strain was developed to express the complete Aβ1–42 peptide, resulting in a more rapid onset of paralysis [72, 73]. Available strains are shown in Fig. 9.3 and listed in Table 9.1.

9.6.2.1 Link Between Aβ Toxicity and Insulin Signalling

The Aβ-induced paralysis in *C. elegans* is strongly linked to insulin signalling, as the paralysis diminishes in the reduced insulin-signalling background. The transcription factors HSF-1 and DAF-16 are required for countering Aβ-induced paralysis. This indicates that Aβ toxicity is influenced by both age-related changes and metabolic regulation. Further studies demonstrated pan-neuronal expression of Aβ, which led to intraneuronal Aβ deposits but did not induce movement-related phenotypic alterations. Instead, worms exhibited behavioural and cognitive impairments, highlighting the neurotoxic effects of Aβ beyond muscle degeneration [22]. Interestingly, the small heat shock proteins (HSP-16 s) are induced in Aβ transgenic worm strains, where they bind with Aβ monomers and soluble oligomers but not with fibrillar aggregates. In contrast, *hsp-70* expression offers a protective effect by suppressing paralysis, suggesting that molecular chaperones play a critical role in modulating Aβ toxicity [31]. Further studies have shown that large Aβ aggregates in the CL2006 worm strain are not directly associated with toxicity, rather the oligomers serve as the primary toxic agents.

9.6.3 APL-1 Models

The *C. elegans* homolog of APP, APL-1 (Amyloid Precursor-Like), shares conservation in the N- and C-terminal sequences, which are highly preserved across species. However, APL-1 lacks the sequence for amyloid-beta (Aβ) peptides, a key factor in APP-associated pathology. Despite APP's central role in AD, with specific mutations linked to AD onset, its physiological functions are unclear [24]. Transgenic mice expressing human or mouse APP and PSEN1 mutations show increased lethality and learning deficits, yet these effects are not directly linked to Aβ deposition, suggesting an association with other underlying mechanisms for the disease progression [40]. This uncertainty makes *C. elegans* a valuable model for further investigation. Loss-of-function mutations in *apl-1*, are lethal during the larval stage, causing defects in cholinergic neurons at the synaptic level. However, reintroducing soluble APL-1 protein (sAPL-1) into the neurons of these mutants restores the normal phenotype, underscoring its essential role in neuronal function and development. Conversely, overexpression of APL-1 results in several developmental, reproductive and movement defects. Remarkably, these overexpression-associated defects can be rescued by the activity of SEL-12, the *C. elegans* ortholog of human Presenilin, highlighting the need for tightly regulated *apl-1* expression. Both overexpression and downregulation of *apl-1* lead to lethality, emphasizing the importance of maintaining optimal levels of this protein [28, 44].

C. elegans relies on sensory neurons to sense and respond to environmental stimuli, including chemoattractive and chemorepulsive cues, by moving toward or away from these signals. Among the sensory neurons, AWC neurons have been identified as key mediators of chemosensory responses. Overexpression of APL-1 in transgenic worms leads to a reduction in chemoattractive responses to compounds such as benzaldehyde and sodium acetate. This occurs despite the fact that the morphology of sensory neurons remains normal, as observed with GFP markers. These findings suggest that APL-1 modulates sensory function through mechanisms unrelated to structural changes, emphasizing the importance of regulating its expression [28]. Interestingly, impaired chemotaxis caused by APL-1 overexpression can be restored when insulin signalling is decreased. However, this restoration does not occur when *daf-16* activity is reduced, suggesting that in APL-1 overexpression worms, DAF-16 is essential for normal chemotaxis. This reveals a complex interplay between APL-1 expression and insulin signalling, providing potential insights into APP-related neurodegeneration and sensory processing in higher organisms.

9.7 *C. elegans* Model for Parkinson's Disease

Parkinson's disease (PD) affects approximately ten million people worldwide, making it the second most common age-related neurodegenerative disease. Despite extensive research, its exact cause remains unknown, with studies identifying both genetic predisposition and environmental influences as major contributors [105]. PD is characterized by α-synuclein protein accumulation, which forms Lewy bodies, a type of abnormal protein aggregate associated with neuronal dysfunction [100]. Misfolded alpha-synuclein disrupts mitochondrial function, induces oxidative stress, and impairs protein degradation pathways, exacerbating neurodegeneration.

Loss of dopaminergic neurons in a brain region that is responsible for movement control is one of the pathological hallmarks of PD. This results in motor impairments, including bradykinesis and hypokinesis (reduced movement), tremors, and postural instability, significantly impacting a patient's ability to coordinate and control movements. However, PD extends beyond motor dysfunction because it affects both the central and peripheral nervous systems, manifests as non-motor symptoms such as cognitive decline, memory impairment, olfactory deficits and depression [47]. Mutations in *LRRK2, PARK2, DJ-1, GBA,* and *UCHL1 genes* have been linked with PD. For example, glucocerebrosidase (*GBA*), a gene involved in lysosomal substrate metabolism, is associated with sporadic (environmental) forms of PD. Likewise, mutations in ubiquitin carboxyl-terminal hydrolase L1 (*UCHL*) enzyme, which is responsible for protein recycling, are also implicated in familial PD, reinforcing the genetic basis of the disease.

9.7.1 Alpha-Synuclein Protein

The Synuclein Alpha (SNCA) gene codes for alpha-synuclein protein. It is a 140 amino acid small protein consisting of 14 kDa in size and comprises three distinct domains. It is a highly conserved cytoplasmic protein, and predominantly localized in the presynaptic neurons of the brain, though expressed in the central and peripheral nervous systems, blood cells, and other tissues [17]. Mutations in the N-terminal region of alpha-synuclein are primarily involved in the development of PD [89]. Although the physiological structure and function of alpha-synuclein remain incompletely understood, existing literature suggests its significance in the storage and recycling of neurotransmitters. Additionally, alpha-synuclein has been observed to regulate various enzymes, enhance dopamine transporter activity, and exhibit molecular chaperone functions that are linked to the release of neurotransmitters [17].

9.7.2 Parkinson's Disease Model

C. elegans possesses eight dopaminergic neurons along with four dopamine receptors. This conserved dopaminergic network makes *C. elegans* a better model to study neurodegeneration and behavioural impairments. To mimic PD pathology, scientists induce dopaminergic neuron degeneration using neurotoxins such as 6-hydroxydopamine (6-OHDA), rotenone, and 1-methyl-4-phenyl-1, 2,3,6-tetrahydropyridine (MPTP). In *C. elegans*, these compounds generate oxidative stress, mitochondrial dysfunction, and protein misfolding, leading to PD-like phenotypes. *C. elegans* does not have a direct homolog for SNCA. However, it contains homologs of key PD-associated genes such as *lrk-1* (LRRK2), *pdr-1* (PARK2), *pink-1* (PINK1), and *djr-1.1*/*djr-1.2* (PARK7), which allow researchers to examine genetic and environmental interactions, providing valuable insights into PD pathogenesis.

9.7.2.1 Alpha-Synuclein Model

The transgenic worm models have been developed by overexpressing human alpha-synuclein. Several *C. elegans* transgenic strains (such as NL5901, OW13, and DDP1) express human alpha-synuclein fused to *YFP* in the body wall muscles. These strains develop inclusion bodies of aggregated alpha-synuclein, closely resembling those found in patients with PD. By analysing movement deficits, behavioural abnormalities, and molecular changes, researchers gain valuable insights into disease progression and genetic modifiers that influence PD pathology [14, 23]. Beyond muscle expression models, additional *C. elegans* strains have been generated carrying wild-type or A53T-mutant alpha-synuclein fused to *GFP*. In these strains, the transgenes are specifically expressed in dopaminergic neurons by the dopamine transporter promoter (*dat-1*) [59]. In these strains, the expression of mutant alpha-synuclein with *GFP* is higher compared to wild-type alpha-synuclein levels. This increased expression leads to greater aggregate formation and progressive cell loss, impacting some or all dopaminergic neurons and mirroring neurodegenerative patterns observed in patients with PD. In addition to alpha-synuclein expression in dopaminergic neurons, several other strains have been created to express pan-neuronally and in motor neurons and have found that neuron loss is induced by the dopaminergic or pan-neuronal specific alpha-synuclein expression. These models have been helping scientists to perform large-scale screens to identify modifiers of aggregation and study neurodegeneration mechanisms [35, 56, 109].

9.7.2.2 LRRK2 Model

The LRRK (leucine-rich repeat kinase) protein is present in two isoforms known as LRRK1 and LRRK2. The LRRK2 is a large protein that contains multiple distinct domains with a molecular weight of 286 kDa, and mutations in LRRK2 are frequently identified in both familial and sporadic PD. The mutation G2019S within the kinase domain is most prevalent in PD that exhibit Lewy bodies [39]. These proteins are primarily localized to intracellular membranous structures, including the endosomal system, Golgi apparatus, endoplasmic reticulum, and mitochondria. The *C. elegans lrk-1* gene encodes a protein homolog of LRRK1/LRRK2. The LRK-1 protein is produced in neurons in addition to other tissues. Importantly, LRRK1 is essential for the synaptic vesicle proteins' localization to presynaptic regions, highlighting its critical role in synaptic function [98]. Two types of LRRK2 models, overexpression and loss-of-function mutants, have been used to study PD.

Mutations in *PINK1* (PTEN-induced putative kinase 1) and *Parkin* genes are also linked to familial forms of PD. Both of these proteins function to maintain the mitochondrial network and homeostasis [112]. Studies in *C. elegans* reveal that LRRK1 mutants can suppress oxidative stress sensitivity and mitochondrial defects caused by *pink-1* loss-of-function. The antagonistic functions of LRRK1 and PINK1 regulate stress response and neuronal development. In the overexpression strains, human wild-type and mutant LRRK2 variants (G2019S and R1441C) are expressed in dopaminergic neurons, co-injected with *dat-1::gfp* for visualization. Overexpression of LRRK2 results in age-dependent dopaminergic neuron degeneration and in reduced dopamine levels, behavioural impairments, and locomotor dysfunction [124]. Available strains are shown in Fig. 9.3 and listed in Table 9.1.

9.8 *C. elegans* Model for Amyotrophic Lateral Sclerosis (ALS)

Amyotrophic lateral sclerosis (ALS) is estimated to affect approximately 1 in 350 individuals, with a notably higher prevalence among military personnel. One of the characteristics of this disorder is a gradual loss of motor neurons, causing a denervated muscle state, movement impairments, and progressive loss of motor function [36]. While 5–10% of ALS cases are linked to inherited genetic mutations referred to as familial ALS (fALS), the majority are sporadic, with no clear hereditary cause. To date, 45 mutations across human genes have been linked to ALS. Both familial and sporadic forms of the disease share a common pathological hallmark, that is, protein inclusions of TDP-43 (Trans-Active Response DNA-binding protein 43) within affected neurons. These aggregates disrupt cellular function, contributing to neurodegeneration and disease progression. In addition, FUS (Fused in Sarcoma) and SOD-1 (Superoxide Dismutase 1) proteins also play pivotal roles in the ALS disease pathogenesis [9, 76].

9.8.1 TDP-43 in ALS

The *TARDBP* gene encodes the TDP-43 protein with a molecular size of 43 kDa. This highly conserved protein is expressed ubiquitously and plays roles in essential cellular processes, such as transcriptional regulation, pre-mRNA processing, alternative splicing, stability, and transport of RNAs. In familial ALS cases, around 30 mutations in the *TARDBP* gene, mostly located in the C-terminal part of the protein, have been identified [111]. At normal physiological conditions, TDP-43 is localized to the nucleus, where it performs its critical functions in RNA metabolism. However, under stress or pathological states, TDP-43 mislocalizes to the cytoplasm, leading to the formation of toxic aggregates, a typical feature observed in patients with ALS [91]. While mutations in *TARDBP* are well-established drivers of ALS, the molecular mechanisms underlying TDP-43-induced pathology are not fully understood. To address this, the *C. elegans* model is frequently employed as a robust system to elucidate these mechanisms.

The *C. elegans* homolog of TDP-43 is named TDP-1. Loss-of-function mutations in *tdp-1* do not naturally lead to neurodegeneration, suggesting that TDP-1 is not inherently toxic in its absence. However, neuronal-specific overexpression of TDP-1 induces motor defects [2] and reduces lifespan when expressed under its endogenous promoter. These effects occur at both 20 °C and 25 °C, highlighting TDP-1's role in promoting neurodegeneration when dysregulated. Interestingly, *tdp-1* loss-of-function mutations render *C. elegans* more sensitive to oxidative stress and DNA damage, suggesting a protective role for TDP-1 under normal conditions. Intriguingly, in *C. elegans*, the TDP-43-induced aggregation and neurotoxicity can be alleviated by reducing the IIS [126].

The human wild-type *TDP-43* expression in *C. elegans* leads to the development of abnormal motor synapses and impaired function, causing uncoordinated responses. This phenotype is exacerbated by mutant *TDP-43* variants (e.g., A315T, M337V, G290A), which escalate toxicity and result in paralysis [2]. Interestingly, when specific domains of TDP-43, such as the RNA recognition domains (RRM1/2) or the C-terminal region, are deleted, the protein forms nuclear inclusions but loses its toxic properties. Conversely, isolated expression of the TDP-43 C-terminal region in neurons is adequate to induce robust aggregation, motor deficits, and neurodegeneration, emphasizing a gain-of-function mechanism behind *TDP-43* mutations. Targeted expression of TDP-43 or its mutant variant (e.g., A315T) in GABAergic neurons promotes aggregate formation, neurodegeneration, and reduced fecundity, closely mimicking ALS phenotypes [107]. Similarly, wild-type and mutant TDP-43 variants (Q331K, M337V) fused to YFP and expressed pan-neuronally result in motor impairments, further demonstrating their role in disease pathology.

9.8.2 FUS in ALS

The FUS protein is composed of 526 amino acids with a molecular weight of 54 kDa. This nuclear protein binds to both DNA and RNA and plays roles in DNA damage repair, mRNA splicing, and RNA transport. Mutations in the *FUS* gene impair its ability to bind RNA effectively, resulting in protein mislocalization to the cytoplasm and the formation of aggregates. These cytoplasmic aggregates disrupt RNA metabolism, destabilize neuronal function, and ultimately contribute to neurodegeneration. Mutations in the FUS gene are linked to several neurodegenerative diseases, including ALS. Mutations in the FUS gene are responsible for 1–5% of familial ALS and 1–2% of sporadic ALS cases, playing a significant role in motor neuron degeneration. Till now, 13 mutations in the C-terminal region of the FUS protein have been identified that contain the nuclear localization sequence [58].

The *C. elegans* ortholog of human *FUS*, known as *fust-1*, is crucial for maintaining neuronal health. Loss-of-function *fust-1* has been implicated in producing ALS-like phenotypes, suggesting a pathogenic role for FUST-1 in neurodegenerative processes. Further studies have examined specific mutations in *fust-1*, such as R446S and P447L, which occur in a conserved region critical for the nuclear localization signal. Expression of these mutations in cholinergic neurons caused FUST-1 mislocalization to the cytoplasm and the formation of aggregates, a phenomenon not observed with the wild-type FUST-1. Under oxidative stress, *fust-1* mutants exhibited reduced survival and impaired autophagy, although their lifespan remained unaffected. These findings highlight the importance of proper FUST-1 localization in mitigating cellular stress and maintaining neuronal function, providing valuable insights into ALS pathology in *C. elegans* [7]. A first *C. elegans* model for studying FUS-related pathology was developed in 2012 by generating neural-specific expression of wild-type (WT) and mutant FUS proteins. Key findings revealed that only the mutant FUS mislocalized to the cytoplasm, disrupting its normal nuclear localization. This mislocalization was linked to motor dysfunction and lifespan reduction, establishing a direct connection between mutant FUS and neurodegeneration. Overexpression of WT FUS failed to rescue the mutant phenotype, supporting the hypothesis that *FUS* mutations exert their effects through a dominant gain-of-function toxic mechanism [80]. However, another study reported that ectopic expression of either wild-type (WT) or mutant FUS in *C. elegans* disrupts neuromuscular junction morphology and function, resulting in impaired motility [71].

9.8.3 Superoxide Dismutase 1 (SOD-1) in ALS

Reactive oxygen species (ROS) cause oxidative damage. Superoxide dismutase 1 (SOD1) functions as an antioxidant enzyme by providing protection against ROS. It is a 32-kDa homo-dimeric protein. Each monomer subunit is 153 amino acids long and is linked by an intramolecular disulfide bond. SOD1 dynamically translocate to mitochondria, safeguarding them from oxidative damage and preserving cellular integrity. The misfolded SOD1 produced by the mutant gene becomes resistant to degradation by the proteasome and autophagy pathways, resulting in the gradual accumulation of toxic aggregates over time. These aggregates exacerbate oxidative stress, disrupt mitochondrial function, and induce motor neuron toxicity [79]. Furthermore, the misfolded SOD1 interferes with synaptic vesicle formation machinery, leading to synaptic vesicle accumulation and impaired neuronal communication. SOD1 mutations were the first genetic cause of amyotrophic lateral sclerosis (ALS) to be identified, with over 150 mutations reported to date. Toxicity arising from SOD1 in ALS models is primarily attributed to a toxic gain-of-function mechanism. Among the known variants, D90A, A4V, and G93A are the most common, collectively accounting for 8–23% of familial ALS cases and 1–4% of sporadic cases.

The *C. elegans* SOD-1 protein shares approximately 70% similarity with human SOD1, making it an important model for studying superoxide dismutase-related mechanisms and oxidative stress responses. Loss-of-function mutation studies in *C. elegans* under oxidative stress conditions revealed selective vulnerability in neurons. Specifically, glutamatergic neurons were shown to degenerate, while cholinergic neurons remained unaffected. Further investigations explored the effects of wild-type (WT) and mutant *sod-1* expression in various tissues of *C. elegans*. Muscle-specific expression of wild-type SOD-1 exhibited resilience and was efficiently degraded, whereas mutant SOD-1 displayed heightened sensitivity, resulting in slower degradation and aggregate formation over time [83]. Pan-neuronal expression of the wild-type SOD-1 and mutant G85R SOD-1 as heterodimers revealed that they formed more aggregates compared to WT homodimers, but fewer aggregates than mutant SOD-1 alone. Animals expressing these heterodimers showed locomotory defects, suggesting a modifier role for mutant SOD-1. This observation aligns with findings from mouse models, in which co-expression of WT and G85R SOD-1 was found to accelerate the disease progression. Another study demonstrated that GABAergic neurons specific expression of the G93A SOD-1 mutant in long-lived *daf-2* mutant worms resulted in reduced aggregate formation [13]. Similarly, expression of the G85R SOD-1 mutant in pan-neurons also showed fewer aggregates, and this reduction was dependent on *daf-16*. These findings underscore the role of longevity-associated pathways, such as the IIS signalling cascade, in mitigating proteotoxicity and aggregate formation.

To study ALS-associated SOD1 pathology in *C. elegans,* researchers utilized mutant *SOD1* genes fused to fluorescence reporters (*GFP* or *YFP*) and expressed either in pan-neurons or GABAergic neurons. Expression promoter. and expressed pan-neuronally, which led to motor deficits. While the G85R mutant exhibited motor deficits, the mutant G93A strain formed larger, more insoluble aggregates within axons, resulting in more severe neurodegeneration compared to the WT strain. This neurodegeneration manifested as pronounced axonal defects and age-dependent paralysis [117].

In neurons, the ALS-associated G85R SOD1 formed both soluble oligomers and insoluble aggregates. These aggregates caused severe locomotor impairments and synaptic dysfunction. A genome-wide RNAi screen identified several chaperones and other cellular components that exacerbated aggregation and further impaired locomotion. The observed neurotoxicity was strongly associated with disruptions in synaptic vesicle biogenesis and trafficking, emphasizing the pivotal role of misfolded SOD1 in ALS pathology. Available strains are shown in Fig. 9.3 and listed in Table 9.1.

9.9 *C. elegans* Model for Huntington's Disease

The populations in the United States and Europe are primarily affected by the Huntington's disease (HD), with an incidence of approximately 1 in 10,000 individuals. Due to its autosomal dominance, HD is strongly linked to familial transmission. The disorder typically manifests in mid-adulthood (age group 30–50) with symptoms like chorea, random, involuntary movements, along with progressive neurodegeneration. Over time, it leads to cognitive decline, psychiatric disturbances, and motor impairment, severely affecting the patient's quality of life. The underlying cause of HD is the CAG repeat expansion in exon 1 of the *HTT* gene, resulting in the production of an altered Huntingtin protein (addition of glutamine or Q repeats). This mutant protein carries extended polyglutamine (polyQ) repeats in its N-terminal region, which disrupt cellular function, contributing to neurotoxicity and degeneration. Molecular genetic studies have determined that Huntington's disease develops when the CAG repeat expansion exceeds a threshold of 35–40 repeats. The longer expansions are associated with early onset and accelerated disease progression [8, 84].

9.9.1 Huntingtin Models in *C. elegans*

C. elegans models have been effectively utilized to replicate polyQ aggregation and toxicity, allowing researchers to study molecular mechanisms at the cellular level underlying polyglutamine-related diseases, including Huntington's disease. These models help unravel how polyQ expansions drive neurodegeneration and cellular dysfunction. Since *C. elegans* lacks an *HTT* gene ortholog, studying polyQ toxicity requires the development of transgenic worms. Researchers have generated worm strains expressing polyQ tracts and human *HTT* fragments with different polyQ repeat lengths in various cell types, offering critical insights into protein aggregation, neuronal integrity, and disease progression.

9.9.1.1 Development of Huntington's Disease Models

The first transgenic *C. elegans* model of Huntington's disease was developed in 1999, where the N-terminal huntingtin (htt) fragment (171 amino acids) was fused with varying glutamine (Q) repeat lengths (Q2, Q23, Q95, and Q150) and expressed under the *osm-10* promoter in ASH chemosensory neurons. Researchers observed progressive neuronal degradation, with Htt-Q150 exhibiting significant aggregate accumulation, reinforcing the correlation between polyQ length and toxicity [29]. Later, in another study, the human htt fragment (first 57 amino acids) was fused with different lengths of glutamines (Q19, Q88, and Q128) with GFP reporter into *C. elegans* mechanosensory neurons under the *nec-3* promoter. These transgenic animals with polyQ128 showed neuronal dysfunction, however, without cell death. These findings validated that polyQ toxicity is length-dependent, with Q128 leading to severe mechanosensory defects in the tail region due to polyQ aggregate deposition [87].

The Morimoto group made a series of worm strains only with glutamine tracks without human htt fragments. Here, various lengths of glutamines were fused to a fluorescence reporter (YFP) to tissue specifically express in body wall muscle cells. A polyQ82 repeat length led to pathological effects, resembling the response observed in neuronal models expressing polyQs with a human htt fragment. Worms with Q82 repeats exhibited significantly more aggregate formation than Q19 worms, confirming that toxicity is length-dependent [94]. To investigate the relationship between polyQ length and ageing, the transgenic worms were observed during ageing. Worms expressing Q35 do not form aggregates at day 1 of adulthood; however, they form aggregates from day 3 of adulthood. Worms expressing Q40 begin to form aggregates from larval stage 4. Worms expressing Q35 become paralyzed in an age-dependent manner [78]. These findings suggest that extended polyQ length enhances aggregation and proteotoxicity, which is age-dependent. The same research group also expressed these transgenes in neurons of the *C. elegans*. In contrast to muscle cells, more polyQ length is needed to induce neuronal toxicity. However, similar to muscle, length-dependent toxicity was observed in neuronal models, reinforcing the idea that polyQ toxicity is directly linked to repeat length [16].

Another research group generated worms with intestinal-specific expression of different lengths of polyQ fused to *YFP*. The threshold length for aggregate formation in the intestine is Q44. This is intermediate length compared to muscle and neurons. This study further showed that pathogen exposure can enhance polyQ aggregation in the intestine [77]. In addition to these strains, worm models expressing polyQ with fluorescent reporters fused to human htt gene fragments were also generated. For example, Wang et al. [116] used htt exon 1 with Q28, Q55, or Q74 in body wall muscle cells. Recently, a new Huntington model was developed that expresses a fragment of the first 513 amino acids of the human htt gene, fused to Q15 or Q128 and containing *YFP* in muscle cells [63]. Similar to observations in neurons, the htt exon 1 fragment also caused aggregation and toxicity in body wall muscle cells in a polyQ-length-dependent manner. Available strains are shown in Fig. 9.3 and listed in Table 9.1.

These worm models have been used as tools to study molecular mechanisms of neurodegeneration. For example, autophagy genes were found to suppress polyQ aggregation and toxicity, making them a key therapeutic target for Huntington's disease. Other studies identified glucose and rutin as neuroprotective factors against polyQ-induced toxicity, which is dependent on the reduced IIS signalling. Various *C. elegans* Huntington models also helped in large-scale screens to identify potential modifiers of the aggregation. For example, forward and reverse genetic screens identified novel regulators of polyQ aggregation and of the proteostasis network.

9.10 *C. elegans* Model for Prion Diseases

Prion diseases, or transmissible spongiform encephalopathies, belong to a rare category of proteinopathies addressed in this chapter. In contrast to the majority of neurodegenerative illnesses, they are contagious and can affect both people and animals. All prion disorders are characterized by significant neuronal loss, accompanied by pronounced astrogliosis and moderate microglial activation. This leads to a gradual and severe spongiform vacuolation and degeneration of the central nervous system, particularly evident in ataxia, behavioural alterations, and, in humans, a markedly progressive decline in cognitive functions [96]. The conversion of normal prion protein PrPC into abnormally folded and aggregated isoform PrPSC is the main cause of prion infection. This fundamentally operates through template-mediated misfolding, wherein PrPC serves as the template. A study using mice indicates that those devoid of the PrP gene exhibit resistance to prion infection, highlighting the significance of PrPc gene expression in prion disease [43]. NMR-based studies gave structural insights indicating that PrPC has a globular domain of approximately 90–230 residues, featuring three α-helices and a short anti-parallel β-sheet. The amino terminus of PrPC comprises oligopeptide repeats, resulting in its flexibility and disorder. The structure of PrPSC remains unsolved; nonetheless, it exhibits an enhanced β-sheet richness. The globular domain of PrPC is susceptible to proteolysis by Proteinase K and is soluble in detergent, but PrPSC exhibits partial resistance to Proteinase K and is insoluble in detergent, indicating that the globular domain of PrP constitutes the core of the PrPSC aggregate [99].

Various experimental models have been investigated to examine the molecular mechanisms of prions; nevertheless, they exhibit some limitations for investigation. *C. elegans* serves as a viable model for investigating the mechanisms of prion-like propagation, as it facilitates real-time observation of intercellular transmission of proteins that are aggregation-prone [81]. Sup35NM is a glutamine/asparagine (Q/N)-rich prion-forming or NM domain of the yeast protein Sup35. This NM domain can adopt different amyloid-like fibrils, generating various prion strains. Three distinct variants of sup35NM were fused to a fluorescence reporter under the *unc-54* promoter to be expressed in the body wall muscle cells of *C. elegans*. Utilization of these *C. elegans* strains demonstrated that NM aggregation resulted in toxicity in both cell-autonomous and non-autonomous manner. Not only did it damage the cellular architecture of neighbouring tissues, but it also had an effect on the muscle cells of the body wall that were expressing the transgene. It was also disclosed that NM, when subjected to autophagy, accumulated in lysosomes rather than undergoing degradation. Furthermore, these NM vesicles did not fuse within a cell but were observed being transported both intracellularly and intercellularly. The results indicate that cytosolic aggregation-prone proteins display prion-like characteristics and can facilitate vesicle-mediated transport [82]. Available strains are shown in Fig. 9.3 and listed in Table 9.1.

In a separate study, the wild-type *PrP* gene and the *PrP* gene with the E200K mutation were used to generate two transgenic *C. elegans* strains that express within mechanosensory neurons. The E200K mutation in the *PrP* gene is responsible for Creutzfeldt-Jakob disease (CJD), the most prevalent form of inherited prion disease. The transgenic lines were constructed using the *mec-7* promoter coupled with both wild-type and mutant PRNP genes, then co-injected with a GFP reporter. They noted that E200K PrP-expressing worms exhibited PrP aggregates and neuronal cell death in GFP-tagged targeted neurons and produced protease-resistant PrP. The researchers demonstrated that the cell loss was mitigated by RNA interference-mediated reduction of PrP E200K [11]. Researchers are leveraging *C. elegans* models to investigate the molecular mechanisms and cellular pathways implicated in prion disorders.

9.11 Neuroprotective Molecules Tested on *C. elegans* Models with Potential for Clinical Trials

C. elegans has proven to be a valuable model organism for identifying potential neuroprotective compounds. It has been in use for early-stage neurodegenerative drug discovery due to its genetic tractability, conserved disease pathways, and suitability for high-throughput screening. Several neuroprotective compounds initially characterised in *C. elegans* have progressed to clinical trials for age-related neurodegenerative diseases [122]. A few examples are given below.

Resveratrol, a polyphenol found in red grapes, has been found to reduce protein aggregation and increase lifespan. In the transgenic CL2006 strain (muscle-specific expression of human $A\beta_{1-42}$), resveratrol (100 μM) reduced Aβ-induced paralysis by about 40%. This effect is through enhanced proteostasis via autophagy and proteasome [90].

Rapamycin, an mTOR inhibitor, was identified in *C. elegans* screens for its ability to extend lifespan and improve stress resistance. In *C. elegans* models of neurodegeneration, rapamycin demonstrated significant neuroprotective effects by enhancing autophagy. This finding in *C. elegans* led to clinical trials investigating rapamycin and other mTOR inhibitors for Alzheimer's disease, particularly focusing on their potential to improve cognitive function by modulating cellular homeostasis and protein clearance [45].

Curcumin, a polyphenol derived from turmeric, was shown in *C. elegans* models to reduce amyloid-beta (Aβ) aggregation and mitigate toxicity in transgenic strains expressing human Aβ peptides. Curcumin and other amyloid-binding compounds have been shown to extend lifespan by maintaining proteostasis [1]. These findings catalyzed multiple clinical trials aimed at evaluating its efficacy in Alzheimer's disease [114].

Trehalose, a disaccharide, has been shown to have health benefits in *C. elegans.* It extended lifespan (30–60%), improved health span, and reduced polyQ aggregation via proteostasis mechanisms [41]. Due to its protective functions, trehalose has been considered as a potential therapeutic for Huntington's and other protein aggregation disorders [55].

Epicatechin and epigallocatechin-3-gallate (EGCG), a flavonoid, are shown to reduce amyloid-β-induced toxicity in *C. elegans* strains expressing human $A\beta_{142}$ peptides. Treatment with epicatechin improved motility, chemotaxis, and survival, and modulated inflammation and cellular stress responses. Epicatechin is being explored for its neuroprotective effects, with studies indicating potential benefits in neurodegenerative diseases. [3]. These compounds are being tested in pre-clinical studies [125].

Methylene blue, a redox-active compound, effectively suppressed paralysis and motor neuron degeneration and reduced TDP-43/FUS-related toxicity in *C. elegans* models expressing mutant human TDP-43 or FUS [108]. Methylene blue has been tested in clinical trials for its neuroprotective and anti-ageing effects [123].

Together, these case studies highlight the translational value of *C. elegans* as a discovery platform for neuroprotective therapeutics. By offering rapid screening capabilities and allowing for high-throughput testing of compounds, *C. elegans* has played an integral role in the drug discovery pipeline, particularly in ageing and age-related neurodegenerative disease research.

9.12 Challenges Working with Transgenic Worms

Many commonly used transgenic *C. elegans* strains are readily available through the Caenorhabditis Genetic Centre (CGC), which provides a central repository of mutant and transgenic strains. However, not all transgenic lines; only the submitted strains are available. [85] recently described how to order strains from CGC. Researchers may encounter difficulties obtaining specific transgenic strains or may need to generate their own lines, which can be a resource-intensive process. In these cases, alternative approaches like requesting directly from available labs or seeking collaborations may be necessary.

The genetic stability of transgenic *C. elegans* strains is a critical consideration for researchers using these models. Over successive generations, transgenic worms may experience loss or alteration of the transgene due to several factors, including genetic drift, silencing of the transgene, or positional effects. These challenges can be especially pronounced in strains where the transgene is randomly integrated into the genome, leading to variability in expression levels or complete loss of the transgene over time. To maintain stable expression, it is often necessary to regularly backcross transgenic worms with wild-type (N2) strains to reduce genetic drift and restore expression levels.

Confirming the transgenic nature of *C. elegans* in resource-limited labs presents a set of unique challenges. Molecular methods, such as genotyping by single-worm PCR or fluorescence microscopy, are typically used to verify transgene integration and expression. However, these techniques often require specialized equipment and expertise, which may not be accessible in low-resource settings. PCR-based genotyping is the most widely accessible method, but it requires specific primers for the transgene and potentially for the worm's genomic DNA. Fluorescence-based methods, such as GFP or other reporter gene expression, are also commonly used but require microscopes capable of detecting specific wavelengths, which may not always be available in resource-constrained labs. In such cases, alternative strategies like using low-cost imaging setups, collaborating with more equipped institutions, or applying simple qualitative assays may provide feasible solutions.

9.13 Conclusions

Proteinopathies represent a diverse group of age-related neurodegenerative diseases driven by protein misfolding and aggregation, which disrupt cellular function and lead to progressive damage. Proteins such as tau, amyloid-beta, and alpha-synuclein play central roles in these conditions, with proteostasis disruption being a critical factor. The *C. elegans* model, with its powerful genetic tools and conserved biological pathways, has proven invaluable in modelling these diseases. Transgenic models expressing human disease proteins has significantly increased the understanding of the molecular mechanisms of proteinopathies and helped identify novel therapeutic approaches.

References

1. Alavez S, Vantipalli MC, Zucker DJ, Klang IM, Lithgow GJ. Amyloid-binding compounds maintain protein homeostasis during ageing and extend lifespan. Nature. 2011;472(7342):226–9.
2. Ash PEA, Zhang Y, Roberts CM, Saldi T, Hutter H, Buratti E, Petrucelli L, Link CD. Neurotoxic effects of TDP-43 overexpression in *C. elegans*. Human Mol Genet. 2010;19(16):3206–18.
3. Ayuda-Durán B, Garzón-García L, González-Manzano S, Santos-Buelga C, González-Paramás AM. Insights into the neuroprotective potential of epicatechin: effects against Aβ-induced toxicity in *Caenorhabditis elegans*. Antioxidants (Basel). 2024;13(1):79.
4. Balch WE, Morimoto RI, Dillin A, Kelly JW. Adapting proteostasis for disease intervention. Science. 2008;319(5865):916–9.
5. Barbieri M, Bonafè M, Franceschi C, Paolisso G. Insulin/IGF-I-signaling pathway: an evolutionarily conserved mechanism of longevity from yeast to humans. Am J Physiol Endocrinol Metab. 2003;285(5):E1064–71.
6. Bard Pohl JA, Goodall EA, Greene ER, Jonsson E, Dong KC, Martin A. Structure and function of the 26S proteasome. Annu Rev Biochem. 2018;87(1):697–724.
7. Baskoylu SN, Chapkis N, Unsal B, Lins J, Schuch K, Simon J, Hart AC. Disrupted autophagy and neuronal dysfunction in *C. elegans* knockin models of FUS amyotrophic lateral sclerosis. Cell Rep. 2022;38(4):110195.
8. Bates GP, Dorsey R, Gusella JF, Hayden MR, Kay C, Leavitt BR, Nance M, Ross CA, Scahill RI, Wetzel R, Wild EJ, Tabrizi SJ. Huntington disease. Nat Rev Dis Primers. 2015;1:15005.
9. Benatar M, Robertson J, Andersen PM. Amyotrophic lateral sclerosis caused by SOD1 variants: from genetic discovery to disease prevention. Lancet Neurol. 2024;24(1):77–86.
10. Biza A, Triantafyllou A, Mavroeidi P, Sklaviadis T. A *Caenorhabditis elegans* model for prion pathogenicity: insights into the role of aggregation state in neurotoxicity. Mol Cell Neurosci. 2010;45(4):306–13.
11. Bizat N, Parrales V, Laoues S, Normant S, Levavasseur E, Roussel J, Privat N, Gougerot A, Ravassard P, Beaudry P, Brandel J, Laplanche J, Haïk S. An in vivo *Caenorhabditis elegans* model for therapeutic research in human prion diseases. Brain. 2021;144(9):2745–58.
12. Blackwell TK, Steinbaugh MJ, Hourihan JM, Ewald CY, Isik M. SKN-1/Nrf, stress responses, and aging in *Caenorhabditis elegans*. Free Radic Biol Med. 2015;88(Pt B):290–301.
13. Boccitto M, Lamitina T, Kalb RG. Daf-2 signaling modifies mutant SOD1 toxicity in *C. elegans*. PLoS One. 2012;7(3):e33494.
14. Bodhicharla R, Nagarajan A, Winter J, Adenle A, Nazir A, Brady D, Vere K, Richens J, O'Shea P, Bell DR, De Pomerai D. Effects of α-synuclein overexpression in transgenic *Caenorhabditis elegans* Strains. CNS Neurol Disord Drug Targets. 2013;11(8):965–75.
15. Brenner S. The genetics of *Caenorhabditis elegans*. Genetics. 1974;77(1):71–94.
16. Brignull HR, Moore FE, Tang SJ, Morimoto RI. Polyglutamine proteins at the pathogenic threshold display neuron-specific aggregation in a pan-neuronal *Caenorhabditis elegans* model. J Neurosci. 2006;26(26):7597–606.
17. Calabresi P, Mechelli A, Natale G, Volpicelli-Daley L, Di Lazzaro G, Ghiglieri V. Alpha-synuclein in Parkinson's disease and other synucleinopathies: from overt neurodegeneration back to early synaptic dysfunction. Cell Death Dis. 2023;14(3):176.
18. Caldwell KA, Willicott CW, Caldwell GA. Modeling neurodegeneration in *Caenorhabditis elegans*. Dis Model Mech. 2020;13(10):dmm046110.
19. Chew YL, Fan X, Götz J, Nicholas HR. PTL-1 regulates neuronal integrity and lifespan in *C. elegans*. J Cell Sci. 2013;126(Pt 9):2079–91.
20. Chiti F, Dobson CM. Protein misfolding, amyloid formation, and human disease: a summary of progress over the last decade. Annu Rev Biochem. 2017;86(1):27–68.
21. Ciechanover A. Proteolysis: from the lysosome to ubiquitin and the proteasome. Nat Rev Mol Cell Biol. 2005;6(1):79–87.

22. Cohen E, Bieschke J, Perciavalle RM, Kelly JW, Dillin A. Opposing activities protect against age-onset proteotoxicity. Science. 2006;313(5793):1604–10.
23. Cooper JF, Van Raamsdonk JM. Modeling Parkinson's Disease in *C. elegans*. J Parkinsons Dis. 2018;8(1):17–32.
24. Daigle I, Li C. *apl-1*, a *Caenorhabditis elegans* gene encoding a protein related to the human beta-amyloid protein precursor. Proc Natl Acad Sci USA. 1993;90(24):12045–9.
25. De Magalhães JP, Stevens M, Thornton D. The business of anti-aging science. Trends Biotechnol. 2017;35(11):1062–73.
26. Dikic I. Proteasomal and autophagic degradation systems. Annu Rev Biochem. 2017;86(1):193–224.
27. Drake J, Link CD, Butterfield DA. Oxidative stress precedes fibrillar deposition of Alzheimer's disease amyloid beta-peptide (1-42) in a transgenic *Caenorhabditis elegans* model. Neurobiol Aging. 2003;24:415–4120.
28. Ewald CY, Cheng R, Tolen L, Shah V, Gillani A, Nasrin A, Li C. Pan-neuronal expression of APL-1, an APP-related protein, disrupts olfactory, gustatory, and touch plasticity in *Caenorhabditis elegans*. J Neurosci. 2012;32(30):10156–69.
29. Faber PW, Alter JR, MacDonald ME, Hart AC. Polyglutamine-mediated dysfunction and apoptotic death of a *Caenorhabditis elegans* sensory neuron. Proc Natl Acad Sci USA. 1999;96(1):179–84.
30. Fatouros C, Pir GJ, Biernat J, Koushika SP, Mandelkow E, Mandelkow E, Schmidt E, Baumeister R. Inhibition of tau aggregation in a novel *Caenorhabditis elegans* model of tauopathy mitigates proteotoxicity. Hum Mol Genet. 2012;21(16):3587–603.
31. Fonte V, Kipp DR, Yerg J, Merin D, Forrestal M, Wagner E, Roberts CM, Link CD. Suppression of in vivo β-amyloid peptide toxicity by overexpression of the HSP-16.2 small chaperone protein. J Biol Chem. 2008;283(2):784–91.
32. Gallrein C, Iburg M, Michelberger T, Koçak A, Puchkov D, Liu F, Ayala Mariscal SM, Nayak T, Kaminski Schierle GS, Kirstein J. Novel amyloid-beta pathology *C. elegans* model reveals distinct neurons as seeds of pathogenicity. Prog Neurobiol. 2021;198:101907.
33. Gidalevitz T, Krupinski T, Garcia S, Morimoto RI. Destabilizing protein polymorphisms in the genetic background direct phenotypic expression of mutant SOD1 toxicity in *Caenorhabditis elegans*. PLoS Genet. 2009;5(3):e1000399.
34. Goedert M, Baur CP, Ahringer J, Jakes R, Hasegawa M, Spillantini MG, Smith MJ, Hill F. PTL-1, a microtubule-associated protein with tau-like repeats from the nematode *Caenorhabditis elegans*. J Cell Sci. 1996;109(11):2661–72.
35. Hamamichi S, Rivas RN, Knight AL, Cao S, Caldwell KA, Caldwell GA. Hypothesis-based RNAi screening identifies neuroprotective genes in a Parkinson's disease model. Proc Natl Acad Sci USA. 2008;105(2):728–33.
36. Hardiman O, Al-Chalabi A, Chio A, Corr EM, Logroscino G, Robberecht W, Shaw PJ, Simmons Z, Van Den Berg LH. Amyotrophic lateral sclerosis. Nat Rev Dis Primers. 2017;3:17071.
37. Hartl FU, Hayer-Hartl M. Molecular chaperones in the cytosol: from nascent chain to folded protein. Science. 2002;295(5561):1852–8.
38. Hipp MS, Kasturi P, Hartl FU. The proteostasis network and its decline in ageing. Nat Rev Mol Cell Biol. 2019;20(7):421–35.
39. Ho DH, Han SJ, Son I. The multifaceted role of LRRK2 in Parkinson's disease. Brain Sci. 2025;15(4):407.
40. Holcomb LA, Gordon MN, Jantzen P, Hsiao K, Duff K, Morgan D. Behavioral changes in transgenic mice expressing both amyloid precursor protein and presenilin-1 mutations: lack of association with amyloid deposits. Behavior Genet. 1999;29(3):177–85.
41. Honda Y, Tanaka M, Honda S. Trehalose extends longevity in the nematode *Caenorhabditis elegans*. Aging Cell. 2010;9(4):558–69.
42. Hong M, Zhukareva V, Vogelsberg-Ragaglia V, Wszolek Z, Reed L, Miller BI, Geschwind DH, Bird TD, McKeel D, Goate A, Morris JC, Wilhelmsen KC, Schellenberg GD, Trojanowski

JQ, Lee VM. Mutation-specific functional impairments in distinct TAU isoforms of hereditary FTDP-17. Science. 1998;282(5395):1914–7.
43. Horiuchi M, Caughey B. Specific binding of normal prion protein to the scrapie form via a localized domain initiates its conversion to the protease-resistant state. EMBO J. 1999;18(12):3193–203.
44. Hornsten A, Liebertha J, Fadia S, Malins R, Ha L, Xu X, Daigle I, Markowitz M, O'Connor G, Plasterk R, Li C. APL-1, a *Caenorhabditis elegans* protein related to the human β-amyloid precursor protein, is essential for viability. Proc Natl Acad Sci USA. 2007;104(6):1971–6.
45. Jahrling JB, Laberge RM. Age-related neurodegeneration prevention through mTOR inhibition: potential mechanisms and remaining questions. Curr Top Med Chem. 2015;15(21):2139–51.
46. Jongsma E, Goyala A, Mateos JM, Ewald CY. Removal of extracellular human amyloid beta aggregates by extracellular proteases in *C. elegans*. eLife. 2023;12:e83465.
47. Kalia LV, Lang AE. Parkinson's disease. Lancet. 2015;386(9996):896–912.
48. Karpinar DP, Balija MB, Kügler S, Opazo F, Rezaei-Ghaleh N, Wender N, Kim HY, Taschenberger G, Falkenburger BH, Heise H, Kumar A, Riedel D, Fichtner L, Voigt A, Braus GH, Giller K, Becker S, Herzig A, Baldus M, Jäckle H, Eimer S, Schulz JB, Griesinger C, Zweckstetter M. Pre-fibrillar alpha-synuclein variants with impaired beta-structure increase neurotoxicity in Parkinson's disease models. EMBO J. 2009;28(20):3256–68.
49. Kennedy BK, Berger SL, Brunet A, Campisi J, Cuervo AM, Epel ES, Franceschi C, Lithgow GJ, Morimoto RI, Pessin JE, Rando TA, Richardson A, Schadt EE, Wyss-Coray T, Sierra F. Geroscience: linking aging to chronic disease. Cell. 2014;159(4):709–13.
50. Kenyon CJ. The genetics of ageing. Nature. 2010;464(7288):504–12.
51. Kenyon C, Chang J, Gensch E, Rudner A, Tabtiang R. A *C. elegans* mutant that lives twice as long as wild type. Nature. 1993;366(6454):461–4.
52. Kidd M. Alzheimer's disease - an electron microscopical study. Brain. 1964;87(2):307–20.
53. Kraemer BC, Zhang B, Leverenz JB, Thomas JH, Trojanowski JQ, Schellenberg GD. Neurodegeneration and defective neurotransmission in a *Caenorhabditis elegans* model of tauopathy. Proc Natl Acad Sci USA. 2003;100(17):9980–5.
54. Krüger L, Mandelkow EM. Tau neurotoxicity and rescue in animal models of human Tauopathies. Curr Opin Neurobiol. 2015;36:52–8.
55. Kuczyńska-Wiśnik D, Stojowska-Swędrzyńska K, Laskowska E. Intracellular protective functions and therapeutical potential of trehalose. Molecules. 2024;29(9):2088.
56. Kuwahara T, Koyama A, Koyama S, Yoshina S, Ren C, Kato T, Mitani S, Iwatsubo T. A systematic RNAi screen reveals involvement of endocytic pathway in neuronal dysfunction in α-synuclein transgenic *C. elegans*. Hum Mol Genetics. 2008;17(19):2997–3009.
57. Kuwahara T, Tonegawa R, Ito G, Mitani S, Iwatsubo T. A systematic RNAi screen reveals involvement of endocytic pathway in α-synuclein toxicity in *Caenorhabditis elegans*. Hum Mol Genet. 2012;21(14):3271–82.
58. Kwiatkowski TJ, Bosco DA, LeClerc AL, Tamrazian E, Vanderburg CR, Russ C, Davis A, Gilchrist J, Kasarskis EJ, Munsat T, Valdmanis P, Rouleau GA, Hosler BA, Cortelli P, De Jong PJ, Yoshinaga Y, Haines JL, Pericak-Vance MA, Yan J, Brown RH. Mutations in the FUS/TLS gene on chromosome 16 cause familial amyotrophic lateral sclerosis. Science. 2009;323(5918):1205–8.
59. Lakso M, Vartiainen S, Moilanen A, Sirviö J, Thomas JH, Nass R, Blakely RD, Wong G. Dopaminergic neuronal loss and motor deficits in *Caenorhabditis elegans* overexpressing human α-synuclein. J Neurochem. 2003;86(1):165–72.
60. Lapierre LR, Hansen M. Lessons from *C. elegans*: signaling pathways for longevity. Trends Endocrinol Metabol. 2012;23(12):637–44.
61. Lazaro-Pena MI, Ward ZC, Yang S, Strohm A, Merrill AK, Soto CA, Samuelson AV. HSF-1: Guardian of the proteome through integration of longevity signals to the proteostatic network. Front Aging. 2022;3:861686.

62. Le WD, Huang KX, Tang Y, Xia Y, Ma Y, Duan WM, Guo J, Xu P, Xu XM, Li J. Establishing a novel *Caenorhabditis elegans* model to investigate the role of autophagy in amyotrophic lateral sclerosis. Acta Pharmacol Sin. 2013;34(5):644–50.
63. Lee AL, Ung HM, Sands LP, Kikis EA. A new *Caenorhabditis elegans* model of human huntingtin 513 aggregation and toxicity in body wall muscles. PLoS One. 2017;12(3):e0173644.
64. Lee RY, Hench J, Ruvkun G. Regulation of *C. elegans* DAF-16 and its human ortholog FKHRL1 by the *DAF-2* insulin-like signaling pathway. Curr Biol. 2001;11(24):1950–7.
65. Liachko NF, Guthrie CR, Kraemer BC. Phosphorylation promotes neurotoxicity in a *Caenorhabditis elegans* model of TDP-43 proteinopathy. J Neurosci. 2010;30(48):16208–19.
66. Link CD. Expression of human beta-amyloid peptide in transgenic *Caenorhabditis elegans*. Proc Natl Acad Sci USA. 1995;92(20):9368–72.
67. Link CD, Fonte V, Roberts CM, Hiester B, Silverman MA, Stein GH. The beta amyloid peptide can act as a modular aggregation domain. Neurobiol Dis. 2008;32(3):420–5.
68. Liu C, Götz J. Profiling Murine Tau with 0N, 1N and 2N isoform-specific antibodies in brain and peripheral organs reveals distinct subcellular localization, with the 1N isoform being enriched in the nucleus. PLoS One. 2013;8(12):e84849.
69. Liu J, Zhang B, Lei H, Feng Z, Liu J, Hsu AL, Xu XZ, Yang Y. Functional analysis of *Caenorhabditis elegans* genes orthologous to human genes associated with Parkinson's disease. Proc Natl Acad Sci USA. 2011;108(26):11106–11.
70. López-Otín C, Blasco MA, Partridge L, Serrano M, Kroemer G. Hallmarks of aging: an expanding universe. Cell. 2023;186(2):243–78.
71. Markert SM, Skoruppa M, Yu B, Mulcahy B, Zhen M, Gao S, Sendtner M, Stigloher C. Overexpression of an ALS-associated FUS mutation in *C. elegans* disrupts NMJ morphology and leads to defective neuromuscular transmission. Biol Open. 2020;9(12):bio055129.
72. McColl G, Roberts BR, Gunn AP, Perez KA, Tew DJ, Masters CL, Barnham KJ, Cherny RA, Bush AI. The *Caenorhabditis elegans* Aβ1–42 model of Alzheimer disease predominantly expresses Aβ 3-42. J Biol Chem. 2009;284(34):22697–702.
73. McColl G, Roberts BR, Pukala TL, Kenche VB, Roberts CM, Link CD, Ryan TM, Masters CL, Barnham KJ, Bush AI, Cherny RA. Utility of an improved model of amyloid-beta (Aβ1-42) toxicity in *Caenorhabditis elegans* for drug screening for Alzheimer's disease. Mol Neurodegener. 2012;7(1):57.
74. Meneely PM, Dahlberg CL, Rose JK. Working with worms: *Caenorhabditis elegans* as a model organism. Curr Protoc Essent Lab Tech. 2019;19(1)
75. Miyasaka T, Ding Z, Gengyo-Ando K, Oue M, Yamaguchi H, Mitani S, Ihara Y. Progressive neurodegeneration in *C. elegans* model of tauopathy. Neurobiol Dis. 2005;20(2):372–83.
76. Moens TG, Da Cruz S, Neumann M, Shelkovnikova TA, Shneider NA, Van Den Bosch L. Amyotrophic lateral sclerosis caused by FUS mutations: advances with broad implications. Lancet Neurol. 2025;24(2):166–78.
77. Mohri-Shiomi A, Garsin DA. Insulin signaling and the heat shock response modulate protein homeostasis in the *Caenorhabditis elegans* intestine during infection. J Biol Chem. 2008;283(1):194–201.
78. Morley JF, Brignull HR, Weyers JJ, Morimoto RI. The threshold for polyglutamine-expansion protein aggregation and cellular toxicity is dynamic and influenced by aging in *Caenorhabditis elegans*. Proc Natl Acad Sci USA. 2002;99(16):10417–22.
79. Müller K, Oh K, Nordin A, Panthi S, Kim SH, Nordin F, Freischmidt A, Ludolph AC, Ki CS, Forsberg K, Weishaupt J, Kim Y, Andersen PM. De novo mutations in SOD1 are a cause of ALS. J Neurol Neurosurg Psychiatry. 2022;93(2):201–6.
80. Murakami T, Yang S, Xie L, Kawano T, Fu D, Mukai A, Bohm C, Chen F, Robertson J, Suzuki H, Tartaglia GG, Vendruscolo M, Schierle GSK, Chan FT, Moloney A, Crowther D, Kaminski CF, Zhen M, St George-Hyslop P. ALS mutations in FUS cause neuronal dysfunction and death in *Caenorhabditis elegans* by a dominant gain-of-function mechanism. Human Mol Genet. 2011;21(1):1–9.

81. Nussbaum-Krammer CI, Morimoto RI. *Caenorhabditis elegans* as a model system for studying non-cell-autonomous mechanisms in protein-misfolding diseases. Dis Model Mech. 2014;7(1):31–9.
82. Nussbaum-Krammer CI, Park K, Li L, Melki R, Morimoto RI. Spreading of a prion domain from cell-to-cell by vesicular transport in *Caenorhabditis elegans*. PLoS Genet. 2013;9(3):e1003351.
83. Oeda T, Shimohama S, Kitagawa N, Kohno R, Imura T, Shibasaki H, Ishii N. Oxidative stress causes abnormal accumulation of familial amyotrophic lateral sclerosis-related mutant SOD1 in transgenic *Caenorhabditis elegans*. Human Mol Genet. 2001;10(19):2013–23.
84. Orr HT, Zoghbi HY. Trinucleotide repeat disorders. Annu Rev Neurosci. 2007;30(1):575–621.
85. Park YJ, Moon KM, Kim K. A practical guide to ordering *C. elegans* strains for biological research. Mol Cells. 2025;48(9):100251.
86. Park KW, Li L. Prion protein in *Caenorhabditis elegans:* distinct models of anti-BAX and neuropathology. Prion. 2011;5(1):28–38.
87. Parker JA, Connolly JB, Wellington C, Hayden M, Dausset J, Neri C. Expanded polyglutamine in *Caenorhabditis elegans* causes axonal abnormalities and progressive motility defects. J Neurochem. 2001;79(4):768–70.
88. Pir GJ, Choudhary B, Mandelkow E, Mandelkow E. Tau mutant A152T, a risk factor for FTD/PSP, induces neuronal dysfunction and reduced lifespan independently of aggregation in a *C. elegans* Tauopathy model. Mol Neurodegener. 2016;11(1):33.
89. Polymeropoulos MH, Lavedan C, Leroy E, Ide SE, Dehejia A, Dutra A, Pike B, Root H, Rubenstein J, Boyer R, Stenroos ES, Chandrasekharappa S, Athanassiadou A, Papapetropoulos T, Johnson WG, Lazzarini AM, Duvoisin RC, Di Iorio G, Golbe LI, Nussbaum RL. Mutation in the α-Synuclein gene identified in families with Parkinson's disease. Science. 1997;276(5321):2045–7.
90. Regitz C, Fitzenberger E, Mahn FL, Dußling LM, Wenzel U. Resveratrol reduces amyloid-beta ($A\beta_{1\text{-}42}$)-induced paralysis through targeting proteostasis in an Alzheimer model of *Caenorhabditis elegans*. Eur J Nutr. 2016;55(2):741–7.
91. Rohn TT. Cytoplasmic inclusions of TDP-43 in neurodegenerative diseases: a potential role for caspases. Histol Histopathol. 2009;24(8):1081–6.
92. Saha S, Guillily MD, Ferree A, Lanceta J, Chan D, Ghosh J, Hsu CH, Segal L, Raghavan K, Matsumoto K, Hisamoto N, Kuwahara T, Iwatsubo T, Moore L, Goldstein L, Cookson M, Wolozin B. LRRK2 modulates vulnerability to mitochondrial dysfunction in *Caenorhabditis elegans*. J Neurosci. 2009;29(29):9210–8.
93. Sandhof CA, Martin N, Tittelmeier J, Schlueter A, Pezzali M, Schoendorf DC, Lange T, Reinhardt P, Ried JS, Liang S, Uzunoglu G, Gasparini L, Jahn TR, Ehrnhoefer DE, Nussbaum-Krammer C. A novel *C. elegans* model for MAPT/Tau spreading reveals genes critical for endolysosomal integrity and seeded MAPT/Tau aggregation. Autophagy. 2025;4:1–19.
94. Satyal SH, Schmidt E, Kitagawa K, Sondheimer N, Lindquist S, Kramer JM, Morimoto RI. Polyglutamine aggregates alter protein folding homeostasis in *Caenorhabditis elegans*. Proc Natl Acad Sci USA. 2000;97(11):5750–5.
95. Sauer RT, Baker TA. AAA+ proteases: ATP-fueled machines of protein destruction. Annu Rev Biochem. 2011;80(1):587–612.
96. Scheckel C, Aguzzi A. Prions, prionoids and protein misfolding disorders. Nat Rev Genet. 2018;19(7):405–18. https://doi.org/10.1038/s41576-018-0011-4.
97. Shcherbakov D, Nigri M, Akbergenov R, Brilkova M, Mantovani M, Petit PI, Grimm A, Karol AA, Teo Y, Sanchón AC, Kumar Y, Eckert A, Thiam K, Seebeck P, Wolfer DP, Böttger EC. Premature aging in mice with error-prone protein synthesis. Sci Adv. 2022;8(9):eabl9051.
98. Shin N, Jeong H, Kwon J, Heo HY, Kwon JJ, Yun HJ, Kim C, Han BS, Tong Y, Shen J, Hatano T, Hattori N, Kim K, Chang S, Seol W. LRRK2 regulates synaptic vesicle endocytosis. Exp Cell Res. 2008;314(10):2055–65.
99. Silveira JR, Raymond GJ, Hughson AG, Race RE, Sim VL, Hayes SF, Caughey B. The most infectious prion protein particles. Nature. 2005;437(7056):257–61.

100. Spillantini MG, Schmidt ML, Lee VM, Trojanowski JQ, Jakes R, Goedert M. α-Synuclein in Lewy bodies. Nat. 1997;388(6645):839–40.
101. Starr LA, McKay LE, Peter KN, Seyfarth LM, Berkowitz LA, Caldwell KA, Caldwell GA. Attenuation of dopaminergic neurodegeneration in a *C. elegans* Parkinson's model through regulation of xanthine dehydrogenase (XDH-1) expression by the RNA Editase, ADR-2. J Dev Biol. 2023;11(2):20.
102. Strang KH, Golde TE, Giasson BI. MAPT mutations, tauopathy, and mechanisms of neurodegeneration. Lab Investig. 2019;99(7):912–28.
103. Sulston J, Schierenberg E, White J, Thomson J. The embryonic cell lineage of the nematode *Caenorhabditis elegans*. Dev Biol. 1983;100(1):64–119.
104. Tabeshmehr P, Eftekharpour E. Tau; one protein, so many diseases. Biology. 2023;12(2):244. https://doi.org/10.3390/biology12020244.
105. Trinh J, Farrer M. Advances in the genetics of Parkinson disease. Nat Rev Neurol. 2013;9(8):445–54.
106. Treusch S, Hamamichi S, Goodman JL, Matlack KE, Chung CY, Baru V, Shulman JM, Parrado A, Bevis BJ, Valastyan JS, Han H, Lindhagen-Persson M, Reiman EM, Evans DA, Bennett DA, Olofsson A, DeJager PL, Tanzi RE, Caldwell KA, Caldwell GA, Lindquist S. Functional links between Aβ toxicity, endocytic trafficking, and Alzheimer's disease risk factors in yeast. Science. 2011;334(6060):1241–5.
107. Vaccaro A, Tauffenberger A, Ash PEA, Carlomagno Y, Petrucelli L, Parker JA. TDP-1/TDP-43 regulates stress signaling and age-dependent proteotoxicity in *Caenorhabditis elegans*. PLoS Genet. 2012a;8(7):e1002806.
108. Vaccaro A, Patten SA, Ciura S, Maios C, Therrien M, Drapeau P, Kabashi E, Parker JA. Methylene blue protects against TDP-43 and FUS neuronal toxicity in *C. elegans* and *D. rerio*. PLoS One. 2012b;7(7):e42117.
109. Van Ham TJ, Thijssen KL, Breitling R, Hofstra RMW, Plasterk RHA, Nollen EAA. *C. elegans* model identifies genetic modifiers of alpha-synuclein inclusion formation during aging. PLoS Genet. 2008;4(3):e1000027.
110. Van Swieten J, Spillantini MG. Hereditary frontotemporal dementia caused by TAU gene mutations. Brain Pathol. 2007;17(1):63–73.
111. Verde EM, Secco V, Ghezzi A, Mandrioli J, Carra S. Molecular mechanisms of protein aggregation in ALS-FTD: focus on TDP-43 and cellular protective responses. Cells. 2025;14(10):680.
112. Vizziello M, Borellini L, Franco G, Ardolino G. Disruption of mitochondrial homeostasis: the role of PINK1 in Parkinson's disease. Cells. 2021;10(11):3022.
113. Volovik Y, Marques FC, Cohen E. The *nematode Caenorhabditis elegans*: a versatile model for the study of proteotoxicity and aging. Methods. 2014;68(3):458–64.
114. Voulgaropoulou SD, van Amelsvoort TAMJ, Prickaerts J, Vingerhoets C. The effect of curcumin on cognition in Alzheimer's disease and healthy aging: a systematic review of preclinical and clinical studies. Brain Res. 2019;1725:146476.
115. Vozdek R, Pramstaller PP, Hicks AA. Functional screening of Parkinson's disease susceptibility genes to identify novel modulators of α-synuclein neurotoxicity in *Caenorhabditis elegans*. Front Aging Neurosci. 2022;14:806000.
116. Wang H, Lim PJ, Yin C, Rieckher M, Vogel BE, Monteiro MJ. Suppression of polyglutamine-induced toxicity in cell and animal models of Huntington's disease by ubiquilin. Hum Mol Genet. 2006;15(6):1025–41.
117. Wang J, Farr GW, Hall DH, Li F, Furtak K, Dreier L, Horwich AL. An ALS-linked mutant SOD1 produces a pocomotor defect associated with aggregation and synaptic dysfunction when expressed in neurons of *Caenorhabditis elegans*. PLoS Genet. 2009;5(1):e1000350.
118. Wang Y, Mandelkow E. Tau in physiology and pathology. Nat Rev Neurosci. 2015;17(1):22–35.
119. Wilson DM, Cookson MR, Van Den Bosch L, Zetterberg H, Holtzman DM, Dewachter I. Hallmarks of neurodegenerative diseases. Cell. 2023;186(4):693–714.
120. Witan H, Kern A, Koziollek-Drechsler I, Wade R, Behl C, Clement AM. Heterodimer formation of wild-type and amyotrophic lateral sclerosis-causing mutant Cu/Zn-superoxide

dismutase induces toxicity independent of protein aggregation. Human Mol Genet. 2008;17(10):1373–85.
121. Wszolek ZK, Tsuboi Y, Ghetti B, Pickering-Brown S, Baba Y, Cheshire WP. Frontotemporal dementia and parkinsonism linked to chromosome 17 (FTDP-17). Orphanet J Rare Dis. 2006;1(1):30.
122. Xu R, Kang Q, Yang X, Yi P, Zhang R. Unravelling molecular targets for neurodegenerative diseases through *Caenorhabditis elegans* models. Int J Mol Sci. 2025;26(7):3030.
123. Xue H, Thaivalappil A, Cao K. The potentials of methylene blue as an anti-aging drug. Cells. 2021;10(12):3379.
124. Yao C, Khoury RE, Wang W, Byrd TA, Pehek EA, Thacker C, Zhu X, Smith MA, Wilson-Delfosse AL, Chen SG. LRRK2-mediated neurodegeneration and dysfunction of dopaminergic neurons in a *Caenorhabditis elegans* model of Parkinson's disease. Neurobiol Dis. 2010;40(1):73–81.
125. Youn K, Ho CT, Jun M. Multifaceted neuroprotective effects of (−)-epigallocatechin-3-gallate (EGCG) in Alzheimer's disease: an overview of pre-clinical studies focused on β-amyloid peptide. Food Sci Human Wellness. 2022;11(3):483–93.
126. Zhang T, Mullane PC, Periz G, Wang J. TDP-43 neurotoxicity and protein aggregation modulated by heat shock factor and insulin/IGF-1 signaling. Human Mol Genet. 2011;20(10):1952–65.
127. Zhang WH, Koyuncu S, Vilchez D. Insights into the links between proteostasis and aging from *C. elegans*. Front Aging. 2022;3:854157.
128. Zhang H, Wei W, Zhao M, Ma L, Jiang X, Pei H, Cao Y, Li H. Interaction between Aβ and Tau in the pathogenesis of alzheimer's disease. Int J Biol Sci. 2021;17(9):2181–92.

Caenorhabditis elegans as a Model Organism in Neurological Research

10

Sakshi Yadav, Sonam Ranwa, Stanzin Nurboo, and Aamir Nazir

10.1 Introduction

The neurological research majorly deals with clinical aspects of brain science. Emerging evidence of age-associated decline in neuronal health and increasing cases of neurological disorders has raised concerns, as a major proportion of the population is globally affected, and the numbers are predicted to go higher. The most commonly occurring NDs include Alzheimer's disease (AD), Parkinson's disease (PD), amyotrophic lateral sclerosis (ALS), and Huntington's disease (HD). As neurons age, they lose their nascent properties and become more vulnerable to decay. The neurological health deteriorates gradually with age, and the conditions become adverse due to the accumulation of DNA damage, dysfunctional mitochondria, protein aggregation, autophagy impairment, and metabolic dysfunction, affecting neuronal viability and integrity. The approach of the scientific community towards finding the cure and understanding the mechanisms of these diseases is breaking conventional barriers by promoting interdisciplinary studies, expanding the inclusivity with the incorporation of different model systems, which will be discussed shortly in later sections [4, 64].

Neurodegeneration progression is mediated by multiple factors, which include genetic and epigenetic alterations, toxic environmental risk factor exposure, and signalling pathway dysregulation. Most of the current ongoing studies aim to decipher the pathogenesis of NDs by primarily focusing on neuroinflammation, a critical event primarily mediated by microglia and astrocytes. The abnormal functioning of the immune system and chronic inflammatory responses further contribute to the worsening of NDs. This suggests that identification of the valid targets could be of

S. Yadav · S. Ranwa · S. Nurboo · A. Nazir (✉)
Division of Toxicology and Experimental Medicine, CSIR-Central Drug Research Institute, Lucknow, India
e-mail: anazir@cdri.res.in

P. Kumar, V. Kothari (eds.), *Model Organisms in Biological Research*,
https://doi.org/10.1007/978-981-92-0038-2_10

therapeutic value against the progression of such diseases. Nevertheless, the progress made in drug discovery is remarkable, but the intricate system of blood-brain and blood-spinal cord barrier systems poses a major challenge to the studies associated with neurological diseases. The incoherent knowledge and complexity of the brain limit the identification of viable targets and biomarkers in neurological disorders, crucial for developing relevant potential therapeutic strategies [31, 67]. Furthermore, this limitation also impacts reproducibility and the outcomes of the preclinical and clinical studies.

The contribution of animal models in laying down evolutionary and mechanistic details of the nervous system and developing therapies against neurological diseases is undeniably significant. Some of the well-established model systems include primates, rodents, zebrafish, *Drosophila*, and *C. elegans* [11]. Initially, evolutionary aspects of the nervous system were studied in animals belonging to the phylum Cnidaria and Ctenophora, and the first theory on neural development came through the cnidarian model system *Hydra*. Later, in the twentieth century, Gastropod molluscs became the most employed organisms for studying the complex mechanisms of neural circuitry, and around the same time, *Drosophila* also gained popularity for conducting genetic and molecular screens. Notably, *C. elegans, Drosophila*, and *Aplysia* share similar memory and learning patterns with higher organisms and have been exploited to explore the same [59]. Rodents have been utilised extensively to study the behavioural and molecular changes associated with NDs. One such successful example from the rodent system is the modelling of diseases in humanized chimeric mouse models [8]. Such models are better suited for understanding disease pathogenesis and conducting screenings because of manifestations of the prominent symptoms, like sleep disorders, locomotory disabilities, and memory loss [6]. The biggest challenge that comes with the employment of alternative animal models in neurological science is the variability in predictive stats from preclinical trials, which ultimately affects the clinical trials [3, 59].

Our primary focus here is to understand the role of *C. elegans* in neurology. The nematode *C. elegans* has earned a credible reputation for its genetic flexibility, screening success, and minimal requirement for maintenance. The transparent body of *C. elegans* makes it compatible for in vivo imaging, and with the incorporation of fluorescent markers, it becomes easier to visualize the expression of genes under different conditions. It is an influential model system that provides a suitable systemic milieu for the emulation of human diseases. Furthermore, the conserved metabolic processes and genetic establishment make the worm relevant for studying various human diseases [68]. The extensive employment of *C. elegans* in conducting studies to understand the basic elements of developmental biology and cell biology has added new perspectives to these domains, and so far, discoveries made using it have enlightened us with some essential life-supporting pathways. The identification of key factors of programmed cell death [19], the discovery of RNA [23], the characterization of signalling pathways linked with longevity, and the complete mapping of its genome and connectome are some of the major highlights from *C. elegans* research. Moreover, the *C. elegans* neuronal architecture is well defined, and it also shares many functional properties with more complex organisms. The neuronal circuitry of worms is capable of processing various complicated behavioural responses, such as aversive responses against mechanical forces and chemical signals, learning and memory patterns, thermotolerance, and defecation. These features make it favourable for conducting neurological research, and currently it is a part of many aging and neurodegenerative disease studies [16, 59].

10.2 *C. elegans* as a Model System

In the realm of biological research, the minuscule nematode *C. elegans* is extensively utilized as a model organism to unravel various human disease mechanisms. This free-living millimetric-sized organism belongs to the Nematoda phylum, commonly called roundworms [79]. *C. elegans* was first isolated by French zoologist Émile Maupas in 1900, who named it *Rhabditis elegans*. However, the potential of *C. elegans* as a model organism was first utilized by Sydney Brenner in the early 1960s to study animal development and behaviour, specifically neuronal development [2]. This tiny system serves as a valuable model to provide remarkable insights into the intricate mechanisms underlying various neurological disorders. Due to the conserved homology (~60% overall gene homology and ~42% human disease-associated genes) of various cellular and molecular pathways with humans, *C. elegans* serves as an excellent model to understand the basic mechanisms of neuronal dysfunction and degeneration [14]. The ease of genetic manipulation makes it amenable to introduce human disease-associated genes (Fig. 10.1), such as amyloid-β (Aβ), tau, or α-synuclein [33, 40, 60]. The creation of these transgenic models can be utilized to study the mechanistic landscape of these proteins, involved in neurological disorders like PD and AD. To study the role of individual genes and gene clusters, techniques like RNA interference (RNAi) and CRISPR/Cas9 (clustered regularly interspaced short palindromic repeats) are widely utilized and well-established in *C. elegans* models.

The experimental advantages of *C. elegans* include its anatomical simplicity, optical transparency, short lifespan, and simple genetic makeup, along with low-cost maintenance. The relatively simple yet functionally rich nervous system with only 302 neurons allows the study of detailed mapping of neuronal circuits and their dysfunctions. In the context of disease research, various cellular processes involved in neurodegenerative diseases, such as protein aggregation, mitochondrial dysfunction, and oxidative stress, can be easily studied using *C. elegans*. Also, the optical transparency of the worms facilitates real-time imaging of neuronal structures by utilizing various fluorescently tagged and genetically modified *C. elegans* strains to visualize specific neurons and synapses in living animals. The fluorescence tagging of these neurons allows the researchers to track the malfunctioning of neuronal cells and their gradual loss throughout the lifespan [68].

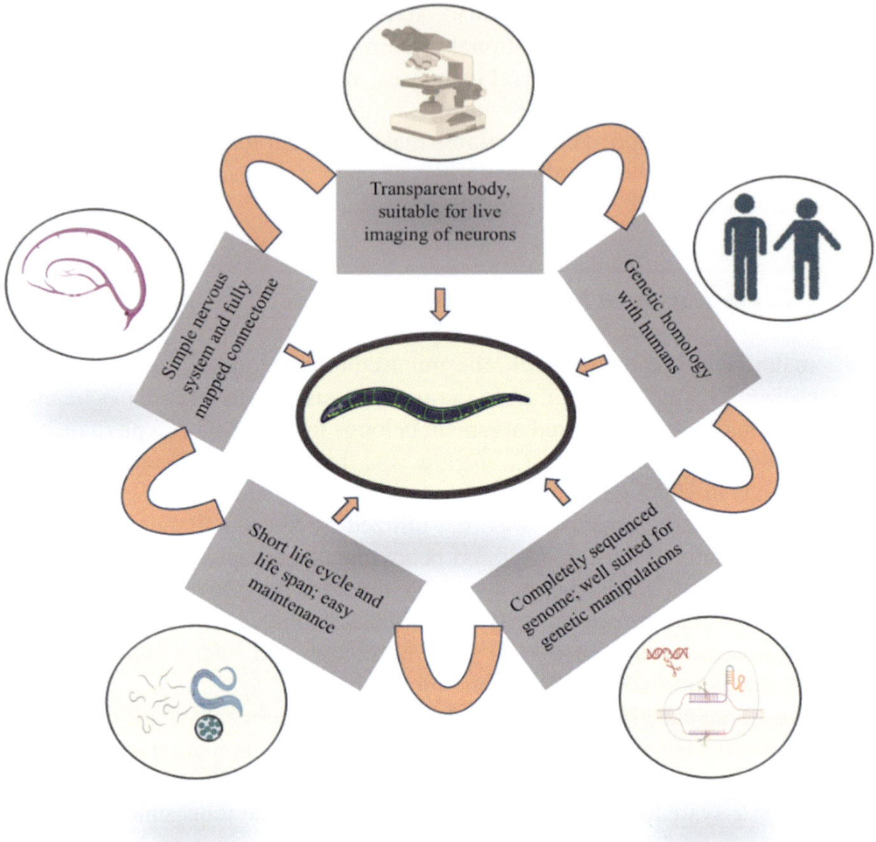

Fig. 10.1 Favourable features of the *C. elegans* model system for neurological investigations

10.2.1 Current Research Hot Topics and *C. elegans* Utility

The versatility of *C. elegans* as a model organism is continuing to accelerate in various fields of biological research to advance our understanding of neurobiology, aging, genetics, and the correlation of the gut-brain axis. In the field of neurodegenerative disease studies, *C. elegans* is extensively utilized to understand the pathophysiology of protein aggregation and disruption in mitochondrial functioning, which are the key factors contributing to the onset of brain disorders. In the case of aging research, the genetic accessibility of the worms is explored to study the role of various transcription factors, epigenetic regulatory factors, and the effects of small molecules on improving healthspan and longevity. The impact of environmental toxicants such as heavy metals and pesticides is assessed by using multi-omics and phenotypic assays. Additionally, the evaluation of neuroprotective nano-formulations is studied to check their efficacy in delivering drugs to the target site to improve neuronal health by reducing oxidative stress. Studies have demonstrated the efficacy of nanoparticles in delivering drugs to target sites, reducing oxidative stress, and improving overall neuronal health. The current research is going on to shed light on the interactions between *C. elegans* and its microbiome since the gut and brain are interconnected through a complex communication system called the microbiota-gut-brain axis. This axis is responsible for the bidirectional communication between the gut microbiome and the central nervous system, which in turn affects brain function, development, and behaviour. In case of dysbiosis or imbalance in the gut microbiota, it leads to the onset of various neurological disorders, including conditions like PD, AD, and ALS [84]. A recent study, which was published in Nature Cell Biology, has demonstrated the influence of specific vitamin B12-producing bacteria on brain signalling. These bacteria were found to negatively modulate the levels of the cognitive neurotransmitter acetylcholine [36].

10.3 Modelling of Neurological Disorders in *C. elegans*

The growing proportion of individuals with neurodegenerative diseases is raising concern, and to overcome the negative consequences of such diseases, cumulative research efforts are required from different organism-based model systems. At present, the rodent model systems are widely utilized as preclinical models for studying neurodegenerative mechanisms and therapies. Nevertheless, the current scenario demands additional efforts to accelerate research with the objective of quickly getting major insights into the prime drivers and interventions against the progression of these diseases. In this lineup, we will explore the valuable perspectives gained through the employment of *C. elegans* in the context of neurodegenerative diseases. It is a powerful tool for fundamental research and can be efficiently employed to identify effective targets and cures in a very short time. This is one of the major advantages of *C. elegans,* whereas mammalian system studies are very time-consuming and expensive [68].

Here, in this section, we describe neurodegenerative diseases modelled in *C. elegans,* which include PD, AD, ALS, and HD. In addition, we will review the *C. elegans* based studies that discuss genetic and environmental factors causing cellular dysfunctions, ultimately leading to neurodegeneration.

10.3.1 Parkinson's Disease (PD)

Parkinson's disease (PD) is the second most frequently occurring neurodegenerative disease, affecting globally around ten million people. PD is mainly idiopathic with a very small number of familial cases. The symptoms associated with PD mostly appear in the later stages of life; early manifestations are usually minimal. The central hallmarks that mark the onset of PD include continuous loss of dopaminergic neurons and abnormal aggregation of α-synuclein protein in the substantia nigra pars compacta, ultimately affecting motor functions associated with impaired

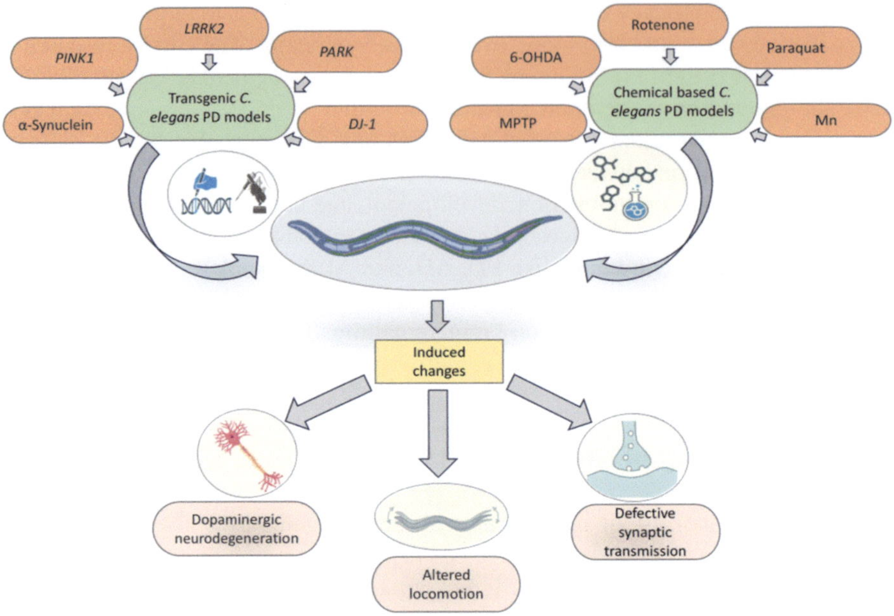

Fig. 10.2 Parkinson's disease models in *C. elegans* and corresponding biological changes

neurons—rigidity, bradykinesia, tremors, and postural instability. The other non-motor PD manifestations consist of cognitive impairment, insomnia, memory loss, and depression [30, 66].

The PD pathogenesis, like other neurodegenerative diseases, is not well outlined, and the cases majorly fall into two categories: sporadic (environmental) and familial (genetic). The genetic factors studied to be involved in the progression of PD include Parkin (PARK2), PARK1/SNCA, PTEN induced putative kinase 1 (PINK1), leucine-rich repeat kinase 2 (LRRK2), glucocerebrosidase (GBA), and many more. Notably, orthologues of some of these genes are also present in *C. elegans,* and the missing ones could be introduced through genome editing techniques to create desired transgenic strains for the purpose of studying any specific phenotype and mechanism (Fig. 10.2) [81].

10.3.1.1 Genetic Models

The α-synuclein is prevalent within the neuronal population and is commonly concentrated near presynaptic terminals, and its expression is governed by the PARK1/*SNCA* locus in humans. *C. elegans* provides an *α*-synuclein null genetic background and allows the generation of α-synuclein transgenic strains, where it could be overexpressed without any hindrance to assess its impact on other healthy parameters affected in the case of PD. The *C. elegans* α-synuclein models rely on both neuronal and non-neuronal expression of human α-synuclein. The phenotypic changes are mainly dependent on the expression pattern of α-synuclein and not on the protein type (wild-type or mutated forms of human α-synuclein) in these models. One of the widely employed transgenic strains of *C. elegans* to conduct α-synuclein studies is NL5901 [unc-54p::alphasynuclein::YFP + unc-119(+)], expressing wild-type human α-synuclein in body wall muscle cells. Transgenic strains harbouring mutant α-synuclein (A53T, E46K, or A30P), causal factors of familial PD, are also employed to examine the PD-associated factors. The neuronal loss is prominent in strains expressing α-synuclein pan-neuronally, ultimately leading to motor defects. The remarkable feature witnessed through the expression of human α-synuclein in *C. elegans* is that it shows a similar pattern of aggregate formation, mimicking the environment of PD neurons. This strengthens the easy assessment of PD-associated molecular players and therapies [41, 42].

Besides, LRRK2 mutations are also prominent in the case of PD, contributing to 4% of autosomal dominant PD cases. Some of the sporadic PD patients have also been reported to exhibit the LRKK2 mutation. The LRKK2 is mainly a cytosolic protein, with not much known about its function, and is reported to interact with the outer mitochondrial membrane. In the case of *C. elegans,* its orthologue is *lrk-1,* which is also expressed in neuronal cells. The transgenic strains harbouring R1441C/G and G2019S mutations within the GTPase and kinase domains exhibit LRRK2 overexpression, dopamine deficiency, and irregular behavioural responses. These LRRK2 *C. elegans* models replicate several aspects of PD (dopaminergic neurodegeneration, behavioural impairments, and mitochondrial dysfunctions), making them an excellent model to conduct LRRK2-linked PD studies [24, 55].

Other mutations used to create a PD-like condition in worms include *pink1* gene mutations (human orthologue of *PINK1*), which hamper mitochondrial functions and dopamine-associated behavioural aspects in worms. Loss of *djr-1.1* (human ortholog of *DJ-1*) in worms also leads to compromised neuronal integrity, increased oxidative stress sensitivity, and mitochondrial inefficiency. Overall, *C. elegans* genetic makeup is favourable for creating PD-like conditions to get mechanistic insights and conduct pharmacological screenings [44].

10.3.1.2 Chemical-Based Models

Several chemical-based models have been created to explore PD conditions in *C. elegans*. The neurotoxin 1-methyl-1, 2, 3, 6-tetrahydropyridine (MPTP) induced-PD model is one of the most widely employed screening models for the assessment of pharmacological agents against PD in worms and rodents. The MPTP forms 1-methyl-4-phenylpyridinium ion (MPP+), which directly affects dopaminergic neurons by inhibiting Mitochondrial Complex I. In the case of *C. elegans*, administration of MPTP impacts its mobility and other dopamine-dependent behaviours, emphasizing the significance of oxidative stress and mitochondrial dysfunction in PD [46, 72].

Similarly, catecholaminergic neurotoxin 6-hydroxydopamine (6-OHDA), structurally analogous to dopamine, also targets dopaminergic circuitry. It is specifically transported to dopaminergic neurons via dopamine transporter (DAT), and once it reaches its target site, it generates ROS through autooxidation, and increases oxidative stress, eventually resulting in degeneration of dopaminergic neurons [61].

In addition to the above-mentioned neurotoxins, pesticides, and herbicides have also been exploited for creating PD-like conditions in *C. elegans*. Rotenone is a broad-spectrum insecticide that inhibits mitochondrial complex I and increases oxidative stress, leading to neuronal damage and altered behaviours. Furthermore, it also induces α-synuclein aggregation in transgenic α-synuclein models of *C. elegans* [35]. On the other hand, paraquat, a commonly employed herbicide, is also used to create a PD model in *C. elegans*. It also affects α-synuclein aggregation, dopaminergic neuronal circuitry, and behavioural responses. Additionally, excess administration of manganese (Mn) also induces PD-like symptoms by disrupting oxidative stress and intracellular homeostasis [76, 94].

10.3.1.3 PD Mechanistic Insights from *C. elegans*

In PD, there is selective degeneration of dopaminergic neurons, and the involvement of dopamine in this context has now been studied under in vitro and in vivo conditions. Dopamine gets quickly oxidized, and this property of it is speculated towards its role in PD pathogenesis. The supply of dopamine to isolated α-synuclein polypeptides enhances the stabilization of the oligomeric state of the protein in vitro. However, this requires the presence of five amino acid residues (^{125}YEMPS^{129}), located near the C-terminal region of α-synuclein. This dopamine and α-synuclein connection was further confirmed through neurodegeneration studies conducted in transgenic *C. elegans* PD models. In *C. elegans*, crossing of a *cat-2* overexpressing strain (which encodes for tyrosine hydroxylase) with an A53T α-synuclein model results in an excessive dopamine-producing strain. The studies have reported that this overproduction of dopamine leads to increased neurodegeneration. Nevertheless, dopamine shows no significant effect on neurodegeneration in worms expressing A53T α-synuclein lacking ^{125}YEMPS^{129}. These results obtained through *C. elegans* were in alignment with other observations made by employing A53T α-synuclein rodent models [5, 56].

Moving forward, the genetic flexibility facilitates the conduction of gene-specific studies in PD models of *C. elegans*. Previous findings have shown the effect of proteins involved in different cellular processes, like protein clearance (ATG7, ULK1, PINK1, PRKN), vesicle trafficking (VPS41), and endolysosomal pathways (cathepsin D, ATP13A2/PARK9, GBA-1, LRKK2), in disease progression and neurodegeneration through these PD models of *C. elegans*. Subsequently, LRRK2 models of worms were used to showcase the effect of reduced glutaredoxin on PD symptoms. In addition, genetic screenings in the α-synuclein model of *daf-2* mutant worms have provided evidence that glycolytic enzyme GPI-1/GPI 1 influences lifespan and neuronal loss, and functions as a neuromodulatory molecule [4, 38]. In summary, *C. elegans* has strengthened the functional understanding of PD through findings made using it.

10.3.2 Alzheimer's Disease (AD)

AD is a rising threat to the global population, characterized by the accumulation of Aβ plaques and tau-associated neurofibrillary tangles (NFT). Currently, AD accounts for 60–80% cases of dementia, and these cases are predicted to ascend multifold in the future. The major AD symptoms include cognitive impairment, behavioural alterations, memory loss, and reduced social skills. The Aβ accumulation starts due to the breakdown of the Amyloid precursor protein (APP) via the amyloidogenic pathway, which involves sequential proteolytic activities of enzymes, β-secretase, and γ-secretase. According to the popular amyloid cascade hypothesis, the accumulation of Aβ acts as a trigger for later events like tau hyperphosphorylation, NFT formation, and neuroinflammation. However, few aspects of this hypothesis are now being questioned due to the failure of amyloid-centric therapies in recent times. Moreover, in the case of AD, cholinergic circuitry is also hampered due to excessive neurodegeneration, leading to a decline in acetylcholine (ACh) levels, which ultimately affects learning, memory, and cognitive health of patients [43, 75].

10.3.2.1 Aβ Models

Aβ accumulation is assumed to be the main culprit behind AD progression and advancement. Many *C. elegans* strains have been constructed based on this idea to simulate similar conditions. Transgenic strains of *C. elegans* expressing human Aβ in muscle cells are popularly utilized for understanding AD pathogenesis. The aggregation of Aβ protein, impaired neuronal function, and motor dysfunctions observed in these transgenic strains make *C. elegans* a suitable model for AD. For specifically studying neurodegeneration, transgenic strains with the expression of Aβ in the neuronal system have also been constructed. AD models that overexpress Aβ in glutamatergic neurons have been found to show age-dependent neurodegeneration [7, 53].

Other factors that contribute to Aβ neurotoxicity in AD have also been modelled in *C. elegans*. Apolipoprotein E (APOE) models, APP models, and presenilin (PSEN) models of *C. elegans* provide useful insights into AD pathology. In this context, in the APOE models, *APOEε2* expression in glutamatergic neurons having Aβ expression rescues the neurodegeneration, whereas *APOEε4* expression shows no such protective effect. The *APOEε4* has been associated with the risk for developing AD and *APOEε2* is known to exhibit the opposite effect. Although in *C. elegans*, expression of *APOE* alleles in neurons without Aβ shows no major effect on structural and functional properties of neurons, suggesting their neuroprotective effect and requirement of Aβ for their counter-regulatory activities [25, 88].

In APP models, single copy expression of human APP in *C. elegans* causes neuronal loss and behavioural impairments. *C. elegans* system has APL-1, an orthologue of the human amyloid beta precursor-like proteins 1 and 2 (APLP1 and APLP2), and overexpression of the *apl-1* gene results in diverse biological outcomes, including neuronal dysfunctions and behavioural alterations, observed in the case of models expressing human APP [20, 21]. AD mainly results from abnormal cleavage of APP, and PSENs are part of the γ-secretase complex, which is known for its key role in APP breakdown and AD disease progression [1]. *C. elegans* has a PSEN orthologue, *sel-12,* which is well studied, and its mutation leads to mitochondrial dysregulation, highlighting the effect of mitochondrial impairments in neurodegeneration [74]. Most of the familial cases of AD show mutations in PSEN1 and PSEN2, but the underlying mechanistic clarity is still lacking.

10.3.2.2 *C. elegans* Tauopathy Models

The tau protein interacts with microtubules and regulates the structural stability and cytoskeletal dynamics, essential for maintaining the neuronal architecture and intracellular functioning. AD is associated with the accumulation of NFT, resulting from hyperphosphorylation and aggregation of tau protein. These molecular changes hamper the neuronal transport activities and disrupt the synaptic connectivity [65]. In *C. elegans*, *ptl-1* is the orthologue of mammalian tau (MAPT), and displays around 50% sequence homology. The loss of *ptl-1* cannot exactly emulate the tauopathy, but it affects the lifespan, mechanosensitivity, and neuronal functions, simulating certain characteristics of AD [12]. Notably, the expression of human tau restores the mechanosensory responses in *ptl-1* mutants.

Various *C. elegans* models expressing tau have been constructed to delineate the tauopathy in AD progression and development. It has been observed that in these transgenic strains, with aging, tau protein phosphorylation and accumulation increase, ultimately affecting the neuronal population and locomotion [54]. Genetic screenings conducted in *C. elegans* models of tauopathy led to the identification of *sut-1* and *sut-2* genes. The knockdown of these genes alleviates the tau-induced neuronal toxicity and restores the normal functioning of neurons [26, 39]. Similarly, the expression of XBP-1, a key factor of unfolded protein response of the ER (UPR^{ER}), reduces tau toxicity and promotes neuronal survival [86].

Mitochondrial dysfunction associated with tau-induced toxicity came to light through a study conducted on *C. elegans* transgenic model expressing wild-type human tau (PIR3 strain), highlighting it as an initial event in disease progression. The study emphasized the effect of compromised mitochondrial health on neuronal functions [63].

10.3.2.3 AD Mechanistic Insights from *C. elegans*

C. elegans has played a crucial role in identifying the genetic factors and molecular mediators that contribute significantly to the development of AD. The early studies on the *sel-12* gene, showing its loss leading to neurodegeneration, provided evidence for the first time that presenilin mutations could be the cause of AD pathology in some of the AD cases [45]. The Aβ models and tau models have enhanced the basic understanding of AD pathogenesis and progression. The reports highlighting the involvement of mitochondrial elements [10] and ER (UPR^{ER}) have been instrumental in providing insights into the AD manifestations (Fig. 10.3) [86].

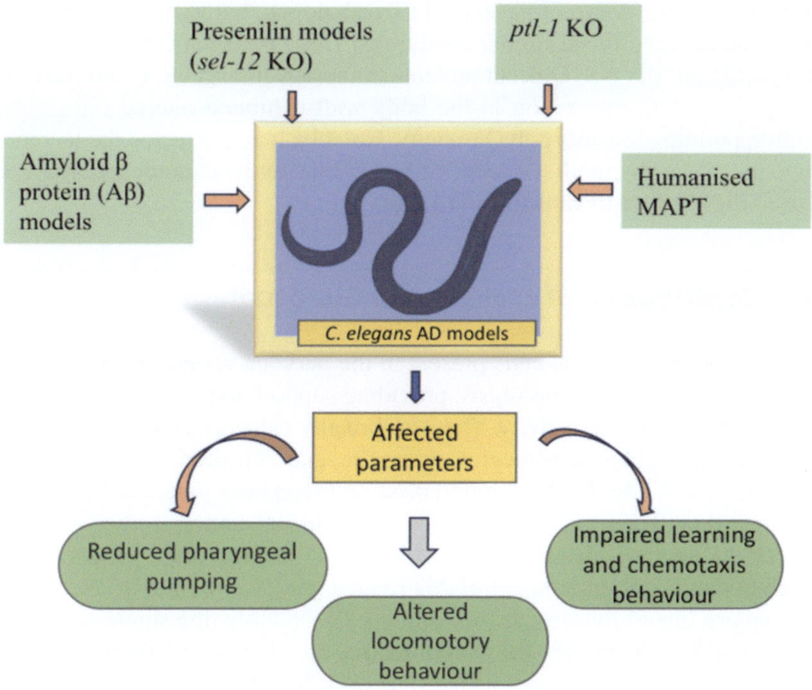

Fig. 10.3 *C. elegans* models for Alzheimer's disease and associated affected parameters

10.3.3 ALS and HD

ALS is a neurodegenerative disorder that attacks motor neurons in the brain and spinal cord, inducing progressive decline in muscle function. The majority of ALS cases are sporadic, and only around 10% of cases show a familial inheritance pattern. The ALS pathology is complex, and a major factor that is known to contribute to its progression is the accumulation of abnormal proteins like TAR DNA-binding protein (TDP-43) and fused in sarcoma (FUS). The mechanistic aspects of ALS are not yet completely deciphered [83, 91]. *C. elegans* has been employed to replicate the four leading mutations observed in the case of ALS: C9orf72-SMCR8 complex subunit (*C9orf72*), superoxide dismutase (*SOD1*), TAR DNA-binding protein (*TARDBP*), RNA-binding protein FUS/TLS (*FUS*), and FTD mutations [microtubule-associated protein tau (*MAPT*)] [4]. The *C. elegans* studies have shed light on the impaired autophagy, mitochondrial defects, and endolysosomal trafficking in ALS-induced neuronal loss and motor dysfunctions [91].

HD is an autosomal dominant disorder marked by the age-associated decline in cognitive abilities and motor disabilities. The causing factor of HD includes CAG repeat expansion in the huntingtin gene (*HTT*), which leads to the formation of abnormal huntingtin protein, carrying extended polyglutamine repeats (polyQs) near its N-terminus. This mutant form of protein induces the degeneration of neurons. *C. elegans* does not have an HTT orthologue, but transgenic strains have been created to recapitulate similar conditions. These transgenic model systems are used to carry out the polyQ accumulation studies. A transgenic *C. elegans* model that displays polyQ expression in the body wall exhibited altered mitochondrial dynamics, pointing towards polyQ toxicity. Notably, it also affected the locomotory behaviours of worms, and silencing of genes that rescued mitochondrial health, improved these motor dysfunctions [13, 70].

10.4 Contribution of *C. elegans* in Glia Studies

Glial cells are non-neuronal cells present in the nervous system and play a significant role in maintaining homeostasis, providing support and protection for neurons, and modulating synaptic activity. They are broadly categorized into neuronal cells such as astrocytes, oligodendrocytes, microglia, and Schwann cells, each having a distinct functional role. Dysfunction in these cells has been linked to several neurological disorders. The most extensive genome-wide association study (GWAS) conducted on Alzheimer's disease (AD), which is the leading cause of dementia, has revealed a significant enrichment of AD-related gene expression in microglia and genes that are linked to astrocyte activation [90]. Notably, the single-cell profiling of Human AD brain tissues has determined a broad range of disease-associated disruptions in glial cells as well as neurons [51]. In the case of PD, the integrated GWAS data have identified the prevalence of disease-associated gene expression in glial cells compared to neurons [37]. The consistency of these findings has been validated through the single-cell sequencing data of PD pathology, which have shown significant upregulation of inflammatory pathways in astrocytes [93]. Apart from PD and AD, HD, which is comparatively less prevalent, the application of conditional mouse genetics in one of its model systems has identified the role of astrocytes in the onset and progression of the disease [92]. Overall, from the above studies focused on uncovering various neurological disorders, the challenge to gain mechanistic insights into the causes and effects of disease-associated cell states remains vague, leaving an open window for future endeavours.

The *C. elegans* glia organization is relatively simpler than higher systems, and it facilitates studies at the single-cell level. The glia-neuron interactions can be studied here in the absence of complexities observed in case of other model systems. Several aspects of glia biology are still unexplored with respect to neurodegeneration and regeneration. Its simple neuronal architecture and restricted number of cells make it a wonderful platform to carry out glia studies in the context of NDs and neuronal repair [77]. The studies in *C. elegans* have unfolded the critical role of glial cells in neuronal cell repair and regeneration. For instance, *C. elegans* cephalic sheath glia can respond to neuronal damage by modulating mitochondrial function and metabolic pathways [29]. These findings can lead to significant advancements for developing neurological therapies to reduce neuronal damage and enhance recovery to treat neurodegenerative disorders. Glial cells also serve as a platform for high-throughput drug screening by testing compounds that can improve glial function. Through this approach, the effect of neuroprotective agents can be studied in glial cell functioning for the development of therapies for treating various NDs.

In conclusion, this transformative nematode with conserved genetic and molecular pathways has provided unparalleled insights into the mechanisms underlying neurodegenerative disorders like Alzheimer's, Parkinson's, ALS, and Huntington's disease. Additionally, the study of glial cells and neuron-glia interactions in *C. elegans* has uncovered critical components of neuronal repair and neuroprotection.

10.5 Approach Towards Establishing Therapeutic Strategies

The *C. elegans* system displays substantial potential for conducting protein aggregation and neurodegeneration studies (Table 10.1). Besides, being a wonderful model for seeking out the questions related to the mechanism and progression of NDs, it also offers the screening assets to develop the cure and therapies against these diseases. A screening conducted against 6-OHDA-induced neurodegeneration for neuroprotective agents in *C. elegans* led to the identification of phytochemicals, Isofloretin, and botulin. Both showed enhanced neuronal responses, offering a protective effect on neurons, and their administration also elevated the lifespan of worms [9, 82].

Similarly, another study utilized *C. elegans* to test the effect of bioactive food nanocomplexes on PD, and it reported that curcumin encapsulated human serum albumin nanoparticles strengthened the neuronal functions and enhanced behavioural performance of worms. Likewise, for AD cases, a screening was conducted in amyloid beta models to evaluate quinoline derivatives, and a few of the molecules showed a positive effect by improving learning and memory [27]. Furthermore, such screenings have also identified different entities (Tetracycline, copper, and noncaffeinated coffee extracts) that offer neuroprotection against tau-induced toxicities [47].

With regard to molecules in clinical trials, several significant case studies exist where *C. elegans*-based screens have directly influenced the progression of compounds into clinical testing. One of the highlights is the screening conducted in a worm model of PMM2-CDG, which played an essential role in the identification of epalrestat [34], a safe, approved drug that boosts PMM2 enzyme activity in patient cells and has progressed into Phase III clinical trials (ClinicalTrials.gov ID: NCT04925960). Popularly known compounds like metformin and resveratrol have also been tested initially in *C. elegans* for their neuroprotective and anti-aging properties. Additionally, compounds like reserpine, trolox, propyl gallate, rolipram, and ethosuximide have also been examined in *C. elegans* for their protective effects against protein toxicity observed in several NDs [80].

Table 10.1 List of commonly used neurodegenerative *C. elegans* models (the strain information has been obtained from Caenorhabditis Genetics Centre (CGC))

Disease models	Genotype	Strain name	References
Parkinson's disease			
α-synuclein expression in the muscles	pkIs2386 [unc54p::alphasynuclein::YFP + unc-119(+)]	NL5901	[28]
Overexpression of α-synuclein in dopaminergic neurons	(P_{dat-1}::α-syn,P_{dat-1}::GFP)	UA44	[69]
α-synuclein expression in the muscles	grk-1(ok1239) X; pkIs2386 [unc54p::alphasynuclein::YFP + unc-119(+)]	OW13	[85]
GFP expression in the soma and dendrites of dopaminergic neurons	egIs1 [dat-1p::GFP]	BZ555	[50]
GFP expression in all dopaminergic neurons	vtIs7 [dat-1p::GFP]	BY250	[57]
Alzheimer's disease			
Chronic Aβ paralysis	dvIs2 [pCL12(unc-54/human Abeta peptide 1–42 minigene) + rol-6(su1006)]	CL2006	[18]
Aβ in muscles	dvIs14 [(pCL12) unc-54::beta 1–42 + (pCL26) mtl-2::GFP]	CL2120	[52]
Neuronal Aβ	dvIs50 [pCL45 (snb-1::Abeta 1–42::3' UTR(long) + mtl-2::GFP]	CL2355	[48]
Acute Aβ paralysis	dvIs27 [myo-3p::A-Beta (1–42)::let-851 3'UTR) + rol-6(su1006)] X. Rollers	CL4176	[49]
Aβ(1–42) in muscles	dvIs100 [unc-54p::A-beta-1-42::unc-54 3'-UTR + mtl-2p::GFP]. mtl-2p::GFP	GMC101	[52]
Pan-neuronal Aβ(1–40)	*Ex*[pPD49.26 + P*aex-3*::Aβ(1–40); *ttx-3*::rfp p*lin-15*(+)]	MT309	[17]
Glutamatergic Aβ	*baIn32*[P*eat-4*::ssAβ42,P*eat-4*::GFP,P*myo-2*::mCherry]	UA198	[73]
Amyotrophic lateral sclerosis			
Constitutive pan-neuronal	Pro-aggregant lines *Is*[rab-3p::F3ΔK280 + myo-2p::mCherry]	BR5270, BR5485, BR 5706, BR5944	[22]
Constitutive pan-neuronal	Anti-aggregant lines: *Is*[P*rab-3*::F3ΔK280(I277P)(I308P) + P*myo-2*::mCherry]	BR5271, BR5486, BR6427, BR6516	[22]

(continued)

Table 10.1 (continued)

Disease models	Genotype	Strain name	References
Huntington's disease			
Expression of polyQ expansions (Q35:YFP) within the body wall muscle cells	rmIs132 [unc54p::Q35::YFP]	AM140	[87]

10.6 Regulatory Acceptance of *C. elegans* for Neurotoxicity Testing and Drug Screening

C. elegans is garnering growing scientific attention in recent times due to its significant contribution to drug discovery and toxicology. It is an ideal model for carrying out large-scale screenings of compounds, facilitating early-stage investigations and exploration of critical pathways, indicative of the toxic effects of compounds. For aging studies, *C. elegans* is extensively utilized and globally acclaimed. Besides, it is also an asset for conducting initial phases of drug discovery studies, and is widely employed for screening purposes. With the rise in the demand for "New Approach Methodologies" (NAMs) that favour 3Rs (Replacement, Reduction, Refinement), *the role of C. elegans* becomes indispensable to overcome the limitations of traditional animal testing [62, 89].

The FDA (U.S. Food and Drug Administration) recognizes *C. elegans* success when it comes to toxicological studies and favours the *C. elegans* studies for drug approval. Similarly, EMA (European Medicines Agency) also acknowledges the importance of *C. elegans* and encourages its use for drug testing studies [58]. This is a major transition observed in recent years, giving researchers freedom to explore beyond the traditional models employed in the routine drug discovery process [32]. In the Indian scenario, the regulatory body, the Central Drugs Standard Control Organization (CDSCO), monitors the clinical trials and new drug approvals, and according to the CDSCO, New Drugs and Clinical Trials Rules, 2019, initial studies are mandatory to be done in at least two mammalian species. The CDSCO stresses significantly over mammalian testing for clinical studies and regulatory approvals. However, recently, due to the introduction of NAMs, researchers have shifted focus towards *C. elegans*. This raises expectations for *C. elegans-based* drug assessments to get their highly awaited recognition from CDSCO. Though nothing is mentioned against employing *C. elegans* in nonclinical studies in the CDSCO guidelines, the regulatory tests required for approvals are incompatible with worms.

Currently, rodents are the dominant model system in preclinical studies and are demanded by all the regulatory bodies mentioned above. They are considered necessary for the pharmacokinetics study, pharmacodynamics study, toxicology, and safety evaluation studies. The major reasons for the success of rodents in preclinical studies include shared homology with humans and a similar complex body environment. The preclinical data obtained from rodent systems have always been used for conducting clinical trials, contributing significantly to its predictive value for clinical outcomes. However, the rodent system is costly, time-consuming, and labour-intensive. It is also not suitable for high-throughput screenings, and the ethical concerns raised over the years for their use in drug assessments have urged the need to look for alternative options [78].

Although the *C. elegans* system cannot completely displace the rodent systems due to differences observed at physiological and metabolic levels with mammalian systems, it still offers a valuable platform for early large-scale drug screenings. The *C. elegans* screenings provide essential findings that could be utilized to reduce mammalian studies for conducting efficient clinical trials [15, 32, 71]. Despite being excessively utilized for fundamental research and toxicological studies, clear guidelines for the requirement of *C. elegans* studies in preclinical trials for drug approval are not yet outlined by major regulatory bodies, raising a critical question for all of us regarding the reliability of *C. elegans* in drug discovery pipelines for future studies.

10.7 Conclusion

One of the chief merits of working with *C. elegans* is that it is free from any kind of bioethical restrictions, and in recent times, *C. elegans* has gained notable popularity as a screening tool for drug discovery. Besides screening, the findings from it have also enlightened the research community with the fundamentals of disease biology and core biological mechanisms. So, constraining its utility to only a screening purpose would not do justice to the other great features it has to offer. A few of the very crucial methods, like RNAi were pioneered in *C. elegans* and are used widely now in other model systems as well. Additionally, from a neuroscience viewpoint, it has a complete neural signal propagation atlas with all possible predictions of interactions for its 302 neurons. The glia-neuron interrelationship is very much like what we observe in mammalian systems, and most of the current *C. elegans* studies are focused on glia-neuron interactions to interrogate the advanced role of glia in age-associated neurological disorders. In conclusion, this 1 mm worm can shorten the time to reach significant research outcomes in the neurological paradigm by bridging the existing research gaps with its A-level merits and fast-track records.

References

1. Beel AJ, Sanders CR. Substrate specificity of γ-secretase and other intramembrane proteases. Cell Mol Life Sci. 2008; https://doi.org/10.1007/s00018-008-7462-2.
2. Brenner S. The Genetics of Caenorhabditis elegans. Genetics. 1974; https://doi.org/10.1093/genetics/77.1.71.
3. Calapai A, Berger M, Niessing M, Heisig K, Brockhausen R, Treue S, Gail A. A cage-based training, cognitive testing and enrichment system optimized for rhesus macaques in neuroscience research. Behav Res Methods. 2017;49(1):35–45. https://doi.org/10.3758/s13428-016-0707-3.

4. Caldwell KA, Willicott CW, Caldwell GA. Modeling neurodegeneration in Caenorhabditis elegans. DMM Dis Models Mech. 2020;13(10). https://doi.org/10.1242/dmm.046110.
5. Cao S, Gelwix CC, Caldwell KA, Caldwell GA. Torsin-mediated protection from cellular stress in the dopaminergic neurons of *Caenorhabditis elegans*. J Neurosci. 2005;25(15):3801–12. https://doi.org/10.1523/JNEUROSCI.5157-04.2005.
6. De Castro Medeiros D, Aguiar CL, Moraes MFD, Fisone G. Sleep disorders in rodent models of Parkinson's disease. Front Pharmacol (Frontiers Media SA). 2019; https://doi.org/10.3389/fphar.2019.01414.
7. Chauhan P, Wadhwa K, Singh G. *Caenorhabditis elegans* as a model system to evaluate neuroprotective potential of nano formulations. Front Nanotechnol (Frontiers Media SA). 2022; https://doi.org/10.3389/fnano.2022.1018754.
8. Chen C, Kim WY, Jiang P. Humanized neuronal chimeric mouse brain generated by neonatally engrafted human iPSC-derived primitive neural progenitor cells. JCI Insight. 2016;1(19):e88632. https://doi.org/10.1172/jci.insight.88632.
9. Chen YM, Liu SP, Lin HL, Chan MC, Chen YC, Huang YL, Tsai MC, Fu RH. Irisflorentin improves alpha-synuclein accumulation and attenuates 6-OHDA-induced dopaminergic neuron degeneration, implication for Parkinson's disease therapy. BioMedicine (Taiwan). 2015;5(1):24–32. https://doi.org/10.7603/s40681-015-0004-y.
10. Cheng Y, Bai F. The association of tau with mitochondrial dysfunction in Alzheimer's disease. Front Neurosci (Frontiers Media SA). 2018; https://doi.org/10.3389/fnins.2018.00163.
11. Chesselet MF, Carmichael ST. Animal models of neurological disorders. Neurotherapeutics. 2012:241–4. https://doi.org/10.1007/s13311-012-0118-9.
12. Chew YL, Fan X, Götz J, Nicholas HR. PTL-1 regulates neuronal integrity and lifespan in C. elegans. J Cell Sci. 2013;126(9):2079–91. https://doi.org/10.1242/jcs.jcs124404.
13. Cordeiro LM, Machado ML, da Silva AF, Obetine Baptista FB, da Silveira TL, Soares FAA, Arantes LP. Rutin protects Huntington's disease through the insulin/IGF1 (IIS) signalling pathway and autophagy activity: Study in Caenorhabditis elegans model. Food Chem Toxicol. 2020;141:111323.
14. Culetto E, Sattelle DB. A role for Caenorhabditis elegans in understanding the function and interactions of human disease genes. Hum Mol Genet. 2000;9(6):869–77. https://doi.org/10.1093/hmg/9.6.869.
15. Deji-Oloruntoba OO, Elufioye TO, Adefegha SA, Jang M. Can *Caenorhabditis elegans* serve as a reliable model for drug and nutraceutical discovery. Appl Biosci. 2025;4(2):23. https://doi.org/10.3390/applbiosci4020023.
16. Dimitriadi M, Hart AC. Neurodegenerative disorders: insights from the nematode *Caenorhabditis elegans*. Neurobiol Dis. 2010:4–11. https://doi.org/10.1016/j.nbd.2010.05.012.
17. Diomede L, Di Fede G, Romeo M, Bagnati R, Ghidoni R, Fiordaliso F, Salio M, Rossi A, Catania M, Paterlini A, Benussi L, Bastone A, Stravalaci M, Gobbi M, Tagliavini F, Salmona M. Expression of A2V-mutated Aβ in *Caenorhabditis elegans* results in oligomer formation and toxicity. Neurobiol Dis. 2014;62:521–32. https://doi.org/10.1016/j.nbd.2013.10.024.
18. Diomede L, Rigacci S, Romeo M, Stefani M, Salmona M. Oleuropein aglycone protects transgenic *C. elegans* strains expressing Aβ42 by reducing plaque load and motor deficit. PLoS One. 2013;8(3). Available at: https://doi.org/10.1371/journal.pone.0058893.
19. Ellis HM, Robert Horvitz H. Genetic control of programmed cell death in the Nematode C. elegans. Cell. 1986; https://doi.org/10.1016/0092-8674(86)90004-8.
20. Ewald CY, Cheng R, Tolen L, Shah V, Gillani A, Nasrin A, Li C. Pan-neuronal expression of APL-1, an APP-related protein, disrupts olfactory, gustatory, and touch plasticity in *Caenorhabditis elegans*. J Neurosci. 2012;32(30):10156–69. https://doi.org/10.1523/JNEUROSCI.0495-12.2012.
21. Ewald CY, Li C. *Caenorhabditis elegans* as a model organism to study APP function. Exp Brain Res. 2012:397–411. https://doi.org/10.1007/s00221-011-2905-7.
22. Fatouros C, Pir GJ, Biernat J, Koushika SP, Mandelkow E, Mandelkow EM, Schmidt E, Baumeister R. Inhibition of Tau aggregation in a novel *Caenorhabditis elegans* model of

tauopathy mitigates proteotoxicity. Hum Mol Genet. 2012;21(16):3587–603. https://doi.org/10.1093/hmg/dds190.
23. Fire A, Xu S, Montgomery MK, Kostas SA, Driver SE, Mello CC. Potent and specific genetic interference by double-stranded RNA in *Caenorhabditis elegans*. Nature. 1998;391:806. https://doi.org/10.1038/35888.
24. Giaime E, Tong Y, Wagner LK, Yuan Y, Huang G, Shen J. Age-dependent dopaminergic neurodegeneration and impairment of the autophagy-lysosomal pathway in LRRK-deficient mice. Neuron. 2017;96(4):796–807.e6. https://doi.org/10.1016/j.neuron.2017.09.036.
25. Griffin EF, Scopel SE, Stephen CA, Holzhauer AC, Vaji MA, Tuckey RA, Berkowitz LA, Caldwell KA, Caldwell GA. ApoE-associated modulation of neuroprotection from Aβ-mediated neurodegeneration in transgenic Caenorhabditis elegans. DMM Dis Models Mech. 2019;12(2):dmm037218. https://doi.org/10.1242/dmm.037218.
26. Guthrie CR, Greenup L, Leverenz JB, Kraemer BC. MSUT2 is a determinant of susceptibility to tau neurotoxicity. Hum Mol Genet. 2011;20(10):1989–99. https://doi.org/10.1093/hmg/ddr079.
27. de Guzman ACV, Razzak MA, Cho JH, Kim JY, Choi SS. Curcumin-loaded human serum albumin nanoparticles prevent Parkinson's disease-like symptoms in *C. elegans*. Nano. 2022;12(5):758. https://doi.org/10.3390/nano12050758.
28. Van Ham TJ, Thijssen KL, Breitling R, Hofstra RMW, Plasterk RHA, Nollen EAA. C. elegans model identifies genetic modifiers of alpha-synuclein inclusion formation during aging. PLoS Genet. 2008;4(3):e1000027. https://doi.org/10.1371/journal.pgen.1000027.
29. Hameed R, Naseer A, Saxena A, Akbar M, Toppo P, Sarkar A, Shukla SK, Nazir A. Functional implications of NHR-210 enrichment in C. elegans cephalic sheath glia: insights into metabolic and mitochondrial disruptions in Parkinson's disease models. Cell Mol Life Sci. 2024;81(1):202. https://doi.org/10.1007/s00018-024-05179-2.
30. Hernandez DG, Reed X, Singleton AB. Genetics in Parkinson disease: Mendelian versus non-Mendelian inheritance. J Neurochem (Blackwell Publishing). 2016:59–74. https://doi.org/10.1111/jnc.13593.
31. Hou Y, Dan X, Babbar M, Wei Y, Hasselbalch SG, Croteau DL, Bohr VA. Ageing as a risk factor for neurodegenerative disease. Nat Rev Neurol (Nature Publishing Group). 2019:565–81. https://doi.org/10.1038/s41582-019-0244-7.
32. Hunt PR. The C. elegans model in toxicity testing. J Appl Toxicol (John Wiley and Sons). 2017;37:50–9. https://doi.org/10.1002/jat.3357.
33. Huang X, et al. Human amyloid beta and α-synuclein co-expression in neurons impair behavior and recapitulate features for Lewy body dementia in Caenorhabditis elegans. Biochim Biophys Acta Mol basis Dis. 2021;1867(10):166203. https://doi.org/10.1016/j.bbadis.2021.166203.
34. Iyer S, Sam FS, DiPrimio N, Preston G, Verheijen J, Murthy K, Parton Z, Tsang H, Lao J, Morava E, Perlstein EO. Repurposing the aldose reductase inhibitor and diabetic neuropathy drug epalrestat for the congenital disorder of glycosylation PMM2-CDG. DMM Dis Models Mech. 2019;12(11). https://doi.org/10.1242/dmm.040584.
35. Jadiya P, Mir S, Nazir A. Effect of various classes of pesticides on expression of stress genes in transgenic C. elegans model of Parkinson's disease. CNS Neurol Disord Drug Targets. 2013;11(8). Available at: https://doi.org/10.2174/187152731121108009.
36. Kang WK, Florman JT, Araya A, Fox BW, Thackeray A, Schroeder FC, Walhout AJM, Alkema MJ. Vitamin B12 produced by gut bacteria modulates cholinergic signalling. Nat Cell Biol. 2024;26(1):72–85. https://doi.org/10.1038/s41556-023-01299-2.
37. Kia DA, Zhang D, Guelfi S, Manzoni C, Hubbard L, Reynolds RH, Botía J, Ryten M, Ferrari R, Lewis PA, Williams N, Trabzuni D, Hardy J, Wood NW. Identification of candidate Parkinson disease genes by integrating genome-wide association study, expression, and epigenetic data sets. JAMA Neurology. 2021;78(4):464–72. https://doi.org/10.1001/jamaneurol.2020.5257.
38. Knight AL, Yan X, Hamamichi S, Ajjuri RR, Mazzulli JR, Zhang MW, Daigle JG, Zhang S, Borom AR, Roberts LR, Lee SK, DeLeon SM, Viollet-Djelassi C, Krainc D, O'Donnell JM, Caldwell KA, Caldwell GA. The glycolytic enzyme, GPI, is a functionally conserved modifier

of dopaminergic neurodegeneration in Parkinson's models. Cell Metab. 2014;20(1):145–57. https://doi.org/10.1016/j.cmet.2014.04.017.

39. Kraemer BC, Burgess JK, Chen JH, Thomas JH, Schellenberg GD. Molecular pathways that influence human tau-induced pathology in Caenorhabditis elegans. Hum Mol Genet. 2006;15(9):1483–96. https://doi.org/10.1093/hmg/ddl067.
40. Kraemer BC, Zhang B, Leverenz JB, Thomas JH, Trojanowski JQ, Schellenberg GD. Neurodegeneration and defective neurotransmission in a Caenorhabditis elegans model of tauopathy. Proc Natl Acad Sci U S A. 2003;100(17):9980–5. https://doi.org/10.1073/pnas.1533448100.
41. Kuwahara T, Koyama A, Gengyo-Ando K, Masuda M, Kowa H, Tsunoda M, Mitani S, Iwatsubo T. Familial Parkinson mutant alpha-synuclein causes dopamine neuron dysfunction in transgenic Caenorhabditis elegans. J Biol Chem. 2006;281(1):334–40. https://doi.org/10.1074/jbc.M504860200.
42. Lakso M, Vartiainen S, Moilanen AM, Sirviö J, Thomas JH, Nass R, Blakely RD, Wong G. Dopaminergic neuronal loss and motor deficits in *Caenorhabditis elegans* overexpressing human alpha-synuclein. J Neurochem. 2003;86(1):165–72. https://doi.org/10.1046/j.1471-4159.2003.01809.x.
43. van der Lee SJ, Wolters FJ, Ikram MK, Hofman A, Ikram MA, Amin N, van Duijn CM. The effect of APOE and other common genetic variants on the onset of Alzheimer's disease and dementia: a community-based cohort study. Lancet Neurol. 2018;17(5):434–44. https://doi.org/10.1016/S1474-4422(18)30053-X.
44. Lee Y, Stevens DA, Kang SU, Jiang H, Lee YI, Ko HS, Scarffe LA, Umanah GE, Kang H, Ham S, Kam TI, Allen K, Brahmachari S, Kim JW, Neifert S, Yun SP, Fiesel FC, Springer W, Dawson VL, Shin JH, Dawson TM. PINK1 primes Parkin-mediated ubiquitination of PARIS in dopaminergic neuronal survival. Cell Rep. 2017;18(4):918–32. https://doi.org/10.1016/j.celrep.2016.12.090.
45. Levitan D, Greenwald I. Facilitation of lin-12-mediated signalling by sel-12, a Caenorhabditis elegans S182 Alzheimer's disease gene. Nature. 1995;377(6547). https://doi.org/10.1038/377351a0.
46. Li H, Feng Y, Chen Z, Jiang X, Zhou Z, Yuan J, Li F, Zhang Y, Huang X, Fan S, Wu X, Huang C. Pepper component 7-ethoxy-4-methylcoumarin, a novel dopamine D2 receptor agonist, ameliorates experimental Parkinson's disease in mice and *Caenorhabditis elegans*. Pharmacol Reseh. 2021;163. https://doi.org/10.1016/j.phrs.2020.105220.
47. Li J, Le W. Modeling neurodegenerative diseases in Caenorhabditis elegans. Exp Neurol. 2013:94–103. https://doi.org/10.1016/j.expneurol.2013.09.024.
48. Lichty JD, Mane H, Yarmey VR, Miguel AS. Amyloid β induces hormetic-like effects through major stress pathways in a C. elegans model of Alzheimer's Disease. PLoS One. 2025;20(4). Available at: https://doi.org/10.1371/journal.pone.0315810.
49. Lin H, Gao Y, Zhang C, Ma B, Wu M, Cui X, Wang H. Autophagy regulation influences β-amyloid toxicity in transgenic Caenorhabditis elegans. Front Aging Neurosci. 2022;14:885145. https://doi.org/10.3389/fnagi.2022.885145.
50. Liu W, Zhou W, Zhao P, Wu T, Gu H, Li Y, Zhong C, Bai H, Zhao N, Huang X. PD-Like Pathogenesis in Caenorhabditis elegans Intestinally Infected with Nocardia farcinica and the Underlying Molecular Mechanisms. Mol Neurobiol [Preprint]. 2024; https://doi.org/10.1007/s12035-024-04076-w.
51. Mathys H, Davila-Velderrain J, Peng Z, Gao F, Mohammadi S, Young JZ, Menon M, He L, Abdurrob F, Jiang X, Martorell AJ, Ransohoff RM, Hafler BP, Bennett DA, Kellis M, Tsai LH. Single-cell transcriptomic analysis of Alzheimer's disease. Nature. 2019;570(7761):332–7. https://doi.org/10.1038/s41586-019-1195-2.
52. Mccoll G, Roberts BR, Pukala TL, Kenche VB, Roberts CM, Link CD, Ryan TM, Masters CL, Barnham KJ, Bush AI, Cherny RA. Utility of an improved model of amyloid-beta (Aβ1–42) toxicity in Caenorhabditis elegans for drug screening for Alzheimer's disease. Mol Neurodegenera. 2012;7(1). https://doi.org/10.1186/1750-1326-7-57.

53. Medeiros R, Baglietto-Vargas D, Laferla FM. The role of tau in Alzheimer's disease and related disorders. CNS Neurosci Ther. 2011:514–24. https://doi.org/10.1111/j.1755-5949.2010.00177.x.
54. Miyasaka T, Shinzaki Y, Yoshimura S, Yoshina S, Kage-Nakadai E, Mitani S, Ihara Y. Imbalanced expression of tau and tubulin induces neuronal dysfunction in C. elegans models of tauopathy. Front Neurosci. 2018;12(JUN). https://doi.org/10.3389/fnins.2018.00415.
55. Moehle MS, Webber PJ, Tse T, Sukar N, Standaert DG, Desilva TM, Cowell RM, West AB. LRRK2 inhibition attenuates microglial inflammatory responses. J Neurosci. 2012;32(5):1602–11. https://doi.org/10.1523/JNEUROSCI.5601-11.2012.
56. Mor DE, Tsika E, Mazzulli JR, Gould NS, Kim H, Daniels MJ, Doshi S, Gupta P, Grossman JL, Tan VX, Kalb RG, Caldwell KA, Caldwell GA, Wolfe JH, Ischiropoulos H. Dopamine induces soluble alpha-synuclein oligomers and nigrostriatal degeneration. Nat Neurosci. 2017;20(11):1560–8. https://doi.org/10.1038/nn.4641.
57. Nass R, Hall DH, Miller DM, Blakely RD. Neurotoxin-induced degeneration of dopamine neurons in Caenorhabditis elegans. Proc Natl Acad Sci USA. 2002;99(5):3264–9. https://doi.org/10.1073/pnas.042497999.
58. Network E-I. New approach methodologies EU-IN horizon scanning report. 2025. Available at: www.ema.europa.euwww.hma.eu.
59. Neziri S, Köseoğlu AE, Deniz Köseoğlu G, Özgültekin B, Özgentürk NÖ. Animal models in neuroscience with alternative approaches: evolutionary, biomedical, and ethical perspectives. Anim Models Exp Med. 2024; [Preprint]. John Wiley and Sons Inc. Available at: https://doi.org/10.1002/ame2.12487.
60. O'Brien TJ, Barlow IL, Feriani L, Brown AEX. High-throughput tracking enables systematic phenotyping and drug repurposing in C. elegans disease models. elife. 2025;12:1–17. https://doi.org/10.7554/elife.92491.4.
61. Offenburger SL, Ho XY, Tachie-Menson T, Coakley S, Hilliard MA, Gartner A. 6-OHDA-induced dopaminergic neurodegeneration in Caenorhabditis elegans is promoted by the engulfment pathway and inhibited by the transthyretin-related protein TTR-33. PLoS Genet. 2017;14(1):e1007125. https://doi.org/10.1371/journal.pgen.1007125.
62. O'Reilly LP, Luke CJ, Perlmutter DH, Silverman GA, Pak SC. C. elegans in high-throughput drug discovery. Adv Drug Deliv Rev (Elsevier). 2014:247–53. https://doi.org/10.1016/j.addr.2013.12.001.
63. Palikaras K, Achanta K, Choi S, Akbari M, Bohr VA. Alteration of mitochondrial homeostasis is an early event in a C. elegans model of human tauopathy. Aging. 2021;13(21):23876. https://doi.org/10.18632/aging.203683.
64. Palmer JE, Wilson N, Son SM, Obrocki P, Wrobel L, Rob M, Takla M, Korolchuk VI, Rubinsztein DC. Autophagy, aging, and age-related neurodegeneration. Neuron (Cell Press). 2024; https://doi.org/10.1016/j.neuron.2024.09.015.
65. Pir GJ, Choudhary B, Mandelkow E. *Caenorhabditis elegans* models of tauopathy. FASEB J. 2017;31(12):5137. https://doi.org/10.1096/fj.201701007.
66. Poewe W, Seppi K, Tanner CM, Halliday GM, Brundin P, Volkmann J, Schrag AE, Lang AE. Parkinson disease. Nat Rev Dis Prim. 2017;3:1–21. https://doi.org/10.1038/nrdp.2017.13.
67. Raggi A, Monasta L, Beghi E, Caso V, Castelpietra G, Mondello S, Giussani G, Logroscino G, Magnani FG, Piccininni M, Pupillo E, Ricci S, Ronfani L, Santalucia P, Sattin D, Schiavolin S, Toppo C, Traini E, Steinmetz J, Nichols E, Ma R, Vos T, Feigin V, Leonardi M. Incidence, prevalence and disability associated with neurological disorders in Italy between 1990 and 2019: an analysis based on the Global Burden of Disease Study 2019. J Neurol. 2022;269(4):2080–98. https://doi.org/10.1007/s00415-021-10774-5.
68. Rapti G. A perspective on C. elegans neurodevelopment: from early visionaries to a booming neuroscience research. J Neurogenet. 2020;34(3–4):259–72. https://doi.org/10.1080/01677063.2020.1837799.
69. Ray A, Zhang S, Rentas C, Caldwell KA, Caldwell GA. RTCB-1 mediates neuroprotection via XBP-1 mRNA splicing in the unfolded protein response pathway. J Neurosci. 2014;34(48):16076–85. https://doi.org/10.1523/JNEUROSCI.1945-14.2014.

70. Reiner A, Dragatsis I, Dietrich P. Genetics and neuropathology of huntington's disease. Int Rev Neurobiol. 2011; https://doi.org/10.1016/B978-0-12-381328-2.00014-6.
71. Roy PJ. Drug screens using the nematode Caenorhabditis elegans. Genetics. 2025;231(1):1–19. https://doi.org/10.1093/genetics/iyaf141.
72. Sakamoto S, Miyara M, Sanoh S, Ohta S, Kotake Y. Mild MPP+ exposure-induced glucose starvation enhances autophagosome synthesis and impairs its degradation. Sci Rep. 2017;7:46668. https://doi.org/10.1038/srep46668.
73. Samara C, Rohde CB, Gilleland CL, Norton S, Haggarty SJ, Yanik MF (2010) Large-scale in vivo femtosecond laser neurosurgery screen reveals small-molecule enhancer of regeneration. https://doi.org/10.1073/pnas.1005372107/-/DCSupplemental.
74. Sarasija S, Laboy JT, Ashkavand Z, Bonner J, Tang Y, Norman KR. Presenilin mutations deregulate mitochondrial Ca2+ homeostasis and metabolic activity causing neurodegeneration in Caenorhabditis elegans. eLife. 2018;7:e33052. https://doi.org/10.7554/eLife.33052.001.
75. Scheltens P, De Strooper B, Kivipelto M, Holstege H, Chételat G, Teunissen CE, Cummings J, van der Flier WM. Alzheimer's disease. The Lancet (Elsevier BV). 2021:1577–90. https://doi.org/10.1016/S0140-6736(20)32205-4.
76. Settivari R, LeVora J, Nass R. The divalent metal transporter homologues SMF-1/2 mediate dopamine neuron sensitivity in Caenorhabditis elegans models of manganism and Parkinson disease. J Biol Chem. 2009;284(51):35758–68. https://doi.org/10.1074/jbc.M109.051409.
77. Singhvi A, Shaham S, Rapti G. Glia development and function in the nematode Caenorhabditis elegans. Cold Spring Harb Perspect Biol. 2024;16:a041346. https://doi.org/10.1101/cshperspect.a041346.
78. Soufizadeh P, Mansouri V, Ahmadbeigi N. A review of animal models utilized in preclinical studies of approved gene therapy products: trends and insights. Lab Anim Res (BioMed Central). 2024; https://doi.org/10.1186/s42826-024-00195-6.
79. Strange K. An overview of C. elegans biology. Methods Mol Biol. 2006; https://doi.org/10.1385/1-59745-151-7:1.
80. Tauffenberger A, Julien C, Parker JA. Evaluation of longevity enhancing compounds against transactive response DNA-binding protein-43 neuronal toxicity. Neurobiol Aging. 2013;34(9):2175–82. https://doi.org/10.1016/j.neurobiolaging.2013.03.014.
81. Trinh J, Farrer M. Advances in the genetics of Parkinson disease. Nat Rev Neurol. 2013;9:445–54. https://doi.org/10.1038/nrneurol.2013.132.
82. Tsai CW, Tsai RT, Liu SP, Chen CS, Tsai MC, Chien SH, Hung HS, Lin SZ, Shyu WC, Fu RH. Neuroprotective effects of Betulin in pharmacological and transgenic Caenorhabditis elegans models of Parkinson's disease. Cell Transplant. 2017;26(12):1903–18. https://doi.org/10.1177/0963689717738785.
83. Turner MR, Hardiman O, Benatar M, Brooks BR, Chio A, De Carvalho M, Ince PG, Lin C, Miller RG, Mitsumoto H, Nicholson G, Ravits J, Shaw PJ, Swash M, Talbot K, Traynor BJ, Van Den Berg LH, Veldink JH, Vucic S, Kiernan MC. Controversies and priorities in amyotrophic lateral sclerosis. Lancet Neurol. 2013; https://doi.org/10.1016/S1474-4422(13)70036-X.
84. Ullah H, Arbab S, Tian Y, Liu CQ, Chen Y, Qijie L, Khan MIU, Hassan IU, Li K. The gut microbiota–brain axis in neurological disorder. Front Neurosci (Frontiers Media SA). 2023; https://doi.org/10.3389/fnins.2023.1225875.
85. Ullah I, Zhao L, Uddin S, Zhou Y, Wang X, Li H. Nicotine-mediated therapy for Parkinson's disease in transgenic Caenorhabditis elegans model. Front Aging Neurosci. 2024;16:1358141. https://doi.org/10.3389/fnagi.2024.1358141.
86. Waldherr SM, Strovas TJ, Vadset TA, Liachko NF, Kraemer BC. Constitutive XBP-1s-mediated activation of the endoplasmic reticulum unfolded protein response protects against pathological tau. Nat Commun. 2019;10(1):4443. https://doi.org/10.1038/s41467-019-12070-3.
87. Walker AC, Bhargava R, Vaziriyan-Sani AS, Pourciau C, Donahue ET, Dove AS, Gebhardt MJ, Ellward GL, Romeo T, Czyż DM. Colonization of the Caenorhabditis elegans gut with human enteric bacterial ltpathogens leads to proteostasis disruption that is rescued by butyrate. PLoS Pathogens. 2021;17(5):e1009510. https://doi.org/10.1371/journal.ppat.1009510.

88. Wang W, Zhao F, Ma X, Perry G, Zhu X. Mitochondria dysfunction in the pathogenesis of Alzheimer's disease: recent advances. Mol Neurodegener (BioMed Central). 2020; https://doi.org/10.1186/s13024-020-00376-6.
89. Wellenberg A, Weides L, Kurzke J, Hennecke T, Bornhorst J, Crone B, Karst U, Brinkmann V, Fritz G, Honnen S. Use of C. elegans as a 3R-compliant in vivo model for the chemoprevention of cisplatin-induced neurotoxicity. Exp Neurol. 2021;341:113705. https://doi.org/10.1016/j.expneurol.2021.113705.
90. Wightman DP, Jansen IE, Savage JE, Shadrin AA, Bahrami S, Holland D, Rongve A, Børte S, Winsvold BS, Drange OK, Martinsen AE, Skogholt AH, Willer C, Bråthen G, Bosnes I, Nielsen JB, Fritsche LG, Thomas LF, Pedersen LM, Gabrielsen ME, Johnsen MB, Meisingset TW, Zhou W, Proitsi P, Hodges A, Dobson R, Velayudhan L, Agee M, Aslibekyan S, Babalola E, Bell RK, Bielenberg J, Bryc K, Bullis E, Cameron B, Coker D, Partida GC, Dhamija D, Das S, Elson SL, Filshtein T, Fletez-Brant K, Fontanillas P, Freyman W, Gandhi PM, Hicks B, Hinds DA, Huber KE, Jewett EM, Jiang Y, Kleinman A, Kukar K, Lane V, Lin KH, Lowe M, Luff MK, McCreight JC, McIntyre MH, McManus KF, Micheletti SJ, Moreno ME, Mountain JL, Mozaffari SV, Nandakumar P, Noblin ES, O'Connell J, Petrakovitz AA, Poznik GD, Schumacher M, Shastri AJ, Shelton JF, Shi J, Shringarpure S, Tian C, Tran V, Tung JY, Wang X, Wang W, Weldon CH, Wilton P, Sealock JM, Davis LK, Pedersen NL, Reynolds CA, Karlsson IK, Magnusson S, Stefansson H, Thordardottir S, Jonsson PV, Snaedal J, Zettergren A, Skoog I, Kern S, Waern M, Zetterberg H, Blennow K, Stordal E, Hveem K, Zwart JA, Athanasiu L, Selnes P, Saltvedt I, Sando SB, Ulstein I, Djurovic S, Fladby T, Aarsland D, Selbæk G, Ripke S, Stefansson K, Andreassen OA, Posthuma D. A genome-wide association study with 1,126,563 individuals identifies new risk loci for Alzheimer's disease. Nature Genetics. 2021;53(9):1276–82. https://doi.org/10.1038/s41588-021-00921-z.
91. Wijesekera LC, Leigh PN. Amyotrophic lateral sclerosis. Orphanet J Rare Dis. 2009;4(1). https://doi.org/10.1186/1750-1172-4-3.
92. Wood TE, Barry J, Yang Z, Cepeda C, Levine MS, Gray M. Mutant huntingtin reduction in astrocytes slows disease progression in the BACHD conditional Huntington's disease mouse model. Hum Mol Genet. 2019;28(3):487–500. https://doi.org/10.1093/hmg/ddy363.
93. Zhong J, Tang G, Zhu J, Wu W, Li G, Lin X, Liang L, Chai C, Zeng Y, Wang F, Luo L, Li J, Chen F, Huang Z, Zhang X, Zhang Y, Liu H, Qiu X, Tang S, Chen D. Single-cell brain atlas of Parkinson's disease mouse model. J Genet Genomics. 2021;48(4):277–88. https://doi.org/10.1016/j.jgg.2021.01.003.
94. Zhou S, Wang Z, Klaunig JE. *Caenorhabditis elegans* neuron degeneration and mitochondrial suppression caused by selected environmental chemicals. Int J Biochem Mol Biol. 2013; Available at: www.ijbmb.org. PMID: 24380023.

Zebrafish (*Danio rerio*): A Transparent Window into Biomedical Research

11

Shaivee Chokshi, Keya Jantrania, and Ashutosh Kumar

11.1 Introduction

Animal models are used to study different biological processes and understand them in the human body. This is possible because development and such processes are regulated by orthologs of the same genes in animals and humans. As this fact was established, a professor named George Streisinger from the nascent Institute of Molecular Biology at the University of Oregon, in the year 1972, chose zebrafish, *Danio rerio*, a small, freshwater, tropical fish, as a model to investigate the genetic basis of vertebrate neural development [37]. It is speculated that Streisinger selected zebrafish over Fundulus and medaka due to the challenges in embryo visibility and manipulation in those species [6, 142]. Zebrafish became popular as a model organism in the 1950s and 1960s when laboratories began using them as a model for studying muscle and nervous system development. Hence, Zebrafish became a popular teleost model organism.

At present, Zebrafish is an excellent model to study development and embryogenesis since they are not seasonal breeders, making it possible to track the development of the embryos, since they do not develop inside the mother and hence can be easily manipulated. Zebrafish larvae are not subject to experimental regulations until 120 hours post-fertilization (hpf), which makes them better for developmental studies [138]. Moreover, the developing embryos also have a transparent covering and have a rapid life cycle [37] (Refer Fig. 11.1).

Shaivee Chokshi and Keya Jantrania contributed equally with all other contributors.

S. Chokshi · K. Jantrania · A. Kumar (✉)
Biological & Life Sciences, School of Arts & Sciences, Ahmedabad University, Central Campus, Ahmedabad, Gujarat, India
e-mail: ashutosh.kumar@ahduni.edu.in

P. Kumar, V. Kothari (eds.), *Model Organisms in Biological Research*, https://doi.org/10.1007/978-981-92-0038-2_11

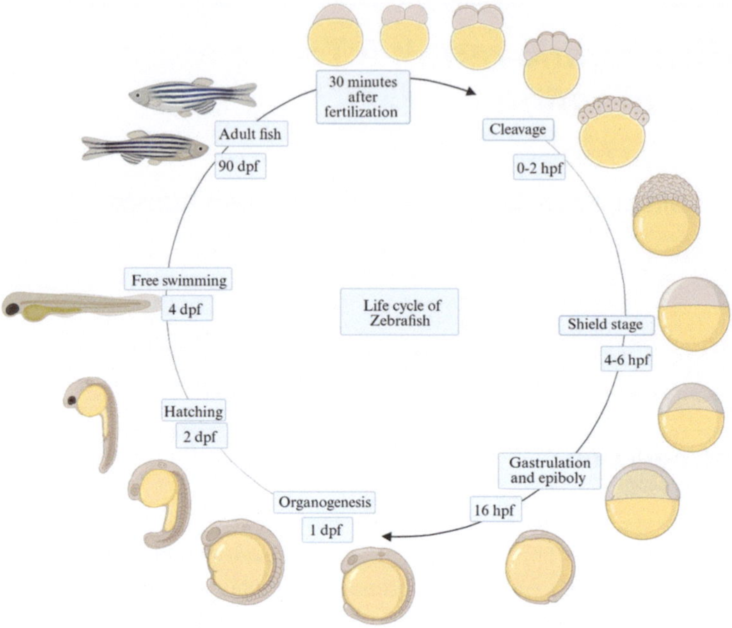

Fig. 11.1 Life cycle of zebrafish: The development begins at 30 minutes after fertilization, at 8 hours post fertilization (hpf) the embryo reaches the 1000 cell stage. At 36 hpf, organs start to develop. The larvae hatch at 48–72 hpf. The swim bladder develops by 3–4 days post-fertilization (dpf). (Image made using BioRender.com)

As models, Zebrafish are maintained in laboratories very feasibly as well. In research laboratories, fish are ideally housed in 20 L tanks which have reverse osmosis/ionised/distilled water and little aquarium salt to maintain salinity. The conductivity is kept within the range of 500 to 1500 μS/cm. The pH of the water is typically maintained using sodium bicarbonate between the range of 7.0–7.6. The following parameters are kept in check along with salinity and pH such as: TDS in the range of 220–320 mg/L, dissolved oxygen should be >6 mg/L, and conductivity is maintained in the range of 250–350 μS [91]. Zebrafish are not housed in tap water because they can be sensitive to chlorine and other chemicals. The temperature is regulated at 28.5 °C, and a photoperiod of 14 hours of light and 10 hours of dark is followed [98]. The fish are fed brine shrimp (Artemia) twice a day, which is cultivated in the laboratory by keeping brine shrimp eggs in 25g/L salt water for 48 hours under aeration, and aquarium flakes once a day. Brine shrimp, once hatched, are stored at 4 °C. There are several other Zebrafish-specific diets that help in optimizing their growth and reproduction [91, 98]. Usually, research laboratories obtain the zebrafish food, such as brine shrimps and flakes, from local suppliers, which are easily available in the area. Zebrafish breed shortly after the light phase of the photoperiod begins. The fish to be bred are set up the night before, with an increased brine shrimp diet. The breeding chambers are set by keeping a mesh- like barrier between the adult fish and the embryos, since the adult fish cannibalize the embryos [98]. The fish are ideally bred in a 4:3 female: male ratio. The embryos settle down beneath the mesh, separating them from the adults. The embryos (eggs) are collected the next day from the breeding box [91]. The adults are removed from the breeding box first, and then the water in the box is sieved so that the eggs collect in it. The eggs could either be fertilized or unfertilized. Healthy fertilized eggs are found at the bottom of the breeding box and appear as transparent balls, while unfertilized or degenerate eggs appear opaque. The eggs are transferred to a sterile Petri plate containing E3 media. The Petri plate is kept in an incubator at 28 °C until further experimentation. The media of these Petri plates is changed every day. The embryo will hatch typically between 48 and 72 hpf at 28.5 °C [98]. The larvae typically begin swimming at 5dpf. Until around 5dpf, the larvae rely on their yolk sac for food supply, but need to be supplied with food by 7dpf. The larvae generally feed on dry diets or live foods like paramecium. In the case where these larvae have to be grown to adulthood, they are shifted to a system that allows feeding and a gradual shift to water. Once larvae begin to grow, they can be shifted to a brine shrimp diet like an adult fish-brine shrimp hatchlings diet [91, 98].

Generally, a zebrafish facility should possess an incubator where the embryos can be kept in ideal conditions until further experimentation. An imaging system should be installed like a stereo microscope with software to conduct behavioural analysis. Moreover, the installation of a microinjector facility would be useful for developmental and exposure studies.

Due to the simple and efficient maintenance techniques, zebrafish have been popularized as a model. Zebrafish are now used as powerful *in vivo* models in studies of nanomaterial interactions with cells and advancing target-based biomedical applications. The zebrafish is used to evaluate biocompatibility, tissue-specific uptake, and intracellular behavior of DNA nanocages [68] and porphyrin-loaded lipid microbubbles [52]. A study on DNA nanocages showed an efficient uptake of tetrahedral DNA nanocages over time without affecting developmental gene expression [68]. Whereas a study on the porphyrin-loaded lipid microbubbles showed concentration-dependent photodynamic and sonodynamic cytotoxicity, along with their potential for near-infrared bioimaging applications [52]. Additionally, another study to evaluate the toxicology of microplastics and phthalates uses zebrafish as a model organism, demonstrating that weathered microplastics (MPs) induce significantly higher toxicity and disrupt key DNA repair pathways in zebrafish larvae compared to virgin MPs, highlighting the zebrafish as a sensitive model for assessing environmentally relevant MP toxicity and phthalate exposure shows cardiac disruption and oxidative stress [89, 90]. In essence, the advantages of teleost models like zebrafish lie in their small size, low maintenance, genetic tractability, and high regenerative ability. Additionally, the transparent nature of zebrafish embryos and larvae allows for high-resolution imaging of osteogenesis and osteoblast function. The availability of well-characterized transgenic and mutant lines specific to particular cell populations or tissues enhances this capability. Together with the possibility of long-term in vivo imaging throughout the embryonic, larval, and adult stages, zebrafish

Table 11.1 Different life stages of zebrafish and the experimental assays associated with it

Life stage	Assays	References
Embryo—0–72 hpf (up to 120 hpf with yolk)	Reactive Oxidative Species (ROS) and Acridine Orange Imaging (AO), developmental toxicity, teratogenicity	[67, 69, 90]
Larva 3–7 dpf	Antioxidant enzyme assays like—SOD, GPx; lipid peroxidation, neurobehavioral assays, toxicity screening	[88, 90]
Juvenile ~30 dpf onward	Behavioral tests, organ specific toxicity, chronic exposure analysis	[123, 145]
Adult >3 months	Cognition and complex behavior analysis, targeted organ assays, histopathological analysis	[91, 159]

offer distinct advantages over other vertebrate models like rodents, where intravital imaging is technically challenging [33]. There is recent research that studied the toxic effects and bioavailability of platinum-based anti-cancer drugs on zebrafish. The research exposed the zebrafish to the drug using various exposure methods, such as direct injection, submerging in the drug added media, as well as the drug dissolved in different solvents, which were further correlated to human health effects [152]. The following sections discuss the popularization of Zebrafish as a model organism to study various diseases, across organ systems, and as a drug screening and toxicity model (Table 11.1).

11.2 Salient Features of Zebrafish as a Model Organism

Zebrafish and medaka are popular small fish models in biomedical science, with zebrafish being the most widely used. These teleost models are ideal for exploring a broad range of human disorders, including cardiovascular disease, bone biology, immunity, and cancer. As vertebrates, they have a conserved chordate body plan, including specific anatomical components such as the vertebral column, neurocranium, neural crest cells, and sensory placodes [13]. Zebrafish offer several advantages as a model organism. They have extracorporeal fertilization, rapid embryonic development, a small size, low maintenance costs, transparent embryos, and allow easy real-time imaging of developmental events [150].

In addition, they have well-established genetic manipulation protocols and are highly amenable to high- throughput screening assays. Their genome is highly homologous to that of higher vertebrates, making them an excellent tool for studying genetic diseases [13]. Zebrafish share a high level of genomic conservation with humans, with nearly 70% of human genes having at least one ortholog in zebrafish. This genetic conservation makes zebrafish an ideal model for studying human genetic diseases. Additionally, tissue-specific transgenic zebrafish can be easily produced using chosen gene promoters. Recent advancements in the Tol2-based transgenic system allow for spatiotemporal regulation of gene expression with high precision, using regulatory systems like GAL4/

UAS and Cre/LoxP. These developments enable real-time *in vivo* imaging of cell dynamics, aiding research into the molecular processes involved in organ development [28]. Zebrafish have also been instrumental in drug discovery and development, with potential therapeutic targets, such as MEK inhibitors, currently being tested in clinical trials for diseases like lymphatic anomalies and melanoma [115]. Zebrafish remain a valuable model for investigating conserved vertebrate molecular mechanisms and are increasingly used in toxicological assessments [23].

Despite their genetic relationship to higher vertebrates, zebrafish display significant biological differences, largely due to their adaptation to life in water. They have specialized structures like gills for respiration and swim bladders to maintain buoyancy. While gills and lungs share certain aspects of development, their physiological functions are fundamentally distinct [13]. Zebrafish have remarkable regenerative capabilities, enabling limb, cardiac tissue, and neuronal regeneration through molecular mechanisms not present or highly limited in higher vertebrates [92]. Other notable differences between zebrafish and higher vertebrates include variations in organogenesis, sex differentiation, immune system function, behaviour (like parental care), and neuroanatomy, such as the absence of a neocortex. Consequently, zebrafish findings must be rigorously validated before translating them to mammals [84].

While mice have historically been the primary model for biomedical studies, due to their well- established genetic tools, human-like physiological responses, and standardized protocols for investigating metabolic syndromes, variations in diet, lifestyles, and microbiota have prompted a shift toward alternative models [28]. Zebrafish provide valuable insights into metabolic disorders and potential therapeutic strategies. Although they have innate limitations—such as being poikilothermic and adapted to marine environments—their metabolic properties complement those of mammalian models, such as rodents, thereby expanding the scope of comparative biomedical research [127].

11.2.1 Key Differences Between Zebrafish and Mice as Model Organisms

Zebrafish (*Danio rerio*) have major advantages over mice (*Mus musculus*) in research studies because they are easy to maintain and have a high reproductive ability. One zebrafish breeding cycle produces about 50 to 300 eggs, and there is frequent and easily inducible spawning, while mice only give 1 to 10 pups per litter and can support a maximum of three litters during their lifetime [21]. The life cycle of Zebrafish is a fast one that can be subjected to repeated experiments, an aspect important in allowing scientific studies to be reproducible and statistically resilient. Therefore, the fecundity and reproductive efficiency of the zebrafish make the animal a potential model organism suitable for genetic as well as developmental studies [84]. Zebrafish showcase external fertilization, which makes early embryonic development observable and possible to make genetic alterations accurately through microinjection of DNA or RNA into the embryo. On the other hand, genetic analysis in mice involves intrusive methods, such as killing the mother to get access to embryos, then embryo transfer to a surrogate to achieve viability [21]. Zebrafish embryos are transparent by nature, allowing one to visualize real-time development under a microscope. Transparency allows non-invasive imaging of all growth stages and is useful for *in vivo* analysis of fluorescently labelled tissues [84]. Mouse embryos, by contrast, are grown inside the mother's uterus and are opaque, requiring invasive techniques for examination and analysis [21]. The optical clarity of zebrafish embryos thus offers an obvious advantage in live imaging and developmental biology research [84]. Zebrafish is an ideal model organism to study cardiovascular diseases, especially because mice depend heavily on the cardiovascular system for obtaining oxygen during early embryonic development, which can lead to poor functions or even death when studying cardiovascular diseases. However, zebrafish embryos can survive for a period even without a fully functional cardiovascular system, obtaining oxygen through passive diffusion [20]. (Refer Fig. 11.2).

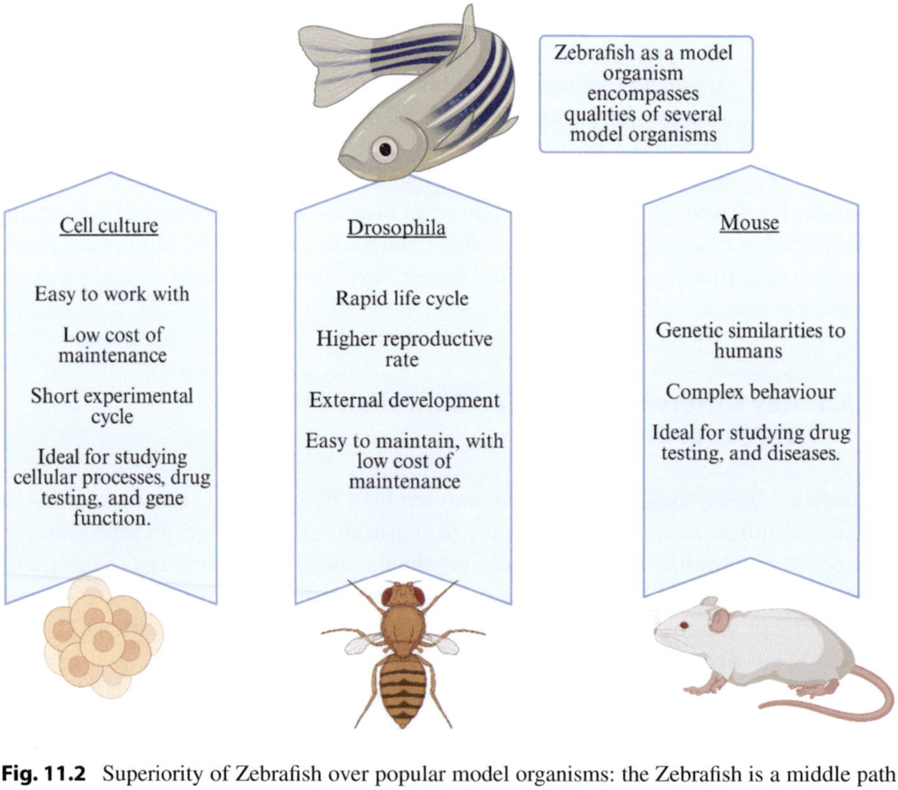

Fig. 11.2 Superiority of Zebrafish over popular model organisms: the Zebrafish is a middle path between vertebrate models and invertebrate models, specifically due to their similarity with humans and rapid life cycle. (Image made using BioRender.com)

11.3 Experimental Applications of the Zebrafish Model

11.3.1 Zebrafish as a Host for Xenograft Models

Zebrafish are an effective model for anticancer drug discovery, screening, and toxicity testing. Zebrafish have an advantage over transgenic mice in this research field, since experiments on mice take longer periods of time than Zebrafish. Moreover, zebrafish are also a valuable model for human cancer cell transplantation. This makes them effective xenograft models. Xenograft transplantation is the process in which there is relocation of living cells from one species to another [76]. Human cells in Zebrafish have been able to interact with the environment and migrate within the body. Furthermore, several successful experiments have been conducted using Zebrafish as a xenograft model. Human cancer cell lines of melanoma, breast, and leukaemia, among numerous other cancer types, have been successfully studied using Zebrafish [30, 118]. The embryo does not have an adaptive immunity in the early stages, until 30 dpf, making it easier to transplant human cells without immune rejection, thereby preventing any need for immunosuppressive drugs or immunocompromised variants to perform xenografts [45]. The zebrafish xenograft model, originally developed with melanoma and human fibroblasts at 3.5–4.5 hours post-fertilization, has been further developed to include additional injection sites, including the yolk sac, duct of Cuvier (common cardinal vein), caudal vein, and brain ventricles [146]. These sites allow for cancer cell behaviour to be studied, including growth, invasion, survival, and metastasis. Yolk sac injections at 48 hours of post-fertilization are most prevalent because the yolk at this stage is relatively large, gastrulation is finished, and there is the onset of rudimentary vasculature and organogenesis. This makes it less likely for catastrophic developmental abnormalities to arise and ensures injected cancer cells cannot diffuse widely [93].

Human cancer cell lines from breast, melanoma, leukaemia, and neuroblastoma, among others, are viable and can grow and expand in the zebrafish, rendering it a suitable organism to model various cancers. Zebrafish patient-derived xenografts (PDXs) recapitulate patient tumour behaviour well, offering a platform for testing treatments on a personalized basis. A proof-of-concept study illustrated the ability of zebrafish avatars to foretell drug responses, e.g., the anti-angiogenic drug bevacizumab. Although no FDA-approved drugs have been directly derived from zebrafish xenograft models, such pathways as YAP1-CXCR2 and PIT1-CXCR4 are being researched clinically, while compounds such as crizotinib against uveal melanoma are also making progress through clinical trials. The model represents enormous potential in drug discovery as well as the personalization of cancer therapy [4, 120].

11.3.2 Application of Transgenic Zebrafish for Functional Genomics

The zebrafish tp53M214K mutated model has been widely utilized to investigate various cancer types, including peripheral nerve sheath tumours, breast cancer, and leukaemia [1, 17]. Zebrafish are favoured as a cancer model due to their small size, which requires fewer cells for transplantation and allows for high-throughput studies. Additionally, early-stage endothelial cell staining in zebrafish aids in monitoring tumour progression and evaluating anti-tumour therapies. Transparent zebrafish embryo and larvae, coupled with fluorescently stained tumour cells, enables precise tracking of metastatic cancer cells at the cellular level [149, 160].

Transgenic expression of human or murine oncogenes has facilitated the development of several cancer models in zebrafish. T-cell acute lymphoblastic leukaemia (T-ALL) was the first cancer induced using transgenic technology, via Myc transgenes [74, 160]. Overexpression of oncogenes such as xmrk, Myc, and KRAS (V12) has led to hepatoma formation in both juvenile and adult zebrafish [82, 83, 104, 161]. Amplification of MYCN and fgf8 expression promotes neuroblastoma formation [162], while upregulation of KRAS (G12D) induces rhabdomyosarcoma [137]. Overexpression of Akt1 enhances lipoma formation. When combined with the p53 mutant, the overexpression of various oncogenes results in diverse tumour phenotypes, including hepatoma (scr) [85], melanoma (BRAF and NRAS) [34, 137] [34, 136], and Ewing's sarcoma (EWS-FIL1) [75]. Furthermore, co-activation of the hedgehog and AKT pathways significantly accelerates tumorigenesis, highlighting the utility of zebrafish as a basis to study oncogene interactions and oncogenic signalling pathways [65, 160]. The zebrafish transgenic lines can be obtained from worldwide repositories like Zebrafish International Resource Centre (ZIRC), the European Zebrafish Resource Centre (EZRC) and the Chinese Zebrafish Resource Centre (CZRC). They house a large number of wild type, mutant and transgenic zebrafish. Moreover, it is also possible to create transgenic lines by microinjection at 2 cell stages and then crossing with a wild type (Table 11.2).

Table 11.2 Different transgenic lines of zebrafish and its applications

Transgenic name	Target	Utility	Reference
Tg(hsp70l:dnBmpr-GFP)^{w30}	BMP Signalling	Study of BMP pathway and developmental biology	[119, 145]
Tg(myl7:EGFP)^{twu34}	Cardiomyocytes	Cardiac development, regeneration and drug testing	[101]
Tg(Hsa.RUNX2-Mmu. Fos:EGFP)^{zf259}	Osteoblasts	Bone biology and skeletal disorders	[99]
Tg(dat:CFP-NTR) / Tg(slc6a3:CFP-NTR)^{ot1413}	Dopaminergic Neurons	Neurodegeneration, Parkinson's disease model	[47]
Tg(−4.9sox10:EGFP)^{ba2}	Neural crest derived cells (Schwann, DRG) and OL lineage cells	Neural crest development and cancer metastasis	[87]
Tg(gsdf:EGFP)^{inr2}	Sertoli and granulosa cells (Gonadal tissues)	Reproductive biology and sex determination	Frétaud et al. [44]
Tg(−2.8fabp10a:EGFP))^{zf235}	Hepatocytes	Liver development, toxicology, metabolism	[58]
Tg(mitfa:GNAQQ209P)	Melanocytes	Uveal melanoma model	[102]
Tg(flk1:EGFP)	Endothelium (endothelial precursor migration and early vascular tubulogenesis)	Angiogenesis and cardiovascular research	[64]
Tg(PT:EGFP)	Enhanced green fluorescence protein (GFP) is expressed in the proximal tubule (PT)	To visualise nephrons	[117]
tp53M214K	Mutant for cancer studies	Cancer biology and tumorigenesis	[17]
Tg(actb1:EWSR1-FLI1), Tg(hsp70:EWSR1-FLI1)	Broad/ inducible fusion oncogene expression	Ewing's sarcoma	[75]
Tg(T2KSAG:GFP-HRAS^{G12V})	Multiple tissues	Rhabdomyosarcoma	[126]

11.4 Zebrafish as a Model for Human Pathologies

Some disorders that are studied using Zebrafish as a model organism are the following:

11.4.1 Neurological Disorders

Zebrafish (*Danio rerio*) are an efficient model for research of neurodegenerative disease, including Parkinson's and Alzheimer's, because they share genetic, anatomical, and physiological features with humans (Refer Fig. 11.3). Although *in vitro* human cell models are invaluable, they cannot mimic complex multi-system interactions. Moreover, conventional mammalian models, i.e., rats and mice, are time and cost- consuming [35]. Zebrafish (*Danio rerio*) offer an affordable and scalable model to investigate neurodegenerative disorders, with some benefits over conventional mammalian models. The zebrafish brain has structural and neurotransmitter system homologies to humans, such as a conserved blood-brain barrier, facilitating translational research in neurological diseases [42]. A study on the ecotoxicological impact of engineered nanoparticles demonstrated TiO_2-induced disruption [67] and CuO nanoparticle's [69] effect on the dorsoventral axis and neurodevelopmental gene expression using zebrafish embryos as an *in vivo* model

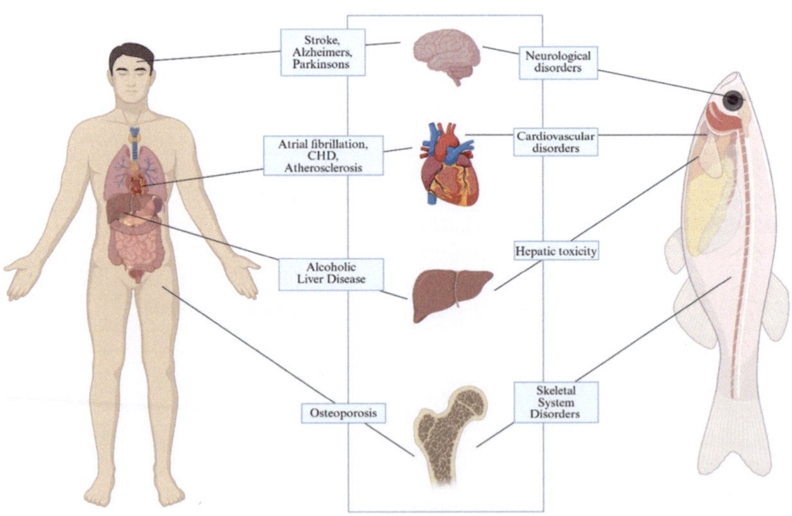

Fig. 11.3 Zebrafish and humans have several homologies. These homologies help in studying specific disorders for specific organs. (Image made using BioRender.com)

[67, 69]. Being susceptible to genetic manipulation, Zebrafish have been used for CRISPR-mediated genome editing and for generating transgenic lines that allow the targeted investigation of disease-causing genes. [148]. Zebrafish display complex behaviours, such as learning, memory, and anxiety-related responses, which offer a functional platform for neuropsychiatric and neurodegenerative studies [84]. Behavioural studies are performed on larvae up to 7dpf. Common behavioural tests include the photo-motor response [PMR], light/dark transition tests, locomotor response test, and moving the object test. These observations are monitored via the automated video systems [79].

Zebrafish is being used as a model based on the known genetic cause of epilepsy. For the model to be reliable, it should capture the key features of the disease, and respond to therapies that are commonly used in patients [8]. The SCN1A mutation is the genetic cause of epilepsy and the zebrafish carry 4 sets of voltage gated sodium channel family genes : *scn1Laa* & *scn1Lab, scn4aa* & *scn4ab, scn5Laa* & *scn5Lab*, and *scn8aa* & *scn8ab* [107].

Screening through the optokinetic response assay has identified a homozygous mutant for this gene, which was called the *didy*^{s552} [128]. Such screens are a result of inducing random point mutations, usually loss-of-function and are recessive, by using the chemical mutagen N-ethyl-N-nitrosourea (ENU). Despite being homozygous, *scn1Lab* zebrafish mutants can still be used for the autosomal dominant Dravet syndrome, since there is the presence of an additional Nav1.1homologue (*scn1Laa*). The study on these mutants proved that they have spontaneous seizures, are resistant to drugs and have been used in novel high-throughput screening. This helps in recognizing compounds that would help in improving the epilepsy phenotype. Clemizole is an FDA-approved drug that has emerged from applying antiseizure treatment to the zebrafish model of Dravet syndrome. Several drugs were tested on a large scale, out of which it was seen that clemizole successfully reversed the effects of Dravet syndrome in the zebrafish model [8].

11.4.1.1 Parkinson's Disease

The zebrafish dopaminergic system is a focus of research for Parkinson's disease (PD). Dopaminergic neurons are mainly found in the posterior tuberculum of the ventral diencephalon and project to the ventral telencephalon, although their direct homology with mammalian midbrain neurons is not agreed upon [151]. These neurons are initially seen at 18 hours post-fertilization and are highly organized at 72 hours. These neurons are also seen in the olfactory bulb, preoptic area, retina, and pretectum. Zebrafish have conserved transcription factors and dopaminergic signalling pathways that are similar to those of mammals and thus are a good model system to study PD. Their neurons are extremely sensitive to oxidative stress, an important cause of neuronal loss in PD, further supporting the zebrafish model's validity for studying disease mechanisms and therapeutic interventions [35]. Furthermore, Zebrafish parkin, essential for neuronal function, is similar to human parkin. This is expressed throughout the development, making it an essential model to study dopaminergic neuron function. LRRK2 is another gene that is linked to Parkinson's. Zebrafish also offer a promising model due to having an LRRK2 orthologue with similarity to the human protein, and allow for rapid observation of developmental effects [125]. Recent research suggests partial LRRK2 knockdown leads to dopaminergic neuron reduction and PD-associated gene upregulation, warranting further investigation into the role of LRRK2 in PD pathogenesis [35, 129].

11.4.1.2 Alzheimer's Disease

Zebrafish (*Danio rerio*) has come up as a promising model for studying Alzheimer's disease (AD) due to it having well-characterized neuroanatomical and neurochemical homologies with the human brain. The cholinergic, glutamatergic, and GABAergic neurotransmitter systems that are involved in cognitive processing and behavioural response are all well-characterized in zebrafish, and therefore, scientists can study crucial aspects of AD-associated cognitive impairments such as spatial and associative memory deficit, anxiety, and neurodegeneration. Compared to other traditional mammalian models, zebrafish possess some valuable benefits, including cost-effectiveness, speed in embryonic development, and transparency for in vivo imaging of the brain structure as well as tracking of disease onset in real-time [130]. Furthermore, zebrafish share genes that are homologous to AD (*presenilin 1* (*PS1*) and *presenilin 2* (*PS2*)) and present a similar structure in the brain, and show cognitive dysfunction, behavioural phenotypes compatible with human patient cases. *PS1* and *PS2* have shown a highly conserved alignment with primary human proteins, allowing for *in vivo* analysis [134].

Their fecundity and genetic tractability also render them perfect for efficient large-scale screening of potential therapeutics. Zebrafish are now a valuable model to study neurodegenerative mechanisms and pharmacological testing, thereby addressing rodent model translational issues such as low predictive validity and limited applicability to human clinical outcomes [10].

11.4.1.3 Stroke and Ischemic Injury

Mainly, two types of strokes have been identified: ischemic stroke and haemorrhagic stroke. In ischemic stroke, blood fails to reach the target site, whereas haemorrhagic stroke is characterised by extravasation of blood outside the lumen of a blood vessel, resulting in direct damage to brain parenchyma [147]. There are four anatomical structures of the zebrafish brain—the telencephalon, the diencephalon, the mesencephalon, and the rhombencephalon. An identical neurotransmitter network to humans is also present, which has noradrenergic, serotonergic, dopaminergic, histaminergic, and cholinergic systems [66, 111]. There are several similarities of cerebral structure between zebrafish and humans, like the presence of an integrated central nervous system, along with homologous brain structures, identical synaptic and cellular structures [96]. There also exists a similar blood-brain barrier, which includes anatomical compositions, cell junctions, enzyme expression, and ion transporters, which enables the study of drugs or substances across the BBB. The pathological change of the BBB in stroke can be visualized with immunofluorescent staining *in vivo* [27, 81]. There are several models to induce stroke in zebrafish at both larval and adult stages. The most common is by inducing hypoxia [31]. A study done by Yu et al. showed that there is a 60% death rate in adult animals when kept for 5–6 minutes in a hypoxic chamber. A bilateral cerebral ischemia was observed in those who survived. They also observed erratic swimming behaviours and eye movements, which were quantified as a general severity measure; however, no correlation was found between lesion volume and behavioural outcomes [155]. In another study done by Matsumoto et al. on 4 days post fertilization (dpf) larvae were given a hypoxic environment by incubating them in a chamber with dissolved oxygen of 3.0 mg/L for 2 hours. The study stated that only 10% of animals survive in hypoxic oxygen chambers, and evidence of global oedema and lessened cerebral blood flow was seen [95]. These stroke models are also used to study potential protective therapies such as increased autophagy [154], brain zinc chelation [22, 156], glucosamine [77], and free radical scavengers [95].

Zebrafish represent a highly effective model for studying neurodegenerative diseases such as Parkinson's, Alzheimer's, and stroke due to their genetic homology, conserved brain architecture, and susceptibility to genetic manipulation. Their rapid development, transparency, and cost-efficiency make them ideal for high-throughput screening and mechanistic studies. Zebrafish models bridge critical translational gaps, offering valuable insights into disease pathology and therapeutic evaluation.

11.4.2 Cancer Models

Cancer remains an interesting topic for research. Cancer is considered to be a collection of genetically variable diseases with a certain commonality at a molecular level, which needs to be explored. Zebrafish come into the picture here. Zebrafish models are easy and convenient models for gene manipulation and for studying various cancers. The transparency in the initial stages helps in studying cancer progression with great ease [1]. It has been shown that melanoma cells can be transplanted into zebrafish embryos, which are immunocompetent and possess functional innate immunity or into immunocompromised adult fish, which can be traced over time. It was shown over the years that there is a direct association between transcriptional states identified through scRNA-seq analyses and their biological function, as revealed by transplantation into early zebrafish embryos [9]. In melanoma subpopulations, BRAF/MEK inhibitor resistance was a key factor for those with a stress-like profile and was effective in generating new tumours. Mutations in the BRAF gene disrupt its function of cell growth and lead to the development of cancer cells [141]. Research using zebrafish models has shown that vemurafenib, an FDA-approved drug, can inhibit the activity of mutated BRAF and reduce tumour growth in melanomas caused by BRAF dysfunction, especially for patients where surgery cannot be carried out or in case of metastatic melanoma (2012).

11.4.2.1 Leukaemia

Leukaemia has also been researched employing zebrafish models for deciphering its mechanism and progression. The CG1-tp53−/− zebrafish model was employed to investigate leukaemia, together with sarcoma. Using this model, the investigation concluded that the ablation of TP53 significantly contributes to the aggressive invasiveness of rhabdomyosarcoma (RMS), which is also associated with leukaemia in the model [59]. The spi1: tel-jak2a zebrafish model has been utilized to investigate T-cell acute lymphoid leukaemia (T-ALL) and chronic myeloid leukaemia (CML). The model offered critical insights into the pathogenesis of these diseases, demonstrating how the alternative fusion of TEL-JAK2 creates oncogenic effects in a cell lineage-specific fashion, most probably as a result of variations in downstream signalling [109]. Also, the spi1:MYST3/NOCA2 zebrafish model was employed to study acute lymphoid myeloma (AML), validating the transforming activity of MYST3/NOCA2 as already noted in other organisms [1, 163].

11.4.2.2 Sarcoma

Sarcoma is also a known cancer that can affect connective, supportive, or soft tissues that originate from mesenchyme. Sarcomas are of different types—rhabdomyosarcoma, malignant peripheral nerve sheath tumours, Ewing's sarcoma, chordoma, hemangiosarcoma, and liposarcoma. There are different models, either transgenic or by gene activation that have been developed in zebrafish. Each of the models is employed to get a unique perspective to identify mutations that drive oncogenesis and their novel drug targets. Zebrafish and human soft tissues are conserved, which makes zebrafish a suitable model for sarcomagenesis [56]. Rag2 promoter in zebrafish is seen to be expressed in both lymphoid as well as non-lymphoid cell populations, which also include mesenchymal progenitor cells [63, 73, 74]. In contrast, in the mammalian system, rag2 shows specific expression in lymphoid cells, though promoter activity is observed in non- lymphoid cells [108, 157]. Rhabdomyosarcoma and liposarcoma models were developed using the ectopic expression of rag2 promoters in mesenchymal progenitor cells of zebrafish [54, 73]. Another model of hemangiosarcoma was developed in zebrafish, where germline mutations in the orthologous tumour suppressor gene PTEN were introduced.

11.4.2.3 Melanoma

Melanoma is a lethal malignant growth of melanocytes. Melanocytes are most commonly found in skin, and are derived from the neural crest and produce melanin. Although melanoma accounts for 1% of skin cancer cases, it is responsible for the most fatalities among skin cancer [19, 43]. Melanocytes of zebrafish are derived from the neural crests and are differentiated into large, dendritic, melanized cells. The dermal melanocytes of zebrafish are responsible for their distinct stripe-like pattern. Along with dermal melanocytes, there are melanocytes associated with scales, which also develop from neural crest and are prone to transformation in adult zebrafish melanoma models. Adult melanocyte stem cells (MCS) were recently identified in zebrafish [60]. There is a conserved melanocyte pathway in zebrafish. Several genes are involved in this developmental pathway, like MITF—master regulator, KIT, and MAPK pathway genes, TYR, TYRP1, and PMEL (melanin biosynthesis). Loss of these orthologs can cause phenotypes identical to those found in mammals [24, 71]. There are several models of melanoma in zebrafish, one of which is injecting p53 loss of function mutant p53(lf), which can give rise to tumours that can mirror human melanoma, histopathologically [114]. In another experiment, Dovey and colleagues were able to successfully model NRAS melanomas, where they introduced oncogenic NRAS Q61K into p53(lf) animals [34]. Even though only 19% of human melanomas have mutated p53, most human melanomas do lose p53 pathway function because of other mutations, which also include loss of the CDKN2A locus. The CDKN2A locus is responsible for encoding the ARF tumour suppressor, which regulates p53 [43, 46, 57]. Mouti and colleagues in 2016 were able to model the first uveal melanoma zebrafish model by expressing the GNAQQ209P oncogene under the mitfa promoter in a loss of functional p53, also known as p53 (lf) background. It was found by them that 33% of TG (mitfa: GNAQQ209P); p53 (lf) animals go on to develop uveal melanoma [102].

Zebrafish serve as a dominant and versatile model for studying various cancers due to their genetic tractability, transparent embryonic stages, and conserved molecular pathways. Specific zebrafish models have elucidated mechanisms of leukaemia, sarcoma, and melanoma, highlighting roles of key oncogenes and tumour suppressors such as TP53, TEL-JAK2, MYST3/NOCA2, and NRAS. These models effectively mirror human cancer progression, enabling insights into tumorigenesis, lineage-specific oncogenesis, and therapeutic target identification. Thus, zebrafish provide a valuable *in vivo* platform for cancer research and drug discovery.

11.4.3 Cardiovascular Disorders

The Zebrafish is often used to study cardiovascular diseases, due to the position of the heart on the ventral side. This allows easy understanding of the mechanisms involved in the development and functioning of the heart. The Zebrafish heart is a WO-chambered organ that consists of one atrium and one ventricle. The atrium receives deoxygenated blood, which is pumped to the gills for oxygenation by the ventricle. The heart is composed of an outer epicardium, a middle myocardium with cardiac muscle fibres, and an inner endocardium lining the chambers [97]. The development of the heart is easy to study because they are not initially dependent on circulation during the development phase. This helps understand the disease progression of various cardiac phenotypes [1]. In a study done by Mansuri et al., zebrafish are also used to assess disruptions in cardiovascular development pathways (krit1, fbn2b), dorsoventral axis formation pathway (chrd, smad5 5), and developmental toxicity caused due to phthalate disruptions [88].

The zebrafish models have been used to study several cardiovascular diseases. However, there are not many that have made it to the clinical trials. For studying ACM pathophysiology, a zebrafish model was developed for using the GAL4/UAS transactivation system for the cardiac myocyte-specific expression of the human 2057del2, affecting the desmosomal protein plakoglobin. This mutation is associated with Naxos in humans and leads to cardiac abnormalities, which develop into cardiomyopathy [5]. After screening for bioactive compounds that could be used as potential disease modifiers, SB216763 was identified as an activator for the Wnt signalling, preventing bradycardia and increasing survival. The application of

SB216763 in patient-derived buccal mucosa cells also resulted in similar observations, demonstrating its potential as a therapeutic intervention [38].

11.4.3.1 Atrial Fibrillation

Atrial fibrillation (AF) is a heart disease researched in zebrafish to determine its causes and treatments. One gene, known as kcip1, is associated with AF. It has been shown that kcip1 influences the atrial rate of the heart in zebrafish, and alterations in the expression of the gene can result in increased heart rates, which may be significant for the development of future AF treatments [143]. A second gene, prrx1, is a transcription factor that is also connected to AF. Experiments in zebrafish demonstrated that even small variations in the amount of prrx1 influence the duration of the atrial potential. Researchers also identified an enhancer of prrx1 that is regulated by a genetic variant (SNP), which alters the expression of prrx1. This implies that prrx1 has an important role in inducing AF [1, 144].

11.4.3.2 Congenital Heart Disease (CHD)

Congenital Heart Disease (CHD) is a common human birth defect, potentially caused by genetic mutations. Zebrafish have certain similarities to human electrophysiology, with transparent embryos and larvae for observation, and are well-suited for forward and reverse genetics technology, which creates various economical and easily controlled zebrafish CHD models [153]. Furthermore, the zebrafish heart develops earlier than other models and can also fully regenerate after an injury [18, 49, 132]. The zebrafish heart has six pairs of aortic arch arteries, whereas mammals possess only five pairs. Nonetheless, the formation and pattern of aortic arch arteries in zebrafish closely resemble those in mammals [3, 7, 139]. The conserved mechanisms and molecular similarities add to the effectiveness of the zebrafish as a genetic model for discovering genes related to cardiac diseases [105]. A key transducer of biochemical signals, ILK, is initiated at the plasma membrane by cell-matrix interactions, including the interaction between laminin and integrin, which supports the notion of efficiency of zebrafish as a CHD model. Zebrafish *ilk* mutant hearts have severe myocardial dysfunction resembling human dilated cardiomyopathy (DCM) [14, 72]. Morpholino-mediated knockdown of laminin α4 (lama4) in zebrafish leads to cardiac dysfunction phenotypically similar to that observed in integrin-linked kinase (ilk) mutants, implicating ILK- dependent cell–cell-cell-extracellular matrix (ECM) signalling as a critical regulator of cardiac development and contractile function. Based on these observations, Knoll et al. conducted a genetic screen in patients with severe dilated cardiomyopathy (DCM) and identified two LAMA4 mutations and one mutation in the coding region of ILK [72]. Although a definitive causal relationship between disruptions in ILK/laminin-mediated signalling and DCM remains to be fully elucidated, these findings support the utility of the zebrafish model in uncovering the genetic underpinnings of human cardiomyopathies.

11.4.3.3 Atherosclerosis

Atherosclerosis, characterized by a large amount of lipid deposition in arterial blood vessels, leading to the formation of plaques, is a cause of several cardiovascular diseases. It is known for causing muscular elastic artery stenosis, which results in lumen occlusion or ruptured haemorrhage. The pathological condition of AS is very complicated; hence, animal models are employed to study it [140]. Larger animal models such as non-human primates and pigs are used for this study owing to their similar vascular lesion morphology and lipid metabolism to humans; however, these models have limitations, including high cost, complex experimental procedures, long modelling times, and difficulties in obtaining huge amounts of data [78]. Using models like mice and rabbits has led to inconsistent results due to the highly varying nature of plasma lipid profile between inbred strains. Furthermore, hepatotoxicity was often reported in well-fed rabbits and mice. These drawbacks required a new model to be established. Zebrafish, which are poikilothermic vertebrates favouring lipids as a source of energy, unlike mammals [135]. Even though there are highly differing plasma lipid profiles in zebrafish and humans, the lipid metabolism process is highly conserved. The following genes are involved in lipoprotein and lipid metabolism, such as *apob, apoe, apoa1, ldlr, apoc2, lpl, lcat, and cetp*, and are conserved in zebrafish [2, 39]. It was observed that the Apob protein in zebrafish corresponds to APOB100, and not APOB48, in humans [110]. Zebrafish can be used to monitor the development of AS dynamically using *in vivo* live imaging because of their embryonic transparency and an abundance of fluorescent reporter lines [40, 53]. There are four models developed for atherosclerosis: high-cholesterol diet model, which mimics the early conditions of atherosclerosis characterized by formation of fatty streaks, ldlr Mutant Zebrafish, a model with a knocked out ldr gene, leading to type 2 atherosclerosis after being subjected to HCD, apoc2 Mutant Zebrafish, with a knocked out gene, results in a phenotype resembling human chylomicronaemia and severe hypertriglyceridemia, and lxr Mutant Zebrafish, a mode which has increased LDL levels and develops severe hypercholesterolemia along with hepatic steatosis when fed either a high-cholesterol or high-fat diet [140].

Zebrafish have emerged as a robust and cost-effective model for studying cardiovascular diseases due to their genetic similarity to humans, optical transparency, and ease of manipulation. Their simple yet functional cardiac structure enables real-time observation of heart development and disease progression. Zebrafish models have provided critical insights into atrial fibrillation through genes like kcip1 and prrx1, congenital heart disease via ILK/laminin signalling, and atherosclerosis through conserved lipid metabolism pathways. These models support dynamic, high-throughput investigations into cardiovascular pathology, offering a valuable platform for genetic screening and therapeutic discovery.

11.4.4 Other Disorders

Zebrafish have been employed as a model organism for several other disorders, apart from the ones discussed here. Some other disease models are: osteoporosis, neuromuscular disorders, and liver disorders. Osteoporosis is a metabolic disease related to aging that is characterized by deterioration of bone architecture, low bone mineral density, and the occurrence of fragility fractures [29]. Teleost and mammalian skeletons share most of the conserved bones, genes, and ossification mechanisms in skeletogenesis [55, 62, 122]. There are several genes associated with osteoporosis modelled in zebrafish and have been validated [15, 50, 158]. For example, the csf1ra—the panther mutant zebrafish line lacks the differentiation-promoting myeloid lineage cells, colony-stimulating factor 1 receptor a [26, 112]. As a result, there is an exhibition of low levels of osteoclast activity and osteoporosis in this model. The integrative recent studies used pedigree analysis to identify mutations in high bone mass genes such as CLCN7 and SMAD9 [33]. Neuromuscular disorders (NMD) can often lead to muscle degeneration, loss of muscle function, skeletal muscle wasting, impaired breathing, lack of mobility, cardiomyopathy, and diaphragm abnormalities [70, 94]. Zebrafish are considered to be an appropriate model for Duchenne muscular dystrophy (DMD). The sapje strain of zebrafish, which is a DMD-deficient strain, has a nonsense mutation in exon 4 [12, 16]. Homozygous sapje embryos at 3dpf exhibit degenerative muscle lesions and many other aspects of human DMD pathology, like skeletal muscle fibrosis and inflammation [11, 16]. Alcoholic liver disease (ALD) is a significant health issue that is known to cause morbidity and mortality worldwide [121, 131]. There are similarities in function seen between the zebrafish liver and the human liver, which include secretion of bile, lipid glycogen storage, response to insulin, ammonia metabolism, and serum protein secretion, like complement, coagulation factors, and albumin-like protein [36, 48]. Upon exposure to ethanol in zebrafish larvae once the liver was developed, there were some acute effects, like: behavioural changes, liver enlargement, and steatosis [113]. However, the effect was reversible within 24 hours. Serum protein deficiency is seen in alcoholic patients because of impaired hepatic secretion [32], which is mirrored in zebrafish by activation of the unfolded protein response to ER stress upon exposure to ethanol.

11.5 Zebrafish in Drug Toxicity and Pharmacological Screening

Zebrafish and humans have conserved disease proteins and processes. This makes them an ideal model for rapid in vivo drug screening, since the drugs that are active in humans will often have the same target in Zebrafish. This specificity is especially noticed in drugs that interact with the active region of the target protein, and drugs designed to target cancer pathways [115]. It has been noted in targeted therapies as MEK inhibitors [80], drugs used daily, like acetaminophen [106], and serotonin (5-HT) modulators [51], that the drug has the same pharmacokinetic properties in Zebrafish and humans.

Zebrafish serve as a valuable vertebrate model for rapid *in vivo* drug screening due to their genetic similarity to humans. Their susceptibility to large-scale mutagenesis allows for the identification of novel disease-relevant targets using a forward genetics approach. Additionally, zebrafish provide a system for validating these targets through rapid gene function analysis [25]. Zebrafish model offers added benefits like high-volume compound screening, target identification, target validation, assay development, and drug toxicity study (Refer Fig. 11.4). Disease proteins and processes are similar in humans and Zebrafish; some drugs have similar pathways, especially for targeting cancer pathways [115]. Although some drugs work in humans and not Zebrafish, or vice versa. However, evidence from drug screening done over more than 20 years in zebrafish shows that, on the whole, molecules that are active in zebrafish are similarly active in human as well as mouse systems with similar pharmacokinetic properties [86].

Alongside target drug screening, phenotypic drug screening is gaining importance. This method helps in identifying drug leads even if they act on multiple targets, as is seen in certain kinase inhibitors. These can be further studied by counter treatments [133]. Zebrafish embryos and larvae are ideal due to their small size, transparency, and aquatic environment, and high-throughput screening is feasible in 96-well plates [124]. The small size of Zebrafish embryos allows phenotypic drug screening in a more efficient way, since the embryo can take in the drug directly from the water and through the gills in adult fish. Even though

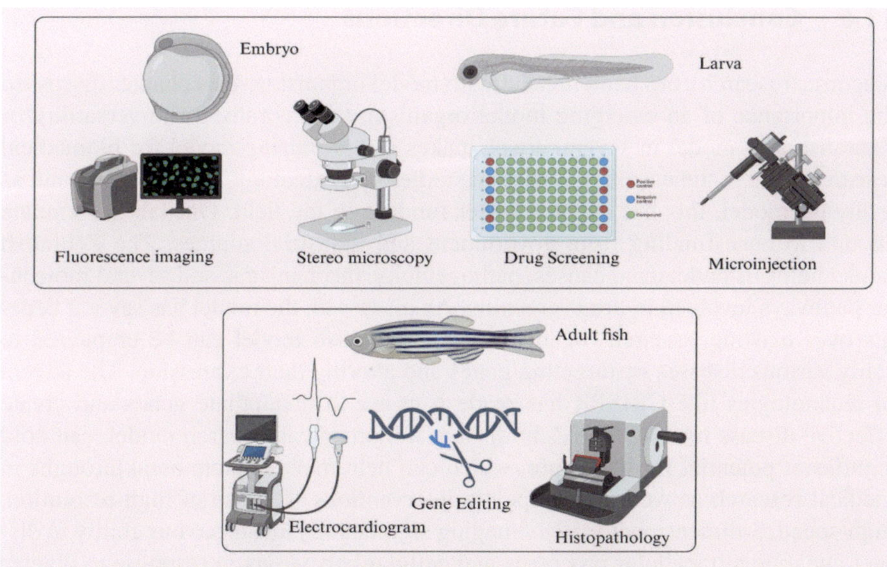

Fig. 11.4 The versatility of Zebrafish across all stages of the life cycle. Several techniques can be employed at different stages to study Zebrafish as a model organism. Although Zebrafish are a popular model, using zebrafish embryos for toxicity and drug testing has some limitations. The protective chorion in the early stages and skin act as barriers, preventing the uptake of compounds with high molecular weights or poor water solubility when compounds are being tested by dissolving them in water, leading to inconsistent results. Environmental factors, such as temperature variations and the lack of adaptive immunity in larvae, further reduce the accuracy and clinical relevance of findings. Alternative drug delivery methods, like gavage or microinjection, are often required but are technically challenging and time-consuming. Overcoming these issues demands a thorough understanding of the drug's properties and careful experimental design. (Image made using BioRender.com)

drug screening is not as high-throughput as in cells, fruit flies, and yeast, Zebrafish surpasses drug screening models in vertebrate systems [115]. Early compound screens have identified phenocopying agents like concentramide [116] and anti-cancer leads targeting tubulin polymerization [100]. Zebrafish responses often mirror human pharmacology's; for instance, atorvastatin, reduces PED6 fluorescence, correlating with disrupted lipid transport [41]. Clotting agents like ADP, thrombin, and collagen produce zebrafish responses similar to mammals. Human antithrombotic blocks these effects, supporting zebrafish predictive validity [61]. Moreover, Zebrafish offer advantages in thrombosis research due to nucleated thrombocytes, unlike anucleate mammalian platelets. This enables functional genomics and the development of thrombocyte-specific cDNA libraries. Transgenic zebrafish with fluorescent thrombocytes streamline isolation for such studies [124]. Zebrafish also facilitate target validation through antisense morpholinos; further knockdown studies, like that of VEGF, demonstrate gene function in angiogenesis, validating potential drug targets [103].

11.6 Conclusion and Future Directions

Scientific research goes hand in hand with model organisms. This chapter discussed the importance of an emerging model organism, the Zebrafish. The versatility of Zebrafish as a model in several areas, makes it a promising model for biomedical research. Due to the use of Zebrafish in studies on toxicology, development, and as a disease model, this has enabled better funding in the field. Our lab is fortunate enough to have funding from government and industrial sources. The Zebrafish model helps us understand causes, pathogenicity, mechanisms, and several molecular pathways involved in drug screening. As discussed, the model has several benefits over existing research organisms. The Zebrafish model can be employed to study various diseases by targeting genes and altering their expression. The advent of technologies like CRISPR has made it easier to manipulate genes and create effective disease models. The Zebrafish, as a biomedical research model, can hold significant potential for the future, which can help make notable breakthroughs in medical research as well as therapeutic interventions. The use of high-resolution, high-speed, 3-dimensional *in vivo* imaging significantly advanced our ability to dissect dynamic intracellular processes and cellular behaviours in response to diverse environmental conditions. This can allow one to predict the physiological conditions at the organism level.

Acknowledgements The authors sincerely thank Gujarat State Biotechnology Mission (GSBTM) (URBASASE222C5/GSBT/23-24_AK_12.26) and Indian Council of Medical Research (ICMR) (URBSASE23B1/ICMR/23-24_AK_02.27) for providing financial support. During the preparation of this work, the author(s) used AI tools like ChatGPT 3.5 to improve the language. After using this tool/service, the author(s) reviewed and edited the content as needed and take(s) full responsibility for the content of the publication.

References

1. Adhish M, Manjubala I. Effectiveness of zebrafish models in understanding human diseases-a review of models. Heliyon. 2023;9:e14557.
2. Anderson JL, Carten JD, Farber SA. Zebrafish lipid metabolism: from mediating early patterning to the metabolism of dietary fat and cholesterol. Methods Cell Biol. 2011;101:111–41.
3. Anderson MJ, Pham VN, Vogel AM, Weinstein BM, Roman BL. Loss of unc45a precipitates arteriovenous shunting in the aortic arches. Dev Biol. 2008;318:258–67.
4. Arnandis T, Monteiro P, Adams SD, Bridgeman VL, Rajeeve V, Gadaleta E, Marzec J, Chelala C, Malanchi I, Cutillas PR, Godinho SA. Oxidative stress in cells with extra centrosomes drives non-cell-autonomous invasion. Dev Cell. 2018;47:409–424.e9.
5. Asimaki A, Kapoor S, Plovie E, Karin Arndt A, Adams E, Liu Z, James CA, Judge DP, Calkins H, Churko J, Wu JC, Macrae CA, Kléber AG, Saffitz JE. Identification of a new modulator of the intercalated disc in a zebrafish model of arrhythmogenic cardiomyopathy. Sci Transl Med. 2014;6:240ra74.
6. Atz JW. Fundulus heteroclitus in the laboratory: a history. Am Zool. 1986;26:111–20.
7. Bamforth SD, Chaudhry B, Bennett M, Wilson R, Mohun TJ, Van Mierop LHS, Henderson DJ, Anderson RH. Clarification of the identity of the mammalian fifth pharyngeal arch artery. Clin Anat. 2013;26:173–82.
8. Baraban SC, Dinday MT, Hortopan GA. Drug screening in Scn1a zebrafish mutant identifies clemizole as a potential Dravet syndrome treatment. Nat Commun. 2013;4:2410.
9. Baron M, Tagore M, Hunter MV, Kim IS, Moncada R, Yan Y, Campbell NR, White RM, Yanai I. The stress-like cancer cell state is a consistent component of tumorigenesis. Cell systems. 2020;11:536–546. e7.
10. Basnet RM, Zizioli D, Taweedet S, Finazzi D, Memo M. Zebrafish larvae as a behavioral model in neuropharmacology. Biomedicine. 2019;7:23.
11. Bassett D, Currie PD. Identification of a zebrafish model of muscular dystrophy. Clin Exp Pharmacol Physiol. 2004;31:537–40.
12. Bassett DI, Bryson-Richardson RJ, Daggett DF, Gautier P, Keenan DG, Currie PD. Dystrophin is required for the formation of stable muscle attachments in the zebrafish embryo. Development. 2003;130:5851–60.
13. Bauer B, Mally A, Liedtke D. Zebrafish embryos and larvae as alternative animal models for toxicity testing. Int J Mol Sci. 2021;22:13417.
14. Bendig G, Grimmler M, Huttner IG, Wessels G, Dahme T, Just S, Trano N, Katus HA, Fishman MC, Rottbauer W. Integrin-linked kinase, a novel component of the cardiac mechanical stretch sensor, controls contractility in the zebrafish heart. Genes Dev. 2006;20:2361–72.
15. Bergen DJM, Kague E, Hammond CL. Zebrafish as an emerging model for osteoporosis: a primary testing platform for screening new osteo-active compounds. Front Endocrinol (Lausanne). 2019;10:6.
16. Berger J, Berger S, Hall TE, Lieschke GJ, Currie PD. Dystrophin-deficient zebrafish feature aspects of the Duchenne muscular dystrophy pathology. Neuromuscul Disord. 2010;20:826–32.
17. Berghmans S, Murphey RD, Wienholds E, Neuberg D, Kutok JL, Fletcher CD, Morris JP, Liu TX, Schulte-Merker S, Kanki JP, Plasterk R, Zon LI, Look AT. tp53 mutant zebrafish develop malignant peripheral nerve sheath tumors. Proc Natl Acad Sci USA. 2005;102:407–12.
18. Bevan L, Lim ZW, Venkatesh B, Riley PR, Martin P, Richardson RJ. Specific macrophage populations promote both cardiac scar deposition and subsequent resolution in adult zebrafish. Cardiovasc Res. 2020;116:1357–71.
19. Bray F, Ferlay J, Soerjomataram I, Siegel RL, Torre LA, Jemal A. Global cancer statistics 2018: GLOBOCAN estimates of incidence and mortality worldwide for 36 cancers in 185 countries. CA Cancer J Clin. 2018;68:394–424.
20. Brown DR, Samsa LA, Qian L, Liu J. Advances in the study of heart development and disease using zebrafish. J Cardiovasc Dev Dis. 2016;3:13.

21. Burke E. Why use zebrafish to study human diseases? [Online]. 2016. Available: https://irp.nih.gov/blog/post/2016/08/why-use-zebrafish-to-study-human-diseases [Accessed].
22. Cappello V, Marchetti L, Parlanti P, Landi S, Tonazzini I, Cecchini M, Piazza V, Gemmi M. Ultrastructural characterization of the lower motor system in a mouse model of Krabbe disease. Sci Rep. 2016;6:1.
23. Cassar S, Adatto I, Freeman JL, Gamse JT, Iturria I, Lawrence C, Muriana A, Peterson RT, van Cruchten S, Zon LI. Use of zebrafish in drug discovery toxicology. Chem Res Toxicol. 2020;33:95–118.
24. Chakrabarti S, Streisinger G, Singer F, Walker C. Frequency of γ-ray induced specific locus and recessive lethal mutations in matur germ cells of the zebrafish, Brachydanio rerio. Genetics. 1983;103:109–23.
25. Chakraborty C, Hsu CH, Wen ZH, Lin CS, Agoramoorthy G. Zebrafish: a complete animal model for in vivo drug discovery and development. Curr Drug Metab. 2009;10:116–24.
26. Chatani M, Takano Y, Kudo A. Osteoclasts in bone modeling, as revealed by in vivo imaging, are essential for organogenesis in fish. Dev Biol. 2011;360:96–109.
27. Chen W, Xie L, Yu F, Li Y, Chen C, Xie W, Huang T, Zhang Y, Zhang S, Li P. Zebrafish as a model for in-depth mechanistic study for stroke. Transl Stroke Res. 2021;12:695–710.
28. Choi T-Y, Choi T-I, Lee Y-R, Choe S-K, Kim C-H. Zebrafish as an animal model for biomedical research. Exp Mol Med. 2021;53:310–7.
29. Compston JE, McClung MR, Leslie WD. Osteoporosis. Lancet. 2019;393:364–76.
30. Corkery DP, Dellaire G, Berman JN. Leukaemia xenotransplantation in zebrafish--chemotherapy response assay in vivo. Br J Haematol. 2011;153:786–9.
31. Crilly S, Emily M, Kasher PR. Zebrafish for modeling stroke and their applicability for drug discovery and development. Expert Opin Drug Discov. 2022;17:559–68.
32. Dara L, Ji C, Kaplowitz N. The contribution of endoplasmic reticulum stress to liver diseases. Hepatology. 2011;53:1752–63.
33. Dietrich K, Fiedler IA, Kurzyukova A, López-Delgado AC, McGowan LM, Geurtzen K, Hammond CL, Busse B, Knopf F. Skeletal biology and disease modeling in zebrafish. J Bone Miner Res. 2021;36:436–58.
34. Dovey M, White RM, Zon LI. Oncogenic NRAS cooperates with p53 loss to generate melanoma in zebrafish. Zebrafish. 2009;6:397–404.
35. Doyle JM, Croll RP. A critical review of zebrafish models of Parkinson's disease. Front Pharmacol. 2022;13:835827.
36. Driessen M, Vitins AP, Pennings JLA, Kienhuis AS, Water BVD, van der Ven LTM. A transcriptomics-based hepatotoxicity comparison between the zebrafish embryo and established human and rodent in vitro and in vivo models using cyclosporine A, amiodarone and acetaminophen. Toxicol Lett. 2015;232:403–12.
37. Eisen JS. Chapter 1 – History of zebrafish research. In: Cartner SC, Eisen JS, Farmer SC, Guillemin KJ, Kent ML, Sanders GE, editors. The zebrafish in biomedical research. Academic Press; 2020.
38. Eldar-Finkelman H, Martinez A. GSK-3 inhibitors: preclinical and clinical focus on CNS. Front Mol Neurosci. 2011;4:32.
39. Fang L, Liu C, Miller YI. Zebrafish models of dyslipidemia: relevance to atherosclerosis and angiogenesis. Transl Res. 2014;163:99–108.
40. Fang L, Miller YI. Emerging applications for zebrafish as a model organism to study oxidative mechanisms and their roles in inflammation and vascular accumulation of oxidized lipids. Free Radic Biol Med. 2012;53:1411–20.
41. Farber SA, Pack M, Ho S-Y, Johnson ID, Wagner DS, Dosch R, Mullins MC, Hendrickson HS, Hendrickson EK, Halpern ME. Genetic analysis of digestive physiology using fluorescent phospholipid reporters. Science. 2001;292:1385–8.
42. Fontana BD, Mezzomo NJ, Kalueff AV, Rosemberg DB. The developing utility of zebrafish models of neurological and neuropsychiatric disorders: a critical review. Exp Neurol. 2018;299:157–71.

43. Frantz WT, Ceol CJ. From tank to treatment: modeling melanoma in zebrafish. Cells. 2020;9:1289.
44. Frétaud M, Rivière L, Job É, Gay S, Lareyre JJ, Joly JS, Affaticati P, Thermes V. High-resolution 3D imaging of whole organ after clearing: taking a new look at the zebrafish testis. Sci Rep. 2017;7:43012. https://doi.org/10.1038/srep43012. PMID: 28211501; PMCID: PMC5314416.
45. Gamble JT, Elson DJ, Greenwood JA, Tanguay RL, Kolluri SK. The zebrafish xenograft models for investigating cancer and cancer therapeutics. Biology (Basel). 2021;10:252.
46. Gembarska A, Luciani F, Fedele C, Russell EA, Dewaele M, Villar S, Zwolinska A, Haupt S, De Lange J, Yip D, Goydos J, Haigh JJ, Haupt Y, Larue L, Jochemsen A, Shi H, Moriceau G, Lo RS, Ghanem G, Shackleton M, Bernal F, Marine J-C. MDM4 is a key therapeutic target in cutaneous melanoma. Nat Med. 2012;18:1239–47.
47. Godoy R, Noble S, Yoon K, Anisman H, Ekker M. Chemogenetic ablation of dopaminergic neurons leads to transient locomotor impairments in zebrafish larvae. J Neurochem. 2015;135:249–60.
48. Goessling W, Sadler KC. Zebrafish: an important tool for liver disease research. Gastroenterology. 2015;149:1361–77.
49. González-Rosa JM. Zebrafish models of cardiac disease: from fortuitous mutants to precision medicine. Circ Res. 2022;130:1803–26.
50. Gregson CL, Bergen DJM, Leo P, Sessions RB, Wheeler L, Hartley A, Youlten S, Croucher PI, Mcinerney-Leo AM, Fraser W, Tang JC, Anderson L, Marshall M, Sergot L, Paternoster L, Davey Smith G, Brown MA, Hammond C, Kemp JP, Tobias JH, Duncan EL. A rare mutation in SMAD9 associated with high bone mass identifies the SMAD-dependent BMP signaling pathway as a potential anabolic target for osteoporosis. J Bone Miner Res. 2020;35:92–105.
51. Griffin A, Hamling KR, Knupp K, Hong S, Lee LP, Baraban SC. Clemizole and modulators of serotonin signaling suppress seizures in Dravet syndrome. Brain. 2017;140:669–83.
52. Guduru A, Manav N, Mansuri A, Gupta I, Bhatia D, Kumar A, Dalvi SV. NIR-active porphyrin-decorated lipid microbubbles for enhanced therapeutic activity enabled by photodynamic effect and ultrasound in 3D tumor models of breast cancer cell line and zebrafish larvae. ACS Appl Bio Mater. 2022;5(9):4270–83.
53. Gut P, Reischauer S, Stainier DYR, Arnaout R. Little fish, big data: zebrafish as a model for cardiovascular and metabolic disease. Physiol Rev. 2017;97:889–938.
54. Gutierrez A, Snyder EL, Marino-Enriquez A, Zhang YX, Sioletic S, Kozakewich E, Grebliunaite R, Ou WB, Sicinska E, Raut CP, Demetri GD, Perez-Atayde AR, Wagner AJ, Fletcher JA, Fletcher CD, Look AT. Aberrant AKT activation drives well-differentiated liposarcoma. Proc Natl Acad Sci USA. 2011;108:16386–91.
55. Hall BK. Bones and cartilage: developmental and evolutionary skeletal biology. Elsevier; 2005.
56. Hayes MN, Langenau DM. Chapter 19 – Discovering novel oncogenic pathways and new therapies using zebrafish models of sarcoma. In: Detrich HW, Westerfield M, Zon LI, editors. Methods in cell biology. Academic Press; 2017.
57. Hodis E, Watson IR, Kryukov GV, Arold ST, Imielinski M, Theurillat J-P, Nickerson E, Auclair D, Li L, Place C, Dicara D, Ramos AH, Lawrence MS, Cibulskis K, Sivachenko A, Voet D, Saksena G, Stransky N, Onofrio RC, Winckler W, Ardlie K, Wagle N, Wargo J, Chong K, Morton DL, Stemke-Hale K, Chen G, Noble M, Meyerson M, Ladbury JE, Davies MA, Gershenwald JE, Wagner SN, Hoon DSB, Schadendorf D, Lander ES, Gabriel SB, Getz G, Garraway LA, Chin L. A landscape of driver mutations in melanoma. Cell. 2012;150:251–63.
58. Huo J, Yu Q, Zhang Y, Liu K, Hsiao CD, Jiang Z, Zhang L. Triptolide-induced hepatotoxicity via apoptosis and autophagy in zebrafish. J Appl Toxicol. 2019;39:1532–40.
59. Ignatius MS, Hayes MN, Moore FE, Tang Q, Garcia SP, Blackburn PR, Baxi K, Wang L, Jin A, Ramakrishnan A, Reeder S, Chen Y, Nielsen GP, Chen EY, Hasserjian RP, Tirode F, Ekker SC, Langenau DM. tp53 deficiency causes a wide tumor spectrum and increases embryonal rhabdomyosarcoma metastasis in zebrafish. Elife. 2018;7:e37202.

60. Iyengar S, Kasheta M, Ceol CJ. Poised regeneration of zebrafish melanocytes involves direct differentiation and concurrent replenishment of tissue-resident progenitor cells. Dev Cell. 2015;33:631–43.
61. Jagadeeswaran P, Sheehan J, Craig F, Troyer D. Identification and characterization of zebrafish thrombocytes. Br J Haematol. 1999;107:731–8.
62. Javidan Y, Schilling TF. Development of cartilage and bone. Methods in cell biology. Academic Press; 2004.
63. Jessen JR, Jessen TN, Vogel SS, Lin S. Concurrent expression of recombination activating genes 1 and 2 in zebrafish olfactory sensory neurons. Genesis. 2001;29:156–62.
64. Jin S-W, Beis D, Mitchell T, Chen J-N, Stainier DYR. Cellular and molecular analyses of vascular tube and lumen formation in zebrafish. Development. 2005;132:5199–209.
65. Ju B, Spitsbergen J, Eden CJ, Taylor MR, Chen W. Co-activation of hedgehog and AKT pathways promote tumorigenesis in zebrafish. Mol Cancer. 2009;8:40.
66. Jurisch-Yaksi N, Yaksi E, Kizil C. Radial glia in the zebrafish brain: functional, structural, and physiological comparison with the mammalian glia. Glia. 2020;68:2451–70.
67. Kansara K, Kumar A, Karakoti AS. Combination of humic acid and clay reduce the ecotoxic effect of TiO2 NPs: a combined physico-chemical and genetic study using zebrafish embryo. Sci Total Environ. 2020;698:134133.
68. Kansara K, Mansuri A, Rajwar A, Vaswani P, Singh R, Kumar A, Bhatia D. Spatiotemporal dynamics of DNA nanocage uptake in zebrafish embryos for targeted tissue bioimaging applications. Nanoscale Adv. 2023;5:2558–64.
69. Kansara K, Paruthi A, Misra SK, Karakoti AS, Kumar A. Montmorillonite clay and humic acid modulate the behavior of copper oxide nanoparticles in aqueous environment and induces developmental defects in zebrafish embryo. Environ Pollut. 2019;255:113313.
70. Karuppasamy M, English KG, Henry CA, Manzini MC, Parant JM, Wright MA, Ruparelia AA, Currie PD, Gupta VA, Dowling JJ, Maves L, Alexander MS. Standardization of zebrafish drug testing parameters for muscle diseases. Dis Model Mech. 2024;17:dmm050339.
71. Kelsh RN, Schmid B, Eisen JS. Genetic analysis of melanophore development in zebrafish embryos. Dev Biol. 2000;225:277–93.
72. Knöll R, Postel R, Wang J, Krätzner R, Hennecke G, Vacaru AM, Vakeel P, Schubert C, Murthy K, Rana BK, Kube D, Knöll G, Schäfer K, Hayashi T, Holm T, Kimura A, Schork N, Toliat MR, Nürnberg P, Schultheiss HP, Schaper W, Schaper J, Bos E, Den Hertog J, Van Eeden FJ, Peters PJ, Hasenfuss G, Chien KR, Bakkers J. Laminin-alpha4 and integrin-linked kinase mutations cause human cardiomyopathy via simultaneous defects in cardiomyocytes and endothelial cells. Circulation. 2007;116:515–25.
73. Langenau DM, Keefe MD, Storer NY, Guyon JR, Kutok JL, Le X, Goessling W, Neuberg DS, Kunkel LM, Zon LI. Effects of RAS on the genesis of embryonal rhabdomyosarcoma. Genes Dev. 2007;21:1382–95.
74. Langenau DM, Traver D, Ferrando AA, Kutok JL, Aster JC, Kanki JP, Lin S, Prochownik E, Trede NS, Zon LI, Look AT. Myc-induced T cell leukemia in transgenic zebrafish. Science. 2003;299:887–90.
75. Leacock SW, Basse AN, Chandler GL, Kirk AM, Rakheja D, Amatruda JF. A zebrafish transgenic model of Ewing's sarcoma reveals conserved mediators of EWS-FLI1 tumorigenesis. Dis Model Mech. 2012;5:95–106.
76. Lee LM, Seftor EA, Bonde G, Cornell RA, Hendrix MJ. The fate of human malignant melanoma cells transplanted into zebrafish embryos: assessment of migration and cell division in the absence of tumor formation. Dev Dyn. 2005;233:1560–70.
77. Lee Y, Lee S, Park JW, Hwang JS, Kim SM, Lyoo IK, Lee CJ, Han IO. Hypoxia-induced neuroinflammation and learning-memory impairments in adult zebrafish are suppressed by glucosamine. Mol Neurobiol. 2018;55:8738–53.
78. Lee YT, Laxton V, Lin HY, Chan YWF, Fitzgerald-Smith S, To TLO, Yan BP, Liu T, Tse G. Animal models of atherosclerosis. Biomed Rep. 2017;6:259–66.
79. Legradi J, El Abdellaoui N, Van Pomeren M, Legler J. Comparability of behavioural assays using zebrafish larvae to assess neurotoxicity. Environ Sci Pollut Res. 2015;22:16277–89.

80. Li D, March ME, Gutierrez-Uzquiza A, Kao C, Seiler C, Pinto E, Matsuoka LS, Battig MR, Bhoj EJ, Wenger TL, Tian L, Robinson N, Wang T, Liu Y, Weinstein BM, Swift M, Jung HM, Kaminski CN, Chiavacci R, Perkins JA, Levine MA, Sleiman PMA, Hicks PJ, Strausbaugh JT, Belasco JB, Dori Y, Hakonarson H. ARAF recurrent mutation causes central conducting lymphatic anomaly treatable with a MEK inhibitor. Nat Med. 2019;25:1116–22.
81. Li Y, Chen T, Miao X, Yi X, Wang X, Zhao H, Lee SM, Zheng Y. Zebrafish: a promising in vivo model for assessing the delivery of natural products, fluorescence dyes and drugs across the blood-brain barrier. Pharmacol Res. 2017;125:246–57.
82. Li Z, Huang X, Zhan H, Zeng Z, Li C, Spitsbergen JM, Meierjohann S, Schartl M, Gong Z. Inducible and repressable oncogene-addicted hepatocellular carcinoma in Tet-on xmrk transgenic zebrafish. J Hepatol. 2012;56:419–25.
83. Li Z, Zheng W, Wang Z, Zeng Z, Zhan H, Li C, Zhou L, Yan C, Spitsbergen JM, Gong Z. A transgenic zebrafish liver tumor model with inducible Myc expression reveals conserved Myc signatures with mammalian liver tumors. Dis Model Mech. 2013;6:414–23.
84. Lieschke GJ, Currie PD. Animal models of human disease: zebrafish swim into view. Nat Rev Genet. 2007;8:353–67.
85. Lu JW, Yang WY, Tsai SM, Lin YM, Chang PH, Chen JR, Wang HD, Wu JL, Jin SL, Yuh CH. Liver-specific expressions of HBx and src in the p53 mutant trigger hepatocarcinogenesis in zebrafish. PLoS One. 2013;8:e76951.
86. Macrae CA, Peterson RT. Zebrafish as tools for drug discovery. Nat Rev Drug Discov. 2015;14:721–31.
87. Malatesta P, Götz M. Radial glia – from boring cables to stem cell stars. Development. 2013;140:483–6.
88. Mansuri A, Kansara K, Raiyani D, Mazmudar D, Kumar A. New insight into long-term effects of phthalates microplastics in developing zebrafish: evidence from genomic alteration and organ development. Environ Toxicol Pharmacol. 2023;99:104087.
89. Mansuri A, Trivedi C, Chokshi S, Jantrania K, Kumar A. Phthalate exposure: prevalence, health effects, regulatory frameworks, and remediation. Chem Res Toxicol. 2025;38:1291–308.
90. Mansuri A, Trivedi C, Kumar A. Impact of virgin and weathered microplastics on zebrafish: bioaccumulation, developmental toxicity and molecular pathway disruptions. Sci Total Environ. 2024a;957:177510.
91. Mansuri A, Trivedi C, Parikh A, Kumar A. Mitigating phthalate toxicity: the protective role of humic acid and clay in zebrafish larvae. Chemosphere. 2024b;354:141756.
92. Marques IJ, Lupi E, Mercader N. Model systems for regeneration: zebrafish. Development. 2019;146:dev167692.
93. Martinez-Ordoñez A, Seoane S, Cabezas P, Eiro N, Sendon-Lago J, Macia M, Garcia-Caballero T, Gonzalez LO, Sanchez L, Vizoso F, Perez-Fernandez R. Breast cancer metastasis to liver and lung is facilitated by Pit-1- CXCL12-CXCR4 axis. Oncogene. 2018;37:1430–44.
94. Mary P, Servais L, Vialle R. Neuromuscular diseases: diagnosis and management. Orthop Traumatol Surg Res. 2018;104:S89–s95.
95. Matsumoto M, Miyamoto M, Sawahata M, Izumi Y, Takada-Takatori Y, Kume T. Establishing a high throughput drug screening system for cerebral ischemia using zebrafish larvae. J Pharmacol Sci. 2021;147:138–42.
96. Menke AL, Spitsbergen JM, Wolterbeek AP, Woutersen RA. Normal anatomy and histology of the adult zebrafish. Toxicol Pathol. 2011a;39:759–75.
97. Menke AL, Spitsbergen JM, Wolterbeek APM, Woutersen RA. Normal anatomy and histology of the adult zebrafish. Toxicol Pathol. 2011b;39:759–75.
98. Meyers JR. Zebrafish: development of a vertebrate model organism. Curr Protoc Essent Lab Tech. 2018;16:e19.
99. Mishra R, Sehring I, Cederlund M, Mulaw M, Weidinger G. NF-κB signaling negatively regulates osteoblast dedifferentiation during zebrafish bone regeneration. Dev Cell. 2020;52:167–182.e7.

100. Moon H-S, Jacobson EM, Khersonsky SM, Luzung MR, Walsh DP, Xiong W, Lee JW, Parikh PB, Lam JC, Kang T-W. A novel microtubule destabilizing entity from orthogonal synthesis of triazine library and zebrafish embryo screening. J Am Chem Soc. 2002;124:11608–9.
101. Moro E, Ozhan-Kizil G, Mongera A, Beis D, Wierzbicki C, Young RM, Bournele D, Domenichini A, Valdivia LE, Lum L, Chen C, Amatruda JF, Tiso N, Weidinger G, Argenton F. In vivo Wnt signaling tracing through a transgenic biosensor fish reveals novel activity domains. Dev Biol. 2012;366:327–40.
102. Mouti MA, Dee C, Coupland SE, Hurlstone AF. Minimal contribution of ERK1/2-MAPK signaling towards the maintenance of oncogenic GNAQQ209P-driven uveal melanomas in zebrafish. Oncotarget. 2016;7:39654–70.
103. Nasevicius A, Larson J, Ekker SC. Distinct requirements for zebrafish angiogenesis revealed by a VEGF-A morphant. Yeast. 2000;17:294–301.
104. Nguyen AT, Emelyanov A, Koh CH, Spitsbergen JM, Parinov S, Gong Z. An inducible kras(V12) transgenic zebrafish model for liver tumorigenesis and chemical drug screening. Dis Model Mech. 2012;5:63–72.
105. Nguyen CT, Lu Q, Wang Y, Chen JN. Zebrafish as a model for cardiovascular development and disease. Drug Discov Today Dis Models. 2008;5:135–40.
106. North TE, Babu IR, Vedder LM, Lord AM, Wishnok JS, Tannenbaum SR, Zon LI, Goessling W. PGE2-regulated wnt signaling and N-acetylcysteine are synergistically hepatoprotective in zebrafish acetaminophen injury. Proc Natl Acad Sci USA. 2010;107:17315–20.
107. Novak AE, Taylor AD, Pineda RH, Lasda EL, Wright MA, Ribera AB. Embryonic and larval expression of zebrafish voltage-gated sodium channel alpha-subunit genes. Dev Dyn. 2006;235:1962–73.
108. Oettinger MA, Schatz DG, Gorka C, Baltimore D. RAG-1 and RAG-2, adjacent genes that synergistically activate V(D)J recombination. Science. 1990;248:1517–23.
109. Onnebo SM, Rasighaemi P, Kumar J, Liongue C, Ward AC. Alternative TEL-JAK2 fusions associated with T-cell acute lymphoblastic leukemia and atypical chronic myelogenous leukemia dissected in zebrafish. Haematologica. 2012;97:1895–903.
110. Otis JP, Zeituni EM, Thierer JH, Anderson JL, Brown AC, Boehm ED, Cerchione DM, Ceasrine AM, Avraham-Davidi I, Tempelhof H, Yaniv K, Farber SA. Zebrafish as a model for apolipoprotein biology: comprehensive expression analysis and a role for ApoA-IV in regulating food intake. Dis Model Mech. 2015;8:295–309.
111. Panula P, Sallinen V, Sundvik M, Kolehmainen J, Torkko V, Tiittula A, Moshnyakov M, Podlasz P. Modulatory neurotransmitter systems and behavior: towards zebrafish models of neurodegenerative diseases. Zebrafish. 2006;3:235–47.
112. Parichy DM, Ransom DG, Paw B, Zon LI, Johnson SL. An orthologue of the kit-related gene fms is required for development of neural crest-derived xanthophores and a subpopulation of adult melanocytes in the zebrafish, Danio rerio. Development. 2000;127:3031–44.
113. Passeri MJ, Cinaroglu A, Gao C, Sadler KC. Hepatic steatosis in response to acute alcohol exposure in zebrafish requires sterol regulatory element binding protein activation. Hepatology. 2009;49:443–52.
114. Patton EE, Widlund HR, Kutok JL, Kopani KR, Amatruda JF, Murphey RD, Berghmans S, Mayhall EA, Traver D, Fletcher CDM, Aster JC, Granter SR, Look AT, Lee C, Fisher DE, Zon LI. BRAF mutations are sufficient to promote nevi formation and cooperate with p53 in the genesis of melanoma. Curr Biol. 2005;15:249–54.
115. Patton EE, Zon LI, Langenau DM. Zebrafish disease models in drug discovery: from preclinical modelling to clinical trials. Nat Rev Drug Discov. 2021;20:611–28.
116. Peterson RT, Mably JD, Chen JN, Fishman MC. Convergence of distinct pathways to heart patterning revealed by the small molecule concentramide and the mutation heart-and-soul. Curr Biol. 2001;11:1481–91.
117. Poureetezadi SJ, Wingert RA. Congenital and acute kidney disease: translational research insights from zebrafish chemical genetics. Gen Med (Los Angel). 2013;1:112.

118. Pruvot B, Jacquel A, Droin N, Auberger P, Bouscary D, Tamburini J, Muller M, Fontenay M, Chluba J, Solary E. Leukemic cell xenograft in zebrafish embryo for investigating drug efficacy. Haematologica. 2011;96:612–6.
119. Ramel MC, Hill CS. The ventral to dorsal BMP activity gradient in the early zebrafish embryo is determined by graded expression of BMP ligands. Dev Biol. 2013;378:170–82.
120. Rebelo De Almeida C, Mendes RV, Pezzarossa A, Gago J, Carvalho C, Alves A, Nunes V, Brito MJ, Cardoso MJ, Ribeiro J, Cardoso F, Ferreira MG, Fior R. Zebrafish xenografts as a fast screening platform for bevacizumab cancer therapy. Commun Biol. 2020;3:299.
121. Rehm J, Gmel GE Sr, Gmel G, Hasan OSM, Imtiaz S, Popova S, Probst C, Roerecke M, Room R, Samokhvalov AV, Shield KD, Shuper PA. The relationship between different dimensions of alcohol use and the burden of disease-an update. Addiction. 2017;112:968–1001.
122. Rosa JT, Laizé V, Gavaia PJ, Cancela ML. Fish models of induced osteoporosis. Front Cell Dev Biol. 2021;9:672424.
123. Roy T, Bhat A. Learning and memory in juvenile zebrafish: what makes the difference—population or rearing environment? Ethology. 2016;122:308–18.
124. Rubinstein AL. Zebrafish: from disease modeling to drug discovery. Curr Opin Drug Discov Devel. 2003;6:218–23.
125. Sager JJ, Bai Q, Burton EA. Transgenic zebrafish models of neurodegenerative diseases. Brain Struct Funct. 2010;214:285–302.
126. Santoriello C, Deflorian G, Pezzimenti F, Kawakami K, Lanfrancone L, D'adda Di Fagagna F, Mione M. Expression of H-RASV12 in a zebrafish model of Costello syndrome causes cellular senescence in adult proliferating cells. Dis Model Mech. 2009;2:56–67.
127. Santoro MM. Zebrafish as a model to explore cell metabolism. Trends Endocrinol Metab. 2014;25:546–54.
128. Schoonheim PJ, Arrenberg AB, Del Bene F, Baier H. Optogenetic localization and genetic perturbation of saccade-generating neurons in zebrafish. J Neurosci. 2010;30:7111–20.
129. Sheng D, Qu D, Kwok KHH, Ng SS, Lim AYM, Aw SS, Lee CWH, Sung WK, Tan EK, Lufkin T. Deletion of the WD40 domain of LRRK2 in zebrafish causes Parkinsonism-like loss of neurons and locomotive defect. PLoS Genet. 2010;6:e1000914.
130. Shenoy A, Banerjee M, Upadhya A, Bagwe-Parab S, Kaur G. The brilliance of the zebrafish model: perception on behavior and Alzheimer's disease. Front Behav Neurosci. 2022;16:861155.
131. Shimizu N, Shiraishi H, Hanada T. Zebrafish as a useful model system for human liver disease. Cells. 2023;12:2246.
132. Simões FC, Cahill TJ, Kenyon A, Gavriouchkina D, Vieira JM, Sun X, Pezzolla D, Ravaud C, Masmanian E, Weinberger M, Mayes S, Lemieux ME, Barnette DN, Gunadasa-Rohling M, Williams RM, Greaves DR, Trinh LA, Fraser SE, Dallas SL, Choudhury RP, Sauka-Spengler T, Riley PR. Macrophages directly contribute collagen to scar formation during zebrafish heart regeneration and mouse heart repair. Nat Commun. 2020;11:600.
133. Sonoshita M, Scopton AP, Ung PMU, Murray MA, Silber L, Maldonado AY, Real A, Schlessinger A, Cagan RL, Dar AC. A whole-animal platform to advance a clinical kinase inhibitor into new disease space. Nat Chem Biol. 2018;14:291–8.
134. Stewart AM, Kalueff AV. The developing utility of zebrafish models for cognitive enhancers research. Curr Neuropharmacol. 2012;10:263–71.
135. Stoletov K, Fang L, Choi SH, Hartvigsen K, Hansen LF, Hall C, Pattison J, Juliano J, Miller ER, Almazan F, Crosier P, Witztum JL, Klemke RL, Miller YI. Vascular lipid accumulation, lipoprotein oxidation, and macrophage lipid uptake in hypercholesterolemic zebrafish. Circ Res. 2009;104:952–60.
136. Storer NY, Zon LI. Zebrafish models of p53 functions. Cold Spring Harb Perspect Biol. 2010 Aug;2(8):a001123. https://doi.org/10.1101/cshperspect.a001123. Epub 2010 May 5. PMID: 20679337; PMCID: PMC2908773.
137. Storer NY, White RM, Uong A, Price E, Nielsen GP, Langenau DM, Zon LI. Zebrafish rhabdomyosarcoma reflects the developmental stage of oncogene expression during myogenesis. Development. 2013;140:3040–50.

138. Strähle U, Scholz S, Geisler R, Greiner P, Hollert H, Rastegar S, Schumacher A, Selderslaghs I, Weiss C, Witters H, Braunbeck T. Zebrafish embryos as an alternative to animal experiments—a commentary on the definition of the onset of protected life stages in animal welfare regulations. Reprod Toxicol. 2012;33:128–32.
139. Sun Y, Kumar SR, Wong CED, Tian Z, Bai H, Crump JG, Bajpai R, Lien CL. Craniofacial and cardiac defects in chd7 zebrafish mutants mimic CHARGE syndrome. Front Cell Dev Biol. 2022;10:1030587.
140. Tang D, Geng F, Yu C, Zhang R. Recent application of zebrafish models in atherosclerosis research. Front Cell Dev Biol. 2021;9:643697.
141. Travnickova J, Patton EE. Deciphering melanoma cell states and plasticity with zebrafish models. J Invest Dermatol. 2021;141:1389–94.
142. Trinkaus JP. Some contributions of research on early teleost embryogenesis to general problems of development. In: Marthy H-J, editor. Experimental embryology in aquatic plants and animals. Boston, MA: Springer US; 1990.
143. Tsai C-T, Hsieh C-S, Chang S-N, Chuang EY, Ueng K-C, Tsai C-F, Lin T-H, Wu C-K, Lee J-K, Lin L-Y, Wang Y-C, Yu C-C, Lai L-P, Tseng C-D, Hwang J-J, Chiang F-T, Lin J-L. Genome-wide screening identifies a KCNIP1 copy number variant as a genetic predictor for atrial fibrillation. Nat Commun. 2016;7:10190.
144. Tucker NR, Dolmatova EV, Lin H, Cooper RR, Ye J, Hucker WJ, Jameson HS, Parsons VA, Weng L-C, Mills RW, Sinner MF, Imakaev M, Leyton-Mange J, Vlahakes G, Benjamin EJ, Lunetta KL, Lubitz SA, Mirny L, Milan DJ, Ellinor PT. Diminished PRRX1 expression is associated with increased risk of atrial fibrillation and shortening of the cardiac action potential. Circ Cardiovasc Genet. 2017;10:e001902.
145. Varga ZK, Zsigmond Á, Pejtsik D, Varga M, Demeter K, Mikics É, Haller J, Aliczki M. The swimming plus-maze test: a novel high-throughput model for assessment of anxiety-related behaviour in larval and juvenile zebrafish (Danio rerio). Sci Rep. 2018;8:16590.
146. Veinotte CJ, Dellaire G, Berman JN. Hooking the big one: the potential of zebrafish xenotransplantation to reform cancer drug screening in the genomic era. Dis Model Mech. 2014;7:745–54.
147. Walcott BP, Peterson RT. Zebrafish models of cerebrovascular disease. J Cereb Blood Flow Metab. 2014;34:571–7.
148. Westermann L, Neubauer B, Köttgen M. Nobel prize 2020 in chemistry honors CRISPR: a tool for rewriting the code of life. Pflugers Arch - Eur J Physiol. 2021;473:1–2.
149. White RM, Sessa A, Burke C, Bowman T, Leblanc J, Ceol C, Bourque C, Dovey M, Goessling W, Burns CE, Zon LI. Transparent adult zebrafish as a tool for in vivo transplantation analysis. Cell Stem Cell. 2008;2:183–9.
150. Wittbrodt J, Shima A, Schartl M. Medaka--a model organism from the far East. Nat Rev Genet. 2002;3:53–64.
151. Xi Y, Noble S, Ekker M. Modeling neurodegeneration in zebrafish. Curr Neurol Neurosci Rep. 2011;11:274–82.
152. Yan J, Takamiya M, Zhang D, Pace G, Rastegar S, Wang H, Schoch S, Köberle B, Hartwig A, Dickmeis T, Weiss C. Zebrafish as a model organism in one health toxicology: impact of solvents and exposure routes on the toxicity of platinum anticancer drugs. Environ Int. 2025;197:109349.
153. Yang D, Jian Z, Tang C, Chen Z, Zhou Z, Zheng L, Peng X. Zebrafish congenital heart disease models: opportunities and challenges. Int J Mol Sci. 2024;25:5943.
154. Yang Z, Lin P, Chen B, Zhang X, Xiao W, Wu S, Huang C, Feng D, Zhang W, Zhang J. Autophagy alleviates hypoxia-induced blood-brain barrier injury via regulation of CLDN5 (claudin 5). Autophagy. 2021;17:3048–67.
155. Yu X, Li YV. Zebrafish as an alternative model for hypoxic-ischemic brain damage. Int J Physiol Pathophysiol Pharmacol. 2011;3:88–96.
156. Yu X, Li YV. Neuroprotective effect of zinc chelator DEDTC in a zebrafish (Danio rerio) model of hypoxic brain injury. Zebrafish. 2013;10:30–5.

157. Zarrin AA, Fong I, Malkin L, Marsden PA, Berinstein NL. Cloning and characterization of the human recombination activating gene 1 (RAG1) and RAG2 promoter regions. J Immunol. 1997;159:4382–94.
158. Zhang Y, Ji D, Li L, Yang S, Zhang H, Duan X. ClC-7 regulates the pattern and early development of craniofacial bone and tooth. Theranostics. 2019;9:1387–400.
159. Zhang X-H, Zhou C-C, Li C-Y, Hua Y, Li K, Wei P, He M-F. Isoliquiritin exert protective effect on telencephalon infarction injury by regulating multi-pathways in zebrafish model of ischemic stroke. Phytomedicine. 2021;83:153469.
160. Zhao S, Huang J, Ye J. A fresh look at zebrafish from the perspective of cancer research. J Exp Clin Cancer Res. 2015;34:80.
161. Zheng W, Li Z, Nguyen AT, Li C, Emelyanov A, Gong Z. Xmrk, kras and myc transgenic zebrafish liver cancer models share molecular signatures with subsets of human hepatocellular carcinoma. PLoS One. 2014;9:e91179.
162. Zhu S, Lee JS, Guo F, Shin J, Perez-Atayde AR, Kutok JL, Rodig SJ, Neuberg DS, Helman D, Feng H, Stewart RA, Wang W, George RE, Kanki JP, Look AT. Activated ALK collaborates with MYCN in neuroblastoma pathogenesis. Cancer Cell. 2012;21:362–73.
163. Zhuravleva J, Paggetti J, Martin L, Hammann A, Solary E, Bastie JN, Delva L. MOZ/TIF2-induced acute myeloid leukaemia in transgenic fish. Br J Haematol. 2008;143:378–82.

12 Zebrafish in Toxicology

Ayelen M. Sosa, M. Jimena Prieto, and Carolina S. Martinez

12.1 Introduction

Toxicity evaluation is crucial for identifying the potential adverse effects of a substance on human health and the environment. These tests can help recognize the relationship between dose and undesired effects as well as the mechanism of toxic action. Toxicity can be evaluated in vitro using cell lines or in vivo in whole organisms, the latter offering a more reliable result since real cell-cell and cell-extracellular matrix interactions guide complex cellular decisions that condition a biological response.

Traditional animal models for toxicological assays have been mammals and higher vertebrates; nevertheless, the rising amount of animal experimentation and the conscientiousness of utilizing living beings demanded consideration about its necessity, feasibility, ethics, and potential harm caused. The formulation of the 3R-principle (replacement, reduction, and refinement) for animal experimentation aimed at minimizing harm and distress to animals while ensuring scientific validity [64], and has led to the adoption of alternative animal models. Among these alternatives are the well-known zebrafish (*Danio rerio*) (Fig. 12.1). Zebrafish exhibit toxicant responses comparable to those observed in other aquatic species, while also sharing organ systems and metabolic pathways with mammals and displaying a high level of homology to the human genome [33, 61]. In addition, due to the simple

A. M. Sosa · M. J. Prieto · C. S. Martinez (✉)
Universidad Nacional de Quilmes, Departamento de Ciencia y Tecnología, Laboratorio de Bio-Nanotecnología (LBN), Bernal, Buenos Aires, Argentina

Grupo de Biología Estructural y Biotecnología, Instituto Multidisciplinario de Biología Celular, Consejo Nacional de Investigaciones Científicas y Técnicas, CIC-PBA, UNLP, La Plata, Buenos Aires, Argentina
e-mail: carolina.martinez@unq.edu.ar

P. Kumar, V. Kothari (eds.), *Model Organisms in Biological Research*,
https://doi.org/10.1007/978-981-92-0038-2_12

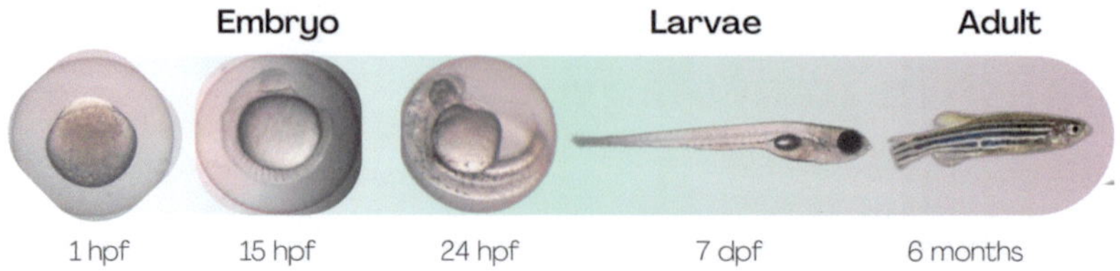

Fig. 12.1 Life cycle of zebrafish. Hpf hours post-fertilization, dpf days post-fertilization

natural environment, it is easy to keep zebrafish conditions to minimize housing stress, which is known to impact the outcome of experiments.

The zebrafish is a distinguished model for toxicity evaluation, offering the possibility to detect from molecular initiating events to physiological and behavioral alterations. Their natural advantages, such as small size, prolific fertilization, and fast growth, enable economic and high-throughput testing of short and long-term effects. Notably, genetic manipulation is considerably simpler in this model, offering various transgenic models to determine toxicity in different organs [9]. Zebrafish have made breakthrough discoveries in toxicology research, particularly in the development and functionality of various organs, underscoring the nervous system [47]. This model is also valuable for understanding the developmental toxicity (birth defects, growth abnormalities, or functional deficits) [28, 29] and the role of the microbiome in toxicity (influence on the absorption, metabolism, and excretion of toxicants) [87]. This chapter presents a summary of the applications of the zebrafish as an animal model in the field of toxicology, focused on different organs, which is very important to predict impacts on health and the environment.

12.2 Early Stages of Zebrafish for Toxicity Assessment

While adult zebrafish are used in research, embryos and larvae are excellent candidates for toxicity assays. Zebrafish embryos develop externally, so the evolution is strikingly well described [38]. By exposing organisms to the target substance, developmental toxicity can be evaluated in a manner that aligns with animal welfare principles [54]. The zebrafish embryo maintains transparency during early ontogeny, with the major organ systems fully developed by 72 hours post-fertilization (hpf). This characteristic allows fast, real-time evaluation with no invasive techniques [27]. Each female can produce around 200 to 300 eggs per week, allowing large-scale studies [27].

Interestingly, zebrafish have a sequenced genome [33], and their gene editing is easy [75], making transgenic lines a suitable in vivo model for "–omics" level evaluations to understand the responses and mechanisms of action of the toxicants [31]. Currently, numerous zebrafish strains are available in the Zebrafish Information Network database (https://zfin.org), including mutants, knockdowns, and transgenics. Toxicological studies reported so far have used transgenic lines expressing various reporter proteins, including EGFP, dsRED, mCherry, KAEDE, mCitrine,

Luciferase, Ceruleus, YFP, and GCaMP, to target specific tissues, cells, or signaling pathways. This chapter will list the main strains used to assess each tissue-specific toxicity.

12.3 Toxicological Research in Zebrafish Embryos and Larvae

The zebrafish embryo and larval stages have been successfully utilized to investigate the toxicological profiles of numerous compounds and to delineate their adverse effects. Acute toxicity evaluations typically include assessment of altered growth trajectories, morphological deformities, systemic functional impairments, and elevated lethality. Organ-directed toxicity is also widely studied; nevertheless, accurate interpretation requires careful consideration of substance identity, administration route, concentration, and exposure timing. Importantly, appropriate control groups must be incorporated to ensure that organ-specific effects are not confounded by developmental toxicity. In this section, we provide a critical overview of the literature addressing zebrafish as a model for organ toxicity testing, with a particular focus on cardiac, hepatic, neurological, and immunological endpoints, which correspond to the most frequently encountered toxicities in human clinical research [72]. Notably, zebrafish are integrated into the internationally accepted standards of the Organization for Economic Co-operation and Development (OECD). The OECD Guidelines for assessing systemic toxicity in fish [57] permit experimentation without regulation until zebrafish are free-swimming and independent feeders, corresponding to 5 days post-fertilization (dpf), when they are considered less prone to pain perception, distress, and suffering [17]. These regulations thus allow toxicological studies in zebrafish at early developmental stages but limit investigations to developmental and acute toxic effects in the countries that have adopted these directives.

Cardiotoxicity This toxicology affects the heart's functionality or cardiac tissues, leading to various types of cardiopathies [15]. The zebrafish heart has two chambers, while humans have four; nevertheless, both species share similar electrophysiological features and a high degree of genetic orthology [43]. In zebrafish, the first organ to develop is the heart during hematopoiesis, and between 22 and 36 hpf it is already fully developed and beating [88]. Most genes involved in the human dilated cardiac pathway have homologous equivalents in zebrafish. Notably, the ERG gene (ether-a-go-go-related gene) is 99% conserved in humans and is expressed during the early stages of embryonic development [88]. The transparency of embryos and larvae allows for the identification of cardiac anomalies, such as the presence of edema, using basic laboratory equipment like a microscope or a magnifying glass (Fig. 12.2). Additionally, at the molecular level, there are toxicity marker genes such as SLC28A3, RARG, and UGT1A6 [77, 84].

Several transgenic zebrafish models have been designed to study cardiac conditions (Table 12.1). Chang et al. [10] utilized the transgenic zebrafish nacre/fli1:EGFP to evaluate the toxicological effects of copper oxide nanoparticles. Additionally, Liu

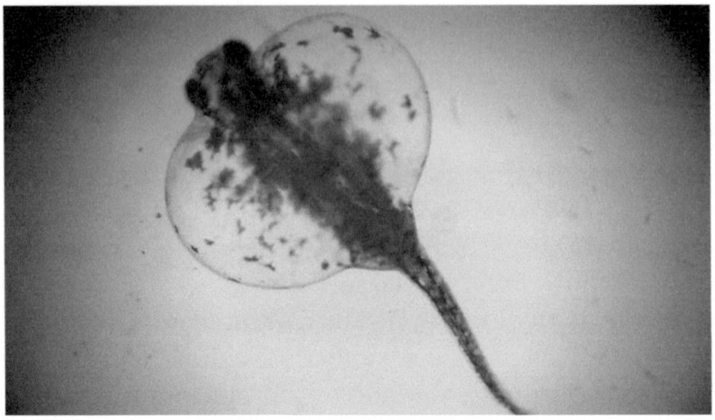

Fig. 12.2 Pericardial edema in zebrafish larvae

Table 12.1 Selected transgenic zebrafish models for investigating cardiotoxicity and cardiomyopathies

Transgenic (Tg)	Characteristics	Bibliography
grl:EGFP	Showed expression in the endothelium of the aorta during embryogenesis	She et al. [68]
BAC(nppa:mCitrine)	A transgenic zebrafish line carrying an nppa reporter will be used to identify these border zone cardiomyocytes	Honkoop et al. [30]
myl7:GFF	Heart-specific mCitrine expression observed pan-myocardial RFP expression	Chávez et al. [11]
ubi: Loxp-stop-Loxp-hmga1a-eGFP	It enables the study of cardiomyocyte proliferation and cardiac regeneration through the hMga1a gene	Brouwman et al. [7]
kdrl:mCherry	Mark endothelial cells with GFP to monitor vascular development	Facchinello et al. [18]

et al. [49] used the cmlc2:eGFP strain, a cardiomyocyte fluorescence-labeled transgenic fish line that enhances heart visualization and enables the detection of beats or abnormalities resulting from incubation with cardiotoxic compounds. Huang et al. [34] studied the p1:d2GFP;myl7:mCherry strain and identified perturbation of early cardiogenesis through disruption of the spatially restricted expression of endocardial Notch signaling. Gu et al. [24] analyzed the number of total cardiomyocytes, endocardial cells, and the atrium and ventricle size in myl7:dsRED2-nuc and fli1a:nGFP. Other useful transgenic lines recently reported are flk1:EGFP and myl7:EGFP [36, 59].

Hepatotoxicity The zebrafish liver is organized into a conserved structure consisting of three lobes, which lack the pedicle separating the mammalian liver's distinct lobes [20]. Despite interspecies differences, zebrafish hepatocytes, biliary epithelial cells, and endothelial cells display morphology and functional properties comparable to those found in humans [40, 70]. While the composition, function, signaling, cellular mechanisms, and cellular response underlying liver diseases are similar between humans and fish, variations in the microenvironment may lead to species-

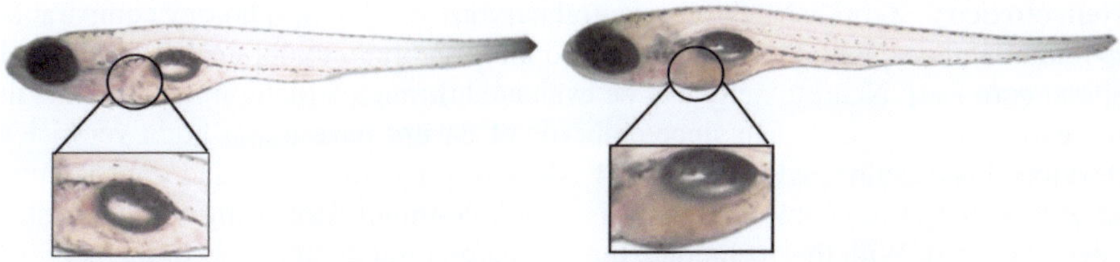

Fig. 12.3 Visualization of a normal and a necrotic liver [73]

Table 12.2 Examples of zebrafish transgenic models for the study of hepatotoxicity

Transgenic (Tg)	Characteristics	Bibliography
lfabp:dsRED	Liver fatty acid binding protein, study the differentiation of the hepatopancreatic ductal system	Yu et al. [83]
fabp10a:DsRed	Reporter line, were used in naphthalene exposure tests	Chen et al. [13]
elastase:eGFP	Study the differentiation, proliferation, and morphogenesis of exocrine cells.	Hong et al. [28], [29]
EPV.Tp1-Mmu. Hbb:EGFP	Marker intrahepatic biliary cells.	Choe et al. [14]

specific responses to injury [86]. During embryonic zebrafish development, the liver begins to form at 28 hpf, with organogenesis completed by 72 hpf, resulting in a functional liver. Full maturation is achieved by 120 hpf. At this stage, the zebrafish liver performs the same functions as the human liver, including lipid, vitamin, and carbohydrate processing, protein synthesis, growth factor production, and gene expression regulation [76].

To analyze the toxicological effects on the liver, various experiments range from anatomical studies to metabolic analyses of the organ, including the assessment of essential genes such as the human CYP gene family. The anatomical study of hepatic tissue can be performed through macroscopic observation, as evidenced by the presence of opacity in the organ, which may result from alterations in hepatic tissue or cellular death (Fig. 12.3). Histological sectioning can also be carried out to examine the damage occurring at the cellular level in detail. In zebrafish, 94 CYP genes have been identified in its genome, with a high degree of conservation between human and zebrafish sequences [76]. As a filtration organ for both exogenous and endogenous compounds, exposure to hepatotoxic substances directly affects liver cells, triggering the activation of the immune system through damage-associated molecular patterns and the subsequent secretion of cytokines and chemokines. Its direct connection with the immune system highlights the liver's critical role in studying hepatotoxic compounds [46, 80]. In addition, the transgenic lines (Table 12.2) are valuable tools to assess hepatotoxicity biomarkers such as mean RFP/GFP intensity in the liver, liver size, and shape [13, 16, 28, 29].

Neurotoxicity Zebrafish exhibit a central nervous system morphology comparable to that of humans, with the brain organized into forebrain, midbrain, hindbrain, and spinal cord [69]. Neurotoxicity can be evaluated through structural impairments in the early stages of zebrafish since embryos of 24 hpf have major brain ventricles developed and delineated visually [63]. Moreover, the small size and transparency of zebrafish larvae enable the analysis of whole-mount fish using immunohistochemistry [65]. With this technique, the main structural components of the nervous system can be observed, encompassing primary and secondary motor neurons, nicotinic acetylcholine receptors, presynaptic vesicles, and muscle fibers [5, 45, 63, 79]. In addition, in situ hybridizations can be applied to genes with region-specific expression in the forebrain, midbrain, hindbrain, or eye-like pax6, otx2, and ephA4 [63].

Neurotoxicity can also be evaluated through behavioral tests, as it correlates with the nervous system's functionality (Fig. 12.4). The advantage of testing behavior is that a wide variety of assays can be done, and depending on the development stage, the information obtained. The application of embryonic and larval zebrafish behavior for toxicity assessment has been validated through exposure to compounds with established mechanisms of action and documented behavioral effects in rodents [52]. Commercial zebrafish behavioral systems have allowed for easier, high-throughput analysis by integrating infrared cameras, contained testing arenas, programmable stimuli, and dedicated analysis software [19]. The first movements of embryos are spontaneous contractions around 17 hpf [78], and from 21 hpf, a response to touch can be detected by a rapid tail coiling [74]. At 3 dpf, the hatched embryo shows an escaping swim burst [37]. Embryos of 30 hpf do not yet see, but they become responsive to light and display a characteristic "photomotor response" upon exposure to light pulses [41]. At 4 dpf, larvae show a spontaneous swimming activity characterized by the tail beating followed by the glide action [37]; from then on, the time the zebrafish spends in motion increases progressively. Free-swimming larvae are regularly used in locomotor testing to study swimming speed, time,

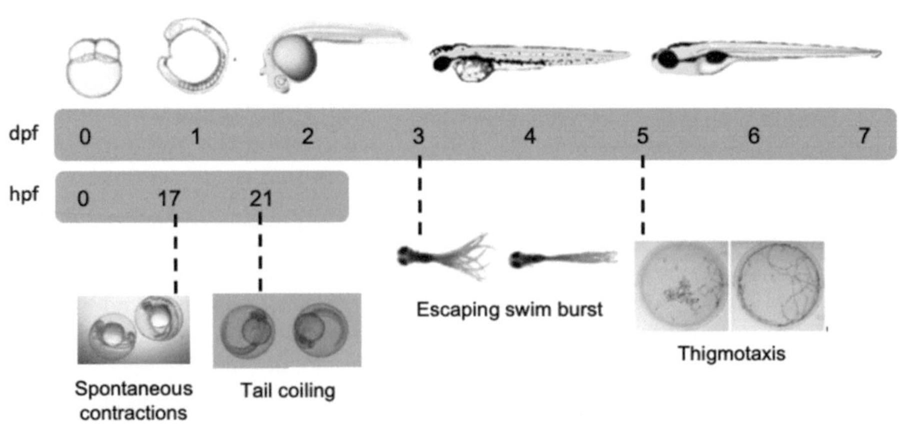

Fig. 12.4 Timeline of zebrafish development and acquired skills over time

Table 12.3 Transgenic zebrafish models employed in neurotoxicity research

Transgenic (Tg)	Characteristics	Bibliography
Oligo2:DsRED	Expresses red fluorescence along olig2-positive cells, including motor neurons and oligodendrocytes, throughout development	An et al. [4]
tg(HUC/D:GFP	These cells express GFP in developing neurons, allowing for their visualization during neurogenesis	Breuer et al. [8]
slc6a3:EGFP	Markers expressing GFP-tagged dopamine transporters were used to study their dynamics during neuronal activity	Siddiqui et al. [71]
mnx1:GFP	Motor neurons were tagged with green fluorescent protein (GFP)	Lee et al. [44]
nkx2.2a:mEGFP	A membrane-anchored enhanced green fluorescent protein (eGFP) is expressed, labeling a specific subset of cells within the oligodendrocyte lineage	Kirby et al. [39]
nkx2.2:eGFP-CAAX	Transgenic line expressing a myelin-specific reporter	Almeida et al. [2]

distance, and pattern [50]. From 5 dpf, when larvae have functioning eyes, an interesting parameter that can be tested is thigmotaxis, a measure of anxiety-like behavior in zebrafish. In this test, the fish's position within the well is tracked as they typically swim around the boundaries, which is attributed to their initial exploration and active avoidance of the central, unprotected region [66]. At this stage and up to 7 dpf, a behavioral variability in zebrafish can be reduced using startling stimuli, including sudden changes in light, noise, or vibration. The resulting startle response is an innate, conserved mechanism that allows swift reactions to perceived dangers [22]. This response typically involves an abrupt turn and rapid swimming away from the eliciting stimulus [58], so deviations in turn dynamics, acceleration, or response latency are commonly used to assess neurotoxicity. Mainly, zebrafish strains designed for neurotoxicity studies are presented in Table 12.3.

Immunotoxicity The immune system is divided into innate and adaptive immunity. In zebrafish, the innate immune system develops on the first day post-fertilization, primarily comprising macrophages and neutrophils [85]. The adaptive immune system fully matures between the fourth and sixth weeks of life, encompassing both T-cells and B-cells [51]. Some components of zebrafish's immunity have been characterized, such as pattern recognition receptors (PRRs), including Toll-like receptors (TLRs), which can detect evolutionarily conserved pathogen-associated molecular patterns present in pathogens. TLRs detect distinct patterns associated with bacteria or xenobiotics and initiate signaling pathways that enhance the expression of cytokines, including tumor necrosis factor-alpha (TNF-α) and interleukin-1β [81]. Conversely, oxidative stress and inflammation are believed to influence this process. The excessive production of reactive oxygen species (ROS) and proinflammatory cytokines results in significant damage to biomacromolecules [60]. An immunoassay used to determine the generation of reactive oxygen species (ROS) is through fluorescence measurements using dihydroethidium (DHE)

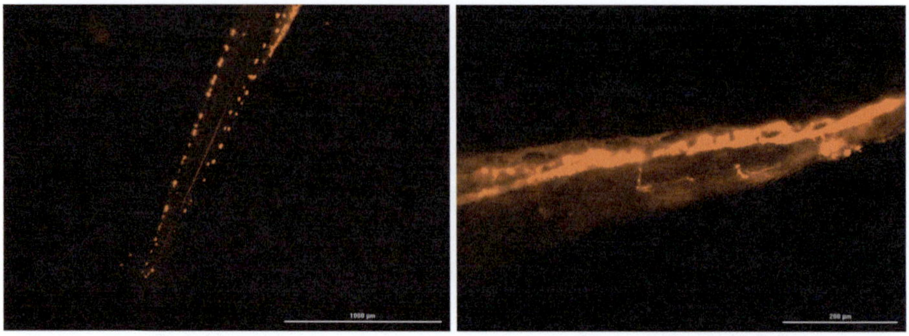

Fig. 12.5 Fluorescence detection of reactive oxygen species (ROS) using dihydroethidium (DHE)

(Fig. 12.5). In untreated control larvae, no orange fluorescent dots were observed in the tail region. In contrast, larvae exposed to hydrogen peroxide exhibited intense orange dots, indicating ROS generation induced by the treatment. The high homology with mammals and the transparency of early zebrafish embryos allow for the convenient use of transgenic lines in immunotoxicity studies. An example is the study by Lu et al. [51], which utilized the transgenic model Tg (rag2:DsRed) that expresses red fluorescent protein in the thymus, allowing for the observation of thymocytes using a confocal or fluorescence microscope. This methodology is used to quantify immune cells, as their numbers may decrease due to an underlying condition or the immunotoxic effects of a specific compound. Other methods include the analysis of gene expression, associated with ROS as an immune response by PCR. Some of these genes include TNF-α, interleukin-1β, and interleukin-6 [12]. Tail fin amputation to observe neutrophil migration through fluorescence is another technique used for these types of assessments [55].

Other Organs' Toxicity The literature about toxicological studies in zebrafish is widely extended, and other tissues not be covered in this chapter are ototoxicity, pacreotoxicity, intestinal toxicity, dermatotoxicity, ophthalmotoxicity, nephrotoxicity, chondrotoxicity, osteotoxicity, systemic toxicity and endocrine disruption, vasotoxiciy, and hematotoxicity.

12.4 Emerging Areas in Toxicology

Environmental Toxicity The environment is continuously exposed to hazardous substances that contribute to pollution. Environmental pollutants and their residues are now commonly found in aquatic environments, where they accumulate in organisms and pose significant risks to ecosystems and human health [82]. During early life stages, organisms are more vulnerable to environmental toxins. As a result, the developmental toxicity of zebrafish embryos is widely studied in environmental pollution research [67]. Since toxicological endpoints in embryos can be assessed through disrupted or abnormal development, monitoring early embryogenesis is a standard method for toxicity screening.

The toxicity of hazardous substances can be evaluated by observing phenotypic changes in zebrafish after a period of exposure. These include developmental abnormalities (such as altered body length and delayed yolk sac absorption), malformations (including spinal curvature, pericardial edema, and liver degeneration), and behavioral irregularities [32, 42]. The development of living color reporter genes, such as the green fluorescent protein (GFP) and other fluorescent proteins, has enabled the creation of transgenic fish that exhibit visible color changes in response to contaminants [21]. This enables real-time monitoring without the need for complex instrumentation, providing a practical and efficient on-site system for detecting water pollution [53].

Developmental Origins of Health and Disease The concept of Developmental Origins of Health and Disease (DOHaD) links early-life developmental stages—including intrauterine growth, lactation, and adolescence—to long-term metabolic health and disease [3]. These critical periods are particularly susceptible to environmental influences, such as exposure to contaminants. This hypothesis proposes that exposure to environmental stressors, including toxicants, during development may result in genetic, epigenetic, or functional modifications within tissues, ultimately increasing disease risk in later life [23, 26].

Zebrafish offer a robust experimental model for studying DOHaD due to its numerous advantages, including the speed of experiments and its high homology with humans. Typically, groups of 48 to 96 zebrafish embryos are exposed to varying chemicals or concentrations, and evolution is monitored through lethality, teratogenic effects, or other phenotypic alterations [31]. The use of zebrafish in chemical screening is well-established, with acute toxicity assays providing greater sensitivity in identifying dangerous chemicals. According to this, the OECD and the Environmental Protection Agency (EPA) have established standard guidelines for toxicity studies involving zebrafish, known as the Zebrafish Embryonic Toxicity Test (ZFET) TG 236 [57], and Fish, Short-term Toxicity Test TG 212 [6, 56].

Recent research has focused on micro and nanoplastics' effects on the environment and human health [89]. Alarmingly, they have been detected in reproductive organs, the placenta, and breast milk, impacting reproductive health [1, 48, 49]. Numerous studies have utilized the zebrafish model to explore the impact of these nanomaterials on health and the environment, these studies suggest that microplastics not only bioaccumulate in the organism but also cause toxicological effects, ranging from generalized and organ-specific necrosis to immunoregulatory responses aimed at minimizing the impact of exposure [25, 35, 62].

12.5 Conclusions

Zebrafish have the potential to become a cornerstone in toxicological analysis. The zebrafish serves as a versatile model, bridging the gap between in vitro and mammalian rodent models of toxicity. The advantages of the zebrafish ensure its continued use in developmental toxicity assays for screening the toxicity of chemicals or drugs. Furthermore, advancements in high-throughput technology will expand the scale, reproducibility, and range of endpoints in these studies. However, one limitation is the need for standardized protocols and robust data quality control measures to ensure reliable and reproducible results across laboratories, thereby strengthening confidence in the model. Additionally, incorporating advanced technologies such as high-throughput screening and sophisticated genetic analysis can significantly enhance efficiency and provide deeper insights into the effects of toxins. This model could also be instrumental in predicting the toxicological impacts of newly identified compounds and facilitating the development of more targeted assessments. Throughout the chapter, it has been emphasized that zebrafish are a valuable model for toxicity studies due to their numerous advantages; however, they have limitations, including a simpler anatomy compared to mammals, the lack of lungs, potential differences in immune system responses, and a lack of maternal-embryo interaction during gestation.

In summary, the zebrafish is a compact yet highly resilient model organism, offering a resource-efficient alternative that bridges the gap between in vitro testing and mammalian experimentation. Its numerous advantages facilitate the seamless integration of new techniques, technologies, and research areas. Given the model has limitations, the use of organoids in conjunction with zebrafish would be an interesting strategy for future research.

References

1. Adjama I, Dave H, Balarabe BY, Masiyambiri V, Marycleopha. Microplastics in dairy products and human breast milk: contamination status and greenness analysis of available analytical methods. J Hazard Mater Lett. 2024;7:100120.
2. Almeida AR, Macklin WB. Early myelination involves the dynamic and repetitive ensheathment of axons which resolves through a low and consistent stabilization rate. elife. 2023;12:e82111.
3. Almeida DL, Pavanello A, Saavedra LP, Pereira TS, de Castro-Prado MA, de Freitas Mathias PC. Environmental monitoring and the developmental origins of health and disease. J Dev Orig Health Dis. 2019;10(6):608–15.
4. An G, Hong T, Park H, Lim W, Song G. Oxamyl exerts developmental toxic effects in zebrafish by disrupting the mitochondrial electron transport chain and modulating PI3K/Akt and p38 Mapk signaling. Sci Total Environ. 2023;859:160458.
5. Babin PJ, Goizet C, Raldúa D. Zebrafish models of human motor neuron diseases: advantages and limitations. Prog Neurobiol. 2014;118:36–58.
6. Bhagat J, Zang L, Nishimura N, Shimada Y. Zebrafish: an emerging model to study microplastic and nanoplastic toxicity. Sci Total Environ. 2020;728:138707.
7. Bouwman M, de Bakker DE, Honkoop H, Giovou AE, Versteeg D, Boender AR, Nguyen PD, Slotboom M, Colquhoun D, Vigil-Garcia M, Kooijman L. Cross-species comparison reveals

that Hmga1 reduces H3K27me3 levels to promote cardiomyocyte proliferation and cardiac regeneration. Nat Cardiovasc Res. 2025:1–9.

8. Breuer M, Guglielmi L, Zielonka M, Hemberger V, Kölker S, Okun JG, Hoffmann GF, Carl M, Sauer SW, Opladen T. QDPR homologues in Danio rerio regulate melanin synthesis, early gliogenesis, and glutamine homeostasis. PLoS One. 2019;14(4):e0215162.
9. Brito RS, Canedo A, Farias D, Rocha TL. Transgenic zebrafish (Danio rerio) as an emerging model system in ecotoxicology and toxicology: historical review, recent advances, and trends. Sci Total Environ. 2022;848:157665.
10. Chang J, Ichihara G, Shimada Y, Tada-Oikawa S, Kuroyanagi J, Zhang B, Suzuki Y, Sehsah R, Kato M, Tanaka T, Ichihara S. Copper oxide nanoparticles reduce vasculogenesis in transgenic zebrafish through down-regulation of vascular endothelial growth factor expression and induction of apoptosis. J Nanosci Nanotechnol. 2015;15(3):2140–7.
11. Chávez MN, Arora P, Meer M, Marques IJ, Ernst A, Castro RA, Mercader N. Spns1-dependent endocardial lysosomal function drives valve morphogenesis through Notch1-signaling. Iscience. 2024;27(12)
12. Chen M, Yin J, Liang Y, Yuan S, Wang F, Song M, Wang H. Oxidative stress and immunotoxicity induced by graphene oxide in zebrafish. Aquat Toxicol. 2016;174:54–60.
13. Chen H, Sheng L, Gong Z, Ru S, Bian H. Investigation of the molecular mechanisms of hepatic injury upon naphthalene exposure in zebrafish (Danio rerio). Ecotoxicology. 2018;27:650–60.
14. Choe CP, Choi SY, Kee Y, Kim MJ, Kim SH, Lee Y, Park HC, Ro H. Transgenic fluorescent zebrafish lines that have revolutionized biomedical research. Lab Anim Res. 2021;37:1–29.
15. Coppola A, Lombari P, Mazzella E, Capolongo G, Simeoni M, Perna AF, Ingrosso D, Borriello M. Zebrafish as a model of cardiac pathology and toxicity: spotlight on uremic toxins. Int J Mol Sci. 2023;24(6):5656.
16. Dey KK, Kamila S, Das T, Chattopadhyay A. Lead induced genotoxicity and hepatotoxicity in zebrafish (Danio rerio) at environmentally relevant concentration: Nrf2-Keap1 regulated stress response and expression of biomarker genes. Environ Toxicol Pharmacol. 2024;107:104396.
17. EFSA. Emergency Food Security Assessment Handbook. World Food Programme (WFP), Emergency Needs Assessment Branch (ODAN); 2005.
18. Facchinello N, Astone M, Audano M, Oberkersch RE, Spizzotin M, Calura E, Marques M, Crisan M, Mitro N, Santoro MM. Oxidative pentose phosphate pathway controls vascular mural cell coverage by regulating extracellular matrix composition. Nat Metab. 2022;4(1):123–40.
19. Fitzgerald JA, Könemann S, Krümpelmann L, Županič A, Vom Berg C. Approaches to test the neurotoxicity of environmental contaminants in the zebrafish model: from behavior to molecular mechanisms. Environ Toxicol Chem. 2021;40(4):989–1006.
20. Goessling W, Sadler KC. Zebrafish: an important tool for liver disease research. Gastroenterology. 2015;149(6):1361–77.
21. Gong Z, Wan H, Tay TL, Wang H, Chen M, Yan T. Development of transgenic fish for ornamental and bioreactor by strong expression of fluorescent proteins in the skeletal muscle. Biochem Biophys Res Commun. 2003;308(1):58–63.
22. Götz T, Janik VM. Repeated elicitation of the acoustic startle reflex leads to sensitisation in subsequent avoidance behaviour and induces fear conditioning. BMC Neurosci. 2011;12:1–3.
23. Goyal D, Limesand SW, Goyal R. Epigenetic responses and the developmental origins of health and disease. J Endocrinol. 2019;242(1):T105–19.
24. Gu J, Wang H, Zhou L, Fan D, Shi L, Ji G, Gu A. Oxidative stress in bisphenol AF-induced cardiotoxicity in zebrafish and the protective role of N-acetyl N-cysteine. Sci Total Environ. 2020;731:139190.
25. Guimarães AT, Charlie-Silva I, Malafaia G. Toxic effects of naturally-aged microplastics on zebrafish juveniles: a more realistic approach to plastic pollution in freshwater ecosystems. J Hazard Mater. 2021;407:124833.
26. Heindel JJ, Vandenberg LN. Developmental origins of health and disease: a paradigm for understanding disease cause and prevention. Curr Opin Pediatr. 2015;27(2):248–53.
27. Hill AJ, Teraoka H, Heideman W, Peterson RE. Zebrafish as a model vertebrate for investigating chemical toxicity. Toxicol Sci. 2005;86(1):6–19.

28. Hong T, Park J, An G, Song J, Song G, Lim W. Evaluation of organ developmental toxicity of environmental toxicants using zebrafish embryos. Mol Cells. 2024a;47:100144.
29. Hong T, Park J, Park H, An G, Lee H, Song G, Lim W. Exposure to acifluorfen induces developmental toxicity in the early life stage of zebrafish. Comp Biochem Physiol C Toxicol Pharmacol. 2024b;281:109909.
30. Honkoop H, de Bakker DE, Aharonov A, Kruse F, Shakked A, Nguyen PD, de Heus C, Garric L, Muraro MJ, Shoffner A, Tessadori F. Single-cell analysis uncovers that metabolic reprogramming by ErbB2 signaling is essential for cardiomyocyte proliferation in the regenerating heart. elife. 2019;8:e50163.
31. Horzmann KA, Freeman JL. Making waves: new developments in toxicology with the zebrafish. Toxicol Sci. 2018;163(1):5–12.
32. Hou Y, Liu X, Qin Y, Hou Y, Hou J, Wu Q, Xu W. Zebrafish as model organisms for toxicological evaluations in the field of food science. Compr Rev Food Sci Food Saf. 2023;22(5):3481–505.
33. Howe K, Clark MD, Torroja CF, Torrance J, Berthelot C, Muffato M, Collins JE, Humphray S, McLaren K, Matthews L, McLaren S. The zebrafish reference genome sequence and its relationship to the human genome. Nature. 2013;496(7446):498–503.
34. Huang M, Zhu F, Jiao J, Wang J, Zhang Y. Exposure to acrylamide disrupts cardiomyocyte interactions during ventricular morphogenesis in zebrafish embryos. Sci Total Environ. 2019;656:1337–45.
35. Huang W, Mo J, Li J, Wu K. Exploring developmental toxicity of microplastics and nanoplastics (MNPS): insights from investigations using zebrafish embryos. Sci Total Environ. 2024;933:173012.
36. Jiang X, Wang J, Liu J, Zhu H, Hu J, Sun X, Zhou W. Resveratrol ameliorates penconazole-induced cardiotoxicity by inhibition of oxidative stress and apoptosis in zebrafish larvae. Ecotoxicol Environ Saf. 2023;256:114865.
37. Kalueff AV, Gebhardt M, Stewart AM, Cachat JM, Brimmer M, Chawla JS, Craddock C, Kyzar EJ, Roth A, Landsman S, Gaikwad S. Towards a comprehensive catalog of zebrafish behavior 1.0 and beyond. Zebrafish. 2013;10(1):70–86.
38. Kimmel CB, Ballard WW, Kimmel SR, Ullmann B, Schilling TF. Stages of embryonic development of the zebrafish. Dev Dyn. 1995;203(3):253–310.
39. Kirby BB, Takada N, Latimer AJ, Shin J, Carney TJ, Kelsh RN, Appel B. In vivo time-lapse imaging shows dynamic oligodendrocyte progenitor behavior during zebrafish development. Nat Neurosci. 2006;9(12):1506–11.
40. Kita S, Shimomura I. Extracellular vesicles as an endocrine mechanism connecting distant cells. Mol Cells. 2022;45(11):771–80.
41. Kokel D, Dunn TW, Ahrens MB, Alshut R, Cheung CY, Saint-Amant L, Bruni G, Mateus R, van Ham TJ, Shiraki T, Fukada Y. Identification of nonvisual photomotor response cells in the vertebrate hindbrain. J Neurosci. 2013;33(9):3834–43.
42. Lai KP, Gong Z, Tse WK. Zebrafish as the toxicant screening model: transgenic and omics approaches. Aquat Toxicol. 2021;234:105813.
43. Lane S, More LA, Asnani A. Zebrafish models of cancer therapy-induced cardiovascular toxicity. J Cardiovasc Dev Dis. 2021;8(2):8.
44. Lee BC, Tsai JC, Huang YH, Wang CC, Lee HC, Tsai HJ. The 419th aspartic acid of neural membrane protein enolase 2 is a key residue involved in the axonal growth of motor neurons mediated by interaction between enolase 2 receptor and extracellular Pgk1 ligand. Int J Mol Sci. 2024;25(19):10753.
45. Li D, Lu C, Wang J, Hu W, Cao Z, Sun D, Ma X. Developmental mechanisms of arsenite toxicity in zebrafish (Danio rerio) embryos. Aquat Toxicol. 2009;91(3):229–237.
46. Li ZQ, Xu Y, Zhao X, Jia Z, Wang ZY, Feng D, Li CB, Ma ZQ, Zhao CJ, Lin RC. Study on liver diseases with zebrafish as an important tool. Zhongguo Zhong yao za zhi= Zhongguo Zhongyao Zazhi= China Journal of Chinese Materia Medica. 2021;46(2):320–32.
47. Li H, Li A, Bian X. Towards zebrafish model applications in drug discovery targeting central nervous system diseases and neurotoxicity assessment. Eur J Pharmacol. 2025;1002:177776.

48. Liu S, Guo J, Liu X, Yang R, Wang H, Sun Y, Chen B, Dong R. Detection of various microplastics in placentas, meconium, infant feces, breastmilk and infant formula: a pilot prospective study. Sci Total Environ. 2023a;854:158699.
49. Liu C, Wang Y, Zeng Y, Kang Z, Zhao H, Qi K, Wu H, Zhao L, Wang Y. Use of deep-learning assisted assessment of cardiac parameters in zebrafish to discover cyanidin chloride as a novel keap1 inhibitor against doxorubicin-induced cardiotoxicity. Adv Sci. 2023b;10(30):2301136.
50. Lu ZG, Li MH, Wang JS, Wei DD, Liu QW, Kong LY. Developmental toxicity and neurotoxicity of two matrine-type alkaloids, matrine and sophocarpine, in zebrafish (Danio rerio) embryos/larvae. Reprod Toxicol. 2014;47:33–41.
51. Lu W, Yang F, Meng Y, An J, Hu B, Jian S, Yang G, Lu H, Wen C. Immunotoxicity and transcriptome analysis of zebrafish embryos exposure to Nitazoxanide. Fish Shellfish Immunol. 2023;141:108977.
52. MacPhail RC, Hunter DL, Irons TD, Padilla S. Locomotion and behavioral toxicity in larval zebrafish: background, methods, and data. Zebrafish Methods Assess Drug Saf Tox. 2011:151–64.
53. MacRae CA, Peterson RT. Zebrafish as a mainstream model for in vivo systems pharmacology and toxicology. Annu Rev Pharmacol Toxicol. 2023;63(1):43–64.
54. Martinez CS, Feas DA, Siri M, Igartúa DE, Chiaramoni NS, Alonso SD, Prieto MJ. In vivo study of teratogenic and anticonvulsant effects of antiepileptics drugs in zebrafish embryo and larvae. Neurotoxicol Teratol. 2018;66:17–24.
55. Martinez CS, Igartúa DE, Czarnowski I, Feas DA, Prieto MJ. Biological response and developmental toxicity of zebrafish embryo and larvae exposed to multi-walled carbon nanotubes with different dimension. Heliyon. 2019;5(8)
56. OECD Test No. 212. Fish, Short-term Toxicity Test on Embryo and Sac-Fry Stages. OECD Guidelines for the Testing of Chemicals; 1998.
57. OECD Test No. 236. Fish embryo acute toxicity (FET) Test. OECD Guidelines for the Testing of Chemicals. Section 2; 2013.
58. Orger MB, Gahtan E, Muto A, Page-McCaw P, Smear MC, Baier H. Behavioral screening assays in zebrafish. In: Methods in cell biology, vol. 77. Academic Press; 2004. p. 53–68.
59. Park J, An G, Park H, Hong T, Lim W, Song G. Developmental defects induced by thiabendazole are mediated via apoptosis, oxidative stress and alteration in PI3K/Akt and MAPK pathways in zebrafish. Environ Int. 2023;176:107973.
60. Peng Y, Li M, Huang Y, Cheng B, Cao Z, Liao X, Xiong G, Liu F, Hu C, Lu H. Bifenazate induces developmental and immunotoxicity in zebrafish. Chemosphere. 2021;271:129457.
61. Prieto MJ, del Rio Zabala NE, Marotta CH, Carreño Gutierrez H, Arevalo Arevalo R, Chiaramoni NS, Alonso SDV. Optimization and in vivo toxicity evaluation of G4. 5 PAMAM dendrimer-risperidone complexes. PloS one. 2014;9(2):e90393.
62. Rainieri S, Conlledo N, Larsen BK, Granby K, Barranco A. Combined effects of microplastics and chemical contaminants on the organ toxicity of zebrafish (Danio rerio). Environ Res. 2018;162:135–43.
63. Roy NM, Carneiro B, Ochs J. Glyphosate induces neurotoxicity in zebrafish. Environ Toxicol Pharmacol. 2016;42:45–54.
64. Russell WM, Burch RL, Hume CW. The principles of humane experimental technique. London: Methuen; 1959.
65. Schacht V, Kern JS. Basics of immunohistochemistry. J Invest Dermatol. 2015;135(3):1–4.
66. Schnörr SJ, Steenbergen PJ, Richardson MK, Champagne D. Measuring thigmotaxis in larval zebrafish. Behav Brain Res. 2012;228(2):367–74.
67. Shao X, Xiao D, Yang Z, Jiang L, Li Y, Wang Y, Ding Y. Frontier of toxicology studies in zebrafish model. J Appl Toxicol. 2024;44(4):488–500.
68. She P, Zhang H, Peng X, Sun J, Gao B, Zhou Y, Zhu X, Hu X, Lai KS, Wong J, Zhou B. The Gridlock transcriptional repressor impedes vertebrate heart regeneration by restricting expression of lysine methyltransferase. Development. 2020;147(18):dev190678.
69. Shenoy A, Banerjee M, Upadhya A, Bagwe-Parab S, Kaur G. The brilliance of the zebrafish model: perception on behavior and Alzheimer's disease. Front Neurosci. 2022;16:861155.

70. Shimizu N, Shiraishi H, Hanada T. Zebrafish as a useful model system for human liver disease. Cells. 2023;12(18):2246.
71. Siddiqui S, Liu F, Kanthasamy AG, McGrail M. Stat3 mediates Fyn kinase-driven dopaminergic neurodegeneration and microglia activation. Dis Model Mech. 2024;17(12):dmm052011.
72. Siramshetty VB, Nickel J, Omieczynski C, Gohlke BO, Drwal MN, Preissner R. WITHDRAWN—a resource for withdrawn and discontinued drugs. Nucleic Acids Res. 2016;44(D1):D1080–6.
73. Sosa AM, Berín BE, Cottet C, Prieto MJ, Martinez CS. Dermatological nanotechnology: gelatin films with O/W emulsions for skin lesion repair. Eur J Pharm Biopharm. 2025;207:114602.
74. Sztal TE, Ruparelia AA, Williams C, Bryson-Richardson RJ. Using touch-evoked response and locomotion assays to assess muscle performance and function in zebrafish. J Visual Exp. 2016;116:54431.
75. Varshney GK, Sood R, Burgess SM. Understanding and editing the zebrafish genome. Adv Genet. 2015;92:1–52.
76. Vliegenthart AB, Tucker CS, Del Pozo J, Dear JW. Zebrafish as model organisms for studying drug-induced liver injury. Br J Clin Pharmacol. 2014;78(6):1217–27.
77. Wei Y, Meng Y, Huang Y, Liu Z, Zhong K, Ma J, Zhang W, Li Y, Lu H. Development toxicity and cardiotoxicity in zebrafish from exposure to iprodione. Chemosphere. 2021;263:127860.
78. Weichert FG, Floeter C, Artmann AS, Kammann U. Assessing the ecotoxicity of potentially neurotoxic substances–evaluation of a behavioural parameter in the embryogenesis of Danio rerio. Chemosphere. 2017;186:43–50.
79. Welsh L, Tanguay RL, Svoboda KR. Uncoupling nicotine mediated motoneuron axonal pathfinding errors and muscle degeneration in zebrafish. Toxicol Appl Pharmacol. 2009;237(1):29–40.
80. Wilkins BJ, Pack M. Zebrafish models of human liver development and disease. Compr Physiol. 2013;3(3):1213.
81. Xiong G, Deng Y, Li J, Cao Z, Liao X, Liu Y, Lu H. Immunotoxicity and transcriptome analysis of zebrafish embryos in response to glufosinate-ammonium exposure. Chemosphere. 2019;236:124423.
82. Yan J, Takamiya M, Zhang D, Pace G, Rastegar S, Wang H, Schoch S, Köberle B, Hartwig A, Dickmeis T, Weiss C. Zebrafish as a model organism in one health toxicology: impact of solvents and exposure routes on the toxicity of platinum anticancer drugs. Environ Int. 2025;197:109349.
83. Yu HC, Bai QR, Guo JJ, Chen MY, Wang L, Tong FC, Zhang SL, Wu J. Elucidating hydroxysafflor yellow A's multi-target mechanisms against alcoholic liver disease through integrative pharmacology. Phytomedicine. 2024;134:155956.
84. Zakaria ZZ, Benslimane FM, Nasrallah GK, Shurbaji S, Younes NN, Mraiche F, Da'as SI, Yalcin HC. Using zebrafish for investigating the molecular mechanisms of drug-induced cardiotoxicity. Biomed Res Int. 2018;2018:1642684.
85. Zeng S, Peng Y, Ma J, Ge Y, Huang Y, Xie S, Yuan W, Lu C, Zhang H, Luo Q, Liao X. Hematopoietic stem cell and immunotoxicity in zebrafish embryos induced by exposure to Metalaxyl-M. Sci Total Environ. 2022;809:152102.
86. Zhang J, Wang K, Guo J, Huang Y, Wei Y, Jia K, Peng Y, Lu H. Study on the mechanism of liver toxicity induced by acenaphthene in zebrafish. Ecotoxicol Environ Saf. 2023;249:114441.
87. Zhong X, Li J, Lu F, Zhang J, Guo L. Application of zebrafish in the study of the gut microbiome. Anim Models Exp Med. 2022;5(4):323–36.
88. Zhu JJ, Xu YQ, He JH, Yu HP, Huang CJ, Gao JM, Dong QX, Xuan YX, Li CQ. Human cardiotoxic drugs delivered by soaking and microinjection induce cardiovascular toxicity in zebrafish. J Appl Toxicol. 2014;34(2):139–48.
89. Zuri G, Karanasiou A, Lacorte S. Human biomonitoring of microplastics and health implications: a review. Environ Res. 2023;237:116966.

Tardigrades as an Emerging Model Animal for Astrobiology Research: A Case Study

13

Soumyadeep Chakraborty, Subhasree Ray, and Soumya Pandit

13.1 Introduction

Astrobiology is a multi-faceted field that seeks to understand the universe's origins, evolution, distribution, and future of life. At the core of astrobiology is the search for extraterrestrial life, and the consideration of the conditions that may support it [17]. An important aspect of astrobiology research is the discovery of life forms that can withstand the extreme environmental conditions that could be encountered in outer space. As the search for life on planets and moons away from Earth grows more advanced, the study of Earth's resilient organisms has gained more importance. An understanding of how life can manage under an extreme set of conditions ranks high in considerations like our evaluation of how likely life will be found beyond Earth and in the understanding of what adaptations life forms need to be able to survive under those conditions. One of the organisms that ranks very highly among these is the tardigrade, tiny microscopic creatures belonging to the Phylum Tardigrada, Class Eutardigrada, Order Parachela, Family Macrobiotidae, Genus *Macrobiotus*, and encompassing various species within this genus. Tardigrades, also known as "water bears," are small, stout, segmented animals (0.3–0.5 mm long) with four pairs of short legs that end in tiny claws to assist with their slow, bear-like crawling. Usually translucent or whitish, but may be pigmented depending on their diet, tardigrades have an uncomplicated body plan with no distinct head; rather, they have a small mouth that is equipped with stylets to allow feeding. Their slow-moving body and claw-like appendages easily identify tardigrades under a microscope [32]. Tardigrades are found in virtually every habitable environment, ranging from the deepest parts of the ocean, to the highest peaks of the mountains, to the moist environments in leaf litter or moss [23].

S. Chakraborty · S. Ray · S. Pandit (✉)
Department of Life Science, School of Bio-Science and Technology, Sharda University, Greater Noida, U.P., India

P. Kumar, V. Kothari (eds.), *Model Organisms in Biological Research*,
https://doi.org/10.1007/978-981-92-0038-2_13

Although they are small in size (0.1–1.5 mm), tardigrades have caught the attention of scholars, especially astrobiologists, for their ability to survive almost anything. Of the many interesting traits of tardigrades, one of the wildest is that they can go through cryptobiosis. This ametabolic state essentially allows them to "turn off" their metabolism and lose most of their water so that they can survive extreme environmental conditions [13]. Through cryptobiosis, tardigrades can endure desiccation (drying out), freezing, extreme heat, high pressure, and even levels of radiation that would kill almost every other life form. They can pull this off because their cells build up sugar (trehalose), stabilizing their structures and protecting them during a desiccated state and from heat, pressure, radiation, and low temperatures. In addition to cryptobiosis, tardigrades are exceptionally resilient to extreme environmental conditions, including exposure to high and low temperatures. Tardigrades can survive temperatures down to −272 °C (almost absolute zero) and up to 150 °C, and probably beyond, which is extraordinary compared to normal limits for most life [26].

This ability to endure temperature extremes is particularly exciting for astrobiologists investigating life's possibilities on biologically hostile planets with a high variance of temperature extremes, such as Mars, and the moons of Jupiter and Saturn, both known to have harsh temperature conditions. Tardigrades can also survive very high amounts of ionizing radiation (5000 to 7500 Gy), or specifically, gamma rays (4000 to 5000 Gy, as compared to humans, who can withstand only up to 10 Gy) and ultraviolet radiation (up to 1000 kJ/m^2), both of which are present in space [22]. Research has demonstrated that tardigrades exposed to very high doses of radiation that would normally kill other life forms show negligible cellular or DNA damage. Tardigrades can also repair severe DNA damage caused by radiation through unique molecular processes. They even create a protein known as Dsup (damage suppressor) that is something like a protective cover for their DNA against oxidative damage caused by radiation [1]. This is also one of the reasons why tardigrades are often thought of as a model organism for studying resistance to space radiation. Probably the most profound research on tardigrades was when they were sent to space and exposed to the vacuum of space. In 2007, the European Space Agency mission, FOTON-M3, sent tardigrades to the cosmos to study them exposed to the cosmic—albeit vacuum—conditions of space, including radiation [30]. The results were incredible: some of the tardigrades not only survived exposure but also, when rehydrated back on Earth, were able to reproduce. This was the first time that any animal (not just a microorganism) was demonstrated to survive the conditions of space and reproduce afterward, providing a legitimate example of a living organism that showed the potential to survive in the extreme conditions of outer space [19]. More evidence was established when additional experiments were conducted aboard the International Space Station (ISS) to again determine whether tardigrades could survive the vacuum and radiation of space. These studies also showed that not only could tardigrades survive both short and long durations in space, but tardigrades evidently withstand the drastic conditions of outer space. It was thought that the discovery of tardigrades surviving the harsh conditions of space represented a significant advance in our knowledge of the possibility of extraterrestrial life [7].

Tardigrades also serve as a practical model for the study of adaptations of life into extraterrestrial habitats and would help in refining how we look for life in other environments throughout the universe.

As far as astrobiology goes, extending past the extremities of life, the use of tardigrades as a model organism becomes pertinent in searching for their existence beyond Earth. Just their very existence in air-drying, non-life conditions under offer a compelling case to argue that life may just be pervasive through the cosmos, from what we used to think, even in environments that seem extreme [3]. Investigating their molecular mechanisms for tolerance to environmental extremes may allow us to continue to develop protective strategies for humans on long missions in space or conserve biological material for space exploration.

13.1.1 Biological Characteristics of Tardigrades

Tardigrades are small, eight-legged microscopic animals that are also called water bears for their slow waddling movements (Fig. 13.1). They range in size from 0.1 to 1.5 millimetres. Tardigrades, despite their size, are incredibly adaptable and can survive some of the harshest environments on the planet as well as in space (Table 13.1). They live in all sorts of environments and even in extreme environments. Tardigrades can be found in the deep-sea trenches, the tallest of the Himalayan Mountains, and extreme polar regions. Tardigrades are exactly what you would consider to be the most adaptable animal on the planet, given their diversity in distribution. Some of the most exciting and interesting characteristics of tardigrades include their survival mechanisms [28]. Tardigrades can enter a reversible ametabolic state called cryptobiosis. Cryptobiosis is a state in which tardigrades can stop their metabolic function and

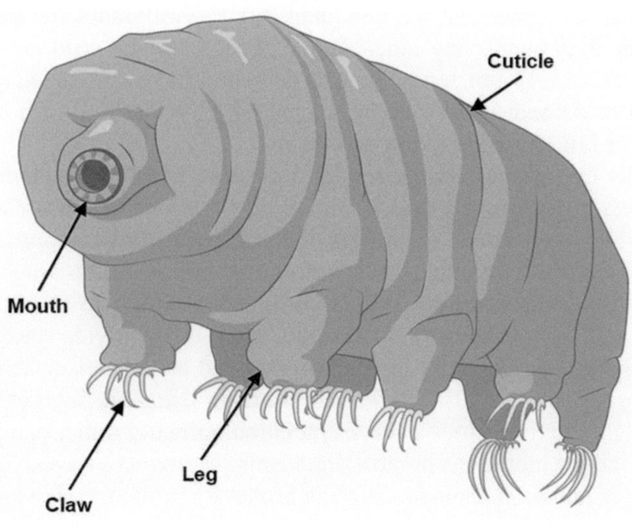

Fig. 13.1 General structure of tardigrade

Table 13.1 Characteristics of tardigrade

Parameters	Description
Common names	Water bears; Moss Piglets
Phylum	Phylum of Ecdysozoa
Occurrence	In mosses and lichens
Food	Bacteria, algae, plants, and small invertebrates
Size	50–2100 μm
Number of cells	Around 40,000
Body	Barrel-shaped, four pairs of legs, and legs without joints
Body cover	Cuticle containing chitin
Reproduction	Oviparous, some species are parthenogenic
Lifespan	3 months–2 years
Resistance	Extreme resistance to various physical and chemical factors
Mechanisms of resistance	Mostly (but not only) cryptobiosis

Adapted from Sadowska-Bartosz and Bartosz [25]

survive extreme environmental stress. Tardigrades can endure environmental extremes, including desiccation, freezing, extreme radiation, and even the vacuum of space, while they are in the cryptobiosis state, which can allow tardigrades to survive when other organisms cannot, and to live under horrible circumstances [12].

As the tardigrades undergo cryptobiosis, they replace most of the water in their bodies with a sugar called trehalose that stabilizes many structures in cellular structures and can allow the cellular structures to maintain cohesion in strict dehydrated environments. The process is a basic requirement for survival and ensures that cells do not collapse upon drying out [20]. Trehalose also helps maintain the integrity of proteins, membranes, and other structures inside the cell, which, in part, accounts for the resilience of the organism. With cryptobiosis as an ability, tardigrades may survive both temperature extremes of −272 °C (which is just above absolute zero) and temperature extremes that are more than 150 °C Tardigrades can endure harsher circumstances than nearly any other animal. They can withstand extreme high or low-cold or extremely hot temperatures, and they have special adaptations that enable them to defend against environmental stress—the kind of conditions that would harm or kill any other living organisms.

Tardigrades can also survive extremes of pressure (6000 atmospheres), up to six times the greatest pressures found in the deepest portions of the ocean, as well as their incredible temperature tolerances. This is particularly important for tardigrades that occupy environments with extremely high pressures, such as deep-sea trenches. The mechanisms involved in tardigrades' ability to withstand extreme pressure are still not completely understood, but it does provide a valuable model for understanding life potential in deep pressurized habitats elsewhere in the universe [6]. Tardigrades are also resistant to ionizing radiation, such as gamma rays and ultraviolet radiation. High levels of radiation normally inflict damage to DNA, kill cells, or induce mutations in most organisms. Tardigrades have evolved mechanisms to repair radiation damage. One can produce a protein, called Dsup (damage suppressor), that shields the tardigrade's damaged DNA from oxidative stress from being irradiated with ionizing radiation. This capability to repair and protect damage to their genetic material is likely fundamental to their ability to survive in environments with high radiation.

Tardigrade's radiation resistance and ability to survive extreme conditions have captured the imagination of many astrobiologists, especially since we are driven by the exploration of space [11]. There is significant evidence that tardigrades can survive not only in extreme terrestrial habitats but also in the vacuum of space. For example, the European Space Agency exposed Tardigrades to the vacuum of space, in addition to radiation and extreme temperatures, during the FOTON-M3 mission in 2007 [30]. When the tardigrades were brought back to Earth and given some water, some survived and were able to reproduce. The fact that they can survive in the vacuum of outer space as well as radiation is just another reason to highlight them as one of the most resilient organisms on Earth. Tardigrades' unique survival methods are something to consider for life, not only on Earth, but also for astrobiology. They can survive in such extreme conditions that some researchers have suggested that by studying tardigrades, we could learn how some life forms could survive in space or extreme conditions on other planets, like Mars or Jupiter, and Saturn's moons. By studying tardigrades, scientists may also shine a light on the extremes of processes required for sustenance, and maybe even life, in extreme environments.

13.2 Tardigrades in Space Research

In 2007, tardigrades became a part of history as the first known animals that could endure outer space. The event happened during the European Space Agency's FOTON-M3 mission when the study exposed tardigrades while in the vacuum of space and exposed them to solar UV radiation that they would experience in outer space [30]. Their results were remarkable: some tardigrades survived the vacuum of space and the intense levels of UV solar radiation, and upon returning to Earth and being rehydrated, some reproduced as well. This work showed that tardigrades can survive extreme conditions in space. It was a significant landmark study for the field of astrobiology as it produced evidence of the survivability of a eukaryotic organism (i.e., a multicelled organism) in an extraterrestrial environment. The findings of the FOTON-M3 mission were an important milestone for science by presenting evidence to suggest that some form of life could exist or survive in environments beyond Earth [30]. The capacity of tardigrades to survive radiation, vacuum, and temperature extremes in outer space has broadened our perspectives on living limits and organisms. This finding supported hypotheses such as panspermia, where life or its building blocks can develop in space and expand throughout the universe by space-faring organisms. The survival of tardigrades in space also challenged views on life's environmental prerequisites and provoked questions about the viability of other life forms living under inhospitable circumstances on remote planets. After the FOTON-M3 mission proved successful, additional studies were conducted on the ISS, investigating the success of tardigrades' survival in space [30]. Because of the microgravity condition on the ISS, researchers were able to observe how well tardigrades could endure long-term exposure to space, since they didn't have to put

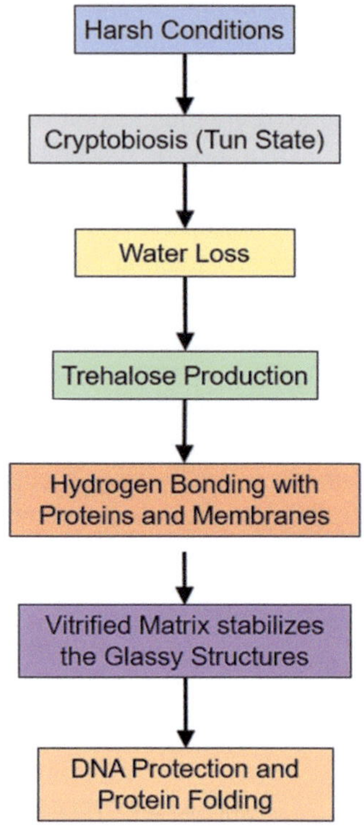

Fig. 13.2 Flowchart explaining the mechanism of cryptobiosis in tardigrades under extreme environmental conditions

the tardigrades into artificial environmental stressors such as space radiation, vacuum, or cosmic rays; that was a great advantage. Tardigrades endured a period of harsh conditions in outer space, and experiments have found that the existence of artifacts further bolsters evidence of their ruggedness and resilience. One of the most novel discoveries that these space studies provided was the identification of multiple molecular mechanisms that could potentially confer resistance against radiation damage and oxidative stress in tardigrades. An interesting protective molecule was Dsup (damage suppressor protein), a protective protein [3].

This protein is necessary in a tardigrade's life for DNA protection against radiation damage. It reduces the oxidative stress caused by oxidative damage to DNA, making it vital in the tardigrades' resiliency. The study of Dsup and other proteins in tardigrades provides significant information about how life can protect itself from radiation-induced genetic damage [10]. Tardigrades also have other molecular adaptations that allow them to endure an array of extreme conditions. Some of the adaptations include changes to the structure of their cell membranes and proteins, which allow tardigrades to survive in cases of extreme radiation and dehydration. Tardigrades also undergo a cryptobiotic state of condensed life and produce protective sugars like trehalose that stabilize cellular structures that allow living cells to enter dried states. These compounds are of interest to astrobiologists, as they may offer insight into how life could survive on other planets and in space [4] (Fig. 13.2).

Exploring the hardiness of tardigrades in space has potential implications for understanding life in other places within the universe, and these organisms can represent a model for the study of life surviving in such extreme environments as planets, moons, and outer space. Continued study of tardigrades' modes of survival may help to more fully understand how life could survive in environments thought to be otherwise uninhabitable [21]. Their ability to endure space radiation and other environmental extremes strengthens the argument that life might exist in unexpected places beyond Earth.

13.3 Implications for Astrobiology

Tardigrades are an outstanding biological model to study the limits of life in extreme conditions. These minute animals surpass what scientists thought could be survivable in extreme conditions, with the ability to withstand extreme hot and cold temperatures, extreme high doses of radiation, extreme desiccation, and even survive in space (the vacuum). Tardigrades are a true testament to the resilience and adaptability of life. For astrobiologists interested in life elsewhere in the universe, the biological model of tardigrades will provide an important basis for understanding how life may exist in regions of the universe with environments that can be vastly different from that of Earth. The ability of tardigrades to survive in space, among others, also provides support for the panspermia hypothesis [28]. The panspermia hypothesis posits that living organisms can exist throughout the universe via space-faring microbes. That means that life, including life on Earth, and possibly life on other planets, came from the journey made by microscopic organisms (and possibly life originated from microorganisms traveling on asteroids, comets, or just dust particles). Even cooler is the fact that tardigrades can survive the vacuum of space, extreme radiation, and extreme temperatures. This suggests that life may be a bit tougher than we give it credit for, and that it may be widespread in the cosmos, maybe still living on other worlds or even the spaces between the stars.

Also, research on tardigrades and how they survive directly benefits the search for life beyond Earth. By understanding how tardigrades survive extreme conditions, we can make life-detection missions better. The ability of tardigrades to survive terrible conditions gives clues about the types of organisms we might find in space, on Mars, or the icy moons of Jupiter and Saturn (for example, Europa) [25]. These clues will inform scientists to make better life-detection techniques in hostile and/or otherwise lifeless environments. The area of research on tardigrades is also being applied to develop bio-inspired protective mechanisms for human space exploration in the future. Knowledge of how tardigrades react to things like starvation, radiation, dehydration, and other stresses could help pursue technologies and techniques that will protect astronauts from similar stresses [2]. For instance, using bio-inspired materials could produce mechanisms for shielding humans from cosmic radiation or dehydration during long-term missions in space. As human space exploration develops, insights from tardigrades could ultimately play important

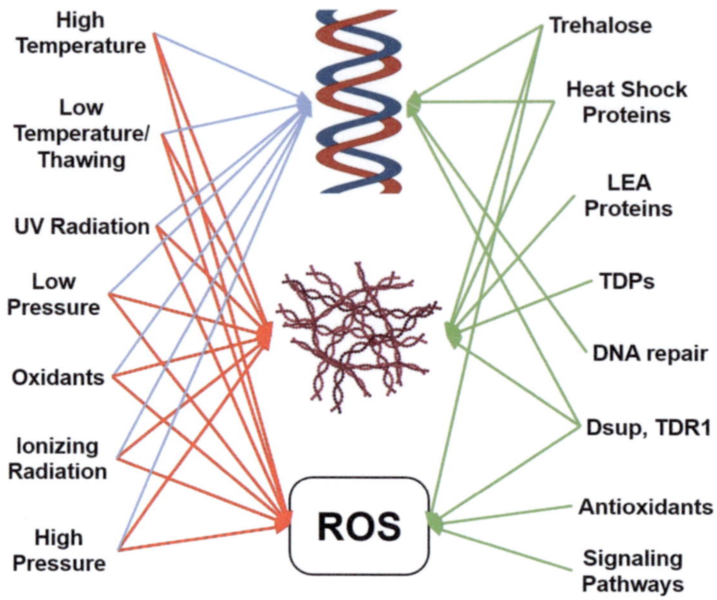

Fig. 13.3 Diagram showing the primary defence systems of tardigrades and how different environmental elements work. All of these elements cause oxidative stress (an increase in the production of reactive oxygen species, or ROS) and damage to proteins and DNA. ROS can harm proteins and DNA. Green indicates the presence of protective systems. Dsup is a damage suppressor protein; TDPs are tardigrade disordered proteins

roles in protecting the health of astronauts as they move towards stops on Mars, the Moon, or beyond.

Ultimately, the study of tardigrades contributes to our understanding of habitable environments on other worlds [27]. These organisms test our ideas about the conditions that could lead to habitable environments by surviving extreme cold, heat, and radiation. The exploration of tardigradology can and will refine our models of what constitutes a habitable environment and present us with new potential sites for future exploration. Whether it is on planets such as Mars, moons such as Europa, or exoplanets orbiting a star, life might be present, and in ways we cannot currently conceive, thrive (Fig. 13.3).

13.4 Significance of Different Studied Cases

13.4.1 FOTON-M3 Mission (2007)

In 2007, one of the first and most significant studies involving tardigrades in space was conducted on the FOTON-M3 mission. The European Space Agency sent tardigrades into space to discover if they could survive the harsh environmental conditions space has to offer, including vacuum, radiation, and solar ultraviolet (UV) radiation [9]. Tardigrades and various microorganisms were placed in a special container that allowed them to undergo such conditions. Upon their return to Earth, researchers found that some of the tardigrades were able to withstand the space conditions and were even able to reproduce after being rehydrated. Still, this research was the tipping point in the study of extremophiles, showing that even eukaryotic organisms could survive space, and lending credence to hypotheses like panspermia [30].

13.4.2 International Space Station (ISS) Experiments (2011–2019)

Following the FOTON-M3 mission's success, additional experiments were conducted aboard the International Space Station (ISS) to assess tardigrades' survival in space. Between 2011 and 2019, Tardigrades were sent to the ISS as part of the BIOEXPOSURE experiment, which exposed tardigrades to microgravity and the space environment. Some key observations include tardigrades being able to withstand the vacuum of space and radiation of space for long durations, thus supporting their extreme resilience [5]. Researchers also gained a better understanding of tardigrade DNA repair and protection thanks to these investigations. They are among the most resilient organisms ever examined in space because, for instance, they generate the Dsup protein, which protects their DNA from radiation damage [31].

13.4.3 Beresheet Lunar Lander (2019)

As it was seen in the Beresheet lunar lander (developed by SpaceIL, with the help of Israel's Ministry of Science and Technology), which accidentally transported tardigrades to the surface of the Moon in the year 2019. While the tardigrades were in a cryptobiotic state, which would keep them dormant, this incident is significant because it was a case of planetary protection and an accidental transportation of life from Earth to another body in the solar system. While the tardigrades were certainly in a sealed container, once the lander crashed, the tardigrades were then on the surface of the Moon [15]. While there's no evidence yet of tardigrade revivification on the Moon, the situation has raised significant ethical and scientific debates about the unintentional contamination of extraterrestrial environments and highlights the need to improve processes to stop contaminating other worlds.

13.4.4 Astrobiology Research in the Search for Life on Mars

In 2018, an experiment involving tardigrades was designed to check their viability within dying Martian extreme conditions, and ultimately their ability to act as a proxy for life on Mars. The tardigrades were exposed to low pressure, low temperature, and high UV light doses when the researchers simulated the conditions of the Martian surface in a controlled experiment. Although the organisms did not survive under those specific conditions as they would have on planet Earth, the information gained from the study to illuminate how life may survive under extreme conditions was useful. Tardigrades demonstrate resiliency to radiation and desiccation, and were ideal candidates to test the viability of life on Mars—not just the possibility, but the possibility of such experiments has continued to foster interest in how we could search for life or viable micro-organisms on Mars in the future [8].

13.4.5 Tardigrades in the "Space Life Sciences" Experiments

In 2019, a series of Space Life Sciences experiments were conducted to investigate how tardigrades can survive long, continuous exposure to space. In order to examine cryptobiosis and evaluate the tardigrade's ability to survive in microgravity, this study was conducted by Ferranti et al., [5] as part of the VITA mission aboard the ISS. The results generated significant information for understanding the potential impacts of long-duration space travel on living organisms, especially circumstantial information about how the mechanisms that allow tardigrades to survive could apply to human space exploration. Additional research was done on how such organisms can be leveraged for bio-regenerative life support systems for future long-duration missions. For example, on missions to Mars [16].

13.5 Future Perspective

13.5.1 Advancing Astrobiology and the Search for Extraterrestrial Life

As space missions to other planets, moons, and asteroids ramp up, tardigrades will continue to be of service in promoting a better understanding of astrobiology. Future research will provide insights for scientists to further understand how life could potentially exist in extreme extraterrestrial conditions, such as those found on Jupiter's moon Europa, Saturn's moon Enceladus, or on exoplanets at vast distances [14]. The researchers may use tardigrades as model organisms to investigate and simulate the possibilities of life existing in subsurface oceans,/over icy and in other extreme environments throughout the solar system. Given their stress-resistance to radiation, freezing, and vacuum conditions, it seems life might not be as rare as previously believed; after all, organisms may exist in areas thought to be unsuitable environments for life.

13.5.2 Improving Space Travel and Protection for Astronauts

Tardigrades' special survival mechanisms, such as their capacity to induce DNA damage repair and withstand extreme radiation, may yield encouraging knowledge for developing bio-inspired technologies for human space exploration. Research may allow scientists to harness tardigrades' biological adaptations into more effective protection for astronauts, such as protective radiation or biological protective suits that mimic the molecular resiliency of tardigrades [25]. There may even be applications of understanding their cryptobiotic state to limit the degradation of biological materials like food or medical supplies for long-duration space travel, such as Martian journeys or beyond.

13.5.3 Planetary Protection and Ethical Considerations

As mankind embraces the expansion of outer space exploration, the ethical issues surrounding planetary protection are bound to be unresolved and pushed into the limelight. The Beresheet incident, which culminated in the unintentional release of tardigrades onto the lunar surface, is a clear example of a future challenge around the contamination of other worlds and the release of Earth life onto other planets or moons [18]. Every upcoming mission to space will need to take the risks associated with space exploration seriously and begin implementing stringent planetary protection policies and protocols to ensure Earth organisms remain on Earth. A focus on the tardigrade as a life form might allow for the evolution of planetary protection policies and guidelines for new decontamination procedures for spacecraft; we want to avoid introducing any terrestrial organisms to unspoiled extraterrestrial environments, ensuring that we preserve the ecological integrity of those ecosystems and don't spoil the potential for discovering extraterrestrial life [24].

13.5.4 Biotechnology and Synthetic Biology Inspired by Tardigrades

The astonishing development capabilities of tardigrades, including cryptobiosis and the expression of protective native proteins such as Dsup that confer different degrees of protection, offer exciting prospects for the realm of biotechnology [29]. Research on these functional mechanisms may lead to synthetic biology applications in the future by harnessing these tardigrade resilience mechanisms and incorporating them into other organisms or materials. Tardigrade proteins like Dsup could be used to create crops that can withstand harsh weather conditions like drought or frost, radiation-resistant biomaterials, and better techniques for boosting DNA repair against oxidative damage in biomedical research. Ultimately, such opportunities would lead to more efficient genetic engineering and genetic therapies and less of a propensity for genetic errors induced by oxidative stress [25].

13.6 Conclusion

Tardigrades are regarded as foundational organisms for the area of astrobiology, which considers life on Earth in relation to the possibility of extraterrestrial life. They have demonstrated an ability to tolerate extreme conditions of radiation, vacuum, temperature extremes, and pressure; they are one of the few organisms known to survive in the vacuum of space. Their approach of behaviourally responding to extreme conditions by entering cryptobiosis (a state of suspended metabolism) provides scientists insight and a framework to understand survival in extreme conditions. Their resilience as an organism contributes to implications for panspermia, the idea that life could move between celestial bodies through comets, asteroids, or even space dust, but it also provides for the possibility of life on Mars, Europa, or one of the many other bodies of our solar system. The advancement in scientific knowledge and exploration of life on other planets becomes more of a reality as exploration and missions to deep space are being planned. Exploration of outer space or Mars will benefit from what we learn about the tardigrade in regard to the viability of life. The tardigrade does not just lend itself as a biological organism, but the adaptations that it presents can lead to bio-inspired novel technologies, such as protecting organisms in space, radiation-resistant materials, enhanced DNA protection, or improved mechanisms to enhance preservation of life. This is why tardigrades serve as a biological model of survival and resilience both on Earth and other locations throughout the universe.

References

1. Baas Becking LG, Raat AJ. The organisms. Geochem Perspect. 2022;11:52–70.
2. Brasch K. Life in the cosmos: when, where, and how? J R Astron Soc Can. 2018;112.
3. Cantara S, Regoli T, Ricci C. Captain tardigrade and its shield to protect DNA. 2025.
4. Cortese F, Klokov D, Osipov A, Stefaniak J, Moskalev A, Schastnaya J, Cantor C, Aliper A, Mamoshina P, Ushakov I. Vive la radiorésistance!: converging research in radiobiology and biogerontology to enhance human radioresistance for deep space exploration and colonization. Oncotarget. 2018;9:14692.
5. Ferranti F, Del Bianco M, Pacelli C. Advantages and limitations of current microgravity platforms for space biology research. Appl Sci. 2020;11:68.
6. Giovannini I, Boothby TC, Cesari M, Goldstein B, Guidetti R, Rebecchi L. Production of reactive oxygen species and involvement of bioprotectants during anhydrobiosis in the tardigrade Paramacrobiotus spatialis. Sci Rep. 2022;12:1938.
7. Grzyb A, Wolna-Maruwka A, Niewiadomska A. Environmental factors affecting the mineralization of crop residues. Agronomy. 2020;10:1951.
8. Guidetti R, Rizzo AM, Altiero T, Rebecchi L. What can we learn from the toughest animals of the Earth? Water bears (tardigrades) as multicellular model organisms in order to perform scientific preparations for lunar exploration. Planet Space Sci. 2012;74:97–102.
9. Gulimova V, Proshchina A, Kharlamova A, Krivova Y, Barabanov V, Berdiev R, Asadchikov V, Buzmakov A, Zolotov D, Saveliev S. Reptiles in space missions: results and perspectives. Int J Mol Sci. 2019;20:3019.
10. Hashimoto T, Kunieda T. DNA protection protein, a novel mechanism of radiation tolerance: lessons from tardigrades. Life. 2017;7:26.
11. Hassan MU, Nawaz M, Shah AN, Raza A, Barbanti L, Skalicky M, Hashem M, Brestic M, Pandey S, Alamri S. Trehalose: a key player in plant growth regulation and tolerance to abiotic stresses. J Plant Growth Regul. 2023;42:4935–57.
12. Hesgrove C, Boothby TC. The biology of tardigrade disordered proteins in extreme stress tolerance. Cell Commun Signal. 2020;18:178. https://doi.org/10.1186/s12964-020-00670-2.
13. Horneck G, Walter N, Westall F, Grenfell JL, Martin WF, Gomez F, Leuko S, Lee N, Onofri S, Tsiganis K, Saladino R, Pilat-Lohinger E, Palomba E, Harrison J, Rull F, Muller C,

Strazzulla G, Brucato JR, Rettberg P, Capria MT. AstRoMap European astrobiology roadmap. Astrobiology. 2016;16:201–43. https://doi.org/10.1089/ast.2015.1441.

14. Kanik I, Paul de Vera J-P. Astrobiology of Mars, Europa, Titan and enceladus-most likely places for alien life. Front Astron Space Sci. 2021;8:643268.
15. Kulu E. In-space economy in 2023-statistical overview and trends. In: 74th International Astronautical Congress (IAC 2023). 2023.
16. Martinez P. Small satellites for science. In: Pelton JN, Madry S, editors. Handbook of small satellites. Cham: Springer International Publishing; 2020. p. 1–25. https://doi.org/10.1007/978-3-030-20707-6_36-1.
17. Martins Z, Cottin H, Kotler JM, Carrasco N, Cockell CS, De La Torre Noetzel R, Demets R, De Vera J-P, d'Hendecourt L, Ehrenfreund P, Elsaesser A, Foing B, Onofri S, Quinn R, Rabbow E, Rettberg P, Ricco AJ, Slenzka K, Stalport F, ten Kate IL, Van Loon JJWA, Westall F. Earth as a tool for astrobiology—a European perspective. Space Sci Rev. 2017;209:43–81. https://doi.org/10.1007/s11214-017-0369-1.
18. McKaig J, Caro T, Burton D, Tavares F, Vidaurri M. Chapter 10: Planetary protection—history, science, and the future. Astrobiology. 2024;24:S-202–15. https://doi.org/10.1089/ast.2021.0112.
19. Möbius K, Savitsky A, Malferrari M, Francia F, Mamedov MD, Semenov AY, Lubitz W, Venturoli G. Soft dynamic confinement of membrane proteins by dehydrated Trehalose matrices: high-field EPR and fast-laser studies. Appl Magn Reson. 2020;51:773–850. https://doi.org/10.1007/s00723-020-01240-y.
20. Nawaz M, Hassan MU, Chattha MU, Mahmood A, Shah AN, Hashem M, Alamri S, Batool M, Rasheed A, Thabit MA. Trehalose: a promising osmo-protectant against salinity stress—physiological and molecular mechanisms and future prospective. Mol Biol Rep. 2022;49:11255–71.
21. Panier S, Wang S, Schumacher B. Genome instability and DNA repair in somatic and reproductive aging. Annu Rev Pathol Mech Dis. 2024;19:261–90. https://doi.org/10.1146/annurev-pathmechdis-051122-093128.
22. Pomeroy KO, Comizzoli P, Rushing JS, Lersten IL, Nel-Themaat L. The ART of cryopreservation and its changing landscape. Fertil Steril. 2022;117:469–76.
23. Prasad B, Richter P, Vadakedath N, Haag FW, Strauch SM, Mancinelli R, Schwarzwälder A, Etcheparre E, Gaume N, Lebert M. How the space environment influences organisms: an astrobiological perspective and review. Int J Astrobiol. 2021;20:159–77.
24. Roger CIT, Roldán ER, Ramos JAS, Pérez SO, González VR. Survival of tardigrade culture in a simulated lunar environment. STAR; 2022.
25. Sadowska-Bartosz I, Bartosz G. Antioxidant defense in the toughest animals on the earth: its contribution to the extreme resistance of tardigrades. Int J Mol Sci. 2024;25:8393.
26. Seager S, Petkowski JJ, Günther MN, Bains W, Mikal-Evans T, Deming D. Possibilities for an aerial biosphere in temperate sub Neptune-sized exoplanet atmospheres. Universe. 2021;7:172.
27. Sithamparam M, Satthiyasilan N, Chen C, Jia TZ, Chandru K. A material-based panspermia hypothesis: the potential of polymer gels and membraneless droplets. Biopolymers. 2022;113:e23486. https://doi.org/10.1002/bip.23486.
28. Swathi P, Sravanti V. Extreme organisms: an Astrobiological perspective. Res J Biotechnol. 2020;15:10.
29. Wang Q, Hu Z, Li Z, Liu T, Bian G. Exploring the application and prospects of synthetic biology in engineered living materials. Adv Mater. 2023:2305828. https://doi.org/10.1002/adma.202305828.
30. Weronika E, Łukasz K. Tardigrades in space research - past and future. Orig Life Evol Biosph. 2017;47:545–53. https://doi.org/10.1007/s11084-016-9522-1.
31. Zarubin M, Murugova T, Ryzhykau Y, Ivankov O, Uversky VN, Kravchenko E. Structural study of the intrinsically disordered tardigrade damage suppressor protein (Dsup) and its complex with DNA. Sci Rep. 2024;14:22910.
32. Zhu A, Stone J. Bearing extremes: impacts from simulated outer space conditions and effective ultraviolet radiation shielding materials on tardigrade life history. Health Phys. 2022;123:238–44.

Part IV

Novel Approaches to Reduce Use of Animals in Research

Alternatives to Animal Testing and the Future of Preclinical Research

14

Subramani Parasuraman, Vijeepallam Kamini, and Aaseer Thamby Sam

14.1 Introduction

The use of animal models in medical research was introduced during the sixth century BC, contributing to the development of surgical techniques, vaccines, antimicrobial agents, and an understanding of human disease processes [1]. Animal models have been used in preclinical research for the last 150 years in pharmacological and toxicological research. Preclinical models were extensively employed in the drug development process in the middle of the twentieth century to confirm the efficacy and safety of pharmaceuticals or medical devices before they were tested in humans [2]. According to the literature, over 20 million animal subjects are used annually in biomedical research, primarily in mouse and rat models. However, the translation of animal models to human subjects has been variable. Ethical concerns have also been raised, such as the study design and data analysis, the inherent heterogeneity of both human and animal subjects, and the translation of preclinical animal trials to human clinical trials [1]. Three key factors undermine confidence in the reliability of animal experimentation, namely the artificial laboratory

Vijeepallam Kamini and Aaseer Thamby Sam contributed equally with all other contributors.

S. Parasuraman (✉) · V. Kamini
Pharmacology, Toxicology and Basic Health Sciences Unit, Faculty of Pharmacy, AIMST University, Bedong, Malaysia
e-mail: parasuraman@aimst.edu.my

A. T. Sam
Pharmacology, Toxicology and Basic Health Sciences Unit, Faculty of Pharmacy, AIMST University, Bedong, Malaysia

Clinical Pharmacy & Pharmacy Practice Unit, Faculty of Pharmacy, AIMST University, Bedong, Malaysia

P. Kumar, V. Kothari (eds.), *Model Organisms in Biological Research*, https://doi.org/10.1007/978-981-92-0038-2_14

environment, differences between animal and human disease models and the differences in physiology and genetics between species [3].

More recently, the scientific community has raised concerns about the clinical validity, usefulness, and ethical implications of using animals in research [4]. However, preclinical research is crucial for human trials, adhering to the ethical principles of the World Medical Association (WMA) Declaration of Helsinki, ensuring scientific principles, are based on literature, and animal welfare [5].

Animal experimentation plays a crucial role in biomedical research due to shared biological processes between animals and humans. Their comparable organ systems and susceptibility to diseases such as diabetes, cancer, and heart disease make them valuable models for study. Animal models have a shorter life cycle, allowing for study throughout their lifespan. The debate revolves around the ethical use of animals in biomedical research and whether speciesism justifies a morally lower value. Animal rights advocates argue that animals should be treated equally to humans, and currently, non-human animal research validates new treatment options before human trials [4]. Animal testing also has several limitations, including ineffective predictions of human responses, higher doses of compounds, poor translation from animal models to human biology, and shorter lifespans. It is expensive, has the potential to cause high mortality rates and consumes time and resources. The reproducibility of animal testing in human biology also raises questions about its applicability in real medical scenarios [6].

In January 2023, changes to the Federal Food, Drug, and Cosmetic Act of 1938 were implemented following the passage of the FDA Modernization Act 2.0 by the United States legislature, allowing alternatives to animal testing in drug development. The new regulation states that animal testing for pharmaceuticals is not necessary prior to human clinical trials. According to the WMA Declaration of Helsinki, clinical trials must be carried out using a comprehensive understanding of the scientific literature, which includes sufficient evidence from animal and laboratory experiments. Alternatives to animal research might be used to ensure the safety of any medications. This chapter summarizes the current state of animals in research and alternatives and directions regarding the use of animal models in research.

14.2 Ethical Considerations and Alternatives

The purpose of animal ethics is to examine both human-animal relationships, as well as how animals ought to be treated. Since the 1980s, proposals for animal research have been subject to approval by animal ethics committees, including Institutional Animal Care and Use Committees (IACUC) in the USA and under Directive 2010/63/EU in Europe, ensuring ethical and scientific requirements. In 1959, the 3R framework was proposed by Russell and Burch to replace animals with alternative methods, reduce animal numbers for statistical validity, and improve animal welfare. This framework is recognized in Directive 2010/63/EU and European national legislation. Animal research regulations in Europe prioritize utilitarianism, minimizing suffering and justifying it with scientific or medical benefits, contrasting with human research ethics, which are dominated by deontological concerns [7].

Table 14.1 The 3Rs principles in animal research: replacement, reduction, and refinement

Principle	Description	Examples/methods	References
Replacement	Replacing the use of animals with modern alternatives	Cell culture In vitro systems Tissue engineering Computer modelling	[11]
Reduction	Decreasing the number of animals used in experiments	Improved experimental design Advanced statistical methods In vitro embryonic stem cell tests Data sharing (e.g., excipient profiles, cumulative findings)	[9, 10]
Refinement	Reducing the pain, distress, and discomfort of experimental animals	Refining animal facilities Minimizing stress to avoid hormonal imbalances and experimental variation Reducing the need for repetition due to stress-induced variability	[9]

In 1981, police raided Edward Taub's laboratory in Silver Spring, Maryland, and removed 17 macaque monkeys amid allegations of animal cruelty. The confrontation lasted over a decade, and the animals became widely known as the "Silver Spring Monkeys." These events were the catalyst for the emergence of the modern animal rights movement. The facility's issues became a catalyst for stronger animal welfare laws and more stringent guidelines for the use and care of animals in research, influencing how such research is conducted and monitored today [8].

Regulatory agencies recommend alternative models for toxicity testing to prevent animal suffering and reduce the use of research animals, guided by the 3Rs principle (replacement, reduction, and refinement). The 3Rs principle is widely accepted as a guideline for most ethical animal research practices [9, 10]. The 3Rs principle in animal research is summarized in Table 14.1. The 3Rs principle is enhanced with the inclusion of responsibility, emphasizing the importance of considering alternative methods in animal research [9].

To reduce the use of animals in research and promote alternative methods, the United States Environmental Protection Agency published two guidelines for waiving acute dermal toxicity tests: *Guidance for waiving or bridging of mammalian acute toxicity tests for pesticides and pesticide products* (2012) and *Guidance for waiving acute dermal toxicity* (2016) [12]. About 45 countries have banned animal testing for cosmetics, including Australia, Brazil, Canada, Chile, Colombia, the European Union, Ecuador, Guatemala, Iceland, India, Israel, Mexico, New Zealand, Norway, South Korea, Switzerland, Taiwan, Turkey, and the United Kingdom [13].

Legislation signed by Joe Biden in December 2022 allows drug developers to use alternatives to animal tesings, rather than requiring animal studies, for approval by U.S. Food and Drug Administration (US FDA) [14]. Animal models are still crucial for drug safety and efficacy approval, and the National Association for

Biomedical Research (NABR) supports alternative models when scientifically feasible and validated. The FDA Modernization Act 2.0 clarifies the definition of nonclinical tests under the Food, Drug, and Cosmetic Act, allowing for adjunct and complementary methods like organs-on-a-chip, micro-physiological systems, and computer simulations, and allows FDA reviewers to determine the best data for new drug applications [15].

14.3 Innovative Alternatives to Animal Testing

Millions of animals are used in experiments worldwide, often experiencing psychological distress, disease, brain damage, paralysis, and exposure to harmful substances. In addition, many animals are bred specifically for genetic modifications [16]. Researchers are developing methods to study diseases and test products that replace animal experiments, focusing on developing alternatives that are relevant to human health, thereby reducing animal usage. The alternative methods include in vitro techniques, in silico models, and studies involving human volunteers.

14.3.1 In Vitro Models (Cellular and Organ-on-a-Chip Technologies)

In vitro models, particularly organ-on-a-chip (OoC) technologies, help to study cellular and organ-level functions in a controlled environment, mimicking human physiology more accurately than traditional conventional Two-dimensional (2D) or Three-dimensional (3D) cell cultures. 2D cell cultures are essential biological models for studying cell functions, signalling pathways, and nuclear processes. However, they often lack the ability to reproduce the unique three-dimensional organization, cell-cell interactions, and physiological microenvironments observed in vivo, limiting their use as predictive tools in drug development and disease modelling [17, 18]. In response to these limitations, 3D cell culture methodologies were developed to recapitulate microenvironments [17]. OoC technology utilizes 3D cell cultures in microfluidic devices, which can better emulate the structural and biochemical behaviour of human organs compared to conventional 2D or static 3D models. The evolution of OoC technology has grown into a game changer for in vitro modelling of human physiology and pathology [19]. These microphysiological systems recreate the interplay of cells, extracellular matrix, and fluid shear in human tissues and organs [20]. Researchers believe that these devices could use to study the ways organs communicate and respond to different drug treatments, leading to new discoveries about how our bodies function, as well as potentially transforming pharmaceutical development and personalized medicine [21, 22].

14.3.2 Computer Modelling and AI-Simulating Biological Systems

Computational models and artificial intelligence (AI) are essential in understanding the fundamentals of biological systems, aiding in novel drug discovery and predicting drug efficacy. The significance of these models is summarized in Table 14.2. These technologies use data from high-throughput systems biology experiments, animal studies, as well as cell culture to measure biomarkers, predict patient responses, and identify novel drug targets [23–26].

Machine learning and AI are transforming drug discovery by enabling the analysis of large and complex datasets, improving prediction accuracy, accelerating research timelines, and reducing the overall cost of identifying potential drug candidates. In addition, AI algorithms can use in vitro data from organoids and OoC systems to assess the potential and potency of a drug [27]. This method helps to resolve biological challenges such as drug repurposing, target identification, and clinical translation.

Machine learning techniques are a subset of AI that can analyse complex datasets, identify patterns, and predict drug properties, thereby facilitating the discovery of new, more effective drugs with fewer side effects [28, 29]. AI and machine learning can significantly reduce drug discovery time, costs, and improve clinical trial success through interdisciplinary collaboration [30].

Table 14.2 Computational approaches to systems biology and drug discovery

Aspect	Details
Methods used	Mathematical algorithms Statistical analysis Machine learning tools
Purpose	Develop computational models of biological systems
Applications	Model molecular content and reactivity of organisms Simulate biological system behaviour Drug discovery optimization
Data types integrated	Genomic Proteomic Metabolomic Imaging
Benefits of simulations	Test biological responses to environmental changes Forecast effects of disruptions Reduce the need for animal testing Shorten drug discovery
Role of machine learning (ML)	Mine large biological databases Pattern recognition Predict outcomes Optimize experiments

14.3.3 3D Bioprinting and Tissue Engineering

Although the term '3D printing' was coined in 1974 by David E. H. Jones, it was not until 1981 that Hideo Kodama produced a functional example of Additive Manufacturing. The field received a huge boost in 1986 when Charles W. Hull introduced a technology he called stereolithography [31]. Taking this concept to the next level, bioprinting was born as an engineering-focused type of 3D printing which employs biological materials (e.g. living cells) to produce tissues that are physiologically functional. In this way, it can more faithfully reproduce how biological systems function and fail, enabling better predictions of human physiology and pathophysiology [27, 32–34].

The scalability, versatility, and cost-effective applications of bioprinting technology in human biofabrication have also furthermore provided promising avenues for functional tissue model generation relevant to artificial grafts, tissue repair, drug testing, and disease modelling [35]. In addition, this technology has enabled personalized medicine by generating patient-specific tissue models reflective of genetic diversity and physiological response variation among individuals to identify pharmacological interventions that can minimize toxicities [34, 36]. However, this is not to diminish the fact that we still have a long way to go in assembling these systems as they are found in nature, and challenges such as cell synchronization and limited availability of cells and tissues are very real today. The models and bioprinting methods are constantly being refined by researchers to address this problem. The success of this process may greatly lessen the dependence on animal testing in biomedical research [33].

14.3.4 Humanized Models: Stem Cells and Organoids

The accuracy of mimicking human physiology is one of the drawbacks of animal models. To address this issue, advancements have been made in in vitro models. However, 2D cultures lack the necessary cell–cell and cell–matrix interactions, compromising drug response and mechanistic investigations [37–39]. Consequently, the current research trend are shifting, focusing on a humanized system. Stem cell research is incorporated in the humanized system research, and 3D structures are designed from stem cells known as organoids. Besides, OoC, physiological systems that mimic human organ physiology, also represent the advancement in these models [19]. The OoC offers a promising alternative that is more efficient and humane compared to animal testing [40].

The disease models with patient-specific genetic backgrounds can be generated for drug screening and personalized therapy using patient-derived pluripotent stem cells as the origin of organoids. Bioengineered organoids, which can replicate organ structural and functional features, have revolutionized the understanding of organogenesis, disease pathogenesis, and drug response. These tissue systems, when combined with microfluidic devices, can provide fine control of the microenvironment, enabling the investigation of dynamic processes like drug absorption, metabolism, and toxicity. Organoid technology faces challenges such as variability in size, shape, and cell composition, and a lack of standardized protocols for formation and characterization [41, 42]. Despite these, limitations organoids are expected to revolutionize biomedical science, as their generation is faster than animal models, and their use can be extended over days, weeks, or months [43].

Organoids offer versatile tools for studying biological processes like development, embryogenesis, cancer metastasis, infectious diseases, genetic disorders, and drug-induced organ damage, particularly neural organoids, which offer unique insights into human brain development [44]. In addition, organoids can be genetically manipulated to incorporate specific mutations or reporter gene, facilitating the investigation of the interaction between a gene and the disease process in greater depth [45]. Also, organoids made from normal tissue are applicable to the study of physiology, as, for example, differentiation programs in the gut can be defined, and they may be infected with various pathogens to model diseases [46]. The possibility of cryopreserving and biobanking the organoids makes it possible to scale up such organoid biobanks, which can be employed for drug screening and toxicology studies.

The development of humanized models, including organoids derived from stem cells, as suitable and reliable models for human biology, requires multidisciplinary approaches involving expertise in stem cell biology, tissue engineering, microfluidics, and computer modelling. Besides, this includes developing a standardized protocol for organoid cultivation and integrating organoids with technologies such as bioreactors and microfluidics [47]. Although human organoids have high potential for application in medical transplantation, the translation of human organoids to clinical use remains challenging [48].

The development of human-like tissue models is crucial for assessing drug efficacy and toxicity, supporting personalized medicine, and reducing reliance on animal testing [49]. Further, pluripotent stem cell-derived organoids offer a valuable platform for disease modelling and drug discovery, as they closely mimic human tissue and retain individual genetic and phenotypic traits, enabling more tailored treatments [50]. Notably, a new phase of the art has been introduced by OoC technology, which integrates organoids with microfluidic devices, showing a better ability to control the microenvironment and monitor organoid function in real-time [51].

14.3.5 Microdosing and Human Trials

A precise amalgamation of pharmacology, pharmaceutical analysis, pharmacokinetics, pharmacodynamics, and pharmacogenomics is the hallmark of proficient drug development. The marketing approval for various novel drugs has experienced a steep decline owing to the burgeoning development costs. The drug development process is a complex and costly endeavour, averaging $500 million to $1 billion per marketed drug over a 10- to 15-year period [52]. Globally, over the past decade, Phases II and III of clinical trials experienced attrition rates of 62% and 45%, respectively [53]. This perilous situation was the precursor for the document "Critical Path" by the FDA, highlighting the numerous hurdles in drug development and encouraging the incorporation of novel approaches within the current drug development paradigm [54, 55].

The steep deterioration in drug development productivity and the need to reduce the expenses and time spent in identifying failed drugs provoked the search for alternative approaches of the first-in-human (FIH) trials, as well as efforts to mitigate the costs and time expended in identifying failed drugs [56]. Exploratory Investigational New Drug applications (eIND), phase 0, and Exploratory Clinical Trials were introduced by regulatory authorities to reduce human risks by minimizing drug exposure during FIH trials [57]. It was at this juncture that the concept of "Microdosing" emerged. Microdosing is a method of using low, non-pharmacologically active doses of a drug to study its pharmacokinetic profile in humans, potentially complementing standard animal-to-human scaling and redefining phase I clinical research by enhancing understanding of drug development processes [58]. The feasibility and consistency of microdosing are highly dependent on drug information, time, costs, alternatives, and validity of modelling data from limited to full-therapeutic dose exposure [59, 60]. Microdosing is also a useful tool for evaluating drug–drug interactions and pharmacokinetic susceptibility, offering advantages such as fewer safety concerns, minimal cross-reaction risk, and reduced variability in results [56].

Microdosing limits exposure to 100 μg or 1/100th of No Observed Adverse Effect Level (NOAEL), ensuring minimal gross effects, therapeutic, toxic, or radiotoxic effects due to low exposures. The International Conference on Harmonization (ICH) lays out two microdosing approaches. Approach-1 comprised a single dose (max. Dose 100 μg and 1/100 of NOAEL), with the preclinical requirement of extended single-dose toxicity in rodents. Approach-2 comprised five doses given at intervals of 6 t½ between doses (each dose 100 μg and 1/100 of NOAEL), with the preclinical requirement of 7-day repeated-dose toxicity studies in rodents. The ICH emphasized that phase 0 trials are First-In-Human (FIH) trials with lower drug exposure than the maximum tolerated dose (MTD), no therapeutic objectives, and no tolerability assessment. Microdosing studies require more sensitive assays than traditional analytical tools, such as liquid chromatography–tandem mass spectrometry (LC–MS/MS), positron emission tomography (PET), and accelerator mass spectrometry (AMS), due to limited systemic exposure [56].

The microdosing method for determining drug concentration has concerns about its predictive accuracy. A CREAM (Consortium for Resourcing and Evaluating AMS Microdosing) study found a 70% approximation between microdose and pharmacological dose pharmacokinetics, but it may not accurately predict clinical drug behaviour due to inconsistent biological responses and nonlinear kinetics [58]. In summary, microdosing uses small, non-therapeutic doses of a drug to study human pharmacokinetics, minimizing side effects, requiring fewer animal studies, and allowing earlier human screening—often within 4–12 months, compared to 12–24 months in traditional Phase I trials. This helps identify unsuitable drug candidates early, reducing unnecessary animal studies and human exposure [56].

However, an ethical question remains: is it right to expose humans to full doses later on if microdosing could have predicted poor drug viability earlier?

14.3.6 CRISPR and Gene Editing as Tools for Disease Modelling

The CRISPR (clustered regularly interspaced short palindromic repeats) technology is used to selectively modify the DNA of living organisms. Yoshizumi Ishino of Osaka University in Japan discovered palindromic repeats in E. coli K-12 cells in 1987, leading to the discovery of new CRISPR-Cas systems and paving the way for gene editing and genomic engineering in the field [61].

The development of the CRISPR-Cas9 technology has enabled the rapid creation of animal models of human diseases, faster and more accurate than the classical techniques [62]. This resource tool allows the generation of precise genetic modifications in specific genes (knock-in or knock-out approaches), and provides the possibility of correcting disease-related variants [63, 64].

The versatile CRISPR-Cas9 technology (Fig. 14.1) is a bacterial adaptive immune system used in biological studies for gene editing. It involves a guide RNA (gRNA)/single guide RNA (sgRNA) guiding the Cas9 endonuclease to a DNA site, causing a double-strand break in the DNA sequence [66]. CRISPR-Cas9 can achieve gene editing through non-homologous end-joining (NHEJ), resulting in insertions or deletions, or through homology-directed repair (HDR) with high precision.

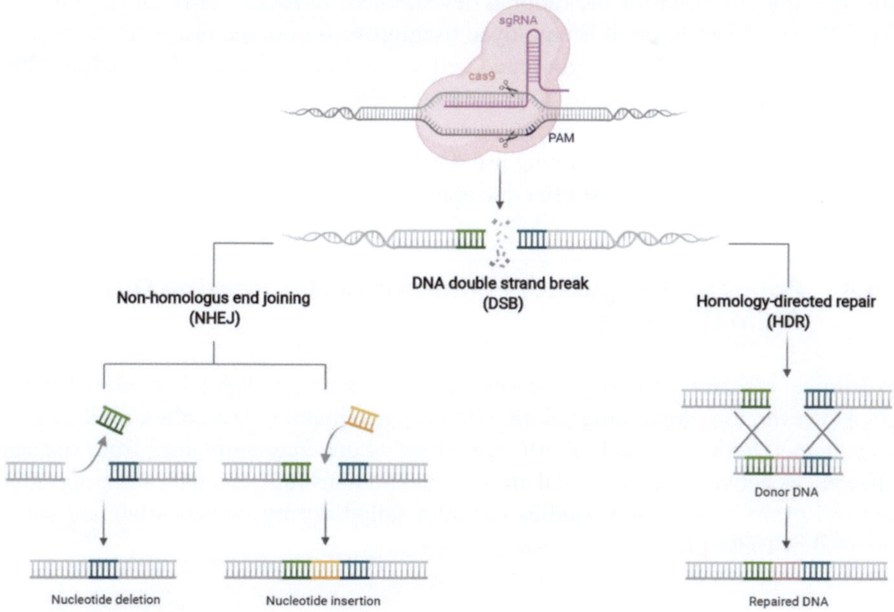

Fig. 14.1 CRISPR-cas9 genome editing through nonhomologous end joining and homology-directed repair pathway [65]

It can be applied to develop a range of disease models as well as modify specific loci in the cell, making it highly valuable for biological and genetic studies [67, 68].

Rapid progress has been made in the development of genetically engineered mouse models, and the application of the CRISPR-Cas9 technology to precisely manipulate small genetic variations known to influence complex traits related to human disease [67, 69, 70]. It can be expanded to multiple targeted mutations, and not only is it capable of replacing mouse genes with human genes, thereby generating humanized mice, which are more robust models for many studies on either gene functions or disease modelling, but also serves as a powerful tool to explore the interaction of human-specific diseases with other cross-species diversity in genomic views [71–73]. The accuracy of CRISPR-Cas9 also allows the creation of in vitro disease models like engineered cell lines and organoids, lowering the dependence on animal models and matching ethical animal research theory with the 3R principles [74]. Well-tuned CRISPR-Cas9 technology is revolutionizing biomedical science, creating new avenues for drug discovery and gene therapy by making the development of sophisticated cellular and animal models much faster than ever before [75]. It can enable the generation of robust, accurate, and physiologically relevant disease models that promise greater efficacy in therapeutic development. CRISPR has been used extensively in ex vivo and in vivo gene editing, with clinical trials for diseases including lung cancer, transthyretin amyloidosis, sickle cell disease, and β-thalassemia showing therapeutic promise [76, 77]. In addition to disease modelling, CRISPR-Cas9 has also contributed to drug discovery as it allows the identification of genes implicated in cancer progression and drug resistance, which are subsequently used for the rational development of molecularly targeted therapies [78, 79]. In addition, it is employed to improve traditional drugs and to enable gene therapies, including programmed cell death-1 based approaches such as chimeric antigen receptor T (CAR T)-cell therapy [79].

Despite its promise, the rapid advancement of CRISPR-Cas9 raises ethical concerns, particularly regarding animal use. However, its application supports the 3Rs, offering a more humane, cost-effective approach to scientific research [80–82].

14.4 Policy and Regulatory Frameworks for Phasing Out Animal Testing

Preclinical research and drug development predominantly depend on animal models. However, the global drug failure rate is approximately 95%, often attributed to unexplained toxicity or lack of efficacy. These issues frequently arise from species differences between experimental animals and humans [83]. Examples of drugs that showed promise in animal studies but later failed during clinical trials are summarised in Table 14.3.

Table 14.3 Examples of drugs with positive results in animal models that failed in human clinical trials

Disease	Drug	Findings in animal models	Clinical trial outcome
Alzheimer's disease	AN1792 vaccine	Cleared amyloid plaques in mice [84]	Caused meningoencephalitis [85]
Stroke	NXY-059	Reduced brain damage in rodents [86]	Ineffective for treatment of acute ischemic stroke (SAINT I & II trials) [87]
Cancer	Matrix metalloproteinase (MMP) inhibitors	Suppressed tumour growth in mice	No benefit; sometimes increased metastasis [88]
HIV infection	AIDSVAX vaccine	Generated protective antibodies in primates [89]	Failed to prevent infection in humans [90]

The use of human biology-based approaches in preclinical drug discovery is expected to be more relevant, predictive, and cost-efficient. Global harmonization, tool standardization, and acceptance of NAMs are needed for effective application [91]. The US FDA announced on April 10, 2025, a strategy to phase out animal testing in monoclonal antibody therapies and other drugs. This includes implementing AI-based models, aiming to enhance safety, reduce research expenses, and lower drug prices [15]. The FDA is introducing a roadmap to replace animal testing in monoclonal antibody safety evaluation with computer modelling, AI, and lab-grown human "organoids" and OoC systems, aiming to promote faster drug development and safety, reduce the use of animals, and promote animal welfare [92].

14.5 Validation of Alternative Methods

The 3Rs principle promotes the continuous exploration of alternative methods to replace animal use. Validation is the key factor in determining the regulatory status of alternative methods, especially for NAMs as alternatives to animal testing [93]. Validation is the process of evaluating an alternative method to ensure it is reliable, relevant, and ready for regulatory use, as a replacement for an existing animal test in a specific regulatory context. Validation bodies like the Interagency Coordinating Committee on the Validation of Alternative Methods (ICCVAM) in the USA and European Centre for the Validation of Alternative Methods (ECVAM) in Europe, along with Japanese Centre for the Validation of Alternative Methods (JaCVAM) and Organization for Economic Co-operation and Development (OECD), are essential in assessing the reliability and suitability of alternative methods to animal testing, ensuring they accurately predict product safety. The scheme for validation of alternative methods is depicted in Fig. 14.2.

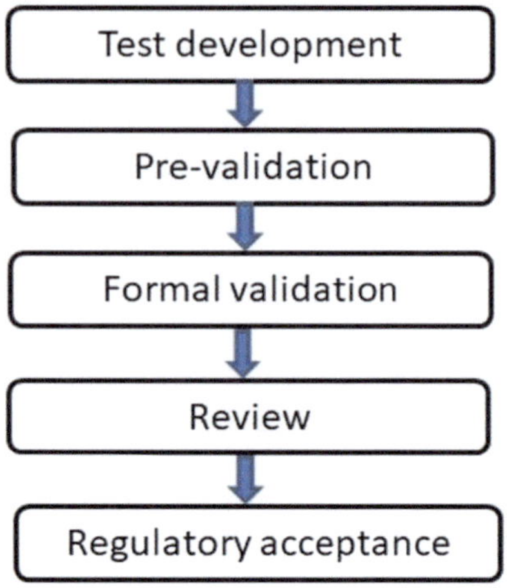

Fig. 14.2 Validation of alternative methods

14.6 Bridging the Gap: Challenges to Adoption

The emerging solution involves the creation of novel research and testing methods that not only avoid animal use but also replace animal experiments. In 2004, the British government established the National Centre for the Replacement, Reduction, and Refinement of Animals in Research to promote the 3Rs and to replace some animal experiments. Non-animal research methods are increasingly recognized as advanced techniques for scientific and medical advancement, with fewer demands on time and funding in countries like Germany, Austria, the Netherlands, and Japan. Replacing animal testing procedures is a well-established concept and is widely accepted among researchers. NAMs are non-animal approaches that replace animal testing for drug toxicity assessment, providing more accurate data and enhancing efficiency. Techniques include organ-on-chip devices using 3D printing to replicate human organs. The OECD is implementing the 3Rs principles through guidance documents and tools, including QSAR models and Adverse Outcome Pathways (AOPs), as part of Integrated Approaches to Testing and Assessment (IATA), to assess pharmaceutical safety. The International Council for Harmonization of Technical Requirements for Pharmaceuticals for Human Use (ICH) also plays an essential role in enhancing the 3Rs principles, reducing drug development resources, and replacing animal testing [94].

The major challenges in implementing alternatives to animal experiments include technological limitations, data interpretation, lack of validated methods, regulatory and ethical barriers, and funding for research. Overcoming these challenges requires a multi-faceted approach thereby enabling the development of effective non-animal models that can be produced cost-effectively. Artificial intelligence (AI), like computational models and machine learning, is providing new opportunities to replace animal experiments with more human-relevant, efficient, and ethical approaches

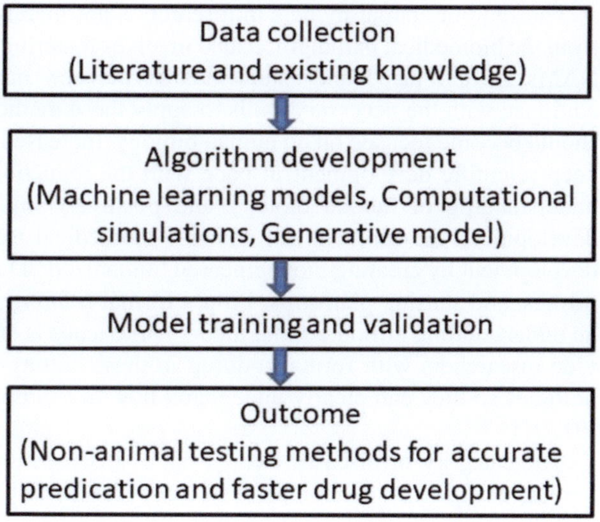

Fig. 14.3 AI workflow for replacing animal models

(Fig. 14.3). AI-based methods can enhance pre-clinical studies by offering virtual platforms for drug screening and optimizing their properties, thereby reducing or replacing the use of animals in drug development [95].

14.7 The Future of Medical Research without Animal Testing

Animal models have been used in research since the eighteenth and nineteenth centuries due to their conservation of biological processes. Advances in genetic engineering have enabled the development of disease-specific animal models; however, they have not been successful in treating complex diseases such as Alzheimer's disease. In recent years, animal-free research alternatives such as computer modelling and human stem cell-derived organoids have gained prominence. Moreover, only about 25% of animal research results are substantial enough to be reported, which negatively impacts the pharmaceutical industry's efficiency and progress [96]. Despite extensive scientific evidence demonstrating the poor quality, publication standards, and limited translatability of many animal studies, the scientific community continues to struggle with making substantial and timely improvements. Addressing these issues requires the active involvement of all stakeholders to ensure the successful implementation of meaningful reforms. The FDA is transitioning to a human-focused approach in drug development, replacing animal testing (phasing out animal testing) with AI-based models. This move is driven by ethical, scientific, and practical reasons, including preventing animal suffering, ensuring accurate results, and meeting public demand for cruelty-free products. Technological advances and evolving regulations are also driving this shift [92].

Phasing out animal models in research is not merely about eliminating animals from the biomedical paradigm; it also involves fostering a scientific culture in which NAMs are embraced and established as the new norm. Researchers should be equipped with the necessary skills to apply these methods effectively, and research should become focused on on human biology. Increased funding for NAMs can help keep scientific development in pace with the reduction in animal use, enhancing understanding of human biology and positively impacting drug discovery and development. Researchers can enhance biomedical research, drug discovery, and development by creating bioengineered humanized 3D models, microfluidics, diagnostics, and sensing platforms. Proper formal training in these advanced tools and an understanding of non-animal model approaches is crucial. It is necessary to provide researchers with formal training in these cutting-edge, human-relevant technologies so they can clearly understand how to use non-animal model approaches [91, 97].

The category of diseases termed "rare" diseases provides an ideal platform to harness genetic and technological advances for identifying new targets, understanding disease progression mechanisms, and facilitating faster drug repurposing. Another feasible approach could be utilizing data mining and patient biological materials for "pre-clinical" testing, instead of expensive and time-consuming animal models. Microphysiological systems such as organ chips are another important system, which can provide efficient drug repurposing and personalized treatments. Human biology-based NAMs are also a useful platform to focus on drug or combination efficacy and safety profiles.

Chou et al. developed successful NAM-based research on human bone marrow chips, which successfully recapitulated haematopoiesis [98]. Medical research is shifting towards human-based technologies sich as 3D bioprinting and computer models, aiming for ethical practices and accurate results. Advancements in AI and personalized medicine could potentially reduce or replace animal testing by 2030, due to more precise health information and the potential for increased computational power and versatility [99].

14.8 Moving Forward: The Road to 2030 and beyond

Medical research is transitioning towards AI, OoC technology, and 3D bioprinting for an ethical understanding of human biology, potentially replacing animal testing in drug development. Regulatory agencies are embracing NAMs, but more validation and regulatory guidance are needed for widespread acceptance [95, 100].

The future of drug development will be shaped by advancements in precision medicine, AI, and machine learning, as well as novel drug modalities such as RNA therapies, cell therapies, and nanomedicines. Global collaboration and open science initiatives will tackle global health challenges. Evolving regulatory frameworks will enable faster approvals without compromising safety, while digital therapeutics will enhance personalized care. Sustainability and ethical considerations will guide the industry's future, with innovations in regenerative medicine, universal vaccines, and decentralized clinical trials (Fig. 14.4).

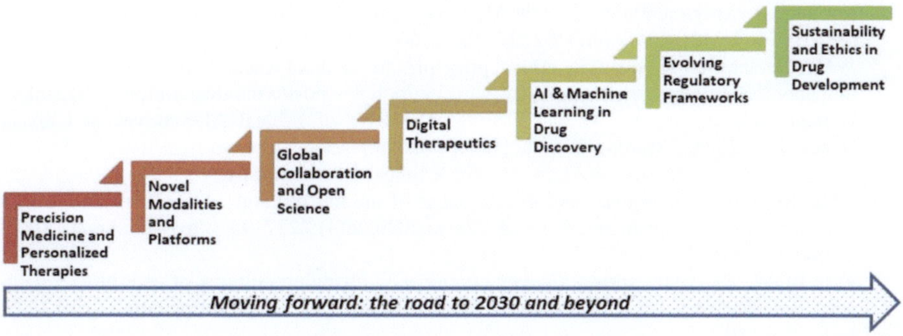

Fig. 14.4 Trends based on global health, AI, and regulatory innovations

14.9 Conclusion

Animal models have been used in preclinical research for over 150 years, primarily relying on mice and rats. However, growing concerns about their clinical relevance and ethical implications have led to significant changes. In January 2023, the US FDA amended the Federal Food, Drug, and Cosmetic Act of 1938, stating that animal testing is no longer required before conducting human clinical trials for pharmaceuticals. However, the World Medical Association's Declaration of Helsinki emphasizes the importance of thorough scientific evidence, including animal and laboratory experiments, in the design and conduct of clinical trials. Phasing out animal models in research is not merely about eliminating animal use; it is about fostering a scientific culture that embraces New Approach Methodologies (NAMs). Increased funding for NAMs can help science keep pace with innovation, enhance our understanding of human biology, and positively impact drug discovery and development.

Acknowledgement Figure 14.1 is reused from Guo et al. [65], licensed under a Creative Commons Attribution 4.0 International License (http://creativecommons.org/licenses/by/4.0/).

References

1. Robinson NB, Krieger K, Khan FM, Huffman W, Chang M, Naik A, Yongle R, Hameed I, Krieger K, Girardi LN, Gaudino M. The current state of animal models in research: a review. Int J Surg (London, England). 2019;72:9–13. https://doi.org/10.1016/j.ijsu.2019.10.015.
2. Agoram BM. Use of pharmacokinetic/pharmacodynamic modelling for starting dose selection in first-in-human trials of high-risk biologics. Br J Clin Pharmacol. 2009;67(2):153–60. https://doi.org/10.1111/j.1365-2125.2008.03297.x.
3. Akhtar A. The flaws and human harms of animal experimentation. Camb Q Healthc Ethics. 2015;24(4):407–19. https://doi.org/10.1017/S0963180115000079.
4. Kiani AK, Pheby D, Henehan G, Brown R, Sieving P, Sykora P, Marks R, Falsini B, Capodicasa N, Miertus S, Lorusso L, Dondossola D, Tartaglia GM, Ergoren MC, Dundar M, Michelini S, Malacarne D, Bonetti G, Dautaj A, et al. Ethical considerations regard-

ing animal experimentation. J Prev Med Hyg. 2022;63(2 Suppl 3):E255–66. https://doi.org/10.15167/2421-4248/jpmh2022.63.2S3.2768.
5. WMA Declaration of helsinki—ethical principles for medical research involving human participants. 2024. [Dataset]. https://www.wma.net/policies-post/wma-declaration-of-helsinki/
6. Deliana Infante. Beyond Animal Testing: The Rise of Ethical Alternatives in Clinical Research. (2024, March 27). https://www.azolifesciences.com/article/Beyond-Animal-Testing-The-Rise-of-Ethical-Alternatives-in-Clinical-Research.aspx
7. Arnason G. The emergence and development of animal research ethics: a review with a focus on nonhuman primates. Sci Eng Ethics. 2020;26(4):2277–93. https://doi.org/10.1007/s11948-020-00219-z.
8. Kirk RGW. The silver spring monkey controversy: changing cultures of care in twentieth-century laboratory animal research. HoST. 2019;13(2):31–59. https://doi.org/10.2478/host-2019-0012.
9. Husain A, Meenakshi DU, Ahmad A, Shrivastava N, Khan SA. A review on alternative methods to experimental animals in biological testing: recent advancement and current strategies. J Pharm Bioallied Sci. 2023;15(4):165–71. https://doi.org/10.4103/jpbs.jpbs_380_23.
10. Tannenbaum J, Bennett BT. Russell and Burch's 3Rs then and now: the need for clarity in definition and purpose. J Am Assoc Lab Anim Sci. 2015;54(2):120–32.
11. Rai J, Kaushik K. Reduction of animal sacrifice in biomedical science & research through alternative design of animal experiments. Saudi Pharm J. 2018;26(6):896–902. https://doi.org/10.1016/j.jsps.2018.03.006.
12. Parasuraman S, Balamurugan S, Vanishya R. Overview of safety assessment and toxicological screening of dermal formulations. SBV J Basic Clin Appl Health Sci. 2020;3(3):96–103. https://doi.org/10.5005/jp-journals-10082-02258.
13. Sreedhar D, Manjula N, Ajay P, Shilpa P, Ligade V. Ban of cosmetic testing on animals: a brief overview. Int J Curr Res Rev. 2020;12(14):113–6. https://doi.org/10.31782/IJCRR.2020.121424.
14. Wadman M. FDA no longer needs to require animal tests before human drug trials. Science, American Association for the Advancement of Science; 2023. https://www.science.org/content/article/fda-no-longer-needs-require-animal-tests-human-drug-trials
15. Reuters. US FDA to phase out animal testing in drug development. 2025, April 11. https://www.reuters.com/world/us/us-fda-phase-out-animal-testing-drug-development-2025-04-10/
16. Animals are not ours to experiment on. n.d.. Retrieved April 14, 2025, from https://www.peta.org.uk/issues/animals-not-experiment-on/
17. Felgueiras J, Ribeiro R, Brevini T a L, Costa PF. State-of-the-art in reproductive bench science: hurdles and new technological solutions. Theriogenology. 2020;150:34–40. https://doi.org/10.1016/j.theriogenology.2020.01.067.
18. Urzì O, Gasparro R, Costanzo E, De Luca A, Giavaresi G, Fontana S, Alessandro R. Three-dimensional cell cultures: the bridge between in vitro and in vivo models. Int J Mol Sci. 2023;24(15):12046. https://doi.org/10.3390/ijms241512046.
19. Edington CD, Chen WLK, Geishecker E, Kassis T, Soenksen LR, Bhushan BM, Freake D, Kirschner J, Maass C, Tsamandouras N, Valdez J, Cook CD, Parent T, Snyder S, Yu J, Suter E, Shockley M, Velazquez J, Velazquez JJ, et al. Interconnected microphysiological systems for quantitative biology and pharmacology studies. Sci Rep. 2018;8(1):4530. https://doi.org/10.1038/s41598-018-22749-0.
20. Kimura H, Sakai Y, Fujii T. Organ/body-on-a-chip based on microfluidic technology for drug discovery. Drug Metab Pharmacokinet. 2018;33(1):43–8. https://doi.org/10.1016/j.dmpk.2017.11.003.
21. Velasco V, Shariati SA, Esfandyarpour R. Microtechnology-based methods for organoid models. Microsyst Nanoeng. 2020;6:76. https://doi.org/10.1038/s41378-020-00185-3.
22. Yang S, Hu H, Kung H, Zou R, Dai Y, Hu Y, Wang T, Lv T, Yu J, Li F. Organoids: the current status and biomedical applications. MedComm. 2023;4(3):e274. https://doi.org/10.1002/mco2.274.

23. Clegg LE, Mac Gabhann F. Molecular mechanism matters: benefits of mechanistic computational models for drug development. Pharmacol Res. 2015;99:149–54. https://doi.org/10.1016/j.phrs.2015.06.002.
24. Han R, Yoon H, Kim G, Lee H, Lee Y. Revolutionizing medicinal chemistry: the application of artificial intelligence (AI) in early drug discovery. Pharmaceuticals (Basel, Switzerland). 2023;16(9):1259. https://doi.org/10.3390/ph16091259.
25. Patel L, Shukla T, Huang X, Ussery DW, Wang S. Machine learning methods in drug discovery. Molecules (Basel, Switzerland). 2020;25(22):5277. https://doi.org/10.3390/molecules25225277.
26. Qureshi R, Irfan M, Gondal TM, Khan S, Wu J, Hadi MU, Heymach J, Le X, Yan H, Alam T. AI in drug discovery and its clinical relevance. Heliyon. 2023;9(7):e17575. https://doi.org/10.1016/j.heliyon.2023.e17575.
27. Wang H, Li X, You X, Zhao G. Harnessing the power of artificial intelligence for human living organoid research. Bioact Mater. 2024;42:140–64. https://doi.org/10.1016/j.bioactmat.2024.08.027.
28. Burki T. A new paradigm for drug development. Lancet Digit Health. 2020;2(5):e226–7. https://doi.org/10.1016/S2589-7500(20)30088-1.
29. Chen W, Liu X, Zhang S, Chen S. Artificial intelligence for drug discovery: resources, methods, and applications. Mol Ther Nucleic Acids. 2023;31:691–702. https://doi.org/10.1016/j.omtn.2023.02.019.
30. Sampene AK, Nyirenda F. Evaluating the effect of artificial intelligence on pharmaceutical product and drug discovery in China. Future J Pharm Sci. 2024;10(1) https://doi.org/10.1186/s43094-024-00632-2.
31. Xie Z, Gao M, Lobo AO, Webster TJ. 3D bioprinting in tissue engineering for medical applications: the classic and the hybrid. Polymers. 2020;12(8):1717. https://doi.org/10.3390/polym12081717.
32. Frankowski J, Kurzątkowska M, Sobczak M, Piotrowska U. Utilization of 3D bioprinting technology in creating human tissue and organoid models for preclinical drug research—state-of-the-art. Int J Pharm. 2023;644:123313. https://doi.org/10.1016/j.ijpharm.2023.123313.
33. Ma J, Wang Y, Liu J. Bioprinting of 3D tissues/organs combined with microfluidics. RSC Adv. 2018a;8(39):21712–27. https://doi.org/10.1039/c8ra03022g.
34. Ma X, Liu J, Zhu W, Tang M, Lawrence N, Yu C, Gou M, Chen S. 3D bioprinting of functional tissue models for personalized drug screening and in vitro disease modeling. Adv Drug Deliv Rev. 2018b;132:235–51. https://doi.org/10.1016/j.addr.2018.06.011.
35. Kolesky DB, Homan KA, Skylar-Scott MA, Lewis JA. Three-dimensional bioprinting of thick vascularized tissues. Proc Natl Acad Sci USA. 2016;113(12):3179–84. https://doi.org/10.1073/pnas.1521342113.
36. Szűcs D, Fekete Z, Guba M, Kemény L, Jemnitz K, Kis E, Veréb Z. Toward better drug development: three-dimensional bioprinting in toxicological research. Int J Bioprinting. 2023;9(2):663. https://doi.org/10.18063/ijb.v9i2.663.
37. Ayuso JM, Park K-Y, Virumbrales-Muñoz M, Beebe DJ. Toward improved in vitro models of human cancer. APL Bioeng. 2021;5(1):010902. https://doi.org/10.1063/5.0026857.
38. Hagenbuchner J, Nothdurfter D, Ausserlechner MJ. 3D bioprinting: novel approaches for engineering complex human tissue equivalents and drug testing. Essays Biochem. 2021;65(3):417–27. https://doi.org/10.1042/EBC20200153.
39. Moysidou C-M, Barberio C, Owens RM. Advances in engineering human tissue models. Front Bioeng Biotechnol. 2020;8:620962. https://doi.org/10.3389/fbioe.2020.620962.
40. Ingber DE. Human organs-on-chips for disease modelling, drug development and personalized medicine. Nat Rev Genet. 2022;23(8):467–91. https://doi.org/10.1038/s41576-022-00466-9.
41. Jensen KB, Little MH. Organoids are not organs: sources of variation and misinformation in organoid biology. Stem Cell Rep. 2023;18(6):1255–70. https://doi.org/10.1016/j.stemcr.2023.05.009.
42. Suhito IR, Kim T-H. Recent advances and challenges in organoid-on-a-chip technology. Organ. 2022;2:e4. https://doi.org/10.51335/organoid.2022.2.e4.

43. Kim J, Koo B-K, Knoblich JA. Human organoids: model systems for human biology and medicine. Nat Rev Mol Cell Biol. 2020;21(10):571–84. https://doi.org/10.1038/s41580-020-0259-3.
44. Andrews MG, Kriegstein AR. Challenges of organoid research. Annu Rev Neurosci. 2022;45:23–39. https://doi.org/10.1146/annurev-neuro-111020-090812.
45. Kaushik G, Ponnusamy MP, Batra SK. Concise review: current status of three-dimensional organoids as preclinical models. Stem Cells (Dayton, Ohio). 2018;36(9):1329–40. https://doi.org/10.1002/stem.2852.
46. Schutgens F, Clevers H. Human organoids: tools for understanding biology and treating diseases. Annu Rev Pathol. 2020;15:211–34. https://doi.org/10.1146/annurev-pathmechdis-012419-032611.
47. Lou Y-R, Leung AW. Next generation organoids for biomedical research and applications. Biotechnol Adv. 2018;36(1):132–49. https://doi.org/10.1016/j.biotechadv.2017.10.005.
48. Hofer M, Lutolf MP. Engineering organoids. Nat Rev Mater. 2021;6(5):402–20. https://doi.org/10.1038/s41578-021-00279-y.
49. Lorenzo-Martín LF, Hübscher T, Bowler AD, Broguiere N, Langer J, Tillard L, Nikolaev M, Radtke F, Lutolf MP. Spatiotemporally resolved colorectal oncogenesis in mini-colons ex vivo. Nature. 2024;629(8011):450–7. https://doi.org/10.1038/s41586-024-07330-2.
50. Pastore M, Giachi A, Spínola-Lasso E, Marra F, Raggi C. Organoids and spheroids: advanced in vitro models for liver cancer research. Front Cell Dev Biol. 2024;12:1536854. https://doi.org/10.3389/fcell.2024.1536854.
51. Balakrishnan S, Atal S, Ray A, Pravin CA, Nanda M. Organoids: an invaluable tool in pharmacology. Indian J Pharmacol. 2020;52(5):422–9. https://doi.org/10.4103/ijp.IJP_137_19.
52. DiMasi JA, Hansen RW, Grabowski HG. The price of innovation: new estimates of drug development costs. J Health Econ. 2003;22(2):151–85. https://doi.org/10.1016/S0167-6296(02)00126-1.
53. Boyd RA, Lalonde RL. Nontraditional approaches to first-in-human studies to increase efficiency of drug development: will microdose studies make a significant impact? Clin Pharmacol Ther. 2007;81(1):24–6. https://doi.org/10.1038/sj.clpt.6100058.
54. Garner R. Less is more: the human microdosing concept☆. Drug Discov Today. 2005;10(7):449–51. https://doi.org/10.1016/S1359-6446(05)03418-5.
55. Kola I, Landis J. Can the pharmaceutical industry reduce attrition rates? Nat Rev Drug Discov. 2004;3(8):711–6. https://doi.org/10.1038/nrd1470.
56. Burt T, Yoshida K, Lappin G, Vuong L, John C, de Wildt SN, Sugiyama Y, Rowland M. Microdosing and other phase 0 clinical trials: facilitating translation in drug development. Clin Transl Sci. 2016;9(2):74–88. https://doi.org/10.1111/cts.12390.
57. Muller PY. Comparative requirements for exploratory clinical trials—eIND, eCTA and microdosing. Adv Drug Deliv Rev. 2011;63(7):511–7. https://doi.org/10.1016/j.addr.2010.10.010.
58. Rani PU, Naidu MUR. Phase 0—microdosing strategy in clinical trials. Indian J Pharmacol. 2008;40(6):240–2. https://doi.org/10.4103/0253-7613.45147.
59. Bergström M, Grahnén A, Långström B. Positron emission tomography microdosing: a new concept with application in tracer and early clinical drug development. Eur J Clin Pharmacol. 2003;59(5–6):357–66. https://doi.org/10.1007/s00228-003-0643-x.
60. Lappin G, Garner RC. Big physics, small doses: the use of AMS and PET in human microdosing of development drugs. Nat Rev Drug Discov. 2003;2(3):233–40. https://doi.org/10.1038/nrd1037.
61. Quansah E, Chen Y, Yang S, Wang J, Sun D, Zhao Y, Chen M, Yu L, Zhang C. CRISPR-Cas13 in malaria parasite: diagnosis and prospective gene function identification. Front Microbiol. 2023;14:1076947. https://doi.org/10.3389/fmicb.2023.1076947.
62. Birling M-C, Herault Y, Pavlovic G. Modeling human disease in rodents by CRISPR/Cas9 genome editing. Mamm Genome. 2017;28(7–8):291–301. https://doi.org/10.1007/s00335-017-9703-x.
63. Cai L, Fisher AL, Huang H, Xie Z. CRISPR-mediated genome editing and human diseases. Genes Dis. 2016;3(4):244–51. https://doi.org/10.1016/j.gendis.2016.07.003.

64. Martinez-Lage M, Torres-Ruiz R, Rodriguez-Perales S. CRISPR/Cas9 technology: applications and human disease modeling. Prog Mol Biol Transl Sci. 2017;152:23–48. https://doi.org/10.1016/bs.pmbts.2017.09.002.
65. Guo M, Chen H, Dong S, Zhang Z, Luo H. CRISPR-Cas gene editing technology and its application prospect in medicinal plants. Chin Med. 2022;17(1) https://doi.org/10.1186/s13020-022-00584-w.
66. Wang JY, Doudna JA. CRISPR technology: a decade of genome editing is only the beginning. Science (New York, NY). 2023;379(6629):eadd8643. https://doi.org/10.1126/science.add8643.
67. Richardson C, Kelsh RN, J Richardson R. New advances in CRISPR/Cas-mediated precise gene-editing techniques. Dis Model Mech. 2023;16(2):dmm049874. https://doi.org/10.1242/dmm.049874.
68. Sander JD, Joung JK. CRISPR-Cas systems for editing, regulating and targeting genomes. Nat Biotechnol. 2014;32(4):347–55. https://doi.org/10.1038/nbt.2842.
69. Daneva T. Crispr/Cas9 system—a revolution in gene editing. Biomed J Sci Tech Res. 2018;10(1) https://doi.org/10.26717/bjstr.2018.10.001902.
70. Rodríguez-Rodríguez DR, Ramírez-Solís R, Garza-Elizondo MA, Garza-Rodríguez MDL, Barrera-Saldaña HA. Genome editing: a perspective on the application of CRISPR/Cas9 to study human diseases (review). Int J Mol Med. 2019;43(4):1559–74. https://doi.org/10.3892/ijmm.2019.4112.
71. Mout R, Ray M, Lee Y-W, Scaletti F, Rotello VM. In vivo delivery of CRISPR/Cas9 for therapeutic gene editing: progress and challenges. Bioconjug Chem. 2017;28(4):880–4. https://doi.org/10.1021/acs.bioconjchem.7b00057.
72. Redman M, King A, Watson C, King D. What is CRISPR/Cas9? Arch Dis Child Educ Pract Ed. 2016;101(4):213–5. https://doi.org/10.1136/archdischild-2016-310459.
73. Xiao-Jie L, Hui-Ying X, Zun-Ping K, Jin-Lian C, Li-Juan J. CRISPR-Cas9: a new and promising player in gene therapy. J Med Genet. 2015;52(5):289–96. https://doi.org/10.1136/jmedgenet-2014-102968.
74. Parsaeimehr A, Ebirim RI, Ozbay G. CRISPR-Cas technology a new era in genomic engineering. Biotechnol Rep (Amst). 2022;34:e00731. https://doi.org/10.1016/j.btre.2022.e00731.
75. Zarei A, Razban V, Hosseini SE, Tabei SMB. Creating cell and animal models of human disease by genome editing using CRISPR/Cas9. J Gene Med. 2019;21(4):e3082. https://doi.org/10.1002/jgm.3082.
76. Abdelnour SA, Xie L, Hassanin AA, Zuo E, Lu Y. The potential of CRISPR/Cas9 gene editing as a treatment strategy for inherited diseases. Front Cell Dev Biol. 2021;9:699597. https://doi.org/10.3389/fcell.2021.699597.
77. Horodecka K, Düchler M. CRISPR/Cas9: principle, applications, and delivery through extracellular vesicles. Int J Mol Sci. 2021;22(11):6072. https://doi.org/10.3390/ijms22116072.
78. Ansori AN, Antonius Y, Susilo RJ, Hayaza S, Kharisma VD, Parikesit AA, Zainul R, Jakhmola V, Saklani T, Rebezov M, Ullah ME, Maksimiuk N, Derkho M, Burkov P. Application of CRISPR-Cas9 genome editing technology in various fields: a review. Narra J. 2023;3(2):e184. https://doi.org/10.52225/narra.v3i2.184.
79. Davis DJ, Yeddula SGR. CRISPR advancements for human health. Mo Med. 2024;121(2):170–6.
80. Dubey AK, Mostafavi E. Biomaterials-mediated CRISPR/Cas9 delivery: recent challenges and opportunities in gene therapy. Front Chem. 2023;11:1259435. https://doi.org/10.3389/fchem.2023.1259435.
81. Guo C, Ma X, Gao F, Guo Y. Off-target effects in CRISPR/Cas9 gene editing. Front Bioeng Biotechnol. 2023;11:1143157. https://doi.org/10.3389/fbioe.2023.1143157.
82. Li T, Yang Y, Qi H, Cui W, Zhang L, Fu X, He X, Liu M, Li P-F, Yu T. CRISPR/Cas9 therapeutics: progress and prospects. Signal Transduct Target Ther. 2023;8(1):36. https://doi.org/10.1038/s41392-023-01309-7.

83. Wong CH, Siah KW, Lo AW. Estimation of clinical trial success rates and related parameters. Biostatistics (Oxford, England). 2019;20(2):273–86. https://doi.org/10.1093/biostatistics/kxx069.
84. Patton RL, Kalback WM, Esh CL, Kokjohn TA, Van Vickle GD, Luehrs DC, Kuo Y-M, Lopez J, Brune D, Ferrer I, Masliah E, Newel AJ, Beach TG, Castaño EM, Roher AE. Amyloid-beta peptide remnants in AN-1792-immunized Alzheimer's disease patients: a biochemical analysis. Am J Pathol. 2006;169(3):1048–63. https://doi.org/10.2353/ajpath.2006.060269.
85. Robinson SR, Bishop GM, Lee H-G, Münch G. Lessons from the AN 1792 Alzheimer vaccine: lest we forget. Neurobiol Aging. 2004;25(5):609–15. https://doi.org/10.1016/j.neurobiolaging.2003.12.020.
86. Zhao Z, Cheng M, Maples KR, Ma JY, Buchan AM. NXY-059, a novel free radical trapping compound, reduces cortical infarction after permanent focal cerebral ischemia in the rat. Brain Res. 2001;909(1–2):46–50. https://doi.org/10.1016/s0006-8993(01)02618-x.
87. Diener H-C, Lees KR, Lyden P, Grotta J, Davalos A, Davis SM, Shuaib A, Ashwood T, Wasiewski W, Alderfer V, Hårdemark H-G, Rodichok L, SAINT I and II Investigators. NXY-059 for the treatment of acute stroke: pooled analysis of the SAINT I and II trials. Stroke. 2008;39(6):1751–8. https://doi.org/10.1161/STROKEAHA.107.503334.
88. Winer A, Adams S, Mignatti P. Matrix metalloproteinase inhibitors in cancer therapy: turning past failures into future successes. Mol Cancer Ther. 2018;17(6):1147–55. https://doi.org/10.1158/1535-7163.MCT-17-0646.
89. Herrera C, Veazey R, Lemke MM, Arnold K, Kim JH, Shattock RJ. Ex vivo evaluation of mucosal responses to vaccination with ALVAC and AIDSVAX of non-human primates. Vaccine. 2022;10(2):187. https://doi.org/10.3390/vaccines10020187.
90. Kim JH, Rerks-Ngarm S, Excler J-L, Michael NL. HIV vaccines: lessons learned and the way forward. Curr Opin HIV AIDS. 2010;5(5):428–34. https://doi.org/10.1097/COH.0b013e32833d17ac.
91. Marshall LJ, Constantino H, Seidle T. Phase-in to phase-out-targeted, inclusive strategies are needed to enable full replacement of animal use in the European Union. Animals. 2022;12(7):863. https://doi.org/10.3390/ani12070863.
92. FDA News Release. FDA announces plan to phase out animal testing requirement for monoclonal antibodies and other drugs. 2025, April 10. https://www.fda.gov/news-events/press-announcements/fda-announces-plan-phase-out-animal-testing-requirement-monoclonal-antibodies-and-other-drugs.
93. Hartung T. The validation of regulatory test methods – conceptual, ethical, and philosophical foundations. ALTEX. 2024;41(4) https://doi.org/10.14573/altex.2409271.
94. Poh WT, Stanslas J. The new paradigm in animal testing—"3Rs alternatives". Regul Toxicol Pharm. 2024;153:105705. https://doi.org/10.1016/j.yrtph.2024.105705.
95. Rudroff T. Artificial intelligence as a replacement for animal experiments in neurology: potential, Progress, and challenges. Neurol Int. 2024;16(4):805–20. https://doi.org/10.3390/neurolint16040060.
96. Ritskes-Hoitinga M. Can animal-free research be the future of medical science? 2021, May 21. https://researchoutreach.org/articles/animal-free-research-future-medical-science/
97. Gharib G, Bütün İ, Muganlı Z, Kozalak G, Namlı İ, Sarraf SS, Ahmadi VE, Toyran E, van Wijnen AJ, Koşar A. Biomedical applications of microfluidic devices: a review. Biosensors. 2022;12(11):1023. https://doi.org/10.3390/bios12111023.
98. Chou DB, Frismantas V, Milton Y, David R, Pop-Damkov P, Ferguson D, MacDonald A, Vargel Bölükbaşı Ö, Joyce CE, Moreira Teixeira LS, Rech A, Jiang A, Calamari E, Jalili-Firoozinezhad S, Furlong BA, O'Sullivan LR, Ng CF, Choe Y, Marquez S, et al. On-chip recapitulation of clinical bone marrow toxicities and patient-specific pathophysiology. Nat Biomed Eng. 2020;4(4):394–406. https://doi.org/10.1038/s41551-019-0495-z.
99. Brownell L. The next decade in science. 2020, January 13
100. Kandarova H. Alternative methods to animal experimentation and their role in modern toxicology. In: Toxicological risk assessment and multi-system health impacts from exposure. Elsevier; 2021. p. 13–22. https://doi.org/10.1016/B978-0-323-85215-9.00047-7.

Ethical Challenges and Barriers in Animal Experimentation 15

Thi Phuong Truc Nguyen and Thi Thu Hoai Nguyen

15.1 Importance of Animal Experimentation

Animal experiments have long played a pivotal role in biomedical research and scientific discovery. They have aided in the development of life-saving vaccines, improved surgical techniques, and elucidated complex physiological mechanisms, contributing to significant advances in human and animal health [18].

Different animal species (rodents, zebrafish, non-human primates, etc.) are selected for each specific research purpose, based on genetic similarity to humans, ease of manipulation, and developed disease models (Fig. 15.1). Animal models play an important role in drug development and safety assessment. Mice and rats are commonly used in preclinical research before clinical trials due to their clear genetic characteristics, ease of handling, and specific disease transgenic models [100]. These rodents not only play a crucial role in preclinical assessment but also help validate the efficacy of chemotherapeutic agents [20], antiretroviral drugs [63], and anti-inflammatory compounds [74]. Drug absorption [67], distribution [41], metabolism [99], and excretion (ADME) [157] are evaluated using large mammals such as rabbits and pigs, while biological products such as monoclonal antibodies and

T. P. T. Nguyen
School of Biotechnology, International University, Vietnam National University Ho Chi Minh City, Ho Chi Minh City, Vietnam

T. T. H. Nguyen (✉)
School of Biotechnology, International University, Vietnam National University Ho Chi Minh City, Ho Chi Minh City, Vietnam

Research Center for Infectious Diseases, International University, Vietnam National University Ho Chi Minh City, Ho Chi Minh City, Vietnam
e-mail: ntthoai@hcmiu.edu.vn

P. Kumar, V. Kothari (eds.), *Model Organisms in Biological Research*,
https://doi.org/10.1007/978-981-92-0038-2_15

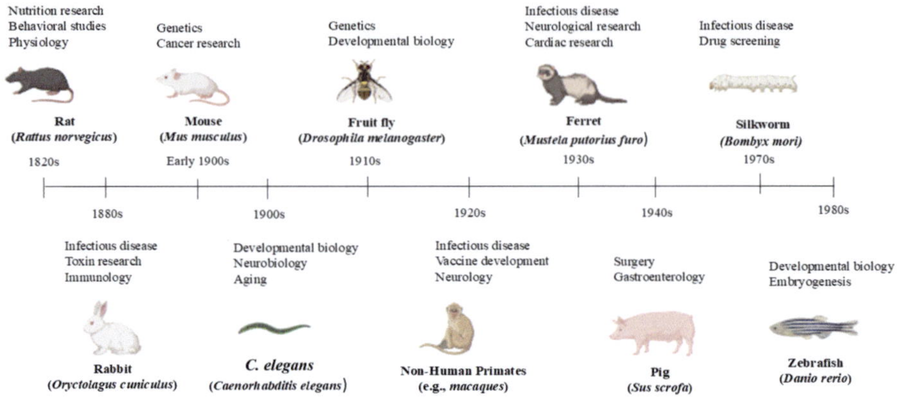

Fig. 15.1 History of animal experimentation and the introduction of different models. The figure is generated using the Biorender free version 2025

vaccines are tested using non-human primates, such as macaques, due to their immune systems being similar to those of humans [19, 124]. As summarized in Table 15.1, the advantages and disadvantages of various animal models are essential for selecting the most appropriate system, balancing scientific goals with ethical responsibility (Table 15.1).

In the early-stage pharmacological research of the 1960s, animal models are essential for investigating disease progression and pathology. These models provide researchers with a dynamic, whole-organism system to explore disease onset, development, and complications under controlled conditions. For example, genetically engineered mice have been used to model human cancers [112], cardiovascular disorders [97], and neurodegenerative diseases such as Alzheimer's [114] and Parkinson's [94]. Zebrafish, owing to their transparent embryos and rapid development, are ideal for studying real-time disease progression at the cellular and tissue level, particularly in fields such as developmental biology and toxicology [65]. Non-human primates remain critical for research into complex disorders such as HIV/AIDS [117] and emerging infectious diseases [108], enabling researchers to replicate human disease pathways and evaluate intervention strategies more accurately.

The rise of molecular biology and genetic engineering in the 1980s–2000s brought forward animal models for exploring gene-disease relationships [12]. The fruit fly (Drosophila melanogaster), though evolutionarily distant from humans, shares a high degree of genetic conservation with human genes and has proven indispensable in uncovering gene-disease associations [56]. Using CRISPR and gene-editing technologies, scientists can create precise mutations in fruit flies [148], mice [131], or zebrafish [47] to observe phenotypes and determine the role of genes, aiding in the modeling of rare genetic diseases and the development of gene therapies.

Table 15.1 Advantages and disadvantages of using different animal models

Models	Advantages	Disadvantages	References
Rat (*Rattus norvegicus*)	Larger size facilitates sampling and surgery Common in behavioral and toxicology studies	*Fewer genetic tools compared to mice* *Still differs from human metabolism*	[25]
Rabbit (*Oryctolagus cuniculus*)	Commonly used in antibody production and ocular research	Sensitive to stress	[17]
Mouse (*Mus musculus*)	Well-characterized genome Availability of transgenic and knockout models Short reproductive cycle Cost-effective	Physiological differences from humans Small size limits surgical procedures	[139]
Nematode *C. elegans* (*Caenorhabditis elegans*)	Simple structure, short life cycle. Easy to culture, low cost. Transparent, easy observation. Fully sequenced genome, many human gene homologs	Invertebrate, far from human physiology. Lacks adaptive immunity, circulatory system. Cannot model complex behavior or human infections	[64]
Fruit fly (*Drosophila melanogaster*)	Fast life cycle and low cost Well-studied genetics High-throughput screening capability	Invertebrate: Limited organ systems Distant evolutionary relation to humans	[144]
Non-human primates (e.g., *macaques*)	Closest to humans in anatomy and physiology Useful for neurological and infectious disease studies	High ethical concerns Expensive and limited availability Long reproductive cycles	[150]
Ferret (*Mustela putorius furo*)	Respiratory physiology close to humans Natural host for human respiratory viruses (e.g., influenza, SARS-CoV-2). Show similar symptoms and transmission Suitable for vaccine and drug studies	High maintenance cost, special housing required. Ethical concerns, strict regulations. Long generation time	[123]
Pig (*Sus scrofa*)	Similar organ size and physiology to humans Useful for cardiovascular and surgical research	Longer gestation and development periods	[162]
Silkworms (*Bombyx mori*)	Low cost and easy to breed Good infection model Genetical well-characterized	Not a vertebrate Limited translational relevance	[96]
Zebrafish (*Danio rerio*)	Transparent embryos for developmental studies High-throughput screening capability	Not ideal for complex organ studies	[142]

Animal models are important tools in nutrition research, helping to understand the physiological effects of diet and metabolic disorders. Rodents have long been used to study obesity, diabetes, and micronutrients [4], while pigs have recently gained popularity due to their similar digestive system to humans [162]. These studies provide the basis for nutritional guidelines, elucidation of the mechanisms of metabolic syndrome, and development of therapeutic interventions.

In psychology and behavioral science, the use of animals began with Pavlov's dog experiments and Skinner's conditioning. Rodents have subsequently been widely used in behavioral paradigms such as the open field test [102], the Morris water maze [161], and fear conditioning to study anxiety [101], memory [38], and stress responses [44], providing insights into the neural mechanisms of depression [95], PTSD [14], and autism spectrum disorders [85]. Primates are used in social cognition research due to their similar brain structures and social systems to humans [149]. Research on enrichment [16], maternal care [37], and group dynamics [98] has also contributed to shaping animal welfare standards in laboratory and natural environments.

Beyond the laboratory, animal experiments have made important contributions to translational and public health research since the 1950s. They have helped to strengthen our understanding of zoonotic transmission and develop responses to outbreaks such as Ebola [48], SARS, and COVID-19 [69]. At the same time, animal research has fueled vaccine development [141] antimicrobial resistance studies [127], and epidemiological modeling [129].

Despite its many contributions, animal experiments pose ethical and logistical challenges, as different species have different costs, care needs, and levels of social acceptance. Species differences sometimes limit the ability to extrapolate results to humans, requiring careful model selection and validation [6]. Additionally, some experiments are conducted primarily due to funding requirements, while review boards may not be fully aware of alternative approaches. For example, in studies evaluating traditional medicine formulations with a history of safe use, animal toxicity testing may not be necessary [146]. Therefore, there is a need for increased awareness among the research community, regulatory agencies, and funding organizations to avoid unnecessary reliance on animal models.

15.2 Ethical Concerns in Animal Experimentation

Despite its critical role, animal research remains a subject of ethical scrutiny, eliciting widespread debate among scientists, ethicists, policymakers, and the general public. At the heart of the debate is the moral dilemma of balancing the potential human benefits derived from animal experimentation against the welfare and rights of the animals involved [73].

In 2013, Oregon passed a law recognizing animals as sentient beings capable of experiencing pain, stress, and fear [111]. Similar legislative actions have been taken globally, reflecting a broader shift in societal values. For example,

Colombia's Law 1774 declares that animals, as sentient beings, are not objects and must be protected from suffering and pain caused by humans [35]. Likewise, countries such as France [109], Sweden [82], the Netherlands [9], Spain [62], New Zealand [8], and the United Kingdom has enacted laws recognizing animal sentience [7]. A prominent example is the United Kingdom's Animal Welfare (Sentience) Act 2022, which formalized the creation of an Animal Sentience Committee tasked with reviewing government policies to ensure that animal welfare considerations are adequately addressed. This recognition has led to growing societal and academic pressure to reevaluate traditional practices in animal-based research. Furthermore, the ethical tension is exacerbated by scientific concerns about the translatability of animal research findings to human conditions [60]. Numerous cases have highlighted the discrepancies between animal models and human physiology, raising doubts about certain animal studies' reliability and predictive value [83, 122].

To address these concerns, ethical review committees and regulatory bodies have adopted the 4Rs: Reduction, Refinement, Replacement, and Responsibility [75], which aim to limit the use and suffering of animals while ensuring the validity of research. However, effective implementation of these principles remains a major challenge. Advances such as organ-on-a-chip systems [79], 3D cell culture [50] and computer modeling [130] are emerging as potential alternatives, but still fall short of fully reproducing the biological complexity of living organisms [137].

15.2.1 Animal Welfare, Moral Status, and the Problem of Speciesism

The treatment of animals as sentient beings, capable of feeling pain, stress, and discomfort, is regarded as a key ethical issue in animal experimentation. Modern research has shown that many species, especially vertebrates, possess neurological structures similar to humans, allowing them to experience suffering. As a result, procedures that cause physical or psychological harm are under increasingly strict ethical scrutiny. Animal rights advocates argue that this ability to feel pain should be considered morally equal, and that animals deserve to be treated in the same way as humans. They challenge the notion of assigning lower moral value to animals solely based on species, a practice referred to as "speciesism"—a form of prejudice that, they argue, is no more justifiable than discrimination based on race or gender [26].

From this standpoint, many argue that the capacity to suffer, not species membership, should determine moral relevance. If animals can experience suffering, then their interests warrant moral consideration [92]. Critics contend that using animals purely as a means to achieve human ends, particularly in research that lacks strong justification or clear societal benefit, represents an ethical shortfall. They call for a rethinking of our responsibilities toward non-human animals, advocating for a more compassionate and rights-based approach [71].

In response to these ethical challenges, many countries have implemented strict animal welfare regulations and oversight mechanisms to ensure the responsible use of animals in science. Ethics committees now require that researchers justify their use of animals, minimize potential harm, and demonstrate the absence of viable alternatives. The 3R principles, which is Replacement, Reduction, and Refinement, form the core of these requirements and serve as a global framework for humane animal research. Researchers actively implement this principle by replacing animals with non-animal models whenever possible, reducing the number of animals used, and refining experimental procedures to alleviate suffering [76].

Despite existing safeguards, debate over the ethics of animal research continues. Some scholars view current welfare standards as a reasonable balance between scientific advancement and animal welfare protection, while others argue that any use of sentient animals is ethically unjustifiable. This tension reflects the struggle to maintain a balance between science and ethics, and underscores the need to continually improve research practices and promote humane, rights-based approaches [58].

15.2.2 Counterarguments and Utilitarian Justifications

Despite growing ethical concerns, many in the scientific community defend animal experimentation by emphasizing the differences between humans and animals in terms of cognitive complexity and moral agency. Proponents argue that animals, while deserving of ethical treatment, do not possess the same degree of self-awareness, autonomy, or reasoning capacities as humans, and thus are not entitled to the same moral status [31]. Within this framework, using animals in research is deemed ethically permissible, so long as harm is minimized and the studies serve a greater good [32].

Utilitarian ethics often underpin this justification, focusing on the overall balance of benefits versus harms [147]. From this perspective, the advancement of medical knowledge, the development of new treatments, and the prevention of human suffering can outweigh the ethical costs associated with animal testing. This pragmatic approach supports the idea that certain sacrifices are acceptable when the outcomes are likely to yield significant benefits to society [32].

Tools such as Bateson's Cube have been developed to assess the ethical acceptability of animal experiments, based on three factors: degree of suffering, potential benefit, and scientific quality. Experiments with high scientific value, clear benefits, and little harm to animals are considered ethically acceptable, while research that causes great suffering and low scientific value is often rejected [77].

Public attitudes toward animal research continue to evolve, fueled by growing awareness of the sentience of animals and advocacy efforts from animal rights groups. At the same time, scientific innovations focus on alternatives to reduce, refine, or replace animal use. As society becomes increasingly concerned about ethics, demands for transparency and accountability in animal research increase, challenging scientists to balance scientific rigor with ethical responsibility.

15.3 The 4Rs Principle: A Framework for Ethical Animal Research

The ethical use of animals in scientific research has long been a source of debate, leading to the development of frameworks to guide responsible practice. One of the most widely accepted models is the 4R Principles, which include Reduction, Refinement, Replacement and Responsibility, providing an ethical foundation for ensuring animal welfare while maintaining scientific integrity [87].

The Reduction focuses on limiting the number of animals used in experiments without reducing the quality or reliability of the data [42]. This is accomplished through careful experimental design, the use of advanced statistical methods, comprehensive literature reviews, and the use of modern technology such as high-resolution imaging or data modeling. Researchers are also encouraged to collaborate, share data, and resources to achieve meaningful results with fewer animals. By optimizing processes and avoiding duplication, the Minimization Principle both protects animal lives and improves scientific efficiency [43].

Refinement emphasizes adjusting experimental procedures to reduce or eliminate potential pain, stress, and discomfort for animals. Measures include providing appropriate housing, improving handling techniques, using anesthesia or pain medication when necessary, training staff, applying humane criteria, and providing environmental enrichment. The goal of Refinement is to both minimize harm and improve the overall well-being of animals, thereby ensuring more reliable and humane research results.[43].

The Replacement uses alternative methods to avoid or reduce the use of live animals. These methods may include in vitro techniques (cell or tissue cultures), in silico (computer simulations), or the use of lower organisms, such as invertebrates, where appropriate [54]. Developments in biotechnology have greatly expanded the range of non-animal models, which sometimes more accurately mimic human biological processes than traditional testing. By replacing animals whenever possible, researchers can both meet ethical requirements and promote scientific innovation [46].

The most recent addition, Responsibility, extends the ethical obligations of researchers beyond the laboratory to animals and society. It encourages scientists to be acutely aware of the ethical implications of their research and to maintain accountability and transparency in their work [55]. It also promotes participation in public discussions about animal ethics and the development of methods to assess animals' sentience, cognition, and emotional states. Commitment to the Principle of Responsibility means continuously improving welfare standards and ensuring animals are treated with respect and dignity [91] (Fig. 15.2).

The 4Rs act as a comprehensive ethical guide, balancing the pursuit of scientific knowledge with the obligation to minimize harm to animals. Adherence to the 4Rs helps researchers meet legal and institutional requirements while promoting humane and ethical scientific practices. While providing a general framework, implementation of the 4Rs can vary between countries and regions depending on specific regulations and policies. Table 15.2 summarizes these differences and highlights how global standards and legal frameworks shape animal experimentation practices.

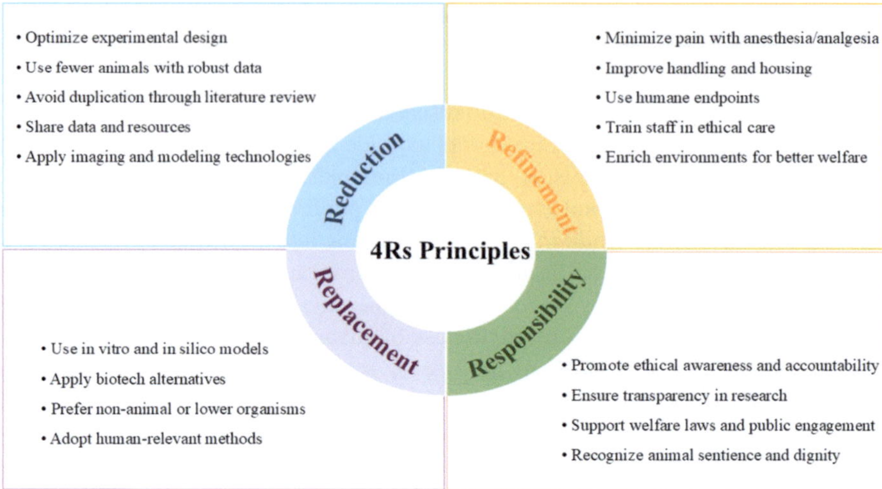

Fig. 15.2 The 4Rs Principle. The figure is generated using the Biorender.com free version 2025

Table 15.2 The regulations on Animal Experimentation in different countries and regions

Country/ region	Regulatory framework	Key requirement	Enforcement and oversight	References
European Union	Directive 2010/63/EU on the protection of animals used for scientific purposes	Mandatory application of the 3Rs (Replacement, Reduction, Refinement), ethical review, licensing of procedures and personnel	National Competent Authorities, inspections by government agencies	[128]
United States	Animal Welfare Act (AWA); Public Health Service (PHS) Policy on Humane Care and Use of Laboratory Animals	Institutional Animal Care and Use Committees (IACUCs), compliance with the 3Rs, annual reporting, accreditation encouraged	USDA (for AWA), Office of Laboratory Animal Welfare (OLAW) under NIH	[10, 126]
Australia	Australian Code for the Care and Use of Animals for Scientific Purposes	Australian code for the care and use of animals for scientific purposes	National Health and Medical Research Council (NHMRC) and state regulators	[119]

(continued)

Table 15.2 (continued)

Country/region	Regulatory framework	Key requirement	Enforcement and oversight	References
Brazil	National Council for the Control of Animal Experimentation (CONCEA)	Adherence to ethical principles and 3Rs, animal ethics committees required in all institutions	CONCEA oversight with regular audits	[39]
Canada	Canadian Council on Animal Care (CCAC) Guidelines	Institutional animal care committees (ACCs), ethics review, adherence to CCAC standards and the 3Rs	CCAC site visits and program assessments	[27]
China	Laboratory Animal Management Regulations (1988, updated 2017); National Guidelines	Ethical approval required, application of the 4Rs encouraged, focus on humane endpoints	Ministry of Science and Technology (MOST), local ethics committees	[72]
India	Committee for the Purpose of Control and Supervision of Experiments on Animals (CPCSEA)	Registration of establishments, animal ethics committees, emphasis on the 3Rs, mandatory training	CPCSEA under Ministry of Environment, Forest and Climate Change	[118]
Japan	Act on Welfare and Management of Animals (amended 2005); Guidelines by Ministry of Education, Culture, Sports, Science and Technology (MEXT)	Promotion of the 3Rs, ethical review by Institutional Animal Care and Use Committees (IACUCs), transparency in research	Institutional self-regulation with MEXT oversight	[80, 107]
South Korea	Laboratory Animal Act (2008)	Animal Ethics Committees required, application of the 3Rs, registration and inspection of facilities	Ministry of Food and Drug Safety (MFDS)	[5]
United Kingdom	Animals (Scientific Procedures) Act 1986 (ASPA), amended to comply with EU Directive	Licensing of institutions, projects, and individuals; harm-benefit analysis; adherence to the 3Rs	Home Office oversight and inspections	[128]

15.4 Barriers to Implementing the 4Rs

15.4.1 Limitations of Alternative Methods

In recent years, considerable progress has been made in developing alternatives to animal experimentation. Technologies such as the organ-on-a-chip system [88], 3D cell culture [81], and computational modeling [13] have emerged as promising tools for simulating human physiology, disease processes, and drug responses. These methods aim to replicate aspects of complex biological systems in a controlled, ethical, and often more cost-effective manner. However, their adoption remains limited due to various scientific and technical limitations (Table 15.3).

3D Cell Culture

Three-dimensional (3D) cell culture models have revolutionized in vitro experimentation by offering more physiologically relevant environments compared to traditional two-dimensional (2D) monolayer cultures. In 3D systems, cells grow within or on a scaffold [70], matrix [134], or self-assemble into spheroids [2], allowing them to interact with their surroundings in a manner that mimics the architecture, gradients, and cell-cell interactions of real tissues. These models provide critical

Table 15.3 Advantages and disadvantages of the new models with year of development and stages of application

Models	Year	Application	Advantages	Disadvantages	Reference
Organ-on-a-Chip (OOC)	2010	Preclinical Drug testing Toxicology Disease modeling	Mimics human organ-level functions and microenvironment Real-time monitoring Reduces interspecies variation	Mainly single-organ systems High technical complexity Limited standardization across labs	[23]
3D cell culture	1970	Cancer research Tissue engineering Regenerative medicine	Better mimics tissue architecture Improves physiological relevance Suitable for drug penetration and tumor modeling	Lack of vasculature limits nutrient diffusion Limited systemic and immune system modeling Reproducibility issues	[105]
Computational modeling	1980	Toxicology screening PK/PD analysis Systems biology Virtual drug trials	High-throughput and scalable Reduces cost and animal use Enables hypothesis generation and risk prioritization	Dependent on data quality May oversimplify biological complexity Requires experimental validation	[89]

insights into processes like tumor growth [93], tissue development [116], drug penetration [115], and cellular differentiation [24].

Various approaches are used in 3D cultures, including scaffold-based methods using hydrogels [49] or biopolymers [136], scaffold-free techniques like hanging drop spheroids [29], and bioprinting technologies [11] that precisely positions cells to build complex structures. 3D cultures are particularly useful in cancer research, as tumor spheres can reproduce the hypoxic core and nutrient gradient of solid tumors, thereby providing a more accurate model for evaluating the efficacy of chemotherapy and radiotherapy [36, 57]. In regenerative medicine, 3D systems also play an important role in studying tissue engineering [158] and stem cell behavior [28].

Computational Modeling

Computational models, often called in silico models, use mathematical frameworks and computer simulations to predict biological behavior and outcomes [33]. These models range from simple statistical correlations to highly complex systems biology approaches that simulate the interactions among genes, proteins, cells, and organs [135]. Computational toxicology [21], pharmacokinetic-pharmacodynamic (PK-PD) modeling [160], and machine learning-based predictive models [143] are some of the major applications designed to reduce reliance on animal experimentation.

One strength of computational modeling lies in its scalability and efficiency. Virtual screening can evaluate thousands of drug candidates for toxicity or efficacy before any laboratory experiment is conducted, saving time, reducing costs, and minimizing ethical concerns [61]. Virtual screening tools, such as AtomNet developed by Atomwise Inc. (USA), can evaluate thousands of drug candidates for toxicity or efficacy before any laboratory experiment is conducted, saving time, reducing costs, and minimizing ethical concerns. AtomNet, a deep convolutional neural network, uses structural information to predict bioactive small molecules and has been applied to screen billions of compounds, significantly accelerating the drug discovery process [51]. Other tools, such as HelixVS, which integrates deep learning-based pose scoring, and LigandScout, which generates 3D pharmacophore models from macromolecule–ligand complexes, have also demonstrated enhanced efficiency and accuracy in virtual screening [151].

Beyond virtual screening, systems biology models can predict how perturbations at the molecular level affect the function of entire biological networks. Additionally, machine learning algorithms trained on large biological datasets are able to identify patterns that traditional testing methods may miss. Regulators are increasingly appreciating the effectiveness of computational models, particularly in assessing initial risk and prioritizing compounds for further testing.

The impact of molecular perturbations on the function of entire biological networks can be predicted using systems biology models. At the same time, patterns that traditional experimental methods may miss can be detected by machine learning algorithms trained on large biological datasets. The value of computational models is increasingly recognized by regulatory agencies, particularly in initial risk assessment and prioritization of compounds for further testing [103].

Organ on a Chip (OOC)

Organ-on-a-chip (OOC) emulation technology marks a breakthrough in biomedical research by reproducing the physiological functions of human organs on a micro-engineering platform. These devices are developed using microfluidic technology, allowing precise control of small volumes of fluid in microchannels. By incorporating living human cells, reproducing the unique structure of each organ, and simulating natural mechanical forces, OOC models recreate a dynamic biological environment that closely resembles the behavior of organs in living organisms [52]. For example, lung-on-a-chip devices can simulate breathing by applying mechanical stretching forces to lung epithelial cells, while heart-on-a-chip can reproduce the natural contraction rhythm of cardiac tissue [59].

The strengths of organ-on-a-chip (OOC) system stem from its ability to realistically reproduce tissue interactions, reflect organ-specific functions, and record biological processes in real time. These platforms are widely used in pharmaceutical toxicity assessment [34], disease modeling [156], cancer metastasis [110], and personalized medicine [125]. Compared to traditional 2D culture models, OOCs provide higher biological validity; at the same time, compared to animal models, they minimize species bias, which often hinders the translation of results to the clinic [104].

15.4.2 Scientific Challenges in Replicating Complex Biological Systems

Although significant advances have been made in alternative models to animal testing, fully reproducing the complexity of biological systems remains a major challenge for scientists. Organisms function as intricate networks where organs, tissues, and cells interact continuously under the influence of genetic factors, biochemical signals, and environmental conditions. Many of these dynamic interactions are difficult to fully simulate in vitro or in silico [140, 159].

A significant limitation lies in the difficulty of reproducing physiological responses at the whole-body level. Processes such as immune modulation [40], endocrine regulation [66], and multi-organ toxicity [15] require a level of integration that current models cannot fully emulate. Advanced tools like three-dimensional (3D) cell cultures and organ-on-a-chip systems have improved the physiological relevance of in vitro models compared to traditional monolayer cultures [121]. However, these innovations still fall short of representing the holistic nature of the human body. For example, although the lung response to a drug can be accurately simulated by an organ-on-a-chip device, hepatic metabolism of the compound, renal drug clearance, and immune system responses cannot be fully reproduced. Systemic phenomena such as pharmacokinetics and biodistribution remain difficult to simulate without the context of an intact organism [78].

In the 3D system, the major limitation noted is the absence of blood vessels, which limits the diffusion of nutrients and oxygen, especially in large structures, leading to the formation of necrotic cores. The size and longevity of the models are therefore limited [132]. Furthermore, although tissue complexity is better represented in 3D models, systemic responses involving multiple organs and immune components are still not fully simulated [113]. The standardization and reproducibility of 3D cultures between different laboratories are still affected by small variations in materials or procedures, which can lead to significant differences in results [145]. Therefore, although 3D cultures are seen as a valuable step forward in reducing animal use, they are currently seen as only a complementary tool rather than a complete replacement for animal models.

In computational models, the accuracy of models is highly dependent on the quality and completeness of input data. When models are built on missing, biased, or species-specific data, misleading conclusions may be drawn when applied to human systems. Furthermore, because biological processes are extremely complex and often context-dependent, comprehensive simulation remains difficult [30]. Computational predictions often need to be validated with real-world experiments, sometimes including animal experiments, to ensure reliability. In addition, many biological problems remain beyond current modeling capabilities due to the complexity of dynamic interactions between tissues, organs, and the environment [22]. Despite these challenges, the power and utility of computational models continue to be enhanced by advances in bioinformatics, artificial intelligence, and high-speed data generation technologies, making them useful tools for humanities and science research.

In OOC, most are designed to represent single organ systems and do not fully simulate the systemic interactions found in the whole body [138]. In addition, the technical complexity of designing, fabricating, and operating these devices may hinder widespread adoption. Standardization across laboratories remains a challenge, making the approval process for use in drug development difficult [84]. However, ongoing research is underway to connect multiple organ chips to create "human-on-a-chip" systems, which are expected to provide an effective alternative to animal testing in the future [70].

Furthermore, current models often do not fully reflect the genetic and physiological diversity of human populations. Although alternative approaches can use human-specific cells or tissues, they often reflect only a limited genetic background and do not capture variation in biological responses. Animal models, despite their limitations, provide controlled systems that provide insight into variation in different genetic and environmental contexts [120].

These challenges become particularly evident in the study of complex, multifactorial diseases such as cancer, autoimmune disorders, and neurodegenerative diseases. For example, the study of Alzheimer's disease requires examining interactions between neurons, glial cells, the immune system, and systemic metabolic processes [3, 133, 154]. Even the most advanced in vitro platforms today struggle to fully reproduce such multidimensional biological interactions [68].

In addition to these limitations, researchers face challenges in scientifically validating alternative models. To gain acceptance in research and regulation, they must demonstrate a high level of relevance to human biology, reproducibility across laboratories, broad applicability, and predictive accuracy. They typically establish this level of confidence through extensive comparative studies with established animal models, and years of refinement and data accumulation [86]. Not only financial constraints but also infrastructural problems are encountered in the development and deployment of alternative systems. Establishing these technologies not only requires significant investments in new equipment but also requires specialized training and interdisciplinary collaboration, while these resources are often limited, especially at academic institutions or in places with limited funding [1].

15.5 Future Development

Despite ongoing ethical debates, animal experimentation remains central to biomedical research and drug development. Animal models continue to provide valuable information about complex physiology, disease progression, and therapeutic efficacy, while alternative platforms remain limited in fully reflecting the interactions and mechanisms of whole organisms [45]. Looking ahead, the ethical and scientific trajectory of animal experimentation is poised to evolve through not only technological innovation but also systemic, educational, and policy-driven transformations. Future strategies must shift from solely developing alternatives to fostering a transdisciplinary research culture that prioritizes integrative approaches and human relevance as core pillars of biomedical science.

A critical future direction is the integration of hybrid models, combining in vitro, in silico, and ex vivo systems to approximate systemic biology [106]. While single-model platforms often fall short of mimicking whole-organism responses, a composite approach where organ-on-a-chip devices are linked with computational models and human-derived biological samples, may bridge existing gaps in complexity, enabling more accurate simulation of multi-organ interactions, immune responses, and metabolic cascades [152, 155]. These integrative systems could serve as intermediate steps in drug development, reducing the dependency on animals at earlier stages and streamlining later in vivo validation.

Another promising area lies in personalized experimental systems, especially with the rising accessibility of induced pluripotent stem cells (iPSCs) and patient-derived organoids. These tools allow the creation of disease models that reflect individual genetic backgrounds, enabling more tailored investigations and reducing the need for generalized animal models [155]. Future research will likely focus on building biobanks and standardized iPSC-based platforms that can represent diverse population groups, addressing the current limitations in modeling human heterogeneity [90]. Another important aspect is the reform of the ethical review system and funding guidelines. Ethics committees need to move from a supervisory role to a facilitative one, actively encouraging the adoption of alternative methods through support, guidance, and financial incentives. This includes increasing funding for the early development of non-animal methods and prioritizing funding for innovative 4R integration projects. In addition, harmonization of international review processes is needed to accelerate the recognition and regulatory acceptance of non-animal models [53]. In particular, capacity building in low- and middle-income countries (LMICs) is key. To date, most efforts to reduce animal testing have focused on countries with strong resources. Future strategies must therefore aim for equitable access, including training programs, open access platforms, and global collaborative alliances. Investment in sustainable infrastructure and supportive policies will help ensure that ethical standards in research are enforced consistently across the globe, rather than being confined to a few regions [153].

15.6 Conclusions

Animal experimentation continues to provoke profound ethical debates, despite its significant contributions to modern scientific and medical progress. The 4Rs framework has provided a foundational guideline for more humane and responsible research, but the full realization of these principles is still hampered by scientific limitations, complex legal requirements, and limited financial resources. Although alternative approaches such as organ-on-a-chip models, 3D culture systems, and computational modeling are rapidly developing and show considerable potential, they still cannot fully reproduce the complexity and physiological interconnectedness of living systems. Therefore, eliminating the reliance on animal models in research remains a significant challenge. However, a combination of sustainable technological innovation, strong ethical commitment, and active support from regulatory bodies is seen as key to gradually reducing the use of animals in research, while promoting the development of more responsible and humane science.

References

1. Abarkan FZ, Wijen AMA, van Eijden RMG, Struijs F, Dennis P, Ritskes-Hoitinga M, Visseren-Hamakers I. Identifying key factors for accelerating the transition to animal-testing-free medical science through co-creative, interdisciplinary learning between students and teachers. Animals. 2022;12(20) https://doi.org/10.3390/ANI12202757.
2. Abuwatfa WH, Pitt WG, Husseini GA. Scaffold-based 3D cell culture models in cancer research. J Biomed Sci. 2024;31(1):7. https://doi.org/10.1186/S12929-024-00994-Y.
3. Afrakoti LGMP, Moghadam SD, Hadinezhad P. Alzheimer's disease and the immune system: a comprehensive overview with a focus on B cells, humoral immunity, and immunotherapy. J Alzheimer's Dis Rep. 2025;9:25424823251329188. https://doi.org/10.1177/25424823251329188.
4. Ahima RS. Editorial: molecular obesity research: lessons learned? Mol Endocrinol. 2014;28(6):785. https://doi.org/10.1210/ME.2014-1140.
5. Ahn N, Roh S, Park J. The status and issues of the institutional animal care and use Committee of Seoul National University: from its establishment to the present day. Exp Anim. 2021;70(4):532. https://doi.org/10.1538/EXPANIM.21-0066.
6. Akhtar A. The flaws and human harms of animal experimentation. Camb Q Healthc Ethics. 2015;24(4):407–19. https://doi.org/10.1017/S0963180115000079.
7. Animal Welfare (Sentience) Thermochim Acta. 2022. n.d.
8. Animal Welfare Act 1999 No 142 (as at 30 April 2023), Public Act – New Zealand Legislation. n.d. Retrieved April 9, 2025. https://www.legislation.govt.nz/act/public/1999/0142/latest/whole.html.

9. Animals Act. | FAOLEX. n.d. Retrieved April 9, 2025. https://www.fao.org/faolex/results/details/en/c/LEX-FAOC213279/.
10. Animals, National Research Council (US) Committee on Guidelines for the Humane Transportation of Laboratory. 2006. 'Summary of the Animal Welfare Act Regulations Pertaining to Transportation'. https://www.ncbi.nlm.nih.gov/books/NBK19631/.
11. Arumugam P, Kaarthikeyan G, Eswaramoorthy R. Three-dimensional bioprinting: the ultimate pinnacle of tissue engineering. Cureus. 2024;16(4):e58029. https://doi.org/10.7759/CUREUS.58029.
12. Asma H, Tieke E, Deem KD, Rahmat J, Dong T, Huang X, Tomoyasu Y, Halfon MS. Regulatory genome annotation of 33 insect species. elife. 2024;13:RP96738. https://doi.org/10.7554/ELIFE.96738.
13. Badyal DK, Modgill V, Kaur J. Computer simulation models are implementable as replacements for animal experiments. Altern Lab Anim. 2009;37(2):191–5. https://doi.org/10.1177/026119290903700208.
14. Balatskyi V, Strilbytska O, Abrat O, Tkachyk A, Lylyk M, Lushchak V, Bayliak M. Behavioral and metabolic effects of escapable electric foot shock stress in male mice. BMC Res Notes. 2025;18(1):127. https://doi.org/10.1186/S13104-025-07189-0.
15. Bassan A, Alves VM, Amberg A, Anger LT, Auerbach S, Beilke L, Bender A, Cronin MTD, Cross KP, Hsieh JH, Greene N, Kemper R, Kim MT, Mumtaz M, Noeske T, Pavan M, Pletz J, Russo DP, Sabnis Y, Schaefer M, Szabo DT, Valentin JP, Wichard J, Williams D, Woolley D, Zwickl C, Myatt GJ. In Silico approaches in organ toxicity Hazard assessment: current status and future needs in predicting liver toxicity. Comput Toxicol (Amsterdam, Netherlands). 2021;20:100187. https://doi.org/10.1016/J.COMTOX.2021.100187.
16. Bayne KAL. Environmental enrichment of nonhuman primates, dogs and rabbits used in toxicology studies. Toxicol Pathol. 2003;31(SUPPL):132–7. https://doi.org/10.1080/01926230390175020.
17. Benedek I, Vilmos Altbcker, Tamás Molnár. Stress reactivity near birth affects nest building timing and offspring number and survival in the European Rabbit (Oryctolagus Cuniculus). PLoS One. 2021;16(1 January) https://doi.org/10.1371/JOURNAL.PONE.0246258.
18. Bhatt LK, Patel JH, Shah CR, Patel SR, Patel SD, Patel VA, Sundar R, Jain MR. Immunization of laboratory animal workers: occupational health and safety aspects. Pathog Glob Health. 2024;118(5):376. https://doi.org/10.1080/20477724.2024.2329376.
19. Brennan FR, Cavagnaro J, McKeever K, Ryan PC, Schutten MM, Vahle J, Weinbauer GF, Marrer-Berger E, Black LE. Safety testing of monoclonal antibodies in non-human primates: case studies highlighting their impact on human risk assessment. MAbs. 2017;10(1):1. https://doi.org/10.1080/19420862.2017.1389364.
20. Bruna J, Alberti P, Calls-Cobos A, Caillaud M, Damaj MI, Navarro X. Methods for in vivo studies in rodents of chemotherapy induced peripheral neuropathy. Exp Neurol. 2019;325:113154. https://doi.org/10.1016/J.EXPNEUROL.2019.113154.
21. Bueso-Bordils JI, Antón-Fos GM, Martín-Algarra R, Alemán-López PA. Overview of computational toxicology methods applied in drug and green chemical discovery. J Xenobiot. 2024;14(4):1901. https://doi.org/10.3390/JOX14040101.
22. Bunne C, Roohani Y, Rosen Y, Gupta A, Zhang X, Roed M, Alexandrov T, AlQuraishi M, Brennan P, Burkhardt DB, Califano A, Cool J, Dernburg AF, Ewing K, Fox EB, Haury M, Herr AE, Horvitz E, Hsu PD, Jain V, Johnson GR, Kalil T, Kelley DR, Kelley SO, Kreshuk A, Mitchison T, Otte S, Shendure J, Sofroniew NJ, Theis F, Theodoris CV, Upadhyayula S, Valer M, Wang B, Xing E, Yeung-Levy S, Zitnik M, Karaletsos T, Regev A, Lundberg E, Leskovec J, Quake SR. How to build the virtual cell with artificial intelligence: priorities and opportunities. ArXiv. 2024;187:arXiv:2409.11654v2. https://pmc.ncbi.nlm.nih.gov/articles/PMC11468656/
23. Busek M, Aizenshtadt A, Amirola Martinez M, Delon L, Krauss S. Academic user view: organ-on-a-Chip Technology. Biosensors. 2022;12(2):126. https://doi.org/10.3390/BIOS12020126/S1.

24. Caballero L, Pasternak T, Riyazuddin R, Pérez-Pérez JM. Connecting high-resolution 3D chromatin maps with cell division and cell differentiation at the root apical meristem. Plant Cell Rep. 2024;43(10):232. https://doi.org/10.1007/S00299-024-03322-8.
25. Carter CS, Richardson A, Huffman DM, Austad S. Bring back the rat! J Gerontol Ser A Biol Med Sci. 2020;75(3):405. https://doi.org/10.1093/GERONA/GLZ298.
26. Caviola L, Charlton Everett JA, Faber NS. The Moral Standing of Animals: Towards a Psychology of Speciesism. 2018; https://doi.org/10.31234/OSF.IO/M5CWQ.
27. CCAC GUIDELINES. Can Vet J. 2003;44(8):631. https://pmc.ncbi.nlm.nih.gov/articles/PMC340226/
28. Chaicharoenaudomrung N, Kunhorm P, Noisa P. Three-dimensional cell culture systems as an in vitro platform for cancer and stem cell modeling. World J Stem Cells. 2019;11(12):1065. https://doi.org/10.4252/WJSC.V11.I12.1065.
29. Chen Z, Park S. Three-dimensional hanging drop spheroid plates for easy chimeric antigen receptor (CAR) T cytotoxicity assay. Methods Mol Biol (Clifton, NJ). 2024;2764:35–42. https://doi.org/10.1007/978-1-0716-3674-9_4.
30. Chen J, Lin A, Jiang A, Qi C, Liu Z, Cheng Q, Yuan S, Luo P. Computational frameworks transform antagonism to synergy in optimizing combination therapies. NPJ Digit Med. 2025;8(1):44. https://doi.org/10.1038/S41746-025-01435-2.
31. Cohen C. The case for the use of animals in biomedical research. N Engl J Med. 1986;315(14):865–70. https://doi.org/10.1056/NEJM198610023151405.
32. Cohen AI, Wellman CH. Contemporary debates in applied ethics. Wiley; 2005.
33. Colquitt RB, Colquhoun DA, Thiele RH. In Silico modelling of physiologic systems. Best Pract Res Clin Anaesthesiol. 2011;25(4):499–510. https://doi.org/10.1016/j.bpa.2011.08.006.
34. Cong Y, Han X, Wang Y, Chen Z, Yao L, Liu T, Zhengzhi W, Jin Y, Luo Y, Zhang X. Drug toxicity evaluation based on organ-on-a-chip technology: a review. Micromachines. 2020;11(4):381. https://doi.org/10.3390/MI11040381.
35. Contreras C. Sentient beings protected by law. Analysis of recent changes in Colombian animal welfare legislation. Glob J Anim Law. 2016;2. https://journal.fi/gjal/article/view/148716
36. Cordeiro S, Oliveira BB, Valente R, Ferreira D, Luz A, Baptista PV, Fernandes AR. Breaking the mold: 3D cell cultures reshaping the future of cancer research. Front Cell Dev Biol. 2024;12 https://doi.org/10.3389/FCELL.2024.1507388.
37. Cox LA, Puppala S, Chan J, Riojas AM, Lange KJ, Birnbaum S, Dick EJ, Comuzzie AG, Nijland MJ, Li C, Nathanielsz PW, Olivier M. Maternal under-nutrition during pregnancy alters the molecular response to over-nutrition in multiple organs and tissues in nonhuman primate juvenile offspring. J Dev Orig Health Dis. 2024;15:e27. https://doi.org/10.1017/S2040174424000163.
38. Cum M, Santiago Pérez JA, Wangia E, Lopez N, Wright ES, Iwata RL, Li A, Chambers AR, Padilla-Coreano N. A Systematic review and meta-analysis of how social memory is studied. Sci Rep. 2024;14(1) https://doi.org/10.1038/S41598-024-52277-Z.
39. de Ávila RI, Valadares MC. Brazil moves toward the replacement of animal experimentation. Altern Lab Anim. 2019;47(2):71–81. https://doi.org/10.1177/0261192919856806.
40. de Moura G, Acácio YM, Rocha FL, Moura D, de Oliveira J, Freitas JP, Rodrigues V, Gonçalves VP, Nicolete R. Immune system cells modulation in patients with reproductive issues: a systematic review approach. Jornal Brasileiro de Reproducao Assistida. 2024;28(1):78–89. https://doi.org/10.5935/1518-0557.20230044.
41. del Amo EM, Hammid A, Tausch M, Toropainen E, Sadeghi A, Valtari A, Puranen J, Reinisalo M, Ruponen M, Urtti A, Sauer A, Honkakoski P. Ocular metabolism and distribution of drugs in the rabbit eye: quantitative assessment after Intracameral and Intravitreal administrations. Int J Pharm. 2022;613 https://doi.org/10.1016/j.ijpharm.2021.121361.
42. Delahaye P. A semiotic methodology for animal studies. 2019. https://link.springer.com/content/pdf/10.1007/978-3-030-28813-6.pdf
43. Di Salvo L. The ethical and moral status of invasive animal research: the dilemma and alternative approaches. 2017

44. Díez-Solinska A, De Miguel Z, Azkona G, Vegas O. Behavioral coping with chronic defeat stress in mice: a Systematic review of current protocols. Neurobiol Stress. 2024;33:100689. https://doi.org/10.1016/J.YNSTR.2024.100689.
45. Domínguez-Oliva A, Hernández-Ávalos I, Martínez-Burnes J, Olmos-Hernández A, Verduzco-Mendoza A, Mota-Rojas D. The importance of animal models in biomedical research: current insights and applications. Anim Open Access J MDPI. 2023;13(7):1223. https://doi.org/10.3390/ANI13071223.
46. Dondossola D, de Falco S, Kersik A, Maggioni M, Di Girolamo L, Biancolilli O, Busana M, Lonati C, Carù F, Zanella A, Gatti S. Procurement and ex-Situ perfusion of isolated slaughterhouse-derived livers as a model of donors after circulatory death. ALTEX. 2020;37(2):243–54. https://doi.org/10.14573/ALTEX.1909131.
47. Dorner L, Stratmann B, Bader L, Podobnik M, Irion U. Efficient genome editing using modified Cas9 proteins in zebrafish. Biol Open. 2024;13(4):bio060401. https://doi.org/10.1242/BIO.060401.
48. Durant O, Marzi A. Ebola virus disease sequelae and viral persistence in animal models: implications for the future. PLoS Pathog. 2024;20(3):e1012065. https://doi.org/10.1371/JOURNAL.PPAT.1012065.
49. Erben J, Jirkovec R, Kalous T, Klicova M, Chvojka J. The combination of hydrogels with 3D fibrous scaffolds based on electrospinning and Meltblown technology. Bioengineering. 2022;9(11):660. https://doi.org/10.3390/BIOENGINEERING9110660.
50. Esposito A, Ferraresi A, Vallino L, Garavaglia B, Dhanasekaran DN, Isidoro C. Three-dimensional in vitro cell cultures as a feasible and promising alternative to two-dimensional and animal models in cancer research. Int J Biol Sci. 2024;20(13):5293–311. https://doi.org/10.7150/IJBS.96469.
51. Eyster C, Matsuzaki S, Pranay A, Giorgione JR, Faakye A, Ahmed M, Humphries KM. Mechanistic studies of PFKFB2 reveal a novel inhibitor of its kinase activity. PLoS One. 2025;20(5) https://doi.org/10.1371/JOURNAL.PONE.0317167.
52. Farhang Doost N, Srivastava SK. A comprehensive review of organ-on-a-chip technology and its applications. Biosensors. 2024;14(5):225. https://doi.org/10.3390/BIOS14050225.
53. Feng J, Song X, Zhang B, Xiang W. Establishing an animal model for post-inflammatory hyperpigmentation following fractional CO_2 laser application. Lasers Med Sci. 2025;40(1):17. https://doi.org/10.1007/S10103-025-04282-5.
54. Fernandes MR, Pedroso AR. Animal experimentation: a look into ethics, welfare and alternative methods. Rev Assoc Med Bras. 2017;63(11):923–8. https://doi.org/10.1590/1806-9282.63.11.923.
55. Field SM, Thompson J, De Rijcke S, Penders B, Munafò MR. Exploring the dimensions of responsible research systems and cultures: a scoping review. R Soc Open Sci. 2024;11(1):230624. https://doi.org/10.1098/RSOS.230624.
56. Fischer FP, Karge RA, Koch H, Voigt A, Weber YG, Wolking S. The fruit fly drosophila melanogaster as a screening model for Antiseizure medications. Front Pharmacol. 2024;15:1489888. https://doi.org/10.3389/FPHAR.2024.1489888/FULL.
57. Foster E, Wardhana O, Zeng Z, Lu X. Comparative analysis of 3D culture methodologies in prostate cancer cells. 2025. BioRxiv 2025.03.11.642589. https://doi.org/10.1101/2025.03.11.642589.
58. Francione GL. Animal welfare and society—part 1, the viewpoints of a philosopher. Anim Front Rev Mag Anim Agric. 2022;12(1):43. https://doi.org/10.1093/AF/VFAC006.
59. Francis I, Shrestha J, Paudel KR, Hansbro PM, Warkiani ME, Saha SC. Recent advances in lung-on-a-Chip models. Drug Discov Today. 2022;27(9):2593–602. https://doi.org/10.1016/j.drudis.2022.06.004.
60. Frühwein H, Paul NW. "Lost in translation?" animal research in the era of precision medicine. J Transl Med. 2025;23(1):152. https://doi.org/10.1186/S12967-025-06084-3.

61. Fu C, Chen Q. The future of pharmaceuticals: artificial intelligence in drug discovery and development. J Pharm Anal. 2025:101248. https://doi.org/10.1016/J.JPHA.2025.101248.
62. Giménez-Candela M. The de-objectification of animals in the Spanish Civil Code. Derecho Animal; 2019. https://portalrecerca.uab.cat/en/publications/the-de-objectification-of-animals-in-the-spanish-civil-code.
63. Goffinet C, Allespach I, Keppler OT. HIV-susceptible transgenic rats allow rapid preclinical testing of antiviral compounds targeting virus entry or reverse transcription. Proc Natl Acad Sci USA. 2007;104(3):1015. https://doi.org/10.1073/PNAS.0607414104.
64. Gottschling DC, Döring F. Is C. Elegans a suitable model for nutritional science? Genes Nutr. 2019;14(1):1. https://doi.org/10.1186/S12263-018-0625-3.
65. Hamm JT, Hsieh JH, Roberts GK, Collins B, Gorospe J, Sparrow B, Walker NJ, Truong L, Tanguay RL, Dyballa S, Miñana R, Schiavone V, Terriente J, Weiner A, Muriana A, Quevedo C, Ryan KR. Interlaboratory study on zebrafish in toxicology: systematic evaluation of the application of zebrafish in toxicology's (SEAZIT'S) evaluation of developmental toxicity. Toxics. 2024;12(1):93. https://doi.org/10.3390/TOXICS12010093/S1.
66. Hart DW, Bennett NC, Mustoe A. Editorial: recent advances in endocrinology of non-traditional mammalian models. Front Endocrinol. 2024;15:1520073. https://doi.org/10.3389/FENDO.2024.1520073.
67. Henze LJ, Koehl NJ, O'Shea JP, Kostewicz ES, Holm R, Griffin BT. The pig as a preclinical model for predicting Oral bioavailability and in vivo performance of pharmaceutical oral dosage forms: a PEARRL review. J Pharm Pharmacol. 2019;71(4):581–602. https://doi.org/10.1111/JPHP.12912.
68. Hirsch C, Schildknecht S. In vitro research reproducibility: keeping up high standards. Front Pharmacol. 2019;10:1484. https://doi.org/10.3389/FPHAR.2019.01484.
69. Hobbs EC, Reid TJ. Animals and SARS-CoV-2: species susceptibility and viral transmission in experimental and natural conditions, and the potential implications for community transmission. Transbound Emerg Dis. 2020;68(4):1850. https://doi.org/10.1111/TBED.13885.
70. Huang W, Chen YY, He FF, Zhang C. Revolutionizing nephrology research: expanding horizons with kidney-on-a-Chip and beyond. Front Bioeng Biotechnol. 2024;12:1373386. https://doi.org/10.3389/FBIOE.2024.1373386/BIBTEX.
71. Jarvis S. Advocating for animal rights. Vet Rec. 2024;194(10):369. https://doi.org/10.1002/VETR.4270.
72. Kang M, Long T, Chang C, Meng T, Ma H, Li Z, Li P, Chen Y. A review of the ethical use of animals in functional experimental research in China based on the "four R" principles of reduction, replacement, refinement, and responsibility. Med Sci Monit. 2022;28:e938807–1. https://doi.org/10.12659/MSM.938807.
73. Kavanagh O, Krebs CE. Mitigating animal methods bias to reduce animal use and improve biomedical translation. Sci Prog. 2024;107(2):00368504241253693. https://doi.org/10.1177/00368504241253693.
74. Kazmi ST, Batool HF, Naz I, Kanwal N, Haq IU. Pre-clinical studies comparing the anti-inflammatory potential of Artemisinic compounds by targeting NFκB/TNF-α/NLRP3 and Nrf2/TRX pathways in Balb/C mice. Front Pharmacol. 2024;15:1352827. https://doi.org/10.3389/FPHAR.2024.1352827.
75. Kiani AK, Pheby D, Henehan G, Brown R, Sieving P, Sykora P, Marks R, Falsini B, Capodicasa N, Miertus S, Lorusso L, Dondossola D, Tartaglia GM, Ergoren MC, Dundar M, Michelini S, Malacarne D, Bonetti G, Dautaj A, Donato K, Medori MC, Beccari T, Samaja M, Connelly ST, Martin D, Morresi A, Bacu A, Herbst KL, Kapustin M, Stuppia L, Lumer L, Farronato G, Bertelli M. Ethical considerations regarding animal experimentation. J Prev Med Hyg. 2022a;63(2 Suppl 3):E255. https://doi.org/10.15167/2421-4248/JPMH2022.63.2S3.2768.
76. Kiani AK, Pheby D, Henehan G, Brown R, Sieving P, Sykora P, Marks R, Falsini B, Capodicasa N, Miertus S, Lorusso L, Dondossola D, Tartaglia GM, Ergoren MC, Dundar M, Michelini S, Malacarne D, Bonetti G, Dautaj A, Donato K, Medori MC, Beccari T, Samaja M, Connelly ST, Martin D, Morresi A, Bacu A, Herbst KL, Kapustin

M, Stuppia L, Lumer L, Farronato G, Bertelli M. Ethical considerations regarding animal experimentation. J Prev Med Hyg. 2022b;63(2 Suppl 3):E255. https://doi.org/10.15167/2421-4248/JPMH2022.63.2S3.2768.

77. Kiani AK, Pheby D, Henehan G, Brown R, Sieving P, Sykora P, Marks R, Falsini B, Capodicasa N, Miertus S, Lorusso L, Dondossola D, Tartaglia GM, Ergoren MC, Dundar M, Michelini S, Malacarne D, Bonetti G, Dautaj A, Donato K, Medori MC, Beccari T, Samaja M, Connelly ST, Martin D, Morresi A, Bacu A, Herbst KL, Kapustin M, Stuppia L, Lumer L, Farronato G, Bertelli M. Ethical considerations regarding animal experimentation. J Prev Med Hyg. 2022c;63(2 Suppl 3):E255. https://doi.org/10.15167/2421-4248/JPMH2022.63.2S3.2768.
78. Kim HJ, Park S, Jeong S, Kim J, Cho YJ. Lung organoid on a chip: a new ensemble model for preclinical studies. Int J Stem Cells. 2023;17(1):30. https://doi.org/10.15283/IJSC23090.
79. Kimura H, Nishikawa M, Kutsuzawa N, Tokito F, Kobayashi T, Kurniawan DA, Shioda H, Cao W, Shinha K, Nakamura H, Doi K, Sakai Y. Advancements in microphysiological systems: exploring organoids and organ-on-a-chip technologies in drug development -focus on pharmacokinetics related organs. Drug Metab Pharmacokinet. 2025;60 https://doi.org/10.1016/J.DMPK.2024.101046.
80. King WW, Paramastri YA, Guillén J. Compliance and regulatory programs. In: Management of animal care and use programs in research, education, and testing: second edition; 2018. p. 105–45. https://doi.org/10.1201/9781315152189-7.
81. Koledova Z. 3D cell culture: an introduction. Methods Mol Biol. 2017;1612 https://doi.org/10.1007/978-1-4939-7021-6_1.
82. Kotzmann J. Sentience and intrinsic worth as a pluralist foundation for Fundamental Animal Rights. Oxf J Leg Stud. 2023;43(2):405–28. https://doi.org/10.1093/OJLS/GQAD003.
83. Kramer LA. The scientific problems with using non-human animals to predict human response to drugs and disease. n.d. https://doi.org/10.1163/9789004391192_018.
84. Kumar D, Nadda R, Repaka R. Advances and challenges in organ-on-chip technology: toward mimicking human physiology and disease in vitro. Med Biol Eng Comput. 2024;62(7):1925–57. https://doi.org/10.1007/S11517-024-03062-7.
85. Kuodza GE, Kawai R, LaSalle JM. Intercontinental insights into autism Spectrum disorder: a synthesis of environmental influences and DNA methylation. Environ Epigenetics. 2024;10(1):dvae023. https://doi.org/10.1093/EEP/DVAE023.
86. Landi M, Everitt J, Berridge B. Bioethical, reproducibility, and translational challenges of animal models. ILAR J. 2021;62(1–2):60–5. https://doi.org/10.1093/ILAR/ILAA027.
87. Lee KH, Lee DW, Kang BC. The "R" principles in laboratory animal experiments. Lab Anim Res. 2020;36(1) https://doi.org/10.1186/S42826-020-00078-6.
88. Leung CM, de Haan P, Ronaldson-Bouchard K, Kim GA, Ko J, Rho HS, Chen Z, Habibovic P, Jeon NL, Takayama S, Shuler ML, Vunjak-Novakovic G, Frey O, Verpoorte E, Toh YC. A guide to the organ-on-a-Chip. Nat Rev Methods Primers. 2022;2(1):1–29.
89. Liang L, Damiani A, Del Brocco M, Rogers ER, Jantz MK, Fisher LE, Gaunt RA, Capogrosso M, Lempka SF, Pirondini E. A systematic review of computational models for the design of spinal cord stimulation therapies: from neural circuits to patient-specific simulations. J Physiol. 2022;601(15):3103. https://doi.org/10.1113/JP282884.
90. Lin JC, Liu YL, Hsiao WWW, Te Fan C. Integrating population-based biobanks: catalyst for advances in precision health. Comput Struct Biotechnol J. 2024;24:690. https://doi.org/10.1016/J.CSBJ.2024.10.049.
91. Liu J, Zai X, Tian X, Li J, Yan S, Wang T. Practical implementation and impact of the 4R principles in ethnopharmacology: pursuing a more humane approach to research. Front Pharmacol. 2025;16:1543316. https://doi.org/10.3389/FPHAR.2025.1543316.
92. Machan TR. Why human beings may use animals. J Value Inq. 2002;36(1):9–16. https://doi.org/10.1023/A:1014993828953.
93. Mangani S, Kremmydas S, Karamanos NK. Mimicking the complexity of solid tumors: how spheroids could advance cancer preclinical transformative approaches. Cancer. 2025;17(7):1161. https://doi.org/10.3390/CANCERS17071161.

94. Medina-Luque J, Piechocinski P, Feyen P, Sgobio C, Herms J. Striatal dopamine neurotransmission is altered in age- and region-specific manner in a Parkinson's disease transgenic mouse. Sci Rep. 2024;14(1) https://doi.org/10.1038/S41598-023-49600-5.
95. Ménard C, Hodes GE, Russo SJ. Pathogenesis of depression: insights from human and rodent studies. Neuroscience. 2015;321:138. https://doi.org/10.1016/J.NEUROSCIENCE.2015.05.053.
96. Meng X, Zhu F, Chen K. Silkworm: a promising model organism in life science. J Insect Sci. 2017;17(5) https://doi.org/10.1093/jisesa/iex064.
97. Miano JM, Zhu QM, Lowenstein CJ. A CRISPR path to engineering new genetic mouse models for cardiovascular research. Arterioscler Thromb Vasc Biol. 2016;36(6):1058. https://doi.org/10.1161/ATVBAHA.116.304790.
98. Milich KM. Male-Philopatric nonhuman primates and their potential role in understanding the evolution of human sociality. Evol Anthropol. 2024;33(1) https://doi.org/10.1002/EVAN.22014.
99. Millecam J, De Baere S, Croubels S, Devreese M. In vivo metabolism of ibuprofen in growing conventional pigs: a pharmacokinetic approach. Front Pharmacol. 2019;10(JUN):712. https://doi.org/10.3389/FPHAR.2019.00712/FULL.
100. Mitchell SJ, MacArthur MR, Kane AE. Optimizing preclinical models of ageing for translation to clinical trials. Br J Clin Pharmacol. 2023;91(1) https://doi.org/10.1111/BCP.15902.
101. Mitchell JR, Vincelette L, Tuberman S, Sheppard V, Bergeron E, Calitri R, Clark R, Cody C, Kannan A, Keith J, Parakoyi A, Pikus MC, Vance V, Ziane L, Brenhouse H, Laine MA, Shansky RM. Behavioral and neural correlates of diverse conditioned fear responses in male and female rats. BioRxiv. 2024;2024(08):20.608817. https://doi.org/10.1101/2024.08.20.608817.
102. Monarca RI, Silva RFB, Gabriel SI, Cerveira AM, von Merten S. The presence of a shelter in an open Field test has differential effects on the behavior and stress response of two mouse species. J Exp Zoo Part A Ecol Integr Physiol. 2025;343(4):480. https://doi.org/10.1002/JEZ.2904.
103. Monfort-Lanzas P, Rungger K, Madersbacher L, Hackl H. Machine learning to dissect perturbations in complex cellular systems. Comput Struct Biotechnol J. 2025;27:832. https://doi.org/10.1016/J.CSBJ.2025.02.028.
104. Morais AS, Mendes M, Cordeiro MA, Sousa JJ, Pais AC, Mihăilă SM, Vitorino C. Organ-on-a-Chip: Ubi Sumus? Fundamentals and design aspects. Pharmaceutics. 2024;16(5):615. https://doi.org/10.3390/PHARMACEUTICS16050615.
105. Moro LG, Guarnier LP, Azevedo MF, Fracasso JAR, Lucio MA, Vidigal M, de Castro M, Dias L, Aparecida dos Reis Lívero F, Ribeiro-Paes JT. A brief history of cell culture: from Harrison to organs-on-a-Chip. Cells. 2024;13(24):2068. https://doi.org/10.3390/CELLS13242068.
106. Moumou M, Tayebi A, Hadini A, Noman OM, Alhalmi A, Ahmoda H, Amrani S, Harnafi H. Combining in vitro, in vivo, and in silico approaches to explore the effect of Ceratonia Siliqua and Ocimum Basilicum rich phenolic formula on lipid metabolism and plasma lipoprotein oxidation in mice fed a high-fat diet: a follow-up study. Meta. 2025;15(1) https://doi.org/10.3390/METABO15010036.
107. Nakayama T, Sakai M, Slingsby BT. Japan's ethical guidelines for epidemiologic research: a history of their development. J Epidemiol. 2005;15(4):107. https://doi.org/10.2188/JEA.15.107.
108. Nederlof RA, Virgilio T, Stemkens HJJ, Luiz CC, da Silva P, Montagna DR, Abdussamad AM, Chipangura J, Bakker J. Yellow fever in non-human primates: a veterinary guide from a one health perspective. Vet Sci. 2025;12(4):339. https://doi.org/10.3390/VETSCI12040339.
109. Neumann J-M. The legal status of animals in the French Civil Code. Glob J Anim Law. 2015;1(1) https://journal.fi/gjal/article/view/148705
110. Nolan J, Pearce OMT, Screen HRC, Knight MM, Verbruggen SW. Organ-on-a-chip and microfluidic platforms for oncology in the UK. Cancer. 2023;15(3):635. https://doi.org/10.3390/CANCERS15030635.

111. ORS 167.305 – Legislative findings. n.d. Retrieved April 9, 2025. https://oregon.public.law/statutes/ors_167.305#google_vignette
112. Oser MG, MacPherson D, Oliver TG, Sage J, Park KS. Genetically-engineered mouse models of small cell lung cancer: the next generation. Oncogene. 2024;43(7):457–69. https://doi.org/10.1038/S41388-023-02929-7.
113. Ozyilmaz ED, Turan A, Comoglu T. An overview on the advantages and limitations of 3D printing of microneedles. Pharm Dev Technol. 2021;26(9):923–33. https://doi.org/10.1080/10837450.2021.1965163.
114. Pádua MS, Guil-Guerrero JL, Prates JAM, Lopes PA. Insights on the use of transgenic mice models in Alzheimer's disease research. Int J Mol Sci. 2024;25(5) https://doi.org/10.3390/IJMS25052805.
115. Paipa-Jabre-Cantu SI, Rodriguez-Salvador M, Castillo-Valdez PF. Revealing three-dimensional printing technology advances for Oral drug delivery: application to central-nervous-system-related diseases. Pharmaceutics. 2025;17(4):445. https://doi.org/10.3390/PHARMACEUTICS17040445.
116. Park CH, Park JH, Suh YJ. Perspective of 3D culture in medicine: transforming disease research and therapeutic applications. Front Bioeng Biotechnol. 2024;12:1491669. https://doi.org/10.3389/FBIOE.2024.1491669.
117. Parsons MS, Bolton DL. The utility of nonhuman primate models for understanding acute HIV-1 infection. Curr Opin HIV AIDS. 2025;20(3):218. https://doi.org/10.1097/COH.0000000000000920.
118. Pereira S, Veeraraghavan P, Ghosh S, Gandhi M. Animal experimentation and ethics in India: the CPCSEA makes a difference. Altern Lab Anim. 2004;32(SUPPL. 1B):411–5. https://doi.org/10.1177/026119290403201S67.
119. Perry M. Revised Australian code of practice for the care and use of animals for scientific purposes. Aust Vet J. 1998;76(4):286. https://doi.org/10.1111/J.1751-0813.1998.TB10161.X.
120. Phillips NLH, Roth TL. Animal models and their contribution to our understanding of the relationship between environments, epigenetic modifications, and behavior. Genes. 2019;10(1):47. https://doi.org/10.3390/GENES10010047.
121. Picchio V, Gaetani R, Chimenti I. Recent advances in 3D cultures. Int J Mol Sci. 2024;25(8):4189. https://doi.org/10.3390/IJMS25084189.
122. Polis B, Samson AO. Addressing the discrepancies between animal models and human Alzheimer's disease pathology: implications for translational research. J Alzheimer's Dis. 2024;98(4):1199–218. https://doi.org/10.3233/JAD-240058.
123. Pramod RK, Atul PK, Pandey M, Anbazhagan S, Mhaske ST, Barathidasan R. Care, management, and use of ferrets in biomedical research. Lab Anim Res. 2024;40(1):10. https://doi.org/10.1186/S42826-024-00197-4.
124. Prescott MJ, Clark C, Dowling WE, Shurtleff AC. Opportunities for refinement of non-human primate vaccine studies. Vaccine. 2021;9(3):284. https://doi.org/10.3390/VACCINES9030284.
125. R N, Aggarwal A, Sravani AB, Mallya P, Lewis S. Organ-on-a-chip: an emerging research platform. Organogenesis. 2023;19(1):2278236. https://doi.org/10.1080/15476278.2023.2278236.
126. Research, National Research Council (US) Institute for Laboratory Animal. Public Health Service Policy and Government Principles Regarding the Care and Use of Animals. 1996. https://www.ncbi.nlm.nih.gov/books/NBK232582/
127. Rhouma M, Archambault M, Butaye P. Antimicrobial use and resistance in animals from a one health perspective. Vet Sci. 2023;10(5):319. https://doi.org/10.3390/VETSCI10050319.
128. Rodríguez-Martínez H. Quality assurance: the key for amendments of the EU-directive/s regulating veterinary training in Europe. Vet Res Commun. 2004;28(SUPPL. 1):29–44. https://doi.org/10.1023/B:VERC.0000045375.12957.22.
129. Rong N, Liu J. Development of animal models for emerging infectious diseases by breaking the barrier of species susceptibility to human pathogens. Emerg Microb Infect. 2023;12(1) https://doi.org/10.1080/22221751.2023.2178242.

130. Rudroff T. Artificial intelligence as a replacement for animal experiments in neurology: potential, progress, and challenges. Neurol Int. 2024;16(4):805. https://doi.org/10.3390/NEUROLINT16040060.
131. Salem AR, Xie X, Griffin SH, Gan L, Miano JM. Genetic modification of mice using prime editing. Curr Protocols. 2024;4(10):e70034. https://doi.org/10.1002/CPZ1.70034.
132. Sano E, Mori C, Nashimoto Y, Yokokawa R, Kotera H, Torisawa YS. Engineering of vascularized 3D cell constructs to model cellular interactions through a vascular network. Biomicrofluidics. 2018;12(4):042204. https://doi.org/10.1063/1.5027183.
133. Seo D o, Holtzman DM. Current understanding of the Alzheimer's disease-associated microbiome and therapeutic strategies. https://doi.org/10.1038/s12276-023-01146-2.
134. Shah D, Dave B, Chorawala MR, Prajapati BG, Singh S, Elossaily GM, Ansari MN, Ali N. An insight on microfluidic organ-on-a-chip models for PM2.5-induced pulmonary complications. ACS Omega. 2024;9(12):13534. https://doi.org/10.1021/ACSOMEGA.3C10271.
135. Sordo Vieira L, Laubenbacher RC. Computational models in systems biology: standards, dissemination, and best practices. Curr Opin Biotechnol. 2022;75 https://doi.org/10.1016/J.COPBIO.2022.102702.
136. Sreepadmanabh M, Arun AB, Bhattacharjee T. Design approaches for 3D cell culture and 3D bioprinting platforms. Biophys Rev. 2024;5(2) https://doi.org/10.1063/5.0188268.
137. Srivastava SK, Foo GW, Aggarwal N, Chang MW. Organ-on-chip technology: opportunities and challenges. Biotechnol Notes. 2024a;5:8–12. https://doi.org/10.1016/J.BIOTNO.2024.01.001.
138. Srivastava SK, Foo GW, Aggarwal N, Chang MW. Organ-on-chip technology: opportunities and challenges. Biotechnol Notes. 2024b;5:8–12. https://doi.org/10.1016/J.BIOTNO.2024.01.001.
139. Suetterlin KJ, Männikkö R, Matthews E, Greensmith L, Hanna MG, Bostock H, Tan SV. Excitability properties of mouse and human skeletal muscle Fibres compared by muscle velocity recovery cycles. Neuromuscul Disord. 2022;32(4):347. https://doi.org/10.1016/J.NMD.2022.02.011.
140. Susilo ME, Schaller S, Jiménez-Franco LD, Kulesza A, de Witte WEA, Chen S-C, Boswell CA, Mandikian D, Li C-C. Whole-body physiologically based pharmacokinetic modeling framework for tissue target engagement of CD3 bispecific antibodies. Pharmaceutics. 2025;17(4):500. https://doi.org/10.3390/PHARMACEUTICS17040500.
141. Tang Z, Wang Z, Ma Z, Jin W, Lin S, Wang L, Min P, Li L, Zhao J, Jia L. Studies on experimental animals immunized with different antigenic vaccine combinations of Neospora Caninum of cattle origin. Parasit Vectors. 2025;18(1):49. https://doi.org/10.1186/S13071-025-06687-1.
142. Teame T, Zhang Z, Ran C, Zhang H, Yang Y, Ding Q, Xie M, Gao C, Ye Y, Duan M, Zhou Z. The use of zebrafish (Danio Rerio) as biomedical models. Anim Front Rev Mag Anim Agric. 2019;9(3):68. https://doi.org/10.1093/AF/VFZ020.
143. Ticliwaerdi X, Manalo K, Abuduweili A, Khan S, Appiah-Kubi E, Williams BA, Oehler AC. Machine learning-based prediction models for healthcare outcomes in patients participating in cardiac rehabilitation a systematic review. J Cardiopulm Rehabil Prev. 2025; https://doi.org/10.1097/HCR.0000000000000943.
144. Tolwinski NS. Introduction: drosophila—a model system for developmental biology. J Dev Biol. 2017;5(3):9. https://doi.org/10.3390/JDB5030009.
145. Urzì O, Gasparro R, Costanzo E, De Luca A, Giavaresi G, Fontana S, Alessandro R. Three-dimensional cell cultures: the bridge between in vitro and in vivo models. Int J Mol Sci. 2023;24(15):12046. https://doi.org/10.3390/IJMS241512046.
146. Van Norman GA. Limitations of animal studies for predicting toxicity in clinical trials: is it time to rethink our current approach? Basic Transl Sci. 2019;4(7):845. https://doi.org/10.1016/J.JACBTS.2019.10.008.

147. Vearrier L, Henderson CM. Utilitarian Principlism as a framework for crisis healthcare ethics. HEC Forum. 2021;33(1–2):45. https://doi.org/10.1007/S10730-020-09431-7.
148. Victor Atoki A, Aja PM, Shinkafi TS, Ondari EN, Adeniyi AI, Fasogbon IV, Dangana RS, Shehu UU, Akin-Adewumi A. Exploring the versatility of drosophila melanogaster as a model organism in biomedical research: a comprehensive review. Fly. 2024;19(1):2420453. https://doi.org/10.1080/19336934.2024.2420453.
149. Völter CJ, Rossano F, Call J. Social manipulation in nonhuman primates: cognitive and motivational determinants. Neurosci Biobehav Rev. 2017;82:76–94. https://doi.org/10.1016/j.neubiorev.2016.09.008.
150. von Bibra C, Hinkel R. Non-human primate studies for Cardiomyocyte transplantation—ready for translation? Front Pharmacol. 2024;15:1408679. https://doi.org/10.3389/FPHAR.2024.1408679.
151. Vu HN, Lee M, Situ AJ, An W, Ley K, Kim C, Ulmer TS. Functional unfolding of the integrin AX transmembrane helix. Proc Natl Acad Sci USA. 2025;122(38) https://doi.org/10.1073/PNAS.2507966122.
152. Vunjak-Novakovic G, Ronaldson-Bouchard K, Radisic M. Organs on a chip models for biological research. Cell. 2021;184(18):4597. https://doi.org/10.1016/J.CELL.2021.08.005.
153. Waechter R, Coomansingh K, Macpherson C, Sarma S, Schweitzer Waechter AR, Coomansingh K, Macpherson C, Sarma S. Ethical challenges in Global Health research. Sustain Dev Goals Ser. 2024;Part F4029:19–34. https://doi.org/10.1007/978-3-031-53793-6_2.
154. Wang Q, Zhu BT, Lei P. Animal models of Alzheimer's disease: current strategies and new directions. Zool Res. 2024;45(6):1385. https://doi.org/10.24272/J.ISSN.2095-8137.2024.274.
155. Wang Y, Marucci L, Homer ME. In Silico modelling of organ-on-a-chip devices: an overview. Front Bioeng Biotechnol. 2025;12:1520795. https://doi.org/10.3389/FBIOE.2024.1520795.
156. Wysoczański B, Świątek M, Wójcik-Gładysz A. Organ-on-a-chip models—new possibilities in experimental science and disease modeling. Biomolecules. 2024;14(12):1569. https://doi.org/10.3390/BIOM14121569.
157. Xu X, Vugmeyster Y. Challenges and opportunities in absorption, distribution, metabolism, and excretion studies of therapeutic biologics. AAPS J. 2012;14(4):781. https://doi.org/10.1208/S12248-012-9388-8.
158. Yoon JK. Advanced 3D cell culture platform for tissue engineering. Tissue Eng Regener Med. 2023;20(4):519. https://doi.org/10.1007/S13770-023-00562-9.
159. Zhang X, Wang Y, Han J, Zhao W, Zhang W, Li X, Chen J, Song W, Wang L. Cardiac-focused multi-organ chips: advanced disease modeling, drug testing, and inter-organ communication. Adv Biol. 2025;9(4) https://doi.org/10.1002/ADBI.202400512.
160. Zhao D, Huang P, Yu L, He Y. Pharmacokinetics–pharmacodynamics modeling for evaluating drug–drug interactions in polypharmacy: development and challenges. Clin Pharmacokinet. 2024;63(7):919–44. https://doi.org/10.1007/S40262-024-01391-2.
161. Zhong JY, Magnusson KR, Swarts ME, Clendinen CA, Reynolds NC, Moffat SD. The application of a rodent-based Morris Water Maze (MWM) protocol to an investigation of age-related differences in human spatial learning. Behav Neurosci. 2017;131(6):470. https://doi.org/10.1037/BNE0000219.
162. Zuo Y. Vascular research of pig knee from the perspective of comparative medicine. Ann Vasc Surg. 2025;113:221–6. https://doi.org/10.1016/J.AVSG.2025.01.022.

Non-animal Models in Research and Teaching: Organoids and Organs-on-Chips

16

Sarah Teakel, Anna Walduck, Joanne H. Connolly, and Thiru Vanniasinkam

16.1 Introduction

Animal models became the standard for regulatory safety testing after significant health crises led governments to mandate comprehensive animal testing before marketing new substances. In 1938, the US Federal Food, Drug, and Cosmetic Act [17] was introduced, and animal models for testing drugs and vaccines, surgery, and dissection have since been considered the gold standard in research and teaching. However, this assumption is increasingly being challenged due to critical ethical and logistical issues [20, 32]. Notably, 90–95% of drugs tested in animals fail in human trials [38], highlighting the significant gap in translational science due primarily to species differences that make animal models unreliable. The 4Rs principle—Reduction, Refinement, Replacement, and Responsibility advocates for minimizing animal use where possible [32]. However, the transition to non-animal models faces challenges, including the need for regulatory changes. Animal testing remains a requirement for drug development in many countries to ensure safety and efficacy, and current practices in teaching and research still heavily rely on animal-based methods.

Interest in non-animal approaches to modelling disease, such as cell culture (2D models) and computational (predictive) analysis, has gained momentum over the past decade. Three-dimensional (3D) models, including organ-on-chip, organoids, and 3D bioprinting, have revolutionized research by providing insights into cellular and molecular mechanisms in normal physiology and pathophysiology, and drug treatments and drug resistance. Organs-on-chips provide environments that mimic the in vivo microenvironment and can be commercially available. Organoids are miniature functional units that mimic organ structures and functions. 3D bioprinting

S. Teakel (✉) · A. Walduck · J. H. Connolly · T. Vanniasinkam
Charles Sturt University, Wagga Wagga, NSW, Australia
e-mail: steakel@csu.edu.au

P. Kumar, V. Kothari (eds.), *Model Organisms in Biological Research*,
https://doi.org/10.1007/978-981-92-0038-2_16

uses bio ink composed of human cells to create tissues such as trachea or neural tissue, offering promising alternatives for regenerative medicine. The potential of these 3D modelling systems is still underexploited.

In this chapter, we cover the concepts and technological developments of organ-on-a-chip and organoid technologies, and discuss their applications. Implementing non-animal models widely involves a coordinated approach between education, practice, and regulation, and a concerted effort to shift the paradigm in research and teaching.

16.2 Technological Developments

To address shortcomings of traditional methods, researchers developed organs-on-chips and organoid technology to simulate the interplay between multiple tissue types, or organ systems.

16.2.1 Organ-on-Chip

Organ-on-chip technology combines micro-manufacturing with tissue engineering to create a system that more precisely simulates microenvironments. These systems enable regulated simulation of physical, mechanical, chemical, and biological parameters, and offer structural, ordered arrangements to simulate tissues or organs with multiple cell types. Design of an organ-on-chip requires consideration of microfluidic technology, biocompatible materials, nutrients, oxygen, and drugs. Cells derived from either cell lines or organoid cultures can be induced to polarize and interact with other cell types. Such systems can be used to investigate tissue physiology, to screen drugs, or to investigate interactions with normal flora or pathogens (Fig. 16.1).

In 2010, the first organ-on-chip simulated the multi-tissue, fluidic human lung alveolus [26]. Advancements have led to an organ-on-chip that enables an air-liquid interface for modelling respiratory viruses [19]. 3D printing is another advancement that has significantly improved organ-on-chip technology, whereby, in addition to micro-bioengineering, cell layers can be 3D printed to assist in structural organization on a minute scale [24, 65]. Modelling of a wide range of systems have been developed, including heart and vasculature [46], liver [60], kidney [41, 61], pancreas [50], gut [63], lung [51], brain neural networking and blood-brain barrier [21, 22, 28, 57], eyes [11], skin [43], and bone [18].

16.2.2 Organoids

Organ-on-chip technologies use small devices to mimic the functions of organs; however, they are not able to entirely replicate the complex structural and cellular organization of organs. In contrast, organoids are 3D cell structures created in vitro from stem cells or include a mix of cell types [44]. These structures closely mimic

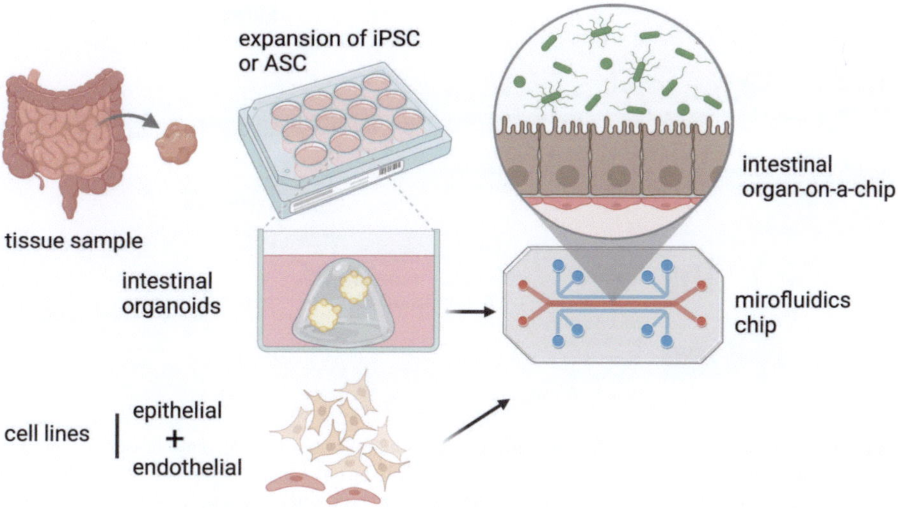

Fig. 16.1 Organ-on-chip technology utilizes microfluidics devices to simulate conditions of normal tissue. Cells derived from expanding induced pluripotent stem cells (iPSC), adult stem cells (ASC), or organoid cultures can be induced to polarize and interact with other cell types. Alternatively, cell lines derived from multiple tissue types can be introduced into a system. (Created in BioRender. Walduck, A. (2025) https://BioRender.com/jto0nu9)

the architecture and function of the target organ or tissue, maintain stable genetic traits, and can be cultivated long-term in a laboratory setting [1, 20]. Organoids can be developed from specific cell types or from stem cells, including induced pluripotent stem cells (iPSCs) or embryonic stem cells (ESCs), to generate specific cell types through targeted induction and differentiation protocols. 3D printing has enabled scaffolding to form organoids with defined structures and functions. The advancement of interdisciplinary methods, including organ-on-chip technology, genome editing, transcriptomics, and microfluidics, has significantly expedited organoid development [12]. Organoids are a promising alternative to animal models, offering researchers the ability to recreate organs and diseases in vitro. This holds significant potential for various translational applications. Applications of organoids in research include disease modelling [8], drug development [56], vaccine development [30], regenerative medicine (transplantation) [13], toxicology [33, 52], and precision medicine [56].

16.3 Applications in Research

16.3.1 Disease Modelling

Organ-on-chip and organoids are revolutionizing disease modelling. Stroke research uses brain organoids to model ischemic conditions and test neuroprotective therapies. For example, translational research from clinical trials for Rheumatoid Arthritis

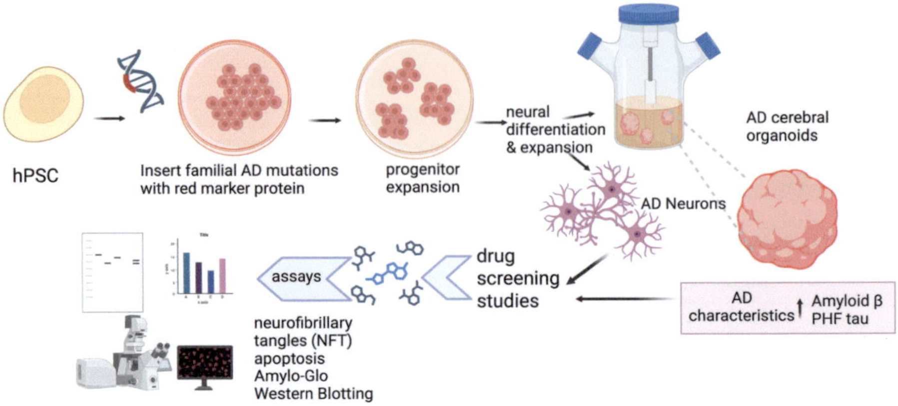

Fig. 16.2 Cerebral organoids for modelling Alzheimer's disease (AD). Patient-derived human pluripotent stem cells (hPSC) can be genetically modified to overexpress genes related to familial AD. Transgenic cells are selected using a fluorescent marker and expanded using specific neural differentiation media to generate either neurons or cerebral organoids. The resulting organoids are grown in spin culture vessels and mature after approximately 70 days. AD cerebral organoids express neuronal marker proteins and display AD characteristics, including increased expression of amyloid β and paired helical filament (PHF)-tau. AD cerebral organoids and neurons are used to screen potential inhibitors, a range of assays are used to identify reductions in AD characteristic proteins, neurofibrillary tangles, β-amyloid deposition, and apoptosis. (Modified from [10]. Created in BioRender. Walduck, A. (2025) https://BioRender.com/o2h75o4)

and Alzheimer's disease using traditional methods of modelling disease and drug treatment have very high failure rates [29, 47]. However, brain organoids replicate amyloid plaques and neurofibrillary tangles (Fig. 16.2), aiding in understanding disease mechanisms and testing treatments [10]. Infectious disease models use organ-on-chip systems to mimic host–pathogen interactions, improving understanding and treatment of infections. Other models of disease include endometriosis, cystic fibrosis, host–pathogen interaction in infections, including viruses, bacteria, and protozoa, and various cancers [8, 13]. Biobanking of tumour-derived organoids is being established as interest in disease modelling using organoids grows [13].

16.3.2 Drug Development

Organoids have been used effectively in the development of drugs and personalized treatments for tumours [56]. For example, head and neck cancer organoids have been used to develop novel treatment approaches that are more targeted to individual patients, as these organoids are developed from their tumours [15]. Additionally, patient-specific combination therapies for pancreatic cancer were identified using organoids, providing a rationale for combination therapies in clinical trials [37]. Using organoids, drug doses and responses can be studied prior to therapy, making it easier and quicker for patients to be provided with effective treatments for potentially life-threatening conditions such as cancer. Organoids have also been used to develop antibiotics. For example, a human epidermis organoid model was used to develop antimicrobials for bacterial biofilms [62]. Biofilms, which are observed in skin and soft tissue infections, are often associated with antimicrobial resistance (AMR) and therefore notoriously difficult to treat. Prior to the use of organoids, research into drugs to treat biofilm-associated infections was limited due to the lack of suitable testing models. With the advent of microfluidics and organoid technology, it has become easier to develop new antibiotics and treatment approaches targeting AMR [55, 66]. Similarly, organs on chips are also being developed to test drugs for various diseases [68].

16.3.3 Vaccine Development

Organoids and organ-on-chips have also been considered as effective alternatives to animal models in vaccine development. Following the COVID pandemic, there have been significant advancements in these technologies [30, 45]. A recent study described the use of a microfluidic lymphoid organ-on-chip (LO chip), which was used to evaluate a COVID vaccine boost [30]. Briefly, peripheral blood mononuclear cells (e.g. lymphocytes) from those previously exposed to SARS-CoV-2 and/or to COVID vaccines were used to construct the LO chip and used to evaluate immune responses to mRNA COVID vaccines (Fig. 16.3). Using the LO chip enabled researchers to investigate antibody and RNA responses in various immune cells. Prior to the advent of organ-on-chip technology, this study would have required an animal model. Another important benefit of the LO chip approach is the ability to evaluate immune responses from multiple individuals, in large cohort studies, using white blood cells obtained from peripheral blood, rather than having to employ invasive procedures to obtain lymphoid tissue from study participants. Organoids have increasingly been used in vaccine research to study immune responses to vaccine antigens and immunological processes [42]. Organoids comprising various white blood cells were used to study effector immune responses to vaccines, including at the level of germinal centres [31]. Overall, organoids have demonstrated success in vaccine development [25, 54, 58].

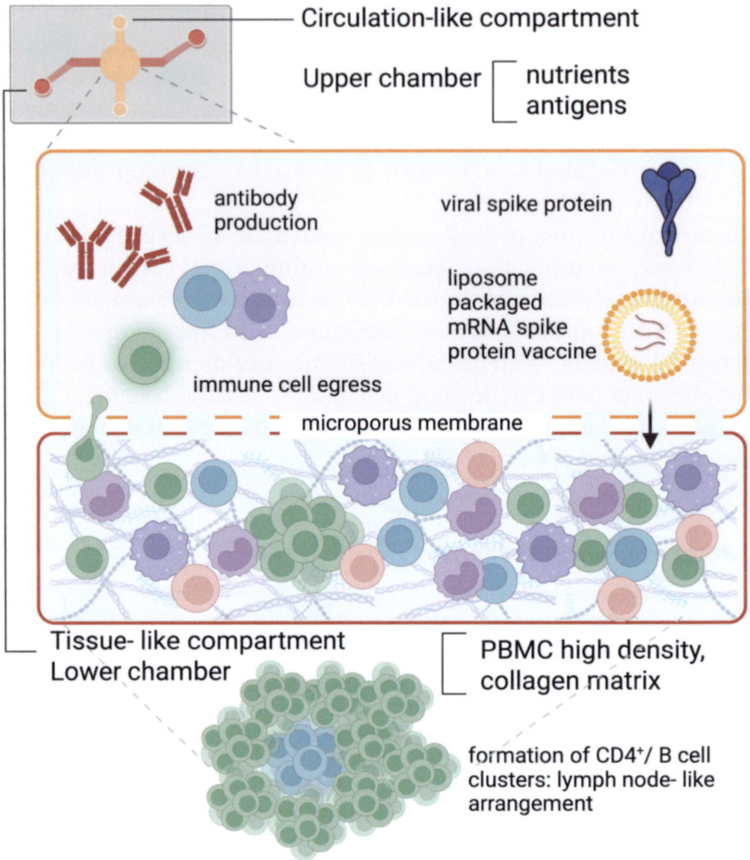

Fig. 16.3 Lymphoid organ-on-chip (LO chip) for investigation of responses to vaccine antigens. Patient peripheral blood mononuclear cells (PBMCs) are cultured at high density in a collagen matrix in the lower chamber of a microfluidics device that is separated from a channel containing media by a microporous membrane, simulating the circulation. On exposure to either antigen or an mRNA vaccine, the immune cells assemble into lymph node–like clusters of CD4 T cells and B cells, permitting antigen presentation and generation of antibody responses. The system can also be used to investigate immune cell egress from the lymph node after various stimuli. Cells can be extracted from both chambers for detailed analysis. (Modified from [30]. Created in BioRender. Walduck, A. (2025) https://BioRender.com/d51ivax)

16.4 Case Study: Using an Organoid to Model Host–Pathogen Interactions Using *Helicobacter pylori* as the Causative Agent of Gastritis, Ulcer, and Gastric Cancer

The gastric pathogen *Helicobacter pylori* causes chronic infection in the stomach and is the cause of gastritis, gastric ulcers, and most cases of gastric cancer [9]. This pathogen is thought to infect approximately 44% of adults and 35% of children

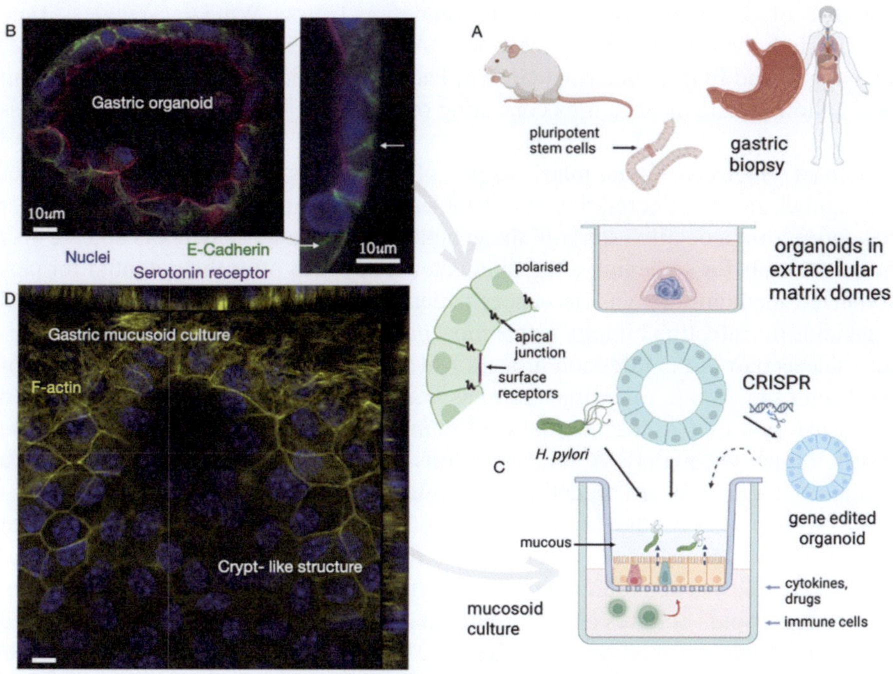

Fig. 16.4 Gastric organoids as effective models to study carcinogenesis, host-pathogen interactions, and immune responses. (**a**) Organoids are generated from primary tissue from patients or experimental animals. Pluripotent stem cells proliferate in vitro and can be passaged. (**b**) Confocal microscopy image of a gastric epithelial spheroid. Cells in organoids are polarized, form normal tight junctions, and express markers found in normal mucosa, including hormone and cytokine receptors. (**c**) 2D mucosoid cultures can be generated by generating air–liquid interface cultures from organoid or primary tissues. (**d**) Confocal Z-series showing form crypt-like 3D structures in a mucosoid culture. Such models are ideal to study the biology of the infection process and the effects of immune cells and inflammatory cytokines. Organoids can also be genetically modified using CRISPR technology, permitting analysis of the role of specific genes in infection and immunological processes

globally [9]. As antibiotic resistance is increasingly reported [64], there is an urgent unmet need for new therapeutics, and preferably a vaccine to prevent *H. pylori* infection and the long-term sequela of infection. An interesting aspect of *H. pylori* biology is that although the pathogen establishes a decades-long infection in the human host, it is non-invasive and remains extracellular. Understanding the bacterial and host factors, and the immune response that leads to the infection being asymptomatic in most patients, is critical for the development of new therapeutics. Gastric organoid models have been developed [5, 6, 39] and applied to study the biology of *H. pylori* infection (Fig. 16.4a).

Organoids developed from PSCs derived from adult tissues can be generated from biopsied tissue and cultured indefinitely, thanks to decades of efforts to refine our understanding of the required tissue growth factors to ensure growth and differentiation of stem cells. Stem cells introduced into an artificial cell matrix, in the presence of essential growth and differentiation factors Wnt3A (Wingless-Type MMTV Integration Site Family, Member 3A), RSPo1 (R-Spondin Family member 1), noggin, and EGF (Epidermal Growth Factor) are able to proliferate and expand into hollow spherical structures consisting of polarized epithelial cells that express characteristic markers of intact gastric epithelia (Fig. 16.4b). The stem cells respond to contact with extracellular matrix as part of the process such that the apical side of the epithelium is oriented inward. Biobanks of organoids from gastric cancer patients permit a detailed study of the genetic changes in these patients and are also excellent tools for screening drug candidates. They also offer the potential for personalized medicine for gastric cancer patients. A limitation of these "outside-in" organoids for infection biology is that most pathogens interact with the apical (luminal) side as part of the infection process which restricts investigating the biology of the bacterium. While microinjection has been used to undertake early studies on *H. pylori* biology [6], the internal environment of a spheroid does not reflect the gastric lumen adequately. A solution to this approach is a 2D air–liquid interface where organoid cells are seeded onto a collagen matrix on a membrane in a transwell system, where the layer receives nutrients from the basal compartment (Fig. 16.4c). Removal of the culture media from the apical side promotes polarization, and media can be modified to promote further differentiation of the monolayer. These "mucosoid" cultures generate a luminal side that contains cell types observed in gastric crypts. Further, over a period of 21–28 days, the cultures develop crypt-like depressions and secrete mucous (Fig. 16.4d). Such a model provides an ideal opportunity to investigate bacteria-epithelial interactions and the impact of immune mediators or cell populations by co-culture. The further opportunity to either generate organoids from KO mice [48] or to gene edit [7] these systems in vitro means that highly focused cell biology studies can be undertaken.

16.5 Advantages and Disadvantages

Evidence supporting organ-on-chip platforms and organoids as superior modelling approaches compared to traditional cell culture methods is increasing in the literature. Researchers are now exploring a synergistic approach that combines the strengths of both models, known as organoid-on-chip [59, 67]. Organoid-on-chip technologies have demonstrated potential for enhancing the complexity of human body modelling by integrating multiple advanced systems. The goal of next-generation platforms is to systematically reconstruct a functional, miniaturized human body in vitro, offering scalable and adaptable advantages [59]. These innovations are expected to accelerate the translation of preclinical research into clinical and industrial applications.

These technologies offer significant advantages over traditional cell culture methods, such as the ability to mimic cellular and tissue structure and function, considering vascularization, perfusion, nutrient requirements, and fluidics [36]. They also use appropriate tissues to increase translatability, addressing interspecies differences and ethical constraints that are unavoidable in animal models. For example, mouse xenograft models are often used to model diseases for other species due to logistical considerations like the expense of feeding and housing animals and animal welfare concerns. The safety of animal models in disease modelling has long been questioned, as substances deemed safe for animals have sometimes led to toxic effects and fatalities in humans [3, 16]. In addition to providing a viable alternative that enhances precision and reproducibility in disease modelling, other advantages include economic benefits, animal welfare, ecological impact, accessibility, time efficiency, social impact, and compliance with legislation. Organoids and organs-on-chips also allow for high-quality data collection when testing vaccine antigens, the ability to test multiple vaccine candidates, reduced costs associated with maintaining animal research facilities, and faster, easier preclinical testing.

Compared to conventional 2D cell lines, organoids and organ-on-chip systems offer several distinct advantages that enhance the reliability and translational relevance of preclinical testing. Traditional cell lines often lack structural complexity, cellular diversity, and microenvironmental factors of tissues, which can lead to misleading or non-predictive results. In contrast, organoids more accurately recapitulate multicellular organization and structure, enabling more physiologically relevant responses to drugs and disease stimuli. Specific advantages of organ-on-chip technology include offering high sensitivity, integration, throughput, and efficiency [20, 34, 35], and microfluidic flow that enables efficient nutrient renewal and waste removal to enhance cell viability and differentiation [14]. These features improve the predictive power of toxicity and efficacy assessments and facilitate earlier identification of promising therapeutic candidates. Additionally, their compatibility with patient-derived cells supports personalized medicine approaches, offering insights into individual variability in drug response that are not possible with immortalized cell lines.

While the issues with traditional cell culture methods, particularly translatability and repeatability, have been highlighted for some time, organoids are still being evaluated for standardization and repeatability. The major limitation of organ-on-chip and organoid technologies is their inability to reproduce the cellular and multi-system communication and transport mechanisms. These systems often lack critical cell and stromal components and can represent immature organs due to time constraints and differentiation timelines. The limited reproducibility of organoid cultures, along with the absence of intricate internal structures and dynamic microenvironments, poses challenges in accurately assessing drug efficacy and toxicity prior to clinical trials [33, 52]. These limitations and the fact that organ-on-chip and organoid technologies are not easily accessible, hinder the broader adoption of these technologies in research.

16.6 Future Applications

Future applications of organ-on-chip and organoid technologies in research are vast and promising. One significant area is studying interactions between organs and microbiomes, which can provide insights into how microbiomes influence health and disease [2]. Additionally, organ-on-chip and organoid technologies can aid in the construction of engineered tissues, offering an alternative to conventional transplantation [53] and regenerative medicine, where organoid technology may in the future replace allogenic transplants and avoid life-long immunosuppression to avoid transplant rejection [13]. Stem cell-derived organoids are poised to revolutionize personalized medicine by providing tailored solutions for individual patients. These organoids can be used to model diseases, test drug responses, and develop personalized treatment plans, advancing the field of personalized medicine [23, 27]. In particular, patient-derived organoids hold great potential for creating more accurate and individualized models of disease, leading to better therapeutic outcomes. However, the widespread adoption of these technologies will require regulatory acceptance and consideration of ethical implications. Ensuring that organ-on-chip and organoid models meet regulatory standards is crucial for their integration into clinical and research settings. Overall, the future of organ-on-chip and organoid technologies has the potential to transform research, personalized medicine, and therapeutic development. Continued advancements in this field will likely lead to more precise, efficient, and ethical teaching and research practices.

Another area for potential high impact is the education of future medical and veterinary professionals and scientists. From an ethical perspective, it is important to consider alternatives that facilitate reducing the number of purpose-killed animals for dissection, animal experimentation, and instrumental animal use [32]. Instructing on ethics, alternatives, and animal welfare while engaging in harmful animal practices is contradictory and may undermine the intended learning outcomes. In veterinary medicine, as in medicine, increasingly, alternatives to living animals or cadavers are being used to teach students. While organoids and organs-on-chips may not be suitable for the delivery of clinically applied anatomy, 3D models, diagnostic images with 3D reconstructions, and plastinated specimens facilitate learning of anatomy without the use of animals. Animal handling skills such as placing a halter on a cow may be practiced on models before students attempt it on a live animal [4]. Other examples include models for blood collection, otoscopic examination, diagnostic imaging, bandaging, surgical skin preparation, incision, and suturing (Fig. 16.5) [4, 40, 49].

An educative approach to raising awareness of non-animal alternatives and implementing replacement alternatives in education and training is a crucial

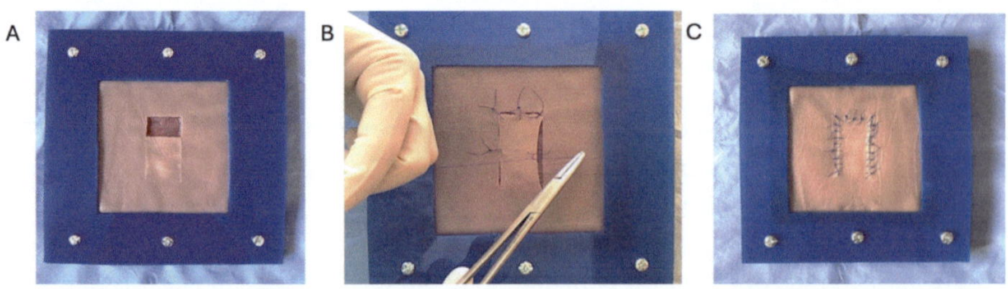

Fig. 16.5 A model for teaching veterinary reconstructive surgical techniques. (**a**) Excision of a mass, leaving a skin deficit. (**b**) A student using the model. (**c**) Completed pedicle flap

component of putting ethics, alternatives, and animal welfare principles into action. These efforts should be aligned. Minimizing the use of animal models should be intentionally embedded into the educational curriculum. Education has the potential to drive forward the research and application of organ-on-chip and organoid technologies, expanding their prospective use and impact.

16.7 Conclusion

The rapid advancements in organoids and organ-on-chip technologies, alongside progress in fluidics, micro-bioengineering, and 3D printing, promise to revolutionize disease modelling and drug development. These technologies offer opportunities to replicate human physiology and pathology in vitro with high fidelity, enabling researchers to study disease mechanisms in a controlled and human-relevant environment. For instance, organoids derived from patient-specific stem cells can mimic the structural and functional properties of native tissues, allowing for personalized disease modelling and therapeutic screening. Similarly, organ-on-chip platforms integrate microfluidic systems to simulate dynamic biological environments, such as blood flow and mechanical stress, which are critical for understanding complex disease processes. This convergence of technologies enhances the predictive power of preclinical models and reduces reliance on animal testing, aligning with regulatory changes and ethical scientific practices. Additionally, the modularity and scalability of these systems support high-throughput screening and real-time monitoring, accelerating drug discovery and development. As research continues to evolve, these systems will enhance medical diagnostics, therapeutic strategies, and educational tools.

Acknowledgements The authors would like to acknowledge Rachel McGowan and Felicity Stanley for providing images of the educational model for teaching veterinary reconstructive surgical techniques.

References

1. Ajmal L, Ajmal S, Ajmal M, Nawaz G. Organ regeneration through stem cells and tissue engineering. Curēus (Palo Alto, CA). 2023;15(1):e34336-e.
2. Allam-Ndoul B, Castonguay-Paradis S, Veilleux A. Gut microbiota and intestinal trans-epithelial permeability. Int J Mol Sci. 2020;21(17):6402.
3. Attarwala H. TGN1412: from discovery to disaster. J Young Pharm. 2010;2(3):332–6.
4. Baillie S, Parkes R, Wood S. Using models as alternatives to live animals. In: Hunt JA, Delcambre J, Baillie S, editors. AAVMC use of animals in veterinary education handbook. Washington, DC: American Association of Veterinary Medical Colleges; 2024. p. 39–43. Available from: https://doi.org/10.35542/osf.io/5k894.
5. Barker N, Es JH, Kuipers J, Kujala P, Born M, Cozijnsen M, et al. Identification of stem cells in small intestine and colon by marker gene Lgr5. Nature. 2007;449(7165):1003.

6. Bartfeld S, Bayram T, Wetering Mvd, Huch M, Begthel H, Kujala P, et al. In vitro expansion of human gastric epithelial stem cells and their responses to bacterial infection. Gastroenterology. 2015;148(1):126–36.e6.
7. Bauer M, Nascakova Z, Mihai A-I, Cheng PF, Levesque MP, Lampart S, et al. The ALPK1/TIFA/NF-κB axis links a bacterial carcinogen to R-loop-induced replication stress. Nat Commun. 2020;11(1):5117.
8. Bloise N, Giannaccari M, Guagliano G, Peluso E, Restivo E, Strada S, et al. Growing role of 3D in vitro cell cultures in the study of cellular and molecular mechanisms: short focus on breast cancer, endometriosis, liver and infectious diseases. Cells (Basel, Switzerland). 2024;13(12):1054.
9. Chen Y-C, Malfertheiner P, Yu H-T, Kuo C-L, Chang Y-Y, Meng F-T, et al. Global prevalence of Helicobacter pylori infection and incidence of gastric cancer between 1980 and 2022. Gastroenterology. 2024;166(4):605–19.
10. Choe MS, Yeo HC, Kim JS, Lee J, Lee HJ, Kim H-R, et al. Simple modeling of familial Alzheimer's disease using human pluripotent stem cell-derived cerebral organoid technology. Stem Cell Res Ther. 2024;15(1):118.
11. Chung M, Lee S, Lee BJ, Son K, Jeon NL, Kim JH. Wet-AMD on a chip: modeling outer blood-retinal barrier in vitro. Adv Healthc Mater. 2018;7(2):n/a.
12. Ciprietti M, Bueds C, Vankelecom H, Vriens J. Organoids as powerful models of endometrium epithelium in transcriptomic, cellular and functional mimicry. Cell Mol Life Sci. 2025;82(1):272.
13. Corrò C, Novellasdemunt L, Li VSW. A brief history of organoids. Am J Physiol Cell Physiol. 2020;319(1):C151–C65.
14. Deng P, Cui K, Shi Y, Zhu Y, Wang Y, Shao X, et al. Fluidic flow enhances the differentiation of placental trophoblast-like 3d tissue from hiPSCs in a perfused macrofluidic device. Front Bioeng Biotechnol. 2022;10:907104.
15. Done AJ, Birkeland AC. Organoids as a tool in drug discovery and patient-specific therapy for head and neck cancer. Cell Rep Med. 2023;4(6):101087.
16. Eddleston M, Cohen AF, Webb DJ. Implications of the BIA-102474-101 study for review of first-into-human clinical trials. Br J Clin Pharmacol. 2016;81(4):582–6.
17. FDA, U. 1938. Part II: 1938, Food, Drug, Cosmetic Act. U.S. Food and Drug Administration. https://www.fda.gov/about-fda/changes-science-law-and-regulatory-authorities/part-ii-1938-food-drug-cosmetic-act.
18. Galván-Chacón VP, Zampouka A, Hesse B, Bohner M, Habibovic P, Barata D. Bone-on-a-chip: a microscale 3D biomimetic model to study bone regeneration. Adv Eng Mater 2022;24(7):n/a.
19. Gard AL, Luu RJ, Miller CR, Maloney R, Cain BP, Marr EE, et al. High-throughput human primary cell-based airway model for evaluating influenza, coronavirus, or other respiratory viruses in vitro. Sci Rep. 2021;11(1):14961.
20. Geng Q, Xu Y, Hu Y, Wang L, Wang Y, Fan Z, et al. Progress in the application of organoids-on-a-chip in diseases. Organogenesis. 2024;20(1):2386727.
21. Hajal C, Offeddu GS, Shin Y, Zhang S, Morozova O, Hickman D, et al. Engineered human blood–brain barrier microfluidic model for vascular permeability analyses. Nat Protocols. 2022;17(1):95–128.
22. Harberts J, Fendler C, Teuber J, Siegmund M, Silva A, Rieck N, et al. Toward brain-on-a-chip: human induced pluripotent stem cell-derived guided neuronal networks in tailor-made 3D nanoprinted microscaffolds. ACS Nano. 2020;14(10):13091–102.
23. Hofer M, Lutolf MP. Engineering organoids. Nat Rev Mater. 2021;6(5):402.
24. Homan KA, Kolesky DB, Skylar-Scott MA, Herrmann J, Obuobi H, Moisan A, et al. Bioprinting of 3D convoluted renal proximal tubules on perfusable chips. Sci Rep. 2016;6(1):34845.
25. Hong K-J, Seo S-H. Organoid as a culture system for viral vaccine strains. Clin Exp Vaccine Res (Seoul). 2018;7(2):145–8.

26. Huh D, Matthews BD, Mammoto A, Montoya-Zavala M, Hsin HY, Ingber DE. Reconstituting organ-level lung functions on a chip. Science. 2010;328(5986):1662–8.
27. Ingber DE. Human organs-on-chips for disease modelling, drug development and personalized medicine. Nat Rev Genetics. 2022;23(8):467–91.
28. Ionescu A, Zahavi EE, Gradus T, Ben-Yaakov K, Perlson E. Compartmental microfluidic system for studying muscle–neuron communication and neuromuscular junction maintenance. Eur J Cell Biol. 2016;95(2):69–88.
29. Jayasundara KS, Keystone EC, Parker JL. Risk of failure of a clinical drug trial in patients with moderate to severe rheumatoid arthritis. J Rheumatol. 2012;39(11):2066.
30. Jeger-Madiot R, Planas D, Staropoli I, Debarnot H, Kervevan J, Mary H, et al. Modeling memory B cell responses in a lymphoid organ-chip to evaluate mRNA vaccine boosting. J Exp Med. 2024;221(10)
31. Kastenschmidt JM, Sureshchandra S, Wagar LE. Leveraging human immune organoids for rational vaccine design. Trends Immunol. 2023;44(12):938–44.
32. Kiani AK, Pheby D, Henehan G, Brown R, Sieving P, Sykora P, et al. Ethical considerations regarding animal experimentation. J Prev Med Hyg. 2022;63(2 Suppl 3):E255–E66.
33. Kim J, Koo B-K, Knoblich JA. Human organoids: model systems for human biology and medicine. Nat Rev Mol Cell Biol. 2020;21(10):571–84.
34. Koyilot MC, Natarajan P, Hunt CR, Sivarajkumar S, Roy R, Joglekar S, et al. Breakthroughs and applications of organ-on-a-chip technology. Cells. 2022;11(11):1828.
35. Li Z, Hui J, Yang P, Mao H. Microfluidic organ-on-a-chip system for disease modeling and drug development. Biosensors (Basel). 2022;12(6):370.
36. Low LA, Mummery C, Berridge BR, Austin CP, Tagle DA. Organs-on-chips: into the next decade. Nat Rev Drug Discov. 2021;20(5):345–61.
37. Mahadevan KK, McAndrews KM, LeBleu VS, Yang S, Lyu H, Li B, Sockwell AM, Kirtley ML, Morse SJ, Moreno Diaz BA, Kim MP, Feng N, Lopez AM, Guerrero PA, Paradiso F, Sugimoto H, Arian KA, Ying H, Barekatain Y, et al. KRASG12D inhibition reprograms the microenvironment of early and advanced pancreatic cancer to promote FAS-mediated killing by CD8+ T cells. Cancer Cell. 2023;41(9):1606–1620.e1608.
38. Marshall LJ, Bailey J, Cassotta M, Herrmann K, Pistollato F. Poor translatability of biomedical research using animals – a narrative review. Altern Lab Anim. 2023;51(2):102–35.
39. McCracken KW, Catá EM, Crawford CM, Sinagoga KL, Schumacher M, Rockich BE, et al. Modeling human development and disease in pluripotent stem cell-derived gastric organoids. Nature. 2014;516(7531):400–4.
40. McGowan RZ, Brookes VJ, Jermyn K, Stanley F, Rotne R. Single pedicle advancement flap models for teaching veterinary reconstructive surgical techniques. J Vet Med Educ. 2023;50(6):703–12.
41. Musah S, Mammoto A, Ferrante TC, Jeanty SSF, Hirano-Kobayashi M, Mammoto T, et al. Mature induced-pluripotent-stem-cell-derived human podocytes reconstitute kidney glomerular-capillary-wall function on a chip. Nat Biomed Eng. 2017;1(5)
42. Perucca A, Llonín AG, Benach OM, Hallopeau C, Rivas EI, Linares J, Garrido M, Sallent-Aragay A, Golde T, Colombelli J, Dalaka E, Linacero J, Cazorla M, Galan T, Pastor Viel J, Badenas X, Recort-Bascuas A, Comerma L, Fernandez-Nogueira P, et al. Micro immune response on-chip (MIRO) models the tumour-stroma interface for immunotherapy testing. Nat Commun. 2025;16(1):1279–20.
43. Ponmozhi J, Dhinakaran S, Varga-Medveczky Z, Fónagy K, Bors LA, Iván K, et al. Development of skin-on-a-chip platforms for different utilizations: factors to be considered. Micromachines (Basel). 2021;12(3):294.
44. Raimondi I, Izzo L, Tunesi M, Comar M, Albani D, Giordano C. Organ-on-a-chip in vitro models of the brain and the blood-brain barrier and their value to study the microbiota-gut-brain axis in neurodegeneration. Front Bioeng Biotechnol. 2020;7:435.

45. Reynaud K, Atwell S, Védrine C, Guyard C, Beloeil L. Innovative micro physiological systems for vaccine development. Hum Vaccin Immunother. 2025;21(1):2518641.
46. Ribas J, Sadeghi H, Manbachi A, Leijten J, Brinegar K, Zhang YS, et al. Cardiovascular organ-on-a-chip platforms for drug discovery and development. Appl In Vitro Toxicol. 2016;2(2):82.
47. Sabbagh MN, Cummings J. Open peer commentary to "failure to demonstrate efficacy of aducanumab: an analysis of the EMERGE and ENGAGE trials as reported by Biogen December 2019". Alzheimer's & Dementia. 2021;17(4):702–3.
48. Sayed IM, Sahan AZ, Venkova T, Chakraborty A, Mukhopadhyay D, Bimczok D, et al. Helicobacter pylori infection downregulates the DNA glycosylase NEIL2, resulting in increased genome damage and inflammation in gastric epithelial cells. J Biol Chem. 2020;295(32):11082–98.
49. Schoenfeld E, Stanley F, Combs M, Callcott E, Williams A, Rotne R. The construction of canine distal limb models used in teaching sonography identification of vegetal foreign bodies. Vet Radiol Ultrasound. 2024;65:486.
50. Shik Mun K, Arora K, Huang Y, Yang F, Yarlagadda S, Ramananda Y, et al. Patient-derived pancreas-on-a-chip to model cystic fibrosis-related disorders. Nat Commun. 2019;10(1):3124–12.
51. Si L, Bai H, Rodas M, Cao W, Oh CY, Jiang A, et al. A human-airway-on-a-chip for the rapid identification of candidate antiviral therapeutics and prophylactics. Nat Biomed Eng. 2021;5(8):815–29.
52. Suarez-Martinez E, Suazo-Sanchez I, Celis-Romero M, Carnero A. 3D and organoid culture in research: physiology, hereditary genetic diseases and cancer. Cell Biosci. 2022;12(1):39.
53. Sugimoto S, Kobayashi E, Fujii M, Ohta Y, Arai K, Matano M, et al. An organoid-based organ-repurposing approach to treat short bowel syndrome. Nature (London). 2021;592(7852):99–104.
54. Suhito IR, Sunil C, Tay A. Engineering human immune organoids for translational immunology. Bioactive Mater. 2025;44:164–83.
55. Tang P-C, Eriksson O, Sjögren J, Fatsis-Kavalopoulos N, Kreuger J, Andersson DI. A microfluidic chip for studies of the dynamics of antibiotic resistance selection in bacterial biofilms. Front Cell Infect Microbiol. 2022;12:896149.
56. Vandana JJ, Manrique C, Lacko LA, Chen S. Human pluripotent-stem-cell-derived organoids for drug discovery and evaluation. Cell Stem Cell. 2023;30(5):571–91.
57. Vatine GD, Barrile R, Workman MJ, Sances S, Barriga BK, Rahnama M, et al. Human iPSC-derived blood-brain barrier chips enable disease modeling and personalized medicine applications. Cell Stem Cell. 2019;24(6):995–1005.e6.
58. Wagar LE. Human immune organoids: a tool to study vaccine responses. Nat Rev Immunol. 2023;23(11):699.
59. Wang H, Ning X, Zhao F, Zhao H, Li D. Human organoids-on-chips for biomedical research and applications. Theranostics. 2024;14(2):788–818.
60. Weng YS, Chang SF, Shih MC, Tseng SH, Lai CH. Scaffold-free liver-on-a-chip with multiscale organotypic cultures. Adv Mater (Weinheim). 2017;29(36):n/a.
61. Wilmer MJ, Ng CP, Lanz HL, Vulto P, Suter-Dick L, Masereeuw R. Kidney-on-a-chip technology for drug-induced nephrotoxicity screening. Trends Biotechnol (Regular ed). 2016;34(2):156–70.
62. Wu B, Haney EF, Akhoundsadegh N, Pletzer D, Trimble MJ, Adriaans AE, et al. Human organoid biofilm model for assessing antibiofilm activity of novel agents. NPJ Biofilms Microbiomes. 2021;7(1):8.
63. Xiang Y, Wen H, Yu Y, Li M, Fu X, Huang S. Gut-on-chip: recreating human intestine in vitro. J Tissue Eng. London, England: SAGE Publications. 2020:2041731420965318.
64. Yu Y, Xue J, Lin F, Liu D, Zhang W, Ru S, et al. Global primary antibiotic resistance rate of Helicobacter pylori in recent 10 years: a systematic review and meta-analysis. Helicobacter. 2024;29(3):e13103.

65. Zhang YS, Arneri A, Bersini S, Shin S-R, Zhu K, Goli-Malekabadi Z, et al. Bioprinting 3D microfibrous scaffolds for engineering endothelialized myocardium and heart-on-a-chip. Biomaterials. 2016;110:45–59.
66. Zhang M, Yuan F, Jia H, Xu Y, Yan L, Zhang T, Xu X, Liu Y, Wang X, Li D. Rapidly in situ forming antibiotic-free injectable hydrogel wound dressing for eradicating drug-resistant bacterial infections in human skin organoids. Int J Biol Macromol. 2024;282(Pt 6):137542.
67. Zhao Y, Landau S, Okhovatian S, et al. Integrating organoids and organ-on-a-chip devices. Nat Rev Bioeng. 2024;2:588–608.
68. Zhou C, Li Z, Lu K, Liu Y, Xuan L, Mao H, et al. Advances in human organs-on-chips and applications for drug screening and personalized medicine. Fundam Res (Beijing). 2024;

MIX
Papier aus verantwortungsvollen Quellen
Paper from responsible sources
FSC® C105338

If you have any concerns about our products,
you can contact us on
ProductSafety@springernature.com

In case Publisher is established outside the EU,
the EU authorized representative is:
Springer Nature Customer Service Center GmbH
Europaplatz 3, 69115 Heidelberg, Germany

Printed by Libri Plureos GmbH
in Hamburg, Germany